Advanced Photonics with Second-order
Optically Nonlinear Processes

T0184271

NATO Science Series

A Series presenting the results of activities sponsored by the NATO Science Committee. The Series is published by IOS Press and Kluwer Academic Publishers, in conjunction with the NATO Scientific Affairs Division.

General Sub-Series

A. Life Sciences	IOS Press
B. Physics	Kluwer Academic Publishers
C. Mathematical and Physical Sciences	Kluwer Academic Publishers
D. Behavioural and Social Sciences	Kluwer Academic Publishers
E. Applied Sciences	Kluwer Academic Publishers
F. Computer and Systems Sciences	IOS Press

Partnership Sub-Series

1. Disarmament Technologies	Kluwer Academic Publishers
2. Environmental Security	Kluwer Academic Publishers
3. High Technology	Kluwer Academic Publishers
4. Science and Technology Policy	IOS Press
5. Computer Networking	IOS Press

The Partnership Sub-Series incorporates activities undertaken in collaboration with NATO's Partners in the Euro-Atlantic Partnership Council – countries of the CIS and Central and Eastern Europe – in Priority Areas of concern to those countries.

NATO-PCO-DATA BASE

The NATO Science Series continues the series of books published formerly in the NATO ASI Series. An electronic index to the NATO ASI Series provides full bibliographical references (with keywords and/or abstracts) to more than 50000 contributions from international scientists published in all sections of the NATO ASI Series.
Access to the NATO-PCO-DATA BASE is possible via CD-ROM "NATO-PCO-DATA BASE" with user-friendly retrieval software in English, French and German (© WTV GmbH and DATAWARE Technologies Inc. 1989).

The CD-ROM of the NATO ASI Series can be ordered from: PCO, Overijse, Belgium.

3. High Technology – Vol. 61

Advanced Photonics with Second-order Optically Nonlinear Processes

edited by

A. D. Boardman

Joule Laboratory,
Department of Physics,
University of Salford,
Salford, U.K.

L. Pavlov

Institute of Electronics,
Bulgarian Academy of Sciences,
Sofia, Bulgaria

and

S. Tanev

Optiwave Corporation,
Nepean,
Ontario, Canada

Kluwer Academic Publishers

Dordrecht / Boston / London

Published in cooperation with NATO Scientific Affairs Division

Proceedings of the NATO Advanced Study Institute on
Advanced Photonics with Second-order Optically Nonlinear Processes
Sozopol, Bulgaria
September 24 - October 3, 1997

A C.I.P. Catalogue record for this book is available from the Library of Congress.

ISBN 0-7923-5315-3 (HB)
ISBN 0-7923-5316-1 (PB)

Published by Kluwer Academic Publishers,
P.O. Box 17, 3300 AA Dordrecht, The Netherlands.

Sold and distributed in North, Central and South America
by Kluwer Academic Publishers,
101 Philip Drive, Norwell, MA 02061, U.S.A.

In all other countries, sold and distributed
by Kluwer Academic Publishers,
P.O. Box 322, 3300 AH Dordrecht, The Netherlands.

Printed on acid-free paper

Printed in the Netherlands

Contents

Preface

Although it took some time to establish the word, *photonics* is both widely accepted and used throughout the world and a major area of activity concerns nonlinear materials. In these the nonlinearity mainly arises from second-order or third-order nonlinear optical processes. A restriction is that second-order processes only occur in media that do not possess a centre of symmetry. Optical fibres, on the other hand, being made of silica glass, created by fusing SiO_2 molecules, are made of material with a centre of symmetry, so the bulk of all processes are governed by third-order nonlinearity. Indeed, optical fibre nonlinearities have been extensively studied for the last thirty years and can be truly hailed as a success story of nonlinear optics. In fact, the fabrication of such fibres, and the exploitation of their nonlinearity, is in an advanced stage - not least being their capacity to sustain envelope solitons. What then of second-order nonlinearity? This is also well-known for its connection to second-harmonic generation. It is an immediate concern, however, to understand how waves can mix and conserve both energy and momentum of the photons involved. The problem is that the wave vectors cannot be made to match without a great deal of effort, or at least some clever arrangement has to be made – a special geometry, or crystal arrangement. The whole business is called phase-matching and an inspection of the state-of-the-art today, reveals the subject to be in an advanced state.

Recently, the community of research workers who are interested in solitons and signal processing, such as the development of the much sought after optical switching, came to the realisation that there is much more to the behaviour of crystals displaying second-order nonlinearity than creating second-harmonic sources. It turns out that there also exists soliton – or, more correctly, solitary wave – solutions to the coupled equations, which involve a balance between the fundamental wave and the co-propagating second-harmonic wave. Although the credit for stimulating this modern activity rightly belongs to the present day community, especially the work of Stegeman and co-workers, the credit for discovering the soliton solution goes to Sukhorukov and co-workers, however, who produced the first paper on this topic some 24 years ago.

It is the case then that second-order nonlinearity is attracting a lot of attention today both in the more traditional phase-matching requirements needed to generate second-harmonic waves very efficiently – see the drive towards the blue laser, or its alternative - and in the new area of solitary waves. In view of this greatly increased interest it was felt that the time was right for NATO Advanced Study Institute, designed to be focused strongly onto these activities.

This type of NATO ASI was created with the title assumed by this book and was held in Sozopol, Bulgaria, September 24 – October 3, 1997. The programme had the main themes – solitons, quasi-phase matching, frequency conversion and parametric interactions. The material was delivered by a distinguished set of lecturers and the range of topics, together with the emphasis put upon them, was selected to give the participants a depth of understanding firmly rooted in the basic material. As always, the blend of topics emphasised the need not only to accumulate some knowledge of basic theory but also a knowledge of materials and what they are capable of. Again the vision of all-optical switching, or some other kind of signal processing application was ever present in our thinking. The ASI was conducted as a school and the Editors felt that the whole school appreciated the presence of so much expertise from all over the world being concentrated in one place. As one participant remarked ' this is like a university' – and indeed it was, for this short period, one of the best in the world.

Nothing like this can take place without organisation and substantial support. First, the organisers are extremely grateful to NATO and the International Center for Scientific Culture – ICSC in Paris for their magnificent support. We are especially indebted to Professor Balkanski, the eminence grise of NATO events in Sozopol. The Directors A D Boardman and L Pavlov are also very indebted to the organising committee member George Stegeman. A D Boardman is also particularly indebted to Mrs L Clarke of Salford for her secretarial work. Finally, we must pay tribute to the school members. Without lively participation there can be no successful school. It is safe to say that the school was very much enjoyed both at work, and at play, by all who attended. We hope that the fine collection of material presented here will be both immediately useful and stand the test of time. We wish everybody well and express a real desire to maintain contact with everybody we met at the school.

INTRODUCTION TO THE PROPAGATION OF CW RADIATION AND SOLITONS IN QUADRATICALLY NONLINEAR MEDIA

A.D. BOARDMAN, P. BONTEMPS and K. XIE
Photonics and Nonlinear Science Group
Joule Laboratory
Department of Physics
University of Salford
Salford, M5 4WT
United Kingdom

Abstract

This chapter provides an introduction to the basic concepts of cw radiation and solitary wave propagation in quadratically nonlinear media. After some introductory comments on the modern motivation for this area of study, a very detailed discussion of polarisation in quadratic media is presented. Through this, an opportunity is taken to present full derivations, with due consideration being given to typical crystal symmetries involved and the accepted notation in the field. Nonlinear plane wave eigenstates are considered both for scalar and vector waves, with thorough discussions being given of how to generate equations that couple the behaviour of fundamental and the harmonic cw radiation. Balanced states are clearly explained and the vector case embraces a discussion of type I, type II and mixed type I-type II phase-matching. The chapter ends with a detailed discussion of scalar solitary waves, in which the cw equations are extended to include diffraction and a Lagrangian approach to stability of single-peak stationary solutions.

1. Introduction

Advanced photonics based upon second-order optical (quadratic) nonlinearity rests, traditionally, upon the efficient generation of second-harmonic radiation [1-5]. To achieve this objective, the emphasis is placed upon linear phase-matching schemes [5]. In recent times, however, the realisation that the coupled equations that model second-harmonic generation admit nonlinear stationary eigenstates in plane wave form, or as solitary waves [6-28], has stimulated a remarkable, and energetic, growth of new interest in this area of work. Historically [29], it turns out that solitary wave solutions to the coupled equations were announced more than two decades ago, but to little obvious acclaim. It is the seminal experimental work of Stegeman and co-workers [11], on cascaded nonlinearity, that has created the modern momentum in this subject. New

1

A. D. Boardman et al. (eds.),
Advanced Photonics with Second-order Optically Nonlinear Processes, 1–57.
© 1999 *Kluwer Academic Publishers. Printed in the Netherlands.*

objectives are being defined all the time and several groups [30-42] are now engaged in this fascinating activity.

To be specific, the topic of second-harmonic generation has blossomed in recent years because there is now a much deeper interest in the properties of the coupled equations that govern the way in which a wave at a fundamental angular frequency ω cooperates with its second-harmonic partner at 2ω. To understand what has opened up a new direction in this area, it is useful to recall, first, that the familiar textbook [1] description of harmonic generation emphasises the question of *linear phase-matching* between *a scalar or vector input* at the fundamental frequency and the second-harmonic that this input generates. This issue arises because, if there is no significant linear phase-matching between the participating waves, there is negligible production of second-harmonic radiation. Furthermore, a lack of matching causes a spatially periodic exchange (amplitude modulation) of energy between the fundamental wave, at frequency ω, and the harmonic wave, at frequency 2ω. As a consequence, in the *usual* study of second-harmonic generation, the possibility of a *steady-state* combination of $(\omega,2\omega)$ waves, i.e. a nonlinear eigenstate, is not addressed. On the other hand, if *both* the fundamental and the harmonic wave form the input to the nonlinear material, some kind of balance between the ω and 2ω waves can be imagined. A balanced, stationary, nonlinear eigenstate is created when the *linear* phase-mismatch is precisely equal to the *nonlinear* phase-mismatch. The latter phase modulation is created by the power of the wave. This type of *plane* wave eigenstate was first noted by Kaplan [6] but, for finite beams, the situation is not so different. If the fundamental beam is above some threshold power value, then the energy will switch backwards and forwards a few times between the ω and 2ω beams, but, eventually, a diffraction-free spatial soliton-like combined beam can be created. This will be a stationary state, and will be subject to precisely the same restraint as that for plane waves viz. that the linear phase-mismatch exactly equals the nonlinear phase-mismatch.

A detailed discussion of second-order optical nonlinearity will be given in the next section but it is useful to state here some of the salient features to be expected. The theory, in section 2, will use a well-known form for the polarisation **P** induced by the passage of an electromagnetic wave, carrying an electric field **E**, through a dielectric material. This is [1]

$$P_i = \varepsilon_0 \{ \chi_{ij}^{(1)} E_j + \chi_{ijk}^{(2)} E_j E_k + \chi_{ijkl}^{(3)} E_j E_k E_l + ... \} \tag{1.1}$$

where $i = x, y, z$ are spatial coordinates, ε_0 is the permittivity of free space, $\chi^{(1)}$ is a *linear* susceptibility tensor, $\chi^{(2)}$ is a *second-order* susceptibility tensor, and finally, $\chi^{(3)}$ is a *third-order* susceptibility tensor. There are other terms in the series, of course, and the form (1.1) carries with it the assumption that there is convergence. It can only do this if the origin of the nonlinearity is non-resonant, so this feature will be commented upon in the next section.

For materials possessing a centre of symmetry [43] all the elements of $\chi^{(2)}$ are zero and the nonlinearity is generally dominated by $\chi^{(3)}$. On the other hand, $\chi^{(2)}$ is non-zero if the material lacks a centre of symmetry and, in general, for materials that lack a centre of symmetry, the nonlinearity arises from a combination of $\chi^{(2)}$ and $\chi^{(3)}$. In many cases $\chi^{(2)}$ dominates, however, and this will be assumed in this chapter even though it is important to realise that this will not always be the case. Indeed, terms beyond $\chi^{(3)}$ may also be significant. Fortunately, there are many materials with a finite, and dominant, $\chi^{(2)}$, including many found in Nature. One example [1], shown in figure 1, is proustite [Ag_3AsS_3], selected here because it is well-known not only in nonlinear optics but to geologists and mineralogists as well.

PROUSTITE Rich scarlet colour

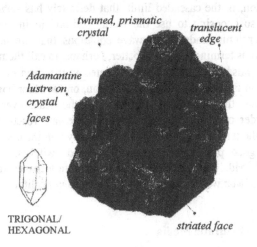

Figure 1 Proustite [Ag_3AsS_3] is 66% silver and is a light red silver ore. It is usually cited as a typical second-order nonlinear crystal and is in the crystal class 3m.

Truncation of the polarisation series (1.1) at the third-order is appropriate for most materials and, furthermore, many elements of the tensors displayed are often zero because of symmetry. In fact, the tensors are often reducible to a single independent parameter or, at most, a small number of independent ones, by the point group symmetry operations. Indeed, it is fortunate that the very materials needed for real applications are, more often than not, isotropic, amorphous, or possess a rather simple crystal structure. The method of processing the tensors to a number of independent elements relies upon the simple fact that all crystal structures are invariant with respect to a set of well-known crystal symmetry operations. The lower the symmetry of the material, the smaller the number of these operations becomes. At the highest level, an isotropic, or amorphous, material is invariant to any symmetry operation. As the symmetry falls from the cubic classes down to the triclinic class, the number of non-zero elements of $\chi_{ijk}^{(2)}$ and $\chi_{ijkl}^{(3)}$, can be easily obtained from the literature.

In a $\chi^{(2)}$ material *two* field components can mix to produce a third one that is, once again, at the fundamental frequency, i.e. backmixing occurs [11], involving the second-harmonic waves and the complex conjugate of the fundamental wave. This theoretical possibility is supported by dramatic [11] experimental evidence showing that this backmixing (cascading) process is clearly observable, even for large linear phase-mismatching. Given this type of wave interaction, the next question is to ask whether solitons can exist under these conditions.

The answer is positive but, a more cautious, *soliton-like*, terminology is appropriate. If the linear phase-mismatch parameter is large, however, the coupled envelope equations for the ω and 2ω waves reduce to a *single* (ω) nonlinear envelope Schrödinger equation and this is called the *cascaded limit* [8,9,11-13,15]. Obviously, it is only this equation, in the cascaded limit, that definitely has stable one-dimensional soliton solutions, so choosing to use the word "soliton" to include solutions of the coupled second-harmonic-fundamental wave equations that are *not* in the cascaded limit is a less rigorous terminology. It is better, perhaps, to call them solitary waves.

$\chi^{(2)}$ materials are likely to be investigated in the form of planar waveguides, for which dispersion is generally weak. Diffraction, on the other hand, is strong and is not dependent upon the material. Furthermore, the scale over which diffraction operates is the order of 1 mm [44-46], so the immediate reaction is that nonlinear beams, for which the tendency to diffract is balanced by nonlinear self-focusing, [spatial solitons] will be good probes for use in $\chi^{(2)}$-based switches, for example. It is interesting, but not widely appreciated that it is possible to have diffraction-free beans as exact solutions of Maxwell's equations but, these are *not* solitons [47.48].

2. Polarisation in Quadratic Media

The presence of electric fields in a material arising from the propagation of an electromagnetic wave, whether in the form of plane wave cw radiation, temporal pulses, or spatial beams polarises it by amount $\mathbf{P(r,t)}$, where $\mathbf{r}$ is a position vector (x,y,z) and t is time. This polarisation then appears as a $\dfrac{\partial^2 \mathbf{P}}{\partial t^2}$ term in Maxwell's equations. The concept of polarisation is a most important one because it links the *material* properties to the electromagnetic field. Indeed, $\mathbf{P}$ must *be modelled to fit the material* under investigation so this task is a major preoccupation of nonlinear optics. Before progression to the full set of coupled field equations, for optically nonlinear media, is obtained, the nature of $\mathbf{P(r,t)}$ has to be explored so that its specific form, in relation to crystal symmetry [43] and a few other considerations, connected with harmonic generation, can be exposed.

At this point of entry into the subject, the literature often implies that the main features of any model of harmonic generation are well-known [1-3]. Various texts and papers have different approaches, however, and the levels of complication needed to reach even simple results can be very confusing. Against this background, important new results concerning nonlinear eigenmodes [6] in quadratically nonlinear media, are now being appreciated [30,32,34-37]. To get to this new perspective, the route taken here starts at the very beginning and moves along pathways laid down in the brilliant early article by Butcher [43]. It also benefits from later expositions, especially the impressive books by Schubert and Wilhelmi, and Boyd [1-3].

First, polarisation is a response of a material to the presence of an electromagnetic field that may be static, or varying with time. This polarisation will involve some charge separation, or movement of some kind, and elementary physics suggest that the electromagnetic field will be modified by this process. In principle, harmonics can be added to the initial driving field so the problem to address is how to formalise the medium response to make this happen.

Before this task is undertaken, it should be pointed out that the language and notation of nonlinear optics can be difficult to compare between one author and another and the reader is strongly recommended to read the excellent textbook by Boyd. Unfortunately, as Boyd points out, a lot of early progress in nonlinear optics was made using Gaussian units, which is why he uses them. To conform to modern practice, however, only SI units will be used here throughout this chapter.

One of the first impressions of nonlinear optics is that processes need not occur just at a single (angular) frequency ω but at several other significant frequencies as well. This is not a feature of nonlinear fibre optics but it is what happens when an electromagnetic wave at ω enters a *quadratically* nonlinear material. In such an experiment, the observer would expect to see waves at 2ω. What does this mean for the total electric field $E(t)$? The answer is that provided the optical field components have frequency spectra that do not overlap, to any extent, the *total electric field* can be expressed as the sum

$$E(t) = E^{\omega_1}(t) + E^{\omega_2}(t) + ... \qquad (2.1)$$

where t is time ω_I are the frequency components of the optical field and are well separated from each other. Note that $E(t)$ is not really only a function of time t; the space coordinate is omitted for convenience. The superscript notation is just a label and has the following definition

$$E^{\omega_i}(t) = \frac{1}{2}\left[\overline{E}(\omega_i)e^{-i\omega_i t} + c.c.\right]$$
$$= \frac{1}{2}\left[E(\omega_i)e^{i(k_{\omega_i}\cdot r - \omega_i t)} + c.c.\right] \qquad (2.2)$$

where c.c. denotes the complex conjugate. In linear physics just the first term in (2.2) will suffice but quadratic nonlinearity will square the field and mix positive and negative frequencies so the c.c., which makes the field real is very important here. Comments on the notation are

- for each frequency ω_i three quantities $E^{\omega_i}(t)$, $\overline{E}(\omega_i)$, $E(\omega_i)$ are defined; provided that consistency is maintained throughout the text no confusion should arise

- this notation is widely used but must not be taken to mean that $\overline{E}(\omega_i)$ as a strict function of ω_i, nor is this the intention; indeed, it would be wrong if it did, since ω_i is a parameter, in this chapter. The notation $\overline{\mathbf{E}}(\omega_i)$, $\mathbf{E}(\omega_i)$ is designed to show ω_i as a parameter

- the transition from $\overline{\mathbf{E}}(\omega_i)$, to $\mathbf{E}(\omega_i)$ is

$$\overline{\mathbf{E}}(\omega_i) = \mathbf{E}(\omega_i)e^{i\mathbf{k}_{\omega_i} \cdot \mathbf{r}} \tag{2.3}$$

which brings in the spatial dependence, which is always there, through the scalar product of the vectors $\mathbf{k}_{\omega_i}$ (wave vector) and $\mathbf{r} = (x,y,z)$, (space vector)

- the complex conjugate c.c. will introduce a starred notation, such as quantities

$$\overline{\mathbf{E}}^*(\omega_i)e^{i\omega_i t}$$

where the frequency has also changed sign.

Generally, this notation expressed by (2.2) and (2.3) will now be maintained both for the polarisations and the electric fields, either explicitly, or implicitly. When it is felt pedagogically necessary, however, the amplitudes of the ω and 2ω waves will be assigned explicit values rather than adhere to this notation. The response of a material to an electric field is quite complex and various methods have been evolved to deal with it but it is an advantage that many materials of interest to nonlinear optics are non-magnetic, do not support a current density and do not generate a space charge. These material features simplify Maxwell's equations and make the modelling of the polarisation considerably easier. The first step is to split the polarisation of the non-magnetic, charge-free, dielectric insulator into a linear part $\mathbf{P}_L$ and a nonlinear part $\mathbf{P}_{NL}$. The next step is to assume that $\mathbf{P}_{NL}$ is the convergent series

$$\mathbf{P}_{NL}^{(t)} = \mathbf{P}^{(2)}(t) + \mathbf{P}^{(3)}(t) + ... \tag{2.4}$$

where the superscript denotes the order of the nonlinearity and t is time. Harmonic processes described by (2.4) involve *virtual levels* which are *not* the actual energy levels of the atoms of the dielectric. If $(\hbar\omega + \hbar\omega)$, where $\hbar = h/2\pi$ and h is Planck's constant, reaches a *real energy level* of the material, resonance will occur and (2.4) will fail to converge and all terms in the series (2.4) will be significant. For (2.4) the physical process involves *bound electrons* and manifests itself as a change in shape of an electron cloud [1]. It is a very fast response indeed. It is important, therefore, for the model adopted here to work, that the harmonic conversion processes are away from any resonances. A simple example of the consequence of resonance is saturation of the nonlinearity with power. It is obvious then that (2.4) will not be valid. Crudely speaking, (2.4) implies that $\mathbf{P}^{(2)}(t) \propto E^2(t)$, and so on, but much more care than this needs to be taken. The precise derivation of $\mathbf{P}^{(2)}(t)$ is important to understand so this is the issue addressed in this section.

An elementary point can be established right away, however. $\mathbf{P}^{(2)}(t)$ *vanishes exactly* for materials possessing a centre of symmetry. For such materials, reversing $\mathbf{E}(t)$ must reverse $\mathbf{P}^{(2)}$ but clearly $\mathbf{E}^2(t)$ cannot change sign with $\mathbf{E}(t)$ so $\mathbf{P}^{(2)}$ must be identically zero in that case. The immediate conclusion is that the materials of interest to this school, do not possess a centre of symmetry and therefore, are quadratically nonlinear with a *non-zero* $\mathbf{P}^{(2)}(t)$ as the focus of interest.

The discussion begins with the *linear* response. It is not hard to imagine that $\mathbf{P}_L(t)$ will not, in general, relate only to $\mathbf{E}(t)$ but that it is somehow related to what the field was at other times. Suppose then that

$$\mathbf{P}_L(t) = \varepsilon_0 \int_{-\infty}^{\infty} dt_1 \kappa^{(1)}(t - t_1) \cdot \mathbf{E}(t_1) \tag{2.5}$$

where ε_0 is the permittivity of free space and $\kappa^{(1)}(t - t_1)$ is a linear susceptibility in the time domain, which must be a second-rank tensor i.e. $\kappa^{(1)} \equiv \kappa_{ij}^{(1)}$ (in the suffix notation). The relationship of $\mathbf{P}_L(t)$ to $\kappa^{(1)}$ embodies the desire to relate $\mathbf{P}_L(t)$ not only to $\mathbf{E}(t)$ but to values of $\mathbf{E}(t)$ that existed at all previous times $t_1 < t$. To do this, it is physically reasonable to introduce a material response $\kappa^{(1)}$ that depends on the time difference $(t - t_1)$ and to use (2.5) without changing its form, provided that $\kappa^{(1)}(t - t_1)$ is zero for times $t_1 > t$ *in the future*. Now make a change of variable $\tau = t - t_1$ to get

$$\mathbf{P}_L(t) = \varepsilon_0 \int_{-\infty}^{\infty} d\tau \kappa^{(1)}(\tau) \cdot \mathbf{E}(t - \tau) \tag{2.6}$$

It must be emphasised that $\kappa^{(1)} = 0$ for $\tau < 0$. Otherwise, $\mathbf{P}_L(t)$ will depend upon field values in the future rather than in the past; this is simply an expression of causality. Another point, alluded to earlier, that will forcibly demand attention later on, is that the field, polarisations are always real functions. This does not matter, at this stage, because only the linear response is being considered. Expressing $\mathbf{E}(t)$ and $\mathbf{P}_L(t)$ as real quantities means the separating of total field and polarisation into two terms; one is the complex conjugate of the other. In the linear case, all the field equations separate into two independent sets determining $\mathbf{E}(\omega)$ and $\mathbf{E}^*(\omega)$. When the nonlinear response is considered, this will not be possible and the field equations then become *coupled*. Clearly, the second-order polarisation has to be approached carefully and a good way to deal with it is in the frequency domain, as will now be demonstrated.

To proceed to the frequency domain the following definitions of the Fourier transform will be adopted

$$\mathbf{E}(t) = \int_{-\infty}^{\infty} d\nu \tilde{\mathbf{E}}(\omega) e^{i2\pi\nu t} \tag{2.7}$$

$$\widetilde{\mathbf{E}}(\omega) = \int\limits_{-\infty}^{\infty} dt \mathbf{E}(t) e^{-i2\pi vt} \tag{2.8}$$

$$\widetilde{\mathbf{P}}^{(2)}(\omega) = \int\limits_{-\infty}^{\infty} dt \mathbf{P}^{(2)}(t) e^{-i2\pi vt} \tag{2.9}$$

where the angular frequency is $\omega = 2\omega v$ and $\mathbf{P}^{(2)}(t)$ is the *total* second-order contribution to the polarisation, in the time domain. The extension of definition (2.6) indicates that the second-order equivalent is

$$P_i^{(2)}(t) = \varepsilon_0 \int\limits_{-\infty}^{\infty} d\tau_1 \int\limits_{-\infty}^{\infty} d\tau_2 \kappa_{ijk}^{(2)}(\tau_1,\tau_2) E_j(t-\tau_1) E_k(t-\tau_2) \tag{2.10}$$

where $\kappa_{ijk}^{(2)}(\tau_1,\tau_2)$ is a third-rank tensor representing the second-order *time-domain* response. Note that the suffix notation is now a much more convenient notation to adopt.

Using the definitions (2.7)-(2.9) results in

$$\widetilde{P}_i^{(2)}(\omega) = \varepsilon_0 \int\limits_{-\infty}^{\infty} dt\, e^{-i2\pi vt} \int\limits_{-\infty}^{\infty} d\tau_1 \int\limits_{-\infty}^{\infty} d\tau_2 \kappa_{ijk}^{(2)}(\tau_1,\tau_2) E_j(t-\tau_1) E_k(t-\tau_2)$$

$$= \varepsilon_0 \int\limits_{-\infty}^{\infty} dt \int\limits_{-\infty}^{\infty} d\tau_1 \int\limits_{-\infty}^{\infty} d\tau_2 \int\limits_{-\infty}^{\infty} dv' \int\limits_{-\infty}^{\infty} dv'' e^{-i2\pi vt} \kappa_{ijk}^{(2)}(\tau_1,\tau_2) \widetilde{E}_j(\omega') \widetilde{E}_k(\omega'') e^{i2\pi v'(t-\tau_1)} \times e^{i2\pi v''(t-\tau_1)}$$

$$= \varepsilon_0 \int\limits_{-\infty}^{\infty} dv' \kappa_{ijk}^{(2)}(\omega;\omega',\omega-\omega') \widetilde{E}_j(\omega') \widetilde{E}_k(\omega-\omega') \tag{2.11}$$

where

$$\widetilde{\kappa}_{ijk}^{(2)}(\omega;\omega',\omega-\omega') = \int\limits_{-\infty}^{\infty} d\tau_1 \int\limits_{-\infty}^{\infty} d\tau_2 \kappa_{ijk}^{(2)}(\tau_1,\tau_2) e^{-i2\pi v'\tau_1} e^{-i2\pi(v-v')\tau_2} \tag{2.12}$$

The notation adopted for the argument of $\widetilde{\kappa}_{ijk}^{(2)}$ is widely accepted and must be read as follows. The semi-colon separates the *sum* $\omega=\omega' + \omega-\omega'$ from the frequencies ω' and $(\omega-\omega')$ of the waves that are mixing in the nonlinear process. This is a convenient, and dramatic, notation that emphasises the physical processes and permits easy tracking of the complex electric field constituents, involved in the producing the second-order polarisation. It is widely used.

Further progress depends upon calculating $\widetilde{E}_j(\omega')$ and $\widetilde{E}_k(\omega-\omega')$. These are transforms of the *total* electric fields but, as mentioned above, when the frequency spectra of the fields at different frequencies do not overlap the total field $E(t)$ can be written as a *sum* of real contributions from frequencies $\omega_1 = 2\pi\nu_1$, $\omega_2 = 2\pi\nu$,... . Hence, for a total field with two contributing frequencies

$$E_i(t) = \frac{1}{2}\left[\overline{E}_i(\omega_1)e^{i2\pi\nu_1 t} + \overline{E}_i(\omega_2)e^{i2\pi\nu_2 t} + c.c.\right] \tag{2.13}$$

The Fourier transforms appearing in (2.11) are then

$$\begin{aligned}
\tilde{E}_j(\omega') = \frac{1}{2}\big[&\overline{E}_j(\omega_1)\delta(-\omega_1+\omega') + \overline{E}_j(\omega_2)\delta(-\omega_2+\omega') \\
&+ \overline{E}_j^*(\omega_1)\delta(\omega_1+\omega') + E_j^*(\omega_2)\delta(\omega_2+\omega')\big]
\end{aligned} \tag{2.14}$$

$$\begin{aligned}
\tilde{E}_k(\omega-\omega') = \frac{1}{2}\big[&\overline{E}_k(\omega_1)\delta(-\omega_1+\omega-\omega') + \overline{E}_k(\omega_2)\delta(-\omega_2+\omega-\omega') \\
&+ \overline{E}_k^*(\omega_1)\delta(\omega_1+\omega-\omega') + \overline{E}_k(\omega_2)\delta(\omega_2+\omega-\omega')\big]
\end{aligned} \tag{2.15}$$

The product $\tilde{E}_k(\omega')\tilde{E}_k(\omega-\omega')$ is

$$\begin{aligned}
\tilde{E}_j(\omega')\tilde{E}_k(\omega-\omega') = \frac{1}{4}\big[&\overline{E}_j(\omega_1)\overline{E}_k(\omega_1)\delta(-\omega_1+\omega')\delta(-\omega_1+\omega-\omega') \\
&+ \overline{E}_j(\omega_1)\overline{E}_k(\omega_2)\delta(-\omega_1+\omega')\delta(-\omega_2+\omega-\omega') \\
&+ \overline{E}_j(\omega_1)\overline{E}_k^*(\omega_1)\delta(-\omega_1+\omega')\delta(\omega_1+\omega-\omega') \\
&+ \overline{E}_j(\omega_1)\overline{E}_k^*(\omega_2)\delta(-\omega_1+\omega')\delta(\omega_2+\omega-\omega') \\
&+ \overline{E}_j(\omega_2)\overline{E}_k(\omega_1)\delta(-\omega_2+\omega')\delta(-\omega_1+\omega-\omega') \\
&+ \overline{E}_j(\omega_2)\overline{E}_k(\omega_2)\delta(-\omega_2+\omega')\delta(-\omega_2+\omega-\omega') \\
&+ \overline{E}_j(\omega_2)\overline{E}_k(\omega_1)\delta(-\omega_2+\omega')\delta(\omega_1+\omega-\omega') \\
&+ E_j(\omega_2)E_k(\omega_2)\delta(-\omega_2+\omega')\delta(\omega_2+\omega-\omega') \\
&+ E_j^*(\omega_1)E_k(\omega_1)\delta(\omega_1+\omega')\delta(-\omega_1+\omega-\omega') \\
&+ E_j^*(\omega_1)E_k(\omega_2)\delta(\omega_1+\omega')\delta(-\omega_2+\omega-\omega') \\
&+ \overline{E}_j^*(\omega_1)\overline{E}_k(\omega_2)\delta(\omega_1+\omega')\delta(\omega_2+\omega-\omega') \\
&+ \overline{E}_j^*(\omega_1)\overline{E}_k(\omega_1)\delta(\omega_1+\omega')\delta(\omega_1+\omega-\omega') \\
&+ \overline{E}_j^*(\omega_2)\overline{E}_k(\omega_2)\delta(\omega_2+\omega')\delta(-\omega_2+\omega-\omega') \\
&+ \overline{E}_j^*(\omega_2)\overline{E}_k(\omega_1)\delta(\omega_2+\omega')\delta(-\omega_1+\omega-\omega') \\
&+ \overline{E}_j^*(\omega_2)\overline{E}_k(\omega_1)\delta(\omega_2+\omega')\delta(\omega_1+\omega-\omega') \\
&+ \overline{E}_j^*(\omega_2)\overline{E}_k^*(\omega_2)\delta(\omega_2+\omega')\delta(\omega_2+\omega-\omega')\big]
\end{aligned} \tag{2.16}$$

The Fourier transform of the second-order polarisation can now be produced using the following property of the delta function

$$\int d\omega' \delta(N - \omega')\delta(\omega' - M) = \delta(N - M) \tag{2.17}$$

where N and M can be identified in the above expression for the field transform product e.g.

$$\int \delta(-\omega_1 + \omega')\delta(-\omega_1 + \omega - \omega')dv'$$
$$= \int dv' \omega \delta(\omega_1 - \omega')\delta(\omega' - (\omega - \omega_1)) = \delta(2\omega_1 - \omega) = \delta(-2\omega_1 + \omega) \tag{2.18}$$

The polarisation $\widetilde{P}_i^{(2)}(v)$ is, therefore,

$$\widetilde{P}_i^{(2)}(v) = \frac{\varepsilon_0}{4} \Big[\widetilde{\kappa}_{ijk}^{(2)}(\omega; \omega_1, \omega - \omega_1)\overline{E}_j(\omega_1)\overline{E}_k(\omega_1)\delta(-2\omega_1 + \omega)$$

$$+ \widetilde{\kappa}_{ijk}^{(2)}(\omega; \omega_1, \omega - \omega_1)\overline{E}_j(\omega_1)\overline{E}_k(\omega_2)\delta(-\omega_1 - \omega_2 + \omega)$$

$$+ \widetilde{\kappa}_{ijk}^{(2)}(\omega; \omega_1, \omega - \omega_1)\overline{E}_j(\omega_1)\overline{E}_k^*(\omega_1)\delta(\omega)$$

$$+ \widetilde{\kappa}_{ijk}^{(2)}(\omega; \omega_1, \omega - \omega_1)\overline{E}_j(\omega_1)\overline{E}_k^*(\omega_2)\delta(-\omega_1 + \omega_2 + \omega)$$

$$+ \widetilde{\kappa}_{ijk}^{(2)}(\omega; \omega_2, \omega - \omega_2)\overline{E}_j(\omega_2)\overline{E}_k(\omega_1)\delta(-\omega_1 - \omega_2 + \omega)$$

$$+ \widetilde{\kappa}_{ijk}^{(2)}(\omega; \omega_2, \omega - \omega_2)\overline{E}_j(\omega_2)\overline{E}_k(\omega_2)\delta(-2\omega_2 + \omega)$$

$$+ \widetilde{\kappa}_{ijk}^{(2)}(\omega; \omega_2, \omega - \omega_2)\overline{E}_j(\omega_2)\overline{E}_k^*(\omega_1)\delta(\omega_1 - \omega_2 + \omega)$$

$$+ \widetilde{\kappa}_{ijk}^{(2)}(\omega; \omega_2, \omega - \omega_2)\overline{E}_j(\omega_2)\overline{E}_k^*(\omega_2)\delta(\omega)$$

$$+ \widetilde{\kappa}_{ijk}^{(2)}(\omega; -\omega_1, \omega + \omega_1)\overline{E}_j^*(\omega_1)\overline{E}_k(\omega_1)\delta(\omega)$$

$$+ \widetilde{\kappa}_{ijk}^{(2)}(\omega; -\omega_1, \omega + \omega_1)\overline{E}_j^*(\omega_1)\overline{E}_k(\omega_2)\delta(-\omega_2 + \omega_1 + \omega)$$

$$+ \widetilde{\kappa}_{ijk}^{(2)}(\omega; -\omega_1, \omega + \omega_1)\overline{E}_j^*(\omega_1)\overline{E}_k^*(\omega_1)\delta(2\omega_1 + \omega)$$

$$+ \widetilde{\kappa}_{ijk}^{(2)}(\omega; -\omega_1, \omega + \omega_1)\overline{E}_j^*(\omega_1)\overline{E}_k^*(\omega_2)\delta(\omega_1 + \omega_2 + \omega)$$

$$+ \widetilde{\kappa}_{ijk}^{(2)}(\omega; -\omega_2, \omega + \omega_2)\overline{E}_j^*(\omega_2)\overline{E}_k(\omega_1)\delta(-\omega_1 + \omega_2 + \omega)$$

$$+ \widetilde{\kappa}_{ijk}^{(2)}(\omega; -\omega_2, \omega + \omega_2)\overline{E}_j^*(\omega_2)\overline{E}_k(\omega_2)\delta(\omega)$$

$$+ \widetilde{\kappa}_{ijk}^{(2)}(\omega; -\omega_2, \omega + \omega_2)\overline{E}_j^*(\omega_2)\overline{E}_k^*(\omega_1)\delta(\omega_1 + \omega_2 + \omega)$$

$$+ \widetilde{\kappa}_{ijk}^{(2)}(\omega; -\omega_2, \omega + \omega_2)\overline{E}_j^*(\omega_2)\overline{E}_k^*(\omega_2)\delta(2\omega_2 + \omega) \Big] \tag{2.19}$$

The second-order polarisation, from the assumption that the total field is the sum of the field components associated with ω_1 and ω_2, must contain contributions $2\omega_1$, 0, $\omega_1 - \omega_2$, $\omega_1 + \omega_2$, $2\omega_2$ and their complex conjugates i.e.

$$\tilde{P}_i^{(2)}(\omega) = \frac{1}{2}\left[\overline{P}_i^{(2)}(2\omega_1)\delta(-2\omega_1 + \omega) + \overline{P}_i^{(2)*}(2\omega_1)\delta(2\omega_1 + \omega) + \overline{P}_i^{(2)}(0)\delta(\omega)\right.$$
$$+ \overline{P}_i^{(2)}(\omega_1 + \omega_2)\delta(-\omega_1 - \omega_2 + \omega) + \overline{P}_i^{(2)*}(\omega_1 + \omega_2)\delta(\omega_1 + \omega_2 + \omega) \qquad (2.20)$$
$$+ \overline{P}_i^{(2)}(\omega_1 - \omega_2)\delta(-\omega_1 + \omega_2 + \omega) + \overline{P}_i^{(2)*}(\omega_1 - \omega_2)\delta(\omega_1 - \omega_2 + \omega)$$
$$\left. + \overline{P}_i^{(2)}(2\omega_2)\delta(\omega - 2\omega_2) + \overline{P}_i^{(2)*}(2\omega_2)\delta(\omega + 2\omega_2\omega)\right]$$

Hence,

$$\overline{P}_i^{(2)}(2\omega_1) = \frac{\varepsilon_0}{2}\tilde{\kappa}_{ijk}^{(2)}(2\omega_1;\omega_1,\omega_1)\overline{E}_j(\omega_1)\overline{E}_k(\omega_2) \qquad (2.21)$$

$$\overline{P}_i^{(2)}(\omega_1 + \omega_2) = \frac{\varepsilon_0}{2}\left[\tilde{\kappa}_{ijk}^{(2)}(\omega_1 + \omega_2;\omega_1,\omega_2)\overline{E}_j(\omega_1)\overline{E}_k(\omega_2)\right.$$
$$\left. + \tilde{\kappa}_{ijk}^{(2)}(\omega_1 + \omega_2;\omega_2,\omega_1)\overline{E}_j(\omega_2)\overline{E}_k(\omega_1)\right] \qquad (2.22)$$

$$\overline{P}_i^{(2)}(0) = \frac{\varepsilon_0}{2}\left[\tilde{\kappa}_{ijk}^{(2)}(0;\omega_1,-\omega_1)\overline{E}_j(\omega_1)\overline{E}_k^*(\omega_1) + \tilde{\kappa}_{ijk}^{(2)}(0;\omega_2,-\omega_2)\overline{E}_j(\omega_2)\overline{E}_k^*(\omega_2)\right.$$
$$\left. + \tilde{\kappa}_{ijk}^{(2)}(0;-\omega_1,\omega_1)\overline{E}_j^*(\omega_1)\overline{E}_k(\omega_1) + \tilde{\kappa}_{ijk}^{(2)}(0;-\omega_2,\omega_2)\overline{E}_j^*(\omega_2)\overline{E}_k(\omega_2)\right] \qquad (2.23)$$

$$\overline{P}_i^{(2)}(-\omega_1 + \omega_2) = \frac{\varepsilon_0}{2}\left[\tilde{\kappa}_{ijk}^{(2)}(-\omega_1 + \omega_2;\omega_2,-\omega_1)\overline{E}_j(\omega_2)\overline{E}_k^*(\omega_1) +\right.$$
$$\left. \tilde{\kappa}_{ijk}^{(2)}(-\omega_1 + \omega_2;-\omega_1,\omega_2)\overline{E}_j^*(\omega_1)\overline{E}_k(\omega_2)\right] \qquad (2.24)$$

$$\overline{P}_i^{(2)}(2\omega_2) = \frac{\varepsilon_0}{2}\tilde{\kappa}_{ijk}^{(2)}(2\omega_2;\omega_2,\omega_2)E_j(\omega_2)E_k(\omega_2) \qquad (2.25)$$

where the superscript still denotes second-order.

This completes the general derivation of the spectral contributions to $\mathbf{P}^{(2)}(t)$, the *total* polarisation. It can be seen that the Fourier transform, at a general frequency ω, has quite a number of contributions and that $\tilde{P}_i^{(2)}(\omega)$ consists of delta-function contributions with complex amplitudes $\overline{P}_i^{(2)}(2\omega_1)$, $\overline{P}_i^{(2)}(2\omega_2)$, $\overline{P}_i^{(2)}(\omega_1 + \omega_2)$, $\overline{P}_i^{(2)}(0)$ and $\overline{P}_i^{(2)}(\omega_2 - \omega_1)$. These are associated with the various frequency combinations, implied by assuming, in the beginning, that $\mathbf{E}(t)$ has two driving frequencies ω_1 and ω_2. In other words, even though the starting point of this calculation constructed $\mathbf{E}(t)$

from two *specific* fields at ω_1 and ω_2, respectively, and it is important to the derivation to take transform of $E(t)$ at some *general* frequency ω. The delta functions then pick out the complex amplitudes for the various combinations of ω_1 and ω_2. Having achieved this objective, it is no longer necessary to maintain the notation ω_1, ω_2 because the focus of attention is going to be on the particular problem of converting some *fundamental* ω to a *second-harmonic* frequency 2ω, and any reverse order, or *cascade*, process that *down-converts* 2ω to ω. The Fourier amplitudes of these processes are contained in the set (2.21) to (2.25) but, because the calculation was done with ω_1 and ω_2, the frequencies are now be set to the values

$$\omega_1 = \omega, \quad \omega_2 = 2\omega \tag{2.26}$$

which permits the Fourier amplitudes of the polarisation contributions involved in the harmonic conversion to be written as

$$\overline{P}_i^{(2)}(2\omega) = \frac{\varepsilon_0}{2} \widetilde{\kappa}_{ijk}^{(2)}(2\omega;\omega,\omega)\overline{E}_j(\omega)\overline{E}_k(\omega) \tag{2.27}$$

$$\overline{P}_i^{(2)}(\omega) = \frac{\varepsilon_0}{2}\left[\widetilde{\kappa}_{ijk}^{(2)}(\omega;2\omega,-\omega)\overline{E}_j(2\omega)\overline{E}_k^*(\omega) + \widetilde{\kappa}_{ijk}^{(2)}(\omega;-\omega,2\omega)\overline{E}_j^*(\omega)\overline{E}_k(\omega_2)\right] \tag{2.28}$$

To be specific, assume that the electromagnetic waves propagate along the z-axis and that the waves are either plane cw radiation or are weakly guided TE or TM waves of a planar structure. Either way, $\overline{E}_z(\omega) \cong 0, \overline{E}_z(2\omega) \cong 0$. The forms of $\overline{P}_i^{(2)}(\omega)$ and $\overline{P}_i^{(2)}(2\omega)$, written, explicitly, in terms of $\overline{E}_x(2\omega), \overline{E}_x(\omega), \overline{E}_y(2\omega), \overline{E}_y(\omega)$, are

$$\begin{aligned}\overline{P}_i^{(2)}(2\omega) = \frac{\varepsilon_0}{2}\Big[&\widetilde{\kappa}_{ixy}^{(2)}(2\omega;\omega,\omega)\overline{E}_x^2(\omega) + \widetilde{\kappa}_{iyy}^{(2)}(2\omega,\omega,\omega)\overline{E}_y^2(\omega)\\ &+\widetilde{\kappa}_{ixy}^{(2)}(2\omega;\omega,\omega)\overline{E}_x(\omega)\overline{E}_y(\omega) + \widetilde{\kappa}_{iyx}^{(2)}(2\omega;\omega,\omega)\overline{E}_y(\omega)\overline{E}_x(\omega)\Big]\end{aligned} \tag{2.29}$$

and

$$\begin{aligned}\widetilde{P}_i^{(2)}(\omega) = \frac{\varepsilon_0}{2}\Big[&\widetilde{\kappa}_{ixx}^{(2)}(\omega;2\omega,-\omega)\overline{E}_x(2\omega)\overline{E}_x^*(\omega)\\ &+\widetilde{\kappa}_{iyy}^{(2)}(\omega;2\omega,-\omega)\overline{E}_y(2\omega)\overline{E}_y^*(\omega) + \widetilde{\kappa}_{ixy}^{(2)}(\omega;2\omega,-\omega)\overline{E}_x(2\omega)\overline{E}_y^*(\omega)\\ &+\widetilde{\kappa}_{iyx}^{(2)}(\omega;2\omega,-\omega)\overline{E}_y(2\omega)\overline{E}_x(\omega) + \widetilde{\kappa}_{ixx}^{(2)}(\omega;-\omega,2\omega)\overline{E}_x(\omega)\overline{E}_x^*(2\omega)\\ &+\widetilde{\kappa}_{iyy}^{(2)}(\omega;-\omega,2\omega)\overline{E}_y^*(\omega)\overline{E}_y(2\omega) + \widetilde{\kappa}_{ixy}^{(2)}(\omega;-\omega,2\omega)\overline{E}_x^*(\omega)\overline{E}_y(2\omega)\\ &+\widetilde{\kappa}_{iyx}^{(2)}(\omega;-\omega,2\omega)\overline{E}_y^*(\omega)\overline{E}_x(2\omega)\Big]\end{aligned} \tag{2.30}$$

A further simplification comes from the fact that the tensor $\widetilde{\kappa}_{ijk}^{(2)}(\omega_s;\omega_1,\omega_2)$ enjoys *intrinsic permutation symmetry*, so that

$$\widetilde{\kappa}_{ixy}^{(2)}(2\omega;\omega,\omega) = \widetilde{\kappa}_{iyx}(2\omega;\omega,\omega) \tag{2.31a}$$

$$\widetilde{\kappa}_{ixx}^{(2)}(\omega;2\omega,-\omega) = \widetilde{\kappa}_{ixx}^{(2)}(\omega;-\omega,2\omega) \tag{2.31b}$$

$$\widetilde{\kappa}_{iyy}^{(2)}(\omega;2\omega,-\omega) = \widetilde{\kappa}_{iyy}^{(2)}(\omega;-\omega,2\omega) \tag{2.31c}$$

$$\widetilde{\kappa}_{iyx}^{(2)}(\omega;2\omega,-\omega) = \widetilde{\kappa}_{ixy}^{(2)}(\omega;-\omega,2\omega) \tag{2.31d}$$

$$\widetilde{\kappa}_{ixy}^{(2)}(\omega;2\omega,-\omega) = \widetilde{\kappa}_{iyx}^{(2)}(\omega;-\omega,2\omega) \tag{2.31e}$$

All of the above involve the *simultaneous* permutation of the *last two* indices and the corresponding *last two* frequencies cited. These symmetry relationships transform $\overline{P}_i^{(2)}(\omega)$ to

$$\widetilde{P}_i^{(2)}(\omega) = \varepsilon[\widetilde{\kappa}_{ixx}^{(2)}(\omega;-\omega,2\omega)\overline{E}_x^*(\omega)\overline{E}_x(2\omega)$$
$$+ \widetilde{\kappa}_{iyy}^{(2)}(\omega;-\omega,2\omega)\overline{E}_y^*(\omega)\overline{E}_y(2\omega) + \widetilde{\kappa}_{ixy}^{(2)}(\omega;-\omega,2\omega)\overline{E}_x^*(\omega)\overline{E}_y(2\omega) \tag{2.32}$$
$$+ \widetilde{\kappa}_{iyx}^{(2)}(\omega;-\omega,2\omega)\overline{E}_y^*(\omega)\overline{E}_x(2\omega)]$$

Clearly, it is now possible to express *both* $\overline{P}_i^{(2)}(\omega)$ and $\overline{P}_i^{(2)}(2\omega)$ in the *same way* i.e.

$$\overline{P}_i^{(2)}(\omega_s) = \varepsilon_0\chi_{ijk}^{(2)}(\omega_s;\omega_1,\omega_2)\overline{E}_j(\omega_1)\overline{E}_k(\omega_2) \tag{2.33}$$

where *another* form of second-order susceptibility tensor $\chi_{ijk}^{(2)}(\omega_s,\omega_1,\omega_2)$ is introduced as [2,3]

$$\chi_{ijk}^{(2)}(\omega_s;\omega_1,\omega_2) = \text{factor x } \widetilde{\kappa}_{ijk}^{(2)}(\omega_s;\omega_1,\omega_2) \tag{2.34}$$

where

$$\omega_s = \omega_1 + \omega_2$$

$$\text{factor} = 1 \quad, \quad \omega_1 \neq \omega_2$$

$$\text{factor} = \frac{1}{2} \quad, \quad \omega_1 = \omega_2$$

The values of the factor are evident from (2.29) and (2.32).

It has already been stated that $\widetilde{\kappa}_{ijk}^{(2)}(\omega_s;\omega_1,\omega_2)$ are controlled by intrinsic symmetry but, in a lossless medium, the components of $\widetilde{\kappa}_{ijk}^{(2)}(\omega_s;\omega_1,\omega_2)$ also have an *overall permutation symmetry* i.e. the nonlinear susceptibility remains invariant if *any* tensor index is interchanged together with its associated frequency. The first frequency in $(\omega_s;\omega_1,\omega_2)$ is the sum of ω_1 and ω_2, so interchanging ω_s with ω_1 for example, means that the sign of *both* must change. If resonance between a pair of real levels in a material is created by an input frequency then the series expansion of $P(t)$ will not be useful as discussed earlier on, examples being dielectric breakdown and saturation. Provided that the driving frequencies are well away from resonances, however, the series will converge and dispersion can be neglected in $\widetilde{\kappa}_{ijk}^{(2)}(\omega_s;\omega_1,\omega_2)$. If this is true, then the indices can be moved *without* moving the frequencies and

$$\widetilde{\kappa}_{ixy}^{(2)}(\omega;-\omega,2\omega) \cong \widetilde{\kappa}_{iyx}^{(2)}(\omega;-\omega,2\omega) \approx \widetilde{\kappa}_{iyx}^{(2)}(2\omega;-\omega,\omega) \approx \widetilde{\kappa}_{iyx}^{(2)}(2\omega;\omega,\omega) \equiv \widetilde{\kappa}_{ixy}^{(2)} \qquad (2.35)$$

$$\widetilde{\kappa}_{ixx}^{(2)}(\omega;-\omega,2\omega) \cong \widetilde{\kappa}_{ixx}^{(2)}(2\omega;\omega,\omega) \equiv \widetilde{\kappa}_{ixx}^{(2)} \qquad (2.36)$$

$$\widetilde{\kappa}_{iyy}^{(2)}(\omega;-\omega,2\omega) \cong \widetilde{\kappa}_{iyy}^{(2)}(2\omega;\omega,\omega) \equiv \widetilde{\kappa}_{iyy}^{(2)} \qquad (2.37)$$

This is called Kleinman's symmetry and leads to

$$\overline{P}_i^{(2)}(2\omega) = \frac{\varepsilon_0}{2}\left[\widetilde{\kappa}_{ixx}^{(2)}\overline{E}_x^2(\omega) + \widetilde{\kappa}_{iyy}^{(2)}\overline{E}_y^2(\omega) + \widetilde{\kappa}_{ixy}^{(2)}\overline{E}_x(\omega)\overline{E}_y(\omega)\right] \qquad (2.38)$$

$$\begin{aligned}\widetilde{P}_i^{(2)}(\omega) = \varepsilon_0\Big[&\widetilde{\kappa}_{ixx}^{(2)}\overline{E}_x(2\omega)\overline{E}_x^*(\omega) + \widetilde{\kappa}_{iyy}^{(2)}\overline{E}_y(2\omega)\overline{E}_y^*(\omega) \\ &+ \widetilde{\kappa}_{ixy}^{(2)}\Big[\overline{E}_x(2\omega)\overline{E}_y^*(\omega) + \overline{E}_x^*(\omega)\overline{E}_y(2\omega)\Big]\Big]\end{aligned} \qquad (2.39)$$

From this, it appears that

$$\widetilde{\kappa}_{ijk}^{(2)}(2\omega;\omega,\omega) = \widetilde{\kappa}_{ijk}^{(2)}(\omega;2\omega,-\omega) \qquad (2.40)$$

so that

$$\chi_{ijk}^{(2)}(2\omega;\omega,\omega) = 2\chi_{ijk}^{(2)}(\omega;2\omega,-\omega) \qquad (2.41)$$

a statement that is often seen in the literature. It arises through symmetry and the use of $\chi_{ijk}^{(2)}(\omega_s;\omega_1,\omega_2)$, as opposed to $\widetilde{\kappa}_{ijk}^{(2)}(\omega_s;\omega_1,\omega_2)$, to produce a *compact* formula that yields both $\overline{P}_i^{(2)}(\omega)$ and $\overline{P}_i^{(2)}(2\omega)$ in a rather elegant manner.

The $\chi_{ijk}^{(2)}(\omega_s;\omega_1,\omega_2)$, or the equivalent $\tilde{\kappa}_{ijk}^{(2)}(\omega_s;\omega_1,\omega_2)$, tensor has, in principle, 27 non-zero elements so, in full, $\chi_{ijk}^{(2)}$ (omitting the frequency argument for clarity of presentation) is the array [43]

$$\begin{bmatrix} xxx & xyy & xzz & xyz & xzy & xzx & xxz & xxy & xyx \\ yxx & yyy & yzz & yyz & yzy & yzx & yxz & yxy & yyx \\ zxx & zyy & zzz & zyz & zzy & zzx & zxz & zxy & zyx \end{bmatrix}$$

Some of these elements are equal to each other by Kleïnman symmetry but the biggest feature of $\chi_{ijk}^{(2)} \equiv \tilde{\kappa}_{ijk}^{(2)}$ is that it also has spatial symmetry. The essential point is that a tensor is a physical property so that even if tensor arrays look like matrices, they are most definitely not. They are tensors and a tensor, being an *observable physical entity*, must be invariant under coordinate transformations.

The transformations are those that take the coordinate system $x_i \equiv (x_1,x_2,x_3)$ [where we use a suffix system of (x,y,z)] to a new coordinate system x_i'. This is effected by a transformation $x_i' = a_{ij}x_j$, where a_{ij} is the set of linear transformations that form a group of symmetry elements. It includes all the elements of the crystal point group and the property is called spatial symmetry. Whenever a crystal has a centre of symmetry its group contains the inversion symmetry operator and $\mathbf{P}^{(2)}(t)$ vanishes exactly! This means that of the 32 crystal classes, 11 of them do not map onto materials that have a finite $\chi_{ijk}^{(2)}$. Interestingly, however, calcite, which is a doubly refracting uniaxial material might be thought to be a $\chi^{(2)}$ candidate. It is one of the few double refractors that has a centre of symmetry, however, i.e. it includes inversion in its list of symmetry operations so it does not have a finite second-order susceptibility. Another well-known uniaxial material is lithium niobate [LiNbO$_3$] but this time, because it is in the 3m crystal class, it does possess a finite second-order nonlinear susceptibility. LiNbO$_3$ is in the trigonal crystal system but there are many other possibilities, such as CdSe and CdS in the 6mm crystal class of the hexagonal system.

For the $\bar{6}$ crystal class, from the hexagonal crystal system, the non-zero elements of $\chi_{ijk}^{(2)}$ are, with the z-axis lying along the single optic axis, [43]

$$\bar{6} \quad \begin{bmatrix} xxx & \overline{xxx} & 0 & 0 & 0 & 0 & 0 & \overline{yyy} & \overline{yyy} \\ (xxx) & (xyy) & & & & & & (xxy) & (xyx) \\ \overline{yyy} & yyy & 0 & 0 & 0 & 0 & 0 & \overline{xxx} & \overline{xxx} \\ (yxx) & (yyy) & & & & & & (yxy) & (yyx) \\ 0 & 0 & 0 & 0 & 0 & 0 & 0 & 0 & 0 \end{bmatrix}$$

where the actual element is shown in brackets and the bar denotes that the element is the negative value of the element indicated. It is clear from this that not many elements survive the crystal symmetry (spatial) operations. For the 3m class (lithium niobate) of the trigonal crystal system

$$
\boxed{3m} \quad
\begin{bmatrix}
0 & 0 & 0 & 0 & 0 & xzx & xxz & \overline{yyy} & \overline{yyy} \\
\overline{yyy} & yyy & 0 & xxz & xzx & 0 & 0 & 0 & 0 \\
zxx & zxx & zzz & 0 & 0 & 0 & 0 & 0 & 0
\end{bmatrix}
$$

The calculation of $\overline{P}_i^{(2)}(\omega)$ and $\overline{P}_i^{(2)}(2\omega)$ results in

$$
\begin{pmatrix} \overline{P}_x^{(2)}(\omega) \\ \overline{P}_y^{(2)}(\omega) \end{pmatrix} = 2\varepsilon_0
\begin{bmatrix}
\frac{1}{2}\widetilde{\kappa}_{xxx}^{(2)} & \frac{1}{2}\widetilde{\kappa}_{xyy}^{(2)} & \frac{1}{2}\widetilde{\kappa}_{xxy}^{(2)} \\
\frac{1}{2}\widetilde{\kappa}_{yxx}^{(2)} & \frac{1}{2}\widetilde{\kappa}_{yyy}^{(2)} & \frac{1}{2}\widetilde{\kappa}_{yxy}^{(2)}
\end{bmatrix}
\begin{bmatrix}
\overline{E}_x(2\omega)\overline{E}_x^*(\omega) \\
\overline{E}_y(2\omega)\overline{E}_y^*(\omega) \\
\overline{E}_x(2\omega)\overline{E}_y^*(\omega) + \overline{E}_x^*(\omega)\overline{E}_y(2\omega)
\end{bmatrix}
\qquad (2.42)
$$

$$
\begin{pmatrix} \overline{P}_x^{(2)}(2\omega) \\ \overline{P}_y^{(2)}(\omega) \end{pmatrix} = \varepsilon_0
\begin{bmatrix}
\frac{1}{2}\widetilde{\kappa}_{xxx}^{(2)} & \frac{1}{2}\widetilde{\kappa}_{xyy}^{(2)} & \frac{1}{2}\widetilde{\kappa}_{xxy}^{(2)} \\
\frac{1}{2}\widetilde{\kappa}_{yxx}^{(2)} & \frac{1}{2}\widetilde{\kappa}_{yyy}^{(2)} & \frac{1}{2}\widetilde{\kappa}_{yxy}^{(2)}
\end{bmatrix}
\begin{bmatrix}
\overline{E}_x^2(\omega) \\
\overline{E}_y^2(\omega) \\
2\overline{E}_x(\omega)\overline{E}_y(\omega)
\end{bmatrix}
\qquad (2.43)
$$

The notation will now be simplified through the definitions

$$
\widetilde{\kappa}_{xxx}^{(2)} = \kappa_1 \, , \quad \widetilde{\kappa}_{xxy}^{(2)} = \kappa_2
\qquad (2.44)
$$

Using the crystal symmetries appropriate to $\overline{6}$ and 3m [see proustite in figure 1] then leads to [15, 30]

$$
\boxed{\overline{6}} \quad
\begin{pmatrix} \overline{P}_x^{(2)}(\omega) \\ \overline{P}_y^{(2)}(\omega) \end{pmatrix} = 2\varepsilon_0
\begin{bmatrix}
\dfrac{\kappa_1}{2} & -\dfrac{\kappa_1}{2} & -\dfrac{\kappa_2}{2} \\
-\dfrac{\kappa_2}{2} & \dfrac{\kappa_2}{2} & -\dfrac{\kappa_1}{2}
\end{bmatrix}
\begin{bmatrix}
\overline{E}_x(2\omega)\overline{E}_x^*(\omega) \\
\overline{E}_y(2\omega)\overline{E}_y^*(\omega) \\
\overline{E}_x(2\omega)\overline{E}_y^*(\omega) + \overline{E}_x^*(\omega)\overline{E}_y(2\omega)
\end{bmatrix}
\qquad (2.45)
$$

$$
\begin{pmatrix} \overline{P}_x^{(2)}(2\omega) \\ \overline{P}_y^{(2)}(2\omega) \end{pmatrix} = \varepsilon_0
\begin{bmatrix}
\dfrac{\kappa_1}{2} & -\dfrac{\kappa_1}{2} & -\dfrac{\kappa_2}{2} \\
-\dfrac{\kappa_2}{2} & \dfrac{\kappa_2}{2} & -\dfrac{\kappa_1}{2}
\end{bmatrix}
\begin{bmatrix}
\overline{E}_x^2(\omega) \\
\overline{E}_y^2(\omega) \\
2\overline{E}_x(\omega)\overline{E}_y(\omega)
\end{bmatrix}
\qquad (2.46)
$$

$$\boxed{3m} \quad \begin{pmatrix} \overline{P}_x^{(2)}(\omega) \\ \overline{P}_y^{(2)}(\omega) \end{pmatrix} = 2\varepsilon_0 \begin{bmatrix} 0 & 0 & -\dfrac{\kappa_2}{2} \\ -\dfrac{\kappa_2}{2} & \dfrac{\kappa_2}{2} & 0 \end{bmatrix} \begin{bmatrix} \overline{E}_x(2\omega)\overline{E}_x(\omega) \\ \overline{E}_y(2\omega)\overline{E}_y(\omega) \\ \overline{E}_x(2\omega)\overline{E}_y^*(\omega) + \overline{E}_x^*(\omega)\overline{E}_y(2\omega) \end{bmatrix} \quad (2.47)$$

$$\begin{pmatrix} \overline{P}_x^{(2)}(2\omega) \\ \overline{P}_y^{(2)}(2\omega) \end{pmatrix} = \varepsilon_0 \begin{bmatrix} 0 & 0 & -\dfrac{\kappa_2}{2} \\ -\dfrac{\kappa_2}{2} & \dfrac{\kappa_2}{2} & 0 \end{bmatrix} \begin{bmatrix} \overline{E}_x^2(\omega) \\ \overline{E}_y^2(\omega) \\ 2\overline{E}_x(\omega)\overline{E}_y(\omega) \end{bmatrix} \quad (2.48)$$

For the hexagonal crystal system

$$\kappa_1 \neq \kappa_2 \neq 0 \quad (\overline{6} \text{ class})$$

$$\kappa_1 = 0 \quad (\overline{6} \text{ m2 class})$$

and for the trigonal system

$$\kappa_2 = 0 \quad (32 \text{ class} : \beta\text{-quartz})$$

$$\kappa_1 = 0 \quad (3m \text{ class} : \text{lithium niobate})$$

The more general $\kappa_1 \neq \kappa_2 \neq 0$ case finally becomes

$$\overline{P}_x^{(2)}(\omega) = \varepsilon_0[\kappa_1\overline{E}_x(2\omega)\overline{E}_x^*(\omega) - \kappa_1\overline{E}_y(2\omega)\overline{E}_y^*(\omega) - \kappa_2(\overline{E}_x(2\omega)\overline{E}_y^*(\omega) + \overline{E}_x^*(\omega)\overline{E}_y(2\omega))]$$

$$(2.49a)$$

$$\overline{P}_y^{(2)}(\omega) = \varepsilon_0[-\kappa_2\overline{E}_x(2\omega)\overline{E}_x^*(\omega) + \kappa_2\overline{E}_y(2\omega)\overline{E}_y^*(\omega) - \kappa_1(\overline{E}_x(2\omega)\overline{E}_y^*(\omega) + \overline{E}_x^*(\omega)\overline{E}_y(2\omega))]$$

$$(2.49b)$$

$$\overline{P}_x^{(2)}(2\omega) = \varepsilon_0\left[\frac{\kappa_1}{2}\overline{E}_x^2(\omega) - \frac{\kappa_1}{2}\overline{E}_x^2(\omega) - \kappa_2\overline{E}_x(\omega)\overline{E}_y(\omega)\right] \quad (2.49c)$$

$$\overline{P}_y^{(2)}(2\omega) = \varepsilon_0\left[-\frac{\kappa_2}{2}\overline{E}_x^2(\omega) + \frac{\kappa_2}{2}\overline{E}_y^2(\omega) - \kappa_1\overline{E}_x(\omega)\overline{E}_y(\omega)\right] \quad (2.49d)$$

For scalar fields propagating along the z-axis TE beams polarised along the x-axis could be considered or, at least, E_y can be set to zero for cw radiation. So for a 32 class crystal, with $E_x \neq 0$, $E_y = 0$, $\kappa_2 = 0$

$$\overline{P}_x(2\omega) = \frac{1}{2}\varepsilon_0\widetilde{\kappa}^{(2)}_{xxx}(2\omega;\omega,\omega)\overline{E}^2_x(\omega) = \varepsilon_0\chi^{(2)}_{xxx}(2\omega;\omega,\omega)\overline{E}^2_x(\omega) = \varepsilon_0\frac{\kappa_1}{2}\overline{E}^2_x \qquad (2.50)$$

$$\overline{P}_x(\omega) = \varepsilon_0\widetilde{\kappa}^{(2)}_{xxx}(\omega;2\omega,-\omega)\overline{E}_x(2\omega)\overline{E}^*_x(\omega)$$
$$= \varepsilon_0\chi^{(2)}_{xxx}(\omega;2\omega,-\omega)\overline{E}_x(2\omega)\overline{E}^*_x(\omega) = \varepsilon_0\kappa_1\overline{E}_x(2\omega)\overline{E}^*_x(\omega) \qquad (2.51)$$

Clearly the x-axis or the y-axis can be selected as the polarisation direction so $E_x = 0$, $E_y \neq 0$, $\kappa_1 = 0$ produces [15, 30]

$$\overline{P}_y(2\omega) = \varepsilon_0\frac{\kappa_2}{2}\overline{E}^2_y(\omega) \qquad (2.52)$$

$$\overline{P}_y(\omega) = \varepsilon_0\kappa_2\overline{E}_y(2\omega)\overline{E}^*_y(\omega) \qquad (2.53)$$

These results will now form the starting point for a study of three-wave interactions in second-order nonlinear materials. This study, which is set up in the next section, is based upon cw plane radiation and looks first at traditional second-harmonic generation and then at the possibility of generating eigenstates, in which a *balanced propagation* of finite ω and 2ω fields can occur, but yet with no energy exchange. This possibility was recently emphasised by Kaplan who talked of the *nonlinear eigenmodes* of $\chi^{(2)}$ nonlinear interactions. The interaction of ω and 2ω waves is more complex, therefore, than the second-harmonic generation story, which is the preoccupation of textbooks on nonlinear optics. The principles will now be exposed in a way developed by the authors of this chapter.

3. Scalar CW Equations: Plane Wave Solutions

In SI units, the four Maxwell equations that describe a non-paramagnetic insulating, charge-free dielectric are

$$\text{curl }\mathbf{E} = -\frac{\partial\mathbf{B}}{\partial t}, \quad \text{curl }\mathbf{H} = -\frac{\partial\mathbf{D}}{\partial t} \qquad (3.1)$$

$$\text{div }\mathbf{D} = 0, \quad \text{div }\mathbf{B} = 0 \qquad (3.2)$$

The medium will be assumed here to be homogeneous and the material properties are brought in through the inclusion of the polarisation P(t) in the displacement vector **D** i.e.

$$\mathbf{D}(t) = \varepsilon_0\mathbf{E}(t) + \mathbf{P}(t) = \varepsilon_0\mathbf{E}(t) + \mathbf{P}_L(t) + \mathbf{P}_{NL}(t) \qquad (3.3)$$

where ε_0 is the permittivity of free space, $\varepsilon_0 \mathbf{E}(t)$ is the free-space displacement and for clarity the *polarisation* of the material has been split into a linear part $\mathbf{P}_L(t)$ and a nonlinear part $\mathbf{P}_{NL}(t)$.

A manipulation of Maxwell's equations, using the identity

$$\text{curl curl } \mathbf{E} = -\nabla^2 \mathbf{E} + \text{grad}(\text{div}\mathbf{E}) \tag{3.4}$$

and, after setting $\text{div}(\mathbf{E}) = 0$, and assuming that the linear polarisation is $\mathbf{P}_L = \varepsilon_0 \, \varepsilon \cdot \mathbf{E}$, the frequency domain equation is

$$\nabla^2 \overline{\mathbf{E}}(\omega_i) + \frac{\omega_i^2}{c^2} \varepsilon^{(i)} \cdot \overline{\mathbf{E}}(\omega_i) + \omega_i^2 \mu_0 \overline{\mathbf{P}}_{NL}(\omega_i) = 0 \tag{3.5}$$

For each frequency $\omega_i \, \varepsilon^{(i)}$, is the linear dielectric tensor. Furthermore $\varepsilon^{(i)}$ is diagonal with respect to the principal axes of the uniaxial crystals to be considered here. Hence there are only principal indices for a uniaxial material associated, respectively, with extraordinary and ordinary waves. Uniaxial crystals are clearly doubly refracting but the treatment given below will not include this effect. Since, it would only serve to cloud the issues being presented, only input radiation incident normally upon the crystal will be addressed.

Given this starting point, for a quadratically nonlinear medium, then the following additional assumptions will be made

- E is polarised along one of the principal axes
- the field envelopes are *slowly varying approximation* so that

$$\left| \frac{\partial^2 E(\omega_i)}{\partial z^2} \right| << \left| k_{\omega_i} \frac{\partial E(\omega_i)}{\partial z} \right|$$

- the wavenumber k_{ω_i} is given by

$$k_{\omega_i}^2 = \frac{\omega_i^2}{c^2} \varepsilon^{(i)} \tag{3.6}$$

where each wave is associated now with a *scalar* principal index $\sqrt{\varepsilon^{(i)}}$ and it should be remembered that, in equation (3.5), $\varepsilon^{(i)}$ is a *diagonal tensor* made up from the principal values.

- the nonlinear polarisation is

$$\overline{P}_{NL}(\omega_i) = P_{NL}(\omega)e^{ik_{\omega_i}^P z}$$

where $k_{\omega_i}^{(P)}$ is the wavenumber of the polarisation. Given these assumptions and definitions the basic equation for the amplitude $E(\omega_i)$ is

$$2ik_{\omega_i}\frac{\partial E(\omega_i)}{\partial z} + \omega_i^2\mu_0 P_{NL}(\omega_i)e^{-i(k_{\omega_i} - k_{\omega_i}^P)z} = 0 \tag{3.7}$$

where $\omega_i = \omega$ or 2ω and $P_{NL}(\omega_i)$ is $P^{(2)}(\omega)$ or $P^{(2)}(2\omega)$. From (2.49)

$$\overline{P}^{(2)}(\omega) = \varepsilon_0\kappa E^*(\omega)E(2\omega)e^{i[k_{2\omega} - k_\omega]} = P^{(2)}(\omega)e^{ik_\omega^P z} \tag{3.8}$$

$$\overline{P}^{(2)}(2\omega) = \varepsilon_0\frac{\kappa}{2}E^2(\omega)e^{2ik_\omega z} = P^{(2)}(2\omega)e^{ik_{2\omega}^P z} \tag{3.9}$$

where $\kappa_1 \equiv \kappa$ has been selected, as an example. The expressions for the polarisations lead, immediately, to the pair of coupled equations [1]

$$2ik_\omega\frac{\partial E(\omega)}{\partial z} + \frac{\omega^2}{c^2}\kappa E^*(\omega)E(2\omega)e^{i\Delta kz} = 0 \tag{3.10}$$

$$2ik_{2\omega}\frac{\partial E(2\omega)}{\partial z} + \left(\frac{2\omega}{c}\right)^2\frac{\kappa}{2}E^2(\omega)e^{-i\Delta kz} = 0 \tag{3.11}$$

where $\Delta k = k_{2\omega} - 2k_\omega$ is the *linear phase-mismatch*. Introducing the *linear* dielectric functions $\varepsilon^{(1)} \equiv \varepsilon(\omega)$ and $\varepsilon^{(2)} \equiv \varepsilon(2\omega)$, at ω and 2ω, and the refractive indices n_ω, $n_{2\omega}$ through $\varepsilon(\omega) = n_\omega^2$, $\varepsilon(2\omega) = n_{2\omega}^2$ implies that the wavenumbers are $k_\omega = \frac{\omega}{c}n_\omega$, $k_{2\omega} = \frac{2\omega}{c}n_{2\omega}$. It will be seen later that, for *type I interactions*, n_ω and $n_{2\omega}$ are refractive indices of ordinary and extraordinary beams, or vice-versa.

The basic coupled equations are, therefore,

$$\frac{dE(\omega)}{dz} = \frac{i\omega\kappa}{2n_\omega c}E(2\omega)E^*(\omega)e^{i\Delta kz} = i\Gamma_1 E(2\omega)E^*(\omega)e^{i\Delta kz} \tag{3.12}$$

$$\frac{dE(2\omega)}{dz} = \frac{i\omega\kappa}{2n_{2\omega}c}E^2(\omega)e^{-i\Delta kz} = i\Gamma_2 E^2(\omega)e^{-i\Delta kz} \tag{3.13}$$

These rather famous equations can be solved in a number of interesting ways. Historically, the driving interest has been in second-harmonic generation, and, quite correctly, still is in many groups. Today, however, there is considerable interest in stationary state [6] eigensolutions of (3.12) and (3.13) that are *balanced* in such a way that there is no energy exchange between the fundamental and the harmonic wave. The extension of these equations to include diffraction or dispersion implies interesting *soliton*-like, or solitary wave, solutions but that question will be left until the end of the chapter. In fact, other chapters in this volume will deal more thoroughly with the soliton issue.

If equation (3.12), for $E(\omega)$, is differentiated with respect to z then

$$\frac{d^2E(\omega)}{dz^2} = i\Gamma_1 \left[\frac{dE(2\omega)}{dz} E^*(\omega)e^{i\Delta kz} - i\Delta kE(2\omega)E^*(\omega)e^{i\Delta kz} + \frac{dE^*(\omega)}{dz} E(2\omega)e^{i\Delta kz} \right] \quad (3.14)$$

Substitution of (3.12) and (3.13) into (3.14) then gives

$$\frac{d^2E(\omega)}{dz^2} = -i\Delta k \frac{dE(\omega)}{dz} - \frac{1}{E^*(\omega)} \left| \frac{dE(\omega)}{dz} \right|^2 + \Gamma_1\Gamma_2 |E(\omega)|^2 E(\omega) = 0 \quad (3.15)$$

At this stage, *nonlinear phases* ϕ_ω, $\phi_{2\omega}$ and amplitudes $F_{2\omega}$, F_ω can be usefully introduced through the definition

$$E(\omega) = F_\omega(z)e^{-i\phi_\omega(z)} \quad (3.16)$$

Using this in equation (3.15), and then separating the equations into real and imaginary parts, gives the pair of equations

$$\frac{d^2F_\omega}{dz^2} - \left(\frac{d\phi}{dz} \right)^2 F_\omega - \Delta k \frac{d\phi_1}{dz} F_\omega - \frac{1}{F_1} \left(\frac{dF_\omega}{dz} \right)^2 F_\omega + \Gamma_1\Gamma_2\Gamma_\omega^3 = 0 \quad (3.17)$$

$$2\frac{d\phi_\omega}{dz} \frac{dF_\omega}{dz} + \frac{d^2\phi_\omega}{dz^2} F_\omega + \Delta k \frac{dF_\omega}{dz} = 0 \quad (3.18)$$

This is a very valuable starting point allowing the focus to be on the role played by the nonlinear phase. There are now a number of ways to proceed based upon the way the interaction is promoted, or seeded, through initial conditions.

The first case is one in which there is *no initial second-harmonic field*. The input conditions are

$$[E(2\omega)]_{z=0} = 0, \quad [E(\omega)]_{z=0} \neq 0 \quad (3.19)$$

From equation (3.12), however

$$\left[\frac{dF_\omega}{dz} - i\frac{d\phi_\omega}{dz}F_\omega\right]_{z=0} = 0 \tag{3.20}$$

so that

$$\left(\frac{dF_\omega}{dz}\right)_{z=0} = 0, \quad \left(\frac{d\phi_\omega}{\partial z}\right)_{z=0} = 0 \tag{3.21}$$

and, from (3.18),

$$\left(\frac{d^2\phi_\omega}{dz^2}\right)_{z=0} = 0 \tag{3.22}$$

Equation (3.18) can be written as

$$\frac{d}{dz}\left[F_\omega^2\frac{d\phi_\omega}{dz} + \frac{\Delta k}{2}F_\omega^2\right] = 0 \tag{3.23}$$

This can be integrated once, to give

$$F_\omega^2\frac{d\phi_\omega}{dz} + \frac{\Delta k}{2}F_\omega^2 = \text{constant} \tag{3.24}$$

But at $z = 0$, $F_\omega(0) = F_\omega(z) \neq 0$ and $\left(\frac{d\phi_\omega}{dz}\right)_{z=0} = 0$ which means that equation (3.24) becomes

$$\frac{d\phi_\omega}{dz} + \frac{\Delta k}{2} = \frac{\Delta k}{2}\frac{F_\omega^2(0)}{F_\omega^2} \tag{3.25}$$

Equation (3.17) is, therefore

$$\frac{d^2F}{dz^2} + \frac{\Delta k^2}{2}\left[-\frac{F_\omega^2(0)}{F_\omega^2} + \frac{F_\omega^2(0)}{F_\omega^4}\right]F_\omega - \frac{1}{F_\omega}\left(\frac{dF_\omega}{dz}\right)^2 + \Gamma_1\Gamma_2\Gamma_\omega^3 = 0 \tag{3.26}$$

The normalised intensity $I_\omega = F_\omega^2$ can now be introduced to change (3.26) to

$$\frac{I_\omega}{2}\frac{d^2I_\omega}{dz^2} - \frac{1}{2}\left(\frac{dI_\omega}{dz}\right)^2 - \frac{\Delta k^2}{2}\left[-I_\omega(0)I_\omega + I_\omega^2(0)\right] + \Gamma_1\Gamma_2I_\omega^3 = 0 \tag{3.27}$$

For a perfectly *linear phase-matched* process, $\Delta k = 0$ and

$$\frac{d\phi_\omega}{dz} = 0 \tag{3.28}$$

$$I_\omega \frac{d^2 I_\omega}{dz^2} - \left(\frac{dI_\omega}{dz}\right)^2 + 2\Gamma_1\Gamma_2 I_\omega^3 = 0 \tag{3.29}$$

for which the solution is

$$F_\omega = \mathrm{sec}\, h\left(\sqrt{\Gamma_1\Gamma_2}\, z\right), \quad \phi_\omega = \pi/4 \tag{3.30}$$

and this implies that

$$F_{2\omega} = \sqrt{\frac{\Gamma_2}{\Gamma_1}} \tanh\left(\sqrt{\Gamma_1\Gamma_2}\, z\right), \quad \phi_{2\omega} = 0 \tag{3.31}$$

From (3.28) we can seen that $\phi_\omega = \pi/4$ is just an arbitrary value, selected to give $\phi_{2\omega} = 0$. Other values constant be used for ϕ_ω but then $\phi_{2\omega}$ would not be zero. The solution $F_\omega = \mathrm{sec}\, h\left(\sqrt{\Gamma_1\Gamma_2}\, z\right)$ can easily be verified by substitution into equation (3.26). The phase is a little more interesting, however. For $\Delta k = 0$ [perfect phase-matching] the division of $E(\omega)$ and $E(2\omega)$ into the products $E(\omega) = F_\omega(z)e^{-i\phi_\omega(z)}$, $E(2\omega) = F_{2\omega}(z)e^{-i\phi_{2\omega}(z)}$ and the substitution into equations (3.12) and (3.13) produces

$$\frac{dF_\omega}{dz} = i\Gamma_1 F_\omega F_{2\omega} e^{i[2\phi_\omega - \phi_{2\omega}]} \tag{3.32}$$

$$\frac{dF_{2\omega}}{dz} = i\Gamma_2 F_\omega^2 e^{i[\phi_{2\omega} - 2\phi_\omega]} \tag{3.33}$$

Now the factor is actually $i = e^{i\pi/2}$ so that

$$\frac{dF_{2\omega}}{dz} = \Gamma_2 F_\omega^2 e^{i[\phi_{2\omega} - 2\phi_\omega + \pi/2]} \tag{3.34}$$

The equality of the real and imaginary parts of (3.34) then requires that

$$\phi_{2\omega} - 2\phi_\omega + \pi/2 = 0 \tag{3.35}$$

Now it can be seen that the choice $\phi_{2\omega} = 0$ that makes $\phi_\omega = \pi/4$. In order to illustrate these second-harmonic generation results $I_\omega = F_\omega^2$ and $I_\omega = F_\omega^2$ are plotted as a function of z in figure 2, for $\Gamma_1 = \Gamma_2 = 1$ and $\Delta k = 0$. The data is selected only to illustrate the points to be made, rather than to show a real scenario. Clearly a 100% conversion of $E(\omega)$ to $E(2\omega)$ takes place and no energy is transferred back (cascaded) to the fundamental. For $\Delta k \neq 0$ [*not a linearly phase-matched process*] the coupled equations can still be solved *exactly* in terms of Jacobean elliptic functions but they are also quite easy to solve numerically.

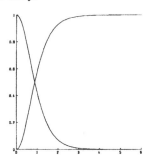

Figure 2 $I_\omega = F_\omega^2$ and $I_{2\omega} = F_{2\omega}^2$, plotted, as a function of z, in the perfectly linearly phase-matched case

$$\Delta k = 0. \quad \Gamma_1 = \Gamma_2 = 1$$

This is the path taken here, with some typical results numerically being shown in figure 3, which shows that the $\Delta k \neq 0$ case is fundamentally different from the phase-matched case because:

- conversion from the fundamental wave (ω) to the second-harmonic [up conversion] (2ω) is not 100%
- when the second-harmonic is accumulated it begins to convert back to the fundamental [down-conversion]
- the interaction exhibits a periodic pattern
- $\phi_\omega(z)$ the phase of the fundamental wave varies considerably with z.

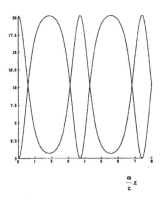

$$\frac{\omega}{c} z$$

Figure 3 Variation of $F_\omega^2(z), F_{2\omega}^2(z)$ in the non-phase matched case, when $\Delta k \neq 0. \ \Gamma_1 = \Gamma_2 = 1$, and

$$\Delta k = -\frac{4\pi}{\lambda} \times 0.05, \ n_{2\omega} = 1.5, n_\omega = 1.55 \ \text{and} \ \lambda = 1.06 \times 10^{-6} \ \text{m}.$$

This process is called *cascading* – like a waterfall or anything resembling this. Cascading is actually well-known in electronics, where pieces of apparatus are so connected that the output of the first constitutes the input of the second, and so on, e.g. a cascade amplifier. In nonlinear optics, it is a process that follows an *up-conversion* of the fundamental to the second-harmonic, i.e. $\omega + \omega \rightarrow 2\omega$, with a *down-conversion* back to the fundamental through $2\omega - \omega \rightarrow \omega$. In principle, the process need not stop at $(2\omega,\omega)$ and can produce other products. All such photon events occur when phase-matching is absent, so, for $\chi^{(2)}$ materials, $\Delta k \neq 0$ focuses attention upon $\omega \underset{\leftarrow}{\overset{\rightarrow}{}} 2\omega$ processes, which are the topic of this school. A significant nonlinear phase shift (modulation) is acquired by the fundamental wave, as the conversions proceed and this is now seen to have realistic device potential, because *both* amplitude and phase modulations can be exploited. Figure 4, for example, shows an interesting 'step-like' structure for the variation of $\phi_\omega(z)$ with z.

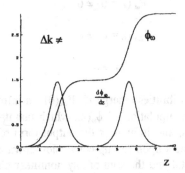

Figure 4 Variation of the nonlinear phase $\phi_\omega(z)$ and $\dfrac{d\phi}{dz}$ of the fundamental wave with propagation distance z.

$$n_\omega = 1.55, n_{2\omega} = 1.5, \lambda = 1.06 \times 10^{-6}.\ \Delta k = \frac{2\omega}{c}(n_{2\omega} - n_\omega) = -0.6 \times 10^6 \text{m}^{-1}.$$

Cascading, on the face of it, looks like a weak process that is not likely to have much of an impact on the fundamental wave. Furthermore, anything that contributes to the fundamental wave adds to the third-order $(\chi^{(3)})$ processes that must already be in operation, however weak they are. Indeed, it is important to remember that the polarisation, being a power series, always has a $\chi^{(3)}$ component, even in $\chi^{(2)}$ materials. The importance of cascading, then, comes down to magnitude. It always occurs in materials that lack a centre of symmetry but it is safe to say that it is only going to be useful if excellent materials are available and $\Delta k \approx 0$. The latter condition can always be achieved through the kind of quasi-phase-matching schemes to be discussed later on in this book. Good materials are available with large $|\chi^{(2)}|$ values, however, and it is the seminal work in this field, by Stegeman and co-workers, using the popular material KTP, that has opened up the field. In addition to KTP, quite a large range of organic materials are now available.

One possible consequence of producing accessible nonlinear phase shifts at the fundamental frequency ω, in a $\chi^{(2)}$ material, is the creation of an effective nonlinear coefficient n_2 so that, broadly speaking, the refractive index is modified from a linear value n_L, to a nonlinear value $n_L + n_2 I$, where I is the intensity of the fundamental. The material used in this way can then, in principle, be used in self-phase modulation applications and a whole range of device possibilities opens up. Stegeman and co-workers, indeed, showed that n_2 has a useable magnitude for KTP, and that its origin was the cascade process. An interesting feature is that n_2 can have positive (self-focusing) signs or negative (self-defocusing) sign. These are but broad conclusions on how cascading can generate an n_2 value; the actual form of n_2 is complicated with some dependence of n_2 upon intensity.

Assanto and co-workers have made great use of equations (3.12) and (3.13), by adopting the more general input conditions

$$\blacksquare \qquad\qquad F_\omega\ (z = 0) \neq 0 \qquad\qquad\qquad (3.36)$$

$$\blacksquare \qquad\qquad F_{2\omega}\ (z = 0) \neq 0 \qquad\qquad\qquad (3.37)$$

$$\blacksquare \qquad\qquad \Delta k \neq 0 \qquad\qquad\qquad (3.38)$$

An amazing number of possibilities emerge if $F_{2\omega}(0)$ is introduced as a small input "seed". This permits the manipulation of $\phi_\omega(z)$, which maintains the interesting step-like nature, as z increases, and influences the output form of the fundamental wave. Maintaining $\Delta k \neq 0$ impacts upon the up-conversion-down-conversion (modulation) behaviour of the fundamental and the sign of any nonlinear phase shift is the same as the sign of Δk. The latter feature means that $\chi^{(2)}$ materials, when $\Delta k \neq 0$, can look like self-focusing or self-defocusing materials. Because of the wealth of opportunities created by the more general input conditions, certain inputs of *both* the fundamental and the second-harmonic will now be investigated.

Initial fundamental and second-harmonic fields

The chapter by Assanto will contain many details of how to manipulate ω and 2ω inputs to a $\chi^{(2)}$ material to get, for example, transistor action. This is not the ground that will be covered here. Instead, the focus is upon the fundamental theory of the generation of *stationary states* in which there is a *balance between the fundamental and the second-harmonic waves*. This balance manifests itself as *restraint* on the linear and nonlinear phase differences.

Assume that the initial conditions are

$$E(\omega) = F_\omega (0)e^{-i\phi_\omega(0)}, \ \ E(2\omega) = F_{2\omega} (0)e^{-i\phi_{2\omega}(0)} \qquad\qquad (3.39)$$

where F_ω, $F_{2\omega}$, ϕ_ω, $\phi_{2\omega}$ are entirely real. Now seek a solution that demands a *balance* between the ω and 2ω waves to be maintained. This *stationary state* is expressed as

$$\frac{dF_\omega(z)}{dz} = 0, \quad \frac{dF_{2\omega}(0)}{dz} = 0 \tag{3.40}$$

which implies that

$$F_\omega(z) = F_\omega(0); \quad F_{2\omega}(z) = F_{2\omega}(0) \tag{3.41}$$

Hence, equations (3.17) and (3.18) become

$$-2\left(\frac{d\phi_\omega}{dz}\right)^2 F_\omega(0) + \Delta k F_\omega(0)\frac{d\phi_\omega}{dz} - \Gamma_1\Gamma_2 F_\omega^3(0) = 0 \tag{3.42}$$

$$-\frac{d^2\phi_\omega}{dz^2} F_\omega(0) = 0 \tag{3.43}$$

The re-organised form of equation (3.42) is

$$2\left(\frac{d\phi_\omega}{dz}\right) = -\frac{\Delta k}{2} \pm \sqrt{\left(\frac{\Delta k}{2}\right)^2 + 2\Gamma_1\Gamma_2 F_\omega^2(0)} \tag{3.44}$$

The substitution of $E(2\omega) = F_{2\omega}(0)e^{-i\phi_{2\omega}(z)}$ into equation (3.13), after setting $\frac{dF_{2\omega}}{dz} = 0$, yields

$$\frac{d\phi_{2\omega}}{dz} = -\Gamma_2 \frac{F_\omega^2(0)}{F_{2\omega}(0)} e^{i[\phi_{2\omega}-2\phi_\omega-\Delta kz]} \tag{3.45}$$

and the fact that $\phi_{2\omega}$ is real, by definition, means that the phase factor, in (3.45), must vanish. Hence

$$\phi_{2\omega} - 2\phi_\omega = \Delta kz \tag{3.46}$$

$$\frac{d\phi_{2\omega}}{dz} = -\Gamma_2 \frac{F_\omega^2(0)}{F_{2\omega}(0)} \tag{3.47}$$

Equations (3.45), (3.46) and (3.47) also give

$$\frac{d}{dz}(\phi_{2\omega} - 2\phi_\omega) = -\Gamma_2 \frac{F_\omega^2(0)}{F_{2\omega}(0)} + \frac{\Delta k}{2} \mp \sqrt{\left(\frac{\Delta k}{2}\right)^2 + 2\Gamma_1\Gamma_2 F_\omega^2(0)} = \Delta k \tag{3.48}$$

so that the amplitude of the second-harmonic wave is

$$F_{2\omega}(0) = \frac{-\Delta k \pm \sqrt{(\Delta k^2) + 8\Gamma_1 \Gamma_2 F_{\omega}^2(0)}}{4\Gamma_1} \tag{3.49}$$

These results mean that if $F_{\omega}(0)$ and $F_{2\omega}(0)$ satisfy equation (3.49) and the *input restraint* $2\phi_{\omega}(0) = \phi_{2\omega}(0)$, then the amplitudes of the fundamental wave and the harmonic wave are stationary. In other words, the nonlinear phase mismatch *compensates* the linear phase mismatch in the following way

$$\underbrace{\frac{d}{dz}(\phi_{2\omega} - 2\phi_{\omega})}_{\text{nonlinear phase mismatch}} = \underbrace{\Delta k}_{\text{linear phase mismatch}} \tag{3.50}$$

In summary, the cw equations for a quadratically nonlinear optical medium can be investigated with two, clearly distinguishable, initial conditions in the scalar case. Each is related to a different physical process. The initial conditions $E(\omega) \neq 0$, $E(2\omega) = 0$ are responsible for *second-harmonic generation* (SHG) and the initial conditions $E(\omega) \neq 0$, $E(2\omega) \neq 0$ with $\frac{dF_{\omega}}{dz} = 0$, $\frac{dF_{2\omega}}{dz} = 0$ lead to *stationary propagation* (SP). It is important to emphasise that SHG and SP *start differently* and *end up differently*. This will not be the case for finite beams but this question will be returned to later on.

4. Vector CW Equations

In the previous section, the electric fields $E(\omega)$ and $E(2\omega)$ are assumed to be scalar quantities. It is well-known for second-harmonic generation, however, that crystals must be used, in which the manipulation of *ordinary* (o) and *extraordinary* (e) rays is, generally, required. If the principal refractive indices seen by o and e rays are n_o and n_e, respectively, then $\Delta n = n_e - n_o$ is called the birefringence and crystals with $\Delta n > 0$ or $\Delta m < 0$ are said to be *positive* or *negative* uniaxial crystals. Type I second-harmonic generation is modelled with the *scalar* equations (3.12) and (3.13) but a vector extension of them is needed to model type II [49] second-harmonic generation which deploys the interaction between ordinary waves, polarised perpendicular to the plane containing the optic axis and the propagation direction, and extraordinary waves that are polarised in the plane containing these axes.

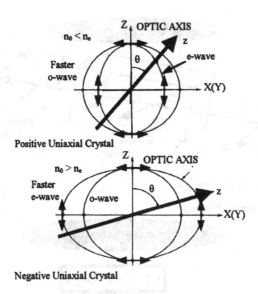

Figure 5 Wave surfaces in a uniaxial crystal. For a source of light at the origin, the wave surfaces connect points of equal phase i.e. the times of arrival of a photon at all points on such a surface are equal. A sphere shows that the velocity is the same in all directions: this is the ordinary wave, polarised perpendicular to the plane (•). An ellipse shows that the velocity varies with θ: this is the extraordinary wave polarised in the plane (↔).

Typical *contours of constant phase* for both positive and negative uniaxial crystals [50,51,52] are shown in figure 5. Briefly, the velocity of an ordinary ray is constant in all directions and, in linear optics, such a ray is controlled by a refractive index n_0. This is not the case for an extraordinary ray. Its velocity varies with the angle θ that the propagation direction sets to the optic axis. In figure 5 the crystal axes are labelled X (or Y) and Z, the propagation direction is called z. The linear refractive index that the extraordinary ray experiences is $n_e(\theta)$, where the argument implies the variation with θ. The wave surface of the extraordinary ray is an ellipsoid and for the ordinary ray it is a sphere. The cross-sections, shown in figure 5 are ellipses and circles and they, in this example, touch on the optic axis. Hence, the velocity of the e and o rays along that direction are equal. For the direction that is perpendicular to the optic axis, the figure shows that the full difference between n_0 and n_e is seen.

Type I phase-matching [1,4,5] is illustrated in figure 6. In this example, the fundamental (pump wave) is entered into the crystal as an ordinary wave, polarised perpendicular to the plane containing the optic axis and the propagation direction (z). The second-harmonic is generated as an extraordinary wave, polarised in the plane containing the optic axis and the propagation direction. The process is phase matched if $k_{2\omega} = 2k_\omega$; otherwise the output consists of a weak second-harmonic pump wave. This is a three-wave (three photon) interaction that proceeds according to the photon energy addition $\hbar\omega + \hbar\omega = 2\hbar\omega$, where $\hbar = h/2\pi$ and h is Planck's constant. This is

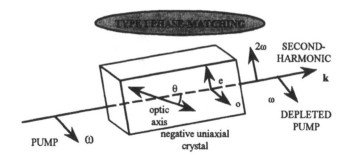

- Optic axis is in vertical plane

- θ_p = phase matching angle

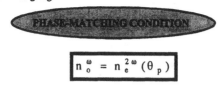

$$n_o^{\omega} = n_e^{2\omega}(\theta_p)$$

- 3–wave interaction in the crystal

$$\frac{\omega}{c}n_o^{\omega} + \frac{\omega}{c}n_o^{\omega} = \frac{2\omega}{c}n_e^{2\omega}(\theta)$$

Figure 6 Illustration of type I phase-matching.

an [ooe] phase-matching interaction and a 100% conversion of the fundamental to the second-harmonic occurs when

$$2\mathbf{k}_{\omega} = \mathbf{k}_{2\omega} \tag{4.1}$$

which, expressed as a three-wave interaction, is [4,5]

$$\frac{\omega}{c}n_o^{\omega} + \frac{\omega}{c}n_o^{\omega} = \frac{2\omega}{c}n_e^{2\omega}(\theta) \tag{4.2}$$

where the superscripts ω and 2ω show which wave is "seeing" this linear index. In other words, Type I phase-matching comes down to

$$n_o^{\omega} = n_e^{2\omega}(\theta_p) \tag{4.3}$$

where θ_p is the *precise* angle the propagation direction sets to the optic axis, to achieve 100% $\omega \rightarrow 2\omega$ conversion. Type I phase-matching is sketched again in figure 7 but, this time, in terms of the dispersion of the refractive indices. It is emphasised in

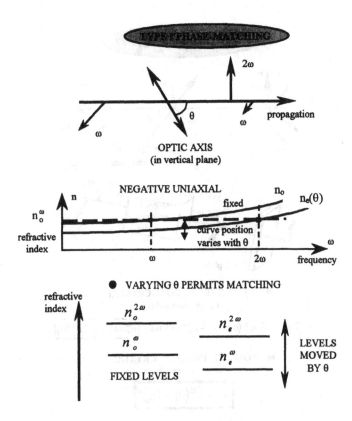

Figure 7 Type I phase-matching as an exploitation of the refractive index dispersion. The refractive index level diagram illustrates how the phase-matching can be achieved by varying θ.

this figure [for a negative uniaxial crystal] that $n_o(\omega)$ (> $n_e(\theta)$) is fixed and that the extraordinary index $n_e(\theta)$ can be moved up and down, by varying θ, to achieve tuning. It is, perhaps, easier to appreciate if the index picture is formed as a set of index levels. $n_o^\omega, n_o^{2\omega}$ levels are then *fixed* and the positions of $n_e^{2\omega}, n_e^\omega$ depend upon θ. Type I phase-matching is not as difficult a process as type II phase-matching [4,5,49], which is shown in figure 8, which admits two pump beams at the input that are orthogonally polarised. This makes one input fundamental wave an ordinary wave and the other one an extraordinary wave. This is no longer a scalar case, because it involves two components of the input fundamental wave and one component of the extraordinary second-harmonic wave. The λ/2 plate, interposed between the crystal and the input beam rotates the ω wave by 45° and thus generates the ordinary and extraordinary ω-wave inputs. In general, the output consists of three waves namely two pump beams, with a phase difference between them, and a second-harmonic e-wave. This is [oee] phase-matching, which is expressed as

$$n_o^\omega + n_e^\omega(\theta) = 2n_e^{2\omega}(\theta) \qquad (4.4)$$

32

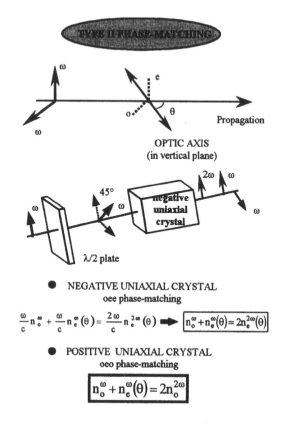

Figure 8. Sketch of type II phase-matching based upon a negative uniaxial crystal

It should be noted that the 2ω wave only comes out as an extraordinary wave because a negative crystal has been selected. For a positive uniaxial crystal

$$n_o^\omega + n_e^\omega(\theta) = n_o^{2\omega} \qquad (4.5)$$

and the second-harmonic is an ordinary wave.

For type II phase-matching figure 9 shows the refractive index plots for a positive uniaxial crystal. In the example selected, $n_o < n_e(\theta)$ and it must be emphasised that n_o does not depend upon frequency. The phase-matching conditions are shown and the refractive index levels are drawn, showing that gaps Δn, Δn_2, Δn_3 and Δn_4 exist. Angle tunability permits the e-levels to be moved up and down with θ. The differences Δn_i will all be used below, as wavenumber differences $\Delta k_i = \dfrac{2\omega}{c}\Delta n_i$ and it will be appreciated that, in principle, all of these index, or wavenumber differences can be important in the most general vector model.

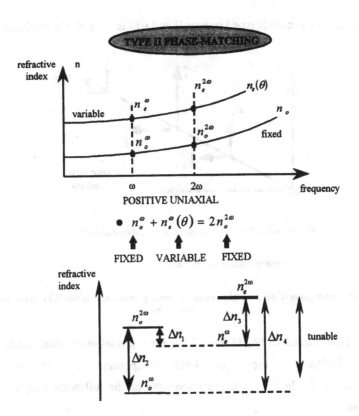

TYPE II PHASE-MATCHING

Figure 9 Type II phase-matching in a positive uniaxial crystal, by exploiting the refractive index dispersion. This is more difficult to do than type I.

The questions posed in figure 10 concern the possibility of the *complete* group $\left(E_x^\omega, E_y^\omega, E_x^{2\omega}, E_y^{2\omega}\right)$ propagating in a quadratically nonlinear crystal. First, the following cases appear to be well-known:

Phase-matching type	Crystal	Field components	Comment
I	negative	$[E_x^\omega, 0, 0, E_y^{2\omega}]$	Scalar
I	positive	$[0, E_y^\omega, E_x^{2\omega}, 0]$	Scalar
II	negative	$[E_x^\omega, E_y^\omega, 0, E_y^{2\omega}]$	Vector
II	positive	$[E_x^\omega, E_y^\omega, E_x^{2\omega}, 0]$	Vector

34

The general case is neither type I or type II and will involve all the field components.

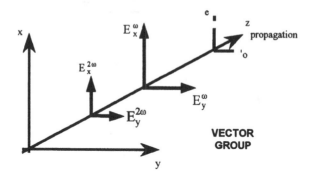

- Are both $E_x^{2\omega}$ and $E_y^{2\omega}$ created in measurable amounts ?

- How many equations do we need ?

Figure 10 Vector group of electric field components showing the existence of two fundamental waves and two harmonic waves.

From equations (2.38) and (2.39) for the nonlinear polarisations $\overline{P}_i^{(2)}(2\omega)$ and $\overline{P}_i^{(2)}(\omega)$ the four field components in the vector group $(\overline{E}_x(\omega), \overline{E}_y(\omega), \overline{E}_x(2\omega), \overline{E}_y(2\omega))$ are solutions of the following coupled differential equations:

$$\frac{\partial^2 \overline{E}_x(\omega)}{\partial z^2} + \left(\frac{\omega}{c}n_x^\omega\right)^2 \overline{E}_x(\omega) + \frac{\omega^2}{c^2}\left[\widetilde{\kappa}_{xxx}^{(2)}\overline{E}_x^*(\omega)\overline{E}_x(2\omega) + \widetilde{\kappa}_{yyy}^{(2)}\overline{E}_y(2\omega)\overline{E}_y^*(\omega)\right.$$
$$\left. + \widetilde{\kappa}_{xxy}^{(2)}\left\{\overline{E}_x^*(\omega)\overline{E}_y(2\omega) + \overline{E}_y^*(\omega)\overline{E}_x(2\omega)\right\}\right] = 0 \tag{4.6}$$

$$\frac{\partial^2 \overline{E}_y(\omega)}{\partial z^2} + \left(\frac{\omega}{c}n_y^\omega\right)^2 \overline{E}_y(\omega) + \frac{\omega^2}{c^2}\left[\widetilde{\kappa}_{yyy}^{(2)}\overline{E}_y^*(\omega)\overline{E}_y(2\omega) + \widetilde{\kappa}_{yxx}^{(2)}\overline{E}_x^*(\omega)\overline{E}_x(2\omega)\right.$$
$$\left. + \widetilde{\kappa}_{yxy}^{(2)}\left\{\overline{E}_y^*(\omega)\overline{E}_x(2\omega) + \overline{E}_x^*(\omega)\overline{E}_y(2\omega)\right\}\right] = 0 \tag{4.7}$$

$$\frac{\partial^2 \overline{E}_x(2\omega)}{\partial z^2} + \left(\frac{2\omega}{c}n_x^{2\omega}\right)^2 \overline{E}_x(2\omega) + \frac{(2\omega)^2}{2c^2}\left[\widetilde{\kappa}_{xxx}^{(2)}\overline{E}_x^2(\omega) + \widetilde{\kappa}_{xyy}^{(2)}\overline{E}_y^2(\omega)\right.$$
$$\left. + 2\widetilde{\kappa}_{xxy}^{(2)}\overline{E}_x(\omega)\overline{E}_y(\omega)\right] = 0 \tag{4.8}$$

$$\frac{\partial^2 \overline{E}_y(2\omega)}{\partial z^2} + \left(\frac{2\omega}{c}n_y^{2\omega}\right)^2 \overline{E}_y(2\omega) + \frac{(2\omega)^2}{2c^2}\left[\widetilde{\kappa}_{yyy}^{(2)}\overline{E}_y^2(\omega) + \widetilde{\kappa}_{yxx}^{(2)}\overline{E}_x^2(\omega)\right.$$
$$\left. + \widetilde{\kappa}_{yxx}^{(2)}\overline{E}_x^2(\omega) + 2\widetilde{\kappa}_{yxy}^{(2)}\overline{E}_y(\omega)\overline{E}_x(\omega)\right] = 0 \tag{4.9}$$

where $n_x^\omega, n_y^\omega, n_x^{2\omega}, n_y^{2\omega}$ are refractive indices. The *linear* phase factors can be factored out by writing

$$(e): \ \overline{E}_x(\omega) = E_x(\omega)e^{i\frac{\omega}{c}n_x^\omega z}, \quad (o): \ \overline{E}_y(\omega) = E_y(\omega)e^{i\frac{\omega}{c}n_y^\omega z} \tag{4.10a}$$

$$(e): \ \overline{E}_x(2\omega) = E_x(2\omega)e^{i\frac{2\omega}{c}n_x^{2\omega}z}, \quad (o): \ \overline{E}_y(2\omega) = E_y(2\omega)e^{i\frac{2\omega}{c}n_y^{2\omega}z} \tag{4.10b}$$

After substituting (4.10) into (4.6) to (4.9) and making a slowly varying approximation, permitting the neglect of $\dfrac{\partial^2 E_{x,y}}{\partial z^2}(\omega), \dfrac{\partial^2 E_{x,y}}{\partial z^2}(2\omega)$ terms, the following set of equations is obtained

$$(e): \ \frac{2i\omega}{c}n_x^\omega \frac{\partial E_x(\omega)}{\partial z} + \left(\frac{\omega}{c}\right)^2 \left[\widetilde{\kappa}_{xxx}^{(2)}E_x^*(\omega)E_x(2\omega)\exp(i\Delta k_{1,1}z)\right.$$
$$+ \widetilde{\kappa}_{xxy}^{(2)}\left\{E_x^*(\omega)E_y(2\omega)\exp(i\Delta k_{2,1}z)\right.$$
$$\left. + E_y^*(\omega)E_x(2\omega)\exp\left(i\frac{[\Delta k_{1,1}+\Delta k_{1,2}]}{2}z\right)\right\} + \widetilde{\kappa}_{xyy}^{(2)}E_y^*(\omega)E_y(2\omega)\exp\left(i\frac{[\Delta k_{2,1}+\Delta k_{2,2}]}{2}z\right)\right] = 0 \tag{4.11}$$

$$(o): \ \frac{2i\omega}{c}n_y^\omega \frac{\partial E_y(\omega)}{\partial z} + \left(\frac{\omega}{c}\right)^2 \left[\widetilde{\kappa}_{yyy}^{(2)}E_y^*(\omega)E_y(2\omega)\exp(i\Delta k_{2,2}z)\right.$$
$$+ \widetilde{\kappa}_{yxy}^{(2)}\left\{E_y^*(\omega)E_x(2\omega)\exp(i\Delta k_{1,2}z)\right.$$
$$\left. + E_x^*(\omega)E_y(2\omega)\exp\left(i\frac{[\Delta k_{2,1}+\Delta k_{2,2}]}{2}z\right)\right\} + \widetilde{\kappa}_{yxx}^{(2)}E_x^*(\omega)E_x(2\omega)\exp\left(i\frac{[\Delta k_{1,1}+\Delta k_{1,2}]}{2}z\right)\right] = 0 \tag{4.12}$$

$$(e): \ 4i\frac{\omega}{c}n_x^{2\omega} \frac{\partial E_x}{\partial z}(2\omega) + \frac{1}{2}\left(\frac{2\omega}{c}\right)^2 \left[\widetilde{\kappa}_{xxx}^{(2)}E_x^2(\omega)\exp(-i\Delta k_{1,1}z)\right.$$
$$\left. + \widetilde{\kappa}_{xyy}^{(2)}E_y^2(\omega)\exp(-i\Delta k_{1,2}z) + 2\widetilde{\kappa}_{xxy}^{(2)}E_x(\omega)E_y(2\omega)\exp\left(-\frac{i[\Delta k_{1,1}+\Delta k_{1,2}]}{2}z\right)\right] = 0 \tag{4.13}$$

$$(o): 4i\frac{\omega}{c}n_y^{2\omega}\frac{\partial E_y}{\partial z}(2\omega)+\frac{1}{2}\left(\frac{2\omega}{c}\right)^2\left[\widetilde{\kappa}_{yyy}^{(2)}E_y^2(\omega)\exp[-i\Delta k_{2,2}z]\right.$$

$$\left.+\widetilde{\kappa}_{yxx}^{(2)}E_y^2(\omega)\exp[-i\Delta k_{2,1}z]+2\widetilde{\kappa}_{yxy}^{(2)}E_x(\omega)E_y(\omega)\exp\left(-\frac{i\left[\Delta k_{2,1}+\Delta k_{2,2}\right]}{2}z\right)\right]=0 \tag{4.14}$$

in which

$$\Delta k_{2,1}=\frac{2\omega}{c}\left[n_y^{2\omega}-n_x^{\omega}\right]=k_y^{2\omega}-2k_x^{\omega}=\frac{2\omega}{c}\left[n_o^{2\omega}-n_e^{\omega}\right]\equiv\Delta k_1 \tag{4.15a}$$

$$\Delta k_{2,2}=\frac{2\omega}{c}\left(n_y^{2\omega}-n_y^{\omega}\right)=k_y^{2\omega}-2k_y^{\omega}=\frac{2\omega}{c}\left(n_o^{2\omega}-n_o^{\omega}\right)\equiv\Delta k_2 \tag{4.15b}$$

$$\Delta k_{1,1}=\frac{2\omega}{c}\left(n_x^{2\omega}-n_x^{\omega}\right)=k_x^{2\omega}-2k_x^{\omega}=\frac{2\omega}{c}\left(n_e^{2\omega}-n_e^{\omega}\right)\equiv\Delta k_3 \tag{4.15c}$$

$$\Delta k_{1,2}=\frac{2\omega}{c}\left(n_x^{2\omega}-n_y^{\omega}\right)=k_x^{2\omega}-2k_y^{\omega}=\frac{2\omega}{c}\left(n_e^{2\omega}-n_o^{\omega}\right)\equiv\Delta k_4 \tag{4.15d}$$

These definitions make it easy to relate to the refractive index level picture shown in figure 9.

This set of coupled equations involves quite a lot of exponentials which could average to zero, as the propagation proceeds, if their arguments do not vanish. Any *visible* argument that vanishes is a linear phase-matching and causes that exponential term to be dominant. Hence, if an experiment is performed in which only a limited number of exponential terms can survive, because of phase-matching, or being near to linear phase-matching, then only three of the equations [21,42,49,52] (4.11) to (4.14) will survive. This is precisely the case for the classic type II phase-matching, used to generate second-harmonic radiation.

More discussion of the *general* vector case is found elsewhere [30,38,53] and in the poster contribution by Bontemps and Boardman, but it is useful to note at this stage, that the variation of the Δk_i with θ, for a given crystal quickly reveals which is the dominant physical process on the basis of a *linear phase-matching* argument. An inspection of the four coupled equations, shows that a typical type II situation is indeed, modelled by the *three equations* (4.11), (4.12) and (4.14). The exponential terms that control the process are $\exp(i\Delta k_1 z)$, $\exp(i\Delta k_2 z)$ and $\exp\left(i\frac{\Delta k_1+\Delta k_2}{2}z\right)$ and equation (4.13) drops out.

The above discussion all looks very sensible, until the equations are investigated under cascading and/or balanced propagation non-linearly phase-matched conditions. In this regime, the separations (4.10a, 4.10b) that take out only the *linear phase* factors hide the fact that *nonlinear phases* are also accumulated. It is not, therefore, only a question of whether the arguments in the exponentials of equations

(4.11-4.14) vanish on linear phase arguments or not but it is the *balance* of the linear phases, or trade-off, with the nonlinear phases, contributed by the amplitude factors that counts. In other words, the linear quantities Δk_i are only part of the argument and once the nonlinear phases accumulate any of the terms in the set of four equations could be important. Indeed, the reasoning for a reduction to the set that is valid for type II SHG phase-matching considerations is no longer appropriate. This interesting line of investigation must be left for the future, in favour of a brief look at soliton propagation.

5. Scalar Spatial Solitons – Diffraction-Free Beams

In the previous section, it was shown, for the cw case, that the initial condition $F_\omega(0) \neq 0$, $F_{2\omega}(0) = 0$ is the starting point for second-harmonic generation, in which a 100% phase matched ($\Delta k = 0$) conversion is the desired outcome. The initial condition $F_\omega(0) \neq 0$, $F_{2\omega}(0) \neq 0$ with $\dfrac{dF_\omega(z)}{dz} = 0, \dfrac{dF_{2\omega}(z)}{dz} = 0$ produces a nonlinear eigenstate in which the *balancing constraint* $\Delta k + \Delta \beta = 0$ is maintained. $\Delta \beta$ is a *nonlinear wavenumber mismatch* where $\Delta \beta = \beta_{2\omega} - 2\beta_\omega$, $\beta_{2\omega} = \dfrac{d\phi_{2\omega}}{dz}$, $\beta_\omega = \dfrac{d\phi_\omega}{dz}$.

If a finite beam, as opposed to cw excitation is entered into a nonlinear crystal then the following questions arise. Will the beam diffract, focus down [self-focusing] or become a *special* diffraction-free entity called a spatial soliton? Note that diffraction-free beams have been found before but as exact solutions for linear media. They are not solitons [47,48] however, or solitary waves, unless special *nonlinear* conditions are met. Physically, for an input condition $E(\omega) \neq 0$, $E(2\omega) = 0$ the outcome will depend upon whether $E(\omega)$ is above, or below, a certain threshold value [a minimum amount]. If, for a given beam width, $E(\omega)$ has an amplitude above a threshold, energy is transferred from this fundamental wave (beam) to a harmonic wave, in the initial state. Then energy switches back and forth between the fundamental and the harmonic beams a few times, and in a process similar to relaxation oscillations, "unwanted energy" is radiated away in the process. Most of the energy, however, redistributes between the fundamental and the harmonic modes to form a composite soliton-like beam - a solitary beam - and a stationary state is finally achieved. Two typical evolutions are shown in figures 11a and 11b. These simulations clearly show that for finite beams [or, indeed, pulses] the emphasis shifts to stationary *fundamental-second-harmonic balanced* soliton-like entities, or solitary waves.

The existence of spatial solitary waves [loosely called solitons] will now be investigated mathematically and in more numerical detail. To deal with finite beams, all that is required is to add *diffraction terms* $\dfrac{\partial^2 E(\omega)}{\partial x^2}, \dfrac{\partial^2 E(2\omega)}{\partial x^2}$ to equations (3.10) and (3.11). The result is

38

$$2ik_\omega \frac{\partial E(\omega)}{\partial z} + \frac{\partial^2 E(\omega)}{\partial x^2} + \frac{\omega^2}{c^2}\kappa E^*(\omega)E(2\omega)e^{i\Delta kz} = 0 \qquad (5.1)$$

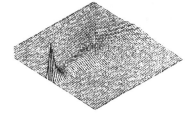

Fundamental below threshold   Second-harmonic beam radiates away

Figure 11(a) Numerical simulation of the fate of an input fundamental beam *below* threshold. F_ω radiates away and although $F_{2\omega}$ is initially created it too radiates away. $\Delta k = 0.2$, $F_{2\omega}(0) = 0$, $F_\omega = 0.5\ \text{sech}(x)$

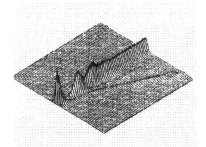
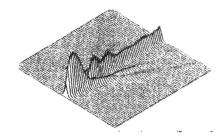

Fundamental beam becomes a solitary wave   Second-harmonic beam becomes a solitary wave
                                                     locked to the fundamental

Figure 11(b) Numerical simulation of an input fundamental beam above a threshold amplitude. After radiating unwanted energy the fundamental beam settles down to a stationary solitary beam. A second-harmonic beam is created. It also radiates but then settles down to a stationary solitary wave state *locked* onto the fundamental. $\Delta k = 0.2$, $F_{2\omega}(0) = 0$, $F_\omega = 2\ \text{sech}(x)$

$$2ik_{2\omega} \frac{\partial E(2\omega)}{\partial z} + \frac{\partial^2 E(2\omega)}{\partial x^2} + \left(\frac{\omega}{c}\right)^2 \frac{\kappa}{2}E^2(\omega)e^{-i\Delta kz} = 0 \qquad (5.2)$$

where $\Delta k = k_{2\omega} - 2k_\omega$ is the *linear* wavenumber mismatch. It should be recalled that to get to (5.1) and (5.2) the linear phases have been factored out by writing the fields as $E(\omega)\exp[i(k_\omega z - \omega t)]$, $E(2\omega)\exp[i(k_{2\omega}z - 2\omega t)]$ in conformity with the notation adopted in (2.2). In addition, the usual slowly varying approximation, with respect to z, has led to the neglect of second derivatives with respect to z.

It is convenient now to factor out the nonlinear phases and introduce them, with the *explicit form* $\beta_\omega z$, for the fundamental beam, and $\beta_{2\omega}z$, for the second-harmonic beam, where β_ω, $\beta_{2\omega}$ are nonlinear wavenumbers. This step is carried out here by writing

$$E(\omega) = \frac{\overline{w}}{\sqrt{2}} \exp(i\beta_\omega z), \quad E(2\omega) = \overline{v} \exp(i\beta_{2\omega} z) \qquad (5.3)$$

where $\overline{w}$, $\overline{v}$ are amplitudes and the usefulness of the $\sqrt{2}$ factor is to get a minor simplification of the equations i.e. to cancel the 4 in the $4\omega^2$ of (5.2) equations. If (5.3) is adopted then equations (5.1) and (5.2) become

$$2ik_\omega \frac{\partial \overline{w}}{\partial z} + \frac{\partial^2 \overline{w}}{\partial x^2} - 2k_\omega \beta_\omega \overline{w} + \left(\frac{\omega}{c}\right)^2 \kappa \overline{w} \, \overline{v} \exp[i(\Delta k + \beta_{2\omega} - 2\beta_\omega)z] = 0 \qquad (5.4)$$

$$2ik_{2\omega} \frac{\partial \overline{v}}{\partial z} + \frac{\partial^2 \overline{v}}{\partial x^2} - 2k_{2\omega} \beta_{2\omega} \overline{v} + \left(\frac{\omega}{c}\right)^2 \kappa \overline{w}^2 \exp[-i(\Delta k + \beta_{2\omega} - 2\beta_\omega)z] = 0 \qquad (5.5)$$

At this stage, $\beta_{2\omega}$, β_ω are chosen to make $\Delta k + \beta_{2\omega} - 2\beta_\omega = 0$ and the further transformations

$$\left(\frac{\omega}{c}\right)^2 \frac{\overline{w}\kappa}{2k_\omega \beta_\omega} \to w, \quad \left(\frac{\omega}{c}\right)^2 \frac{\overline{v}\kappa}{2k_\omega \beta_\omega} \to v, \qquad (5.6a)$$

$$\sqrt{2k_\omega \beta_\omega} \, x \to x, \quad \beta_\omega z \to z \qquad (5.6b)$$

are made, the basic coupled differential equations reduce to the simple *scalar* forms [15]

$$i\frac{\partial w}{\partial z} + \frac{\partial^2 w}{\partial x^2} - w + w^* v = 0 \qquad (5.7a)$$

$$i2\alpha \frac{\partial v}{\partial z} + \frac{\partial^2 v}{\partial x^2} - \alpha\beta v + w^2 = 0 \qquad (5.7b)$$

where

$$\alpha = \frac{k_{2\omega}}{2k_\omega} = \sqrt{\frac{\varepsilon(2\omega)}{\varepsilon(\omega)}}, \quad \beta = 2\frac{\beta_{2\omega}}{\beta_\omega} \qquad (5.8)$$

The forms of (5.7a) and (5.7b) are to be found in Boardman and Xie. The alternative notation of Buryak and Kivshar [18] is reached by the identifications $\alpha\beta \to \alpha$ and $\sqrt{k_\omega \beta_\omega} \, x \to x$.

This resolution into *fast* and *slow* dependences is very convenient and w and v have the following interpretation. Usually, for *stationary solutions*,

$\dfrac{\partial |w|}{\partial z} = 0, \dfrac{\partial |v|}{\partial z} = 0$, with the phases of w and v being allowed to change during the propagation. In this particular case, however, any phase change can be included in the transformations. In fact, for stationary states, w, v are real and independent of z and β_ω, $\beta_{2\omega}$ are *exactly* the nonlinear phase shifts.

Consider now a quantity M that can be thought of as an *"effective mass"*, where

$$M = \frac{1}{\sqrt{2\pi}} \int \left(\frac{|w|^2}{2} + \alpha |v|^2 \right) dx \tag{5.9}$$

$$\frac{dM}{dz} = \frac{i}{2\sqrt{2\pi}} \int \left[w^* \frac{\partial^2 w}{\partial x^2} - w \frac{\partial^2 w^*}{\partial x^2} + v^* \frac{\partial^2 v}{\partial x^2} - v \frac{\partial^2 v^*}{\partial x^2} \right] dx = 0 \tag{5.10}$$

so that M is a constant of the motion. M, called here an *effective mass*, is related to the total power carried by the coupled ω and 2ω beams. To see this note, first of all, that the time-averaged value of the Poynting vector is

$$S = \frac{1}{2} Re(\mathbf{E} \times \mathbf{H}^*) \text{ watts/meter}^2 \tag{5.11}$$

where Re denotes real part, $\mathbf{E}$ is electric field, $\mathbf{H}$ is magnetic field intensity. Writing $H = \varepsilon_0(c/n)E$ - as is easily deduced for plane waves - and noting that n is refractive index and c the vacuum velocity of light, the power in the system is

$$P = \frac{1}{2} \sqrt{\frac{\varepsilon_0}{\mu_0}} \int \left[n_\omega \frac{|\overline{w}|^2}{2} + n_{2\omega} |\overline{v}|^2 \right] dx \tag{5.12}$$

Using the transformations of $\overline{w} \to w$, $\overline{v} \to v$ and, finally, $\sqrt{2k_\omega \beta_\omega} x \to x$, yields

$$P = \left(\frac{c}{w} \right)^4 \frac{4k_\omega^2 \beta_\omega^2}{2\kappa^2 n_\omega} \sqrt{\frac{\varepsilon_0}{\mu_0}} \int \left(\frac{1}{2} |w|^2 + \alpha |v|^2 \, dx \right)$$

$$= \left(\frac{c}{w} \right)^4 (2k_\omega \beta_\omega)^{3/2} \frac{\sqrt{2\pi}}{2\kappa^2 n_\omega} \sqrt{\frac{\varepsilon_0}{\mu_0}} M \tag{5.13}$$

$$= 2\sqrt{\pi} \frac{c}{\omega} \frac{\varepsilon(\omega)}{\kappa^2} \sqrt{\frac{\varepsilon_0}{\mu_0}} \frac{(\alpha - 1)^{3/2}}{(1 - \beta/4)} M$$

Hence M is related to the total power in the system.

The "mass centre" of the system is x_0, where

$$2Mx_0 = \sqrt{\frac{2}{\pi}} \int x \left[\frac{1}{2} |w|^2 + \alpha |v|^2 \right] dx \qquad (5.14)$$

The second conserved quantity of the coupled equations is the momentum

$$P = \frac{i}{\sqrt{2\pi}} \int \left[w \frac{\partial w^*}{\partial x} - w^* \frac{\partial w}{\partial x} + \alpha v \frac{\partial v^*}{\partial x} - \alpha v^* \frac{\partial v}{\partial x} \right] dx \qquad (5.15)$$

and the third conserved quantity is the Hamiltonian [15]

$$H = \sqrt{\frac{2}{\pi}} \int \left[\left| \frac{\partial w}{\partial x} \right|^2 + \frac{1}{2} \left| \frac{\partial v}{\partial x} \right|^2 + |w|^2 + \frac{\alpha\beta}{2} |v|^2 - \frac{1}{2} (w^{*2}v + w^2 v^*) \right] dx \qquad (5.16)$$

Stationary solutions are w, v which are solutions of the equations

$$\frac{d^2 w}{dx^2} - w + wv = 0 \qquad (5.17a)$$

$$\frac{d^2 v}{dx^2} - \alpha\beta v + w^2 = 0 \qquad (5.17b)$$

Integrating these once gives

$$\frac{1}{2} \left(\frac{dw}{dx} \right)^2 + \frac{1}{4} \left(\frac{dv}{dx} \right)^2 + \frac{1}{2} \left(w^2 v - \frac{\alpha\beta}{2} v^2 - w^2 \right) = C \qquad (5.18)$$

where C is the constant of integration. $C = 0$ corresponds to a soliton-like solution. It is interesting that the above coupled equations do not yield any more conservation laws. Furthermore, analytical solutions cannot generally be found but one does exist for $\alpha\beta = 1$, and these are

$$w = v = \frac{3}{2} \text{sech}^2 (x/2) \qquad (5.19)$$

Solitary wave solutions of the coupled equations, for other parameter values, have to be calculated numerically. Figure 12 shows some possibilities, including, symmetric and antisymmetric field profiles. It is extremely interesting that both single peak and multipeak solutions [15] exist. The stability of such solutions is still a topic for investigation. It is important to note that they have finite binding energy, unlike solutions of the nonlinear Schrödinger equation. In addition to the soliton-like solitons

42

(solitary waves) the coupled equations also have a lot of periodic solutions, which exist for all different values of C.

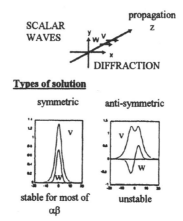

Figure 12 Representative numerical solutions of equations (5.7a) and (5.7b). Note that both single peak and multi-peak solutions exist. In the multi-peak solution, w has a crossing point on the axis. Multi-peak solutions have a finite bonding energy, however, unlike higher-order soliton solutions of the nonlinear Schrödinger equation.

Since the original coupled equations are *not integrable*, many aspects of soliton properties, such as the stability of the solutions and what kind of initial conditions will produce a soliton, can not be deduced mathematically. Numerical simulation can reveal some of this information, however. For example, numerical simulations have shown [15] that multipeak structures may well be unstable for most values of $\alpha\beta$, while single-peak solitary waves are rather stable for large α and unstable for small α. A reliable conclusion can not be drawn from such discrete numerical experiments, however. Another method is needed. In the following, therefore, an outline will be given of an *approximate but versatile method*, called the variational method. Using this elegant analysis, it is the stability of fundamental solitary, single-peak waves can be deduced that will be addressed.

5.1 THE VARIATIONAL APPROACH

The original coupled equations can be derived from the following Lagrangian [15,44,45]

$$
L = \frac{i}{2}\left(w^* \frac{\partial w}{\partial z} - w \frac{\partial w^*}{\partial z} \right) + \frac{i\alpha}{2}\left(v^* \frac{\partial v}{\partial z} - v \frac{\partial v^*}{\partial z} \right)
$$
$$
- \left| \frac{\partial w}{\partial x} \right|^2 - |w|^2 - \frac{1}{2}\left| \frac{\partial v}{\partial x} \right|^2 - \frac{\alpha\beta}{2}|v|^2 + \frac{1}{2}\left(w^{*2}v + w^2 v^* \right)
$$

(5.20)

At this point trial functions are selected and it is common practice to use Gaussian forms, rather than the more obvious sech-hyperbolic secant – forms. The reason is that Gaussians are much easier to deal with mathematically. Only a modicum of accuracy is lost but lots of easy to handle analytical results [54] are gained that give an excellent representation of the solitary wave dynamics. The presentation here has been published by Boardman and Xie [15] but more detail is given here to enable the reader to derive *all* the results and to fill in the gaps inevitably appearing in a paper.

The Gaussian trial functions selected here are

$$w = \eta_1 \exp\left[-\rho_1^2(x - x_1)^2 + i\frac{\xi_1}{2}(x - x_1) + i\frac{\theta_1}{2}\right] \tag{5.21a}$$

$$v = \eta_2 \exp\left[-\rho_2^2(x - x_2)^2 + i\xi_2(x - x_2) + i\theta_2\right] \tag{5.21b}$$

where η_1, η_2 are the amplitudes of the beams, $\rho_{1,2}$ are the inverse beams, $\xi_{1,2}$ are the propagation directions, $x_{1,2}$ are the beam centres and $\theta_{1,2}$ are phases. The *reduced* Lagrangian [15,44,45] is

$$\mathcal{L} = \sqrt{\frac{2}{\pi}} \int L dx = \mathcal{L}_1 + \mathcal{L}_2 + \mathcal{L}_{12} \tag{5.22}$$

where

$$\mathcal{L}_1 = \frac{\eta_1^2}{2\rho_1}\xi_1\frac{\partial x_1}{\partial z} - \frac{\eta_1^2}{2\rho_1}\frac{\partial \theta_1}{\partial z} - \rho_1\eta_1^2 - \frac{\eta_1^2}{4\rho_1}\xi_1^2 - \frac{\eta_1^2}{\rho_1} \tag{5.23a}$$

$$\mathcal{L}_2 = \alpha\frac{\eta_2^2}{\rho_2}\xi_2\frac{\partial x_2}{\partial z} - \alpha\frac{\eta_2^2}{\rho_2}\frac{\partial \theta_2}{\partial z} - \frac{1}{2}\rho_2\eta_2^2 - \frac{\eta_2^2}{2\rho_2}\xi_2^2 - \frac{\alpha\beta}{2}\frac{\eta_2^2}{\rho_2} \tag{5.23b}$$

$$\mathcal{L}_{12} = \sqrt{\frac{2}{\pi}}\eta_1^2\eta_2 \int \exp\left[-2\rho_1^2(x - x_1)^2 - \rho_2^2(x - x_2)^2\right]$$
$$\cos\left[\theta_2 - \theta_1 + \xi_2(x - x_2) - \xi_1(x - x_1)\right]dx \tag{5.23c}$$

The next step is to apply the Euler-Lagrange equations

$$\frac{\partial \mathcal{L}}{\partial g} - \frac{d}{dz}\frac{\partial \mathcal{L}}{\partial(dg/dz)} = 0 \tag{5.24}$$

in which the variable g is $g = \eta_j, \rho_j, \theta_j, \xi_j, x_j$. Equations (5.24) then leads to a set of coupled, nonlinear, ordinary, differential equations, namely

$$\frac{\eta_1^2}{2\rho_1}\frac{dx_1}{dz} - \frac{\eta_1^2}{2\rho_1}\xi_1 + \frac{\partial \mathcal{L}_{12}}{\partial \xi_1} = 0 \tag{5.25a}$$

$$\alpha \frac{\eta_2^2}{\rho_2}\frac{dx_2}{dz} - \frac{\eta_2^2}{\rho_2}\xi_2 + \frac{\partial \mathcal{L}_{12}}{\partial \xi_2} = 0 \tag{5.25b}$$

$$\frac{d}{dz}\left(\frac{\eta_1^2}{2\rho_1}\xi_1\right) = \frac{\partial \mathcal{L}_{12}}{\partial x_1} \tag{5.25c}$$

$$\alpha \frac{d}{dz}\left(\frac{\eta_2^2}{\rho_2}\xi_2\right) = \frac{\partial \mathcal{L}_{12}}{\partial x_2} \tag{5.25d}$$

$$\frac{d}{dz}\left(\frac{\eta_1^2}{2\rho_2}\right) = -\frac{\partial \mathcal{L}_{12}}{\partial \theta_1} \tag{5.25e}$$

$$\alpha \frac{d}{dz}\left(\frac{\eta_2^2}{\rho_2}\right) = -\frac{\partial \mathcal{L}_{12}}{\partial \theta_2} \tag{5.25f}$$

$$\frac{\eta_1}{\rho_1}\xi_1\frac{dx_1}{dz} - \frac{\eta_1}{\rho_1}\frac{d\theta_1}{dz} - 2\rho_1\eta_1 - \frac{\eta_1}{2\rho_1}\xi_1^2 - 2\frac{\eta_1}{\rho_1} + \frac{\partial \mathcal{L}_{12}}{\partial \eta_1} = 0 \tag{5.25g}$$

$$2\alpha\frac{\eta_2}{\rho_2}\xi_2\frac{dx_2}{dz} - 2\alpha\frac{\eta_2}{\rho_2}\frac{d\theta_2}{dz} - \rho_2\eta_2 - \frac{\eta_2}{\rho_2}\xi_2^2 - \alpha\beta\frac{\eta_2}{\rho_2} + \frac{\partial \mathcal{L}_{12}}{\partial \eta_2} = 0 \tag{5.25h}$$

$$-\frac{\eta_1^2}{2\rho_1^2}\xi_1\frac{dx_1}{dz} + \frac{\eta_1^2}{2\rho_1^2}\frac{d\theta_1}{dz} - \eta_1^2 + \frac{\eta_1^2}{4\rho_1^2}\xi_1^2 + \frac{\eta_1^2}{\rho_1^2} + \frac{\partial \mathcal{L}_{12}}{\partial \rho_1} = 0 \tag{5.25i}$$

$$-\alpha\frac{\eta_2^2}{\rho_2^2}\xi_2\frac{dx_2}{dz} + \alpha\frac{\eta_2^2}{\rho_2^2}\frac{d\theta_2}{dz} - \frac{1}{2}\eta_2^2 + \frac{\eta_2^2}{2\rho_2^2}\xi_2^2 + \frac{\alpha\beta}{2}\frac{\eta_2^2}{\rho_2^2} + \frac{\partial \mathcal{L}_{12}}{\partial \rho_2} = 0 \tag{5.25j}$$

The law conserving "mass" is

$$\frac{d}{dz}\left(\frac{\eta_1^2}{2\rho_1} + \alpha\frac{\eta_2^2}{\rho_2}\right) = -\frac{\partial \mathcal{L}_{12}}{\partial \theta_1} - \frac{\partial \mathcal{L}_{12}}{\partial \theta_2} = 0 \tag{5.26}$$

Hence,

$$\frac{\eta_1^2}{2\rho_1} + \alpha\frac{\eta_2^2}{\rho_2} = 2M = \text{constant} \qquad (5.27)$$

where it should be noted that M has the previous definition $M = \frac{1}{\sqrt{2\pi}}\int\left(\frac{1}{2}|w|^2 + \alpha|v|^2\right)dx$. The law conserving "momentum" is

$$\frac{d}{dz}\left(\frac{\eta_1^2}{2\rho_1}\xi_1 + \alpha\frac{\eta_2^2}{\rho_2}\xi_2\right) = \frac{\partial\mathcal{L}_{12}}{\partial x_1} + \frac{\partial\mathcal{L}_{12}}{\partial x_2} = -\int\frac{d}{dx}L_{12}dx\sqrt{\frac{2}{\pi}} = 0 \qquad (5.28)$$

where

$$\mathcal{L}_{12} = \sqrt{\frac{2}{\pi}}\int\mathcal{L}_{12}dx, \quad L_{12} = \frac{1}{2}(w^{*2}v + w^2v^*) = L_{12}(x - x_1, x - x_2) \qquad (5.29)$$

and

$$\frac{dL_{12}}{dx} = -\frac{\partial L_{12}}{\partial x_1} - \frac{\partial L_{12}}{\partial x_2} \qquad (5.30a)$$

To obtain (5.30) set $A_1 = x - x_1$, $A_2 = x - x_2$ and use

$$\frac{\partial L_{12}}{\partial x} = \frac{\partial L_{12}}{\partial A_1}\frac{\partial A_1}{\partial x} + \frac{\partial L_{12}}{\partial A_2}\frac{\partial A_2}{\partial x} = \frac{\partial L_{12}}{\partial A_1} + \frac{\partial L_{12}}{\partial A_2}; \frac{\partial L_{12}}{\partial x_{1,2}} = \frac{\partial L_{12}}{\partial A_{1,2}}\frac{\partial A_{1,2}}{\partial x_{1,2}} = -\frac{\partial L_{12}}{\partial A_{1,2}} \qquad (5.30b)$$

The conclusion is that

$$\frac{\eta_1^2}{2\rho_1}\xi_1 + \alpha\frac{\eta_2^2}{\rho_2}\xi_2 = \text{constant} = 0 \text{ (chosen initially)} \qquad (5.31)$$

It also can be easily confirmed that

$$\frac{\eta_1^2}{2\rho_1}\xi_1 + \alpha\frac{\eta_2^2}{\rho_2}\xi_2 - P = \frac{i}{\sqrt{2\pi}}\int\left(w\frac{\partial w^*}{\partial x} - w^*\frac{\partial w}{\partial x} + \alpha v\frac{\partial v^*}{\partial x} - \alpha v^*\frac{\partial v}{\partial x}\right)dx \qquad (5.32)$$

Considering now the movement of the "centre of mass"

$$\frac{d}{dz}\left(\frac{\eta_1^2}{2\rho_1}x_1 + \alpha\frac{\eta_2^2}{\rho_2}x_2\right) = -x_1\frac{\partial \mathcal{L}_{12}}{\partial\theta_1} - x_2\frac{\partial \mathcal{L}_{12}}{\partial\theta_2} + \frac{\eta_1^2}{2\rho_1}\xi_1 - \frac{\partial \mathcal{L}_{12}}{\partial\xi_1} + \frac{\eta_2^2}{\rho_2}\xi_2 - \frac{\partial \mathcal{L}_{12}}{\partial\xi_2}$$

$$= \frac{\eta_1^2}{2\rho_1}\xi_1 + \frac{\eta_2^2}{\rho_2}\xi_2 = (1-\alpha)\frac{\eta_2^2}{\rho_2}\xi_2$$

(5.33)

Hence,

$$\frac{\eta_1^2}{2\rho_1}x_1 + \frac{\eta_2^2}{\rho_2}x_2 \neq \text{constant}$$

(5.34)

It can also be confirmed that

$$\frac{\eta_1^2}{2\rho_1}x_1 + \alpha\frac{\eta_2^2}{\rho_2}x_2 = 2Mx_0 = \sqrt{\frac{2}{\pi}}\int x\left(\frac{1}{2}|w|^2 + \alpha|v|^2\right)dx$$

(5.35)

Finally, the Hamiltonian of the system is

$$H = \rho_1\eta_1^2 + \frac{\eta_1^2}{4\rho_1}\xi_1^2 + \frac{\eta_1^2}{\rho_1} + \frac{1}{2}\rho_2\eta_2^2 + \frac{\eta_2^2}{2\rho_2}\xi_2^2 + \frac{\alpha\beta}{2}\frac{\eta_2^2}{\rho_2}$$

$$-\sqrt{\frac{2}{\pi}}\eta_1^2\eta_2\int \exp\left[-2\rho_1^2(x-x_1)^2 - \rho_2^2(x-x_2)^2\right]\cos\left[\theta_2 - \theta_1 + \xi_2(x-x_2) - \xi_1(x-x_1)\right]dx$$

(5.36)

5.2 STATIONARY SOLITARY SOLUTIONS

Stationary solutions occur when all the z derivatives are zero, i.e. $\dfrac{dx_1}{dz} = \dfrac{dx_2}{dz} = 0$,

$\dfrac{d\xi_1}{dz} = \dfrac{d\xi_2}{dz} = 0$, $\dfrac{d\theta_1}{dz} = \dfrac{d\theta_2}{dz} = 0$, $\dfrac{d}{dz}\left(\dfrac{\eta_1^2}{2\rho_1}\right) = \dfrac{d}{dz}\left(\dfrac{\eta_2^2}{\rho_2}\right) = 0$. This requires $x_1 = x_2 = 0$,

$\xi_1 = \xi_2 = 0$, $\theta_1 = \theta_2 = 0$, $\eta_1 = \eta_{10}$, $\eta_2 = \eta_{20}$, $\rho_1 = \rho_{10}$, $\rho_2 = \rho_{20}$. The stationary values can be found from the analytical evaluation of the Euler-Lagrange equations (5.25a)=(5.25j). In fact ρ_{10}, ρ_{20}, η_{10}, η_{20} are solutions of the following

$$\rho_{10}\eta_{10}^2 + \frac{\eta_{10}^2}{\rho_{10}} - \frac{\sqrt{2}\eta_{10}^2\eta_{20}}{\sqrt{2\rho_{10}^2 + \rho_{20}^2}} = 0$$

(5.37)

$$\rho_{20}\eta_{20}^2 + \alpha\beta\frac{\eta_{20}^2}{\rho_{20}} - \frac{\sqrt{2}\eta_{10}^2\eta_{20}}{\sqrt{2\rho_{10}^2 + \rho_{20}^2}} = 0$$

(5.38)

$$-\rho_{10}\eta_{10}^2 + \frac{\eta_{10}^2}{\rho_{10}} - \frac{2\sqrt{2}\eta_{10}^2\eta_{20}}{\sqrt{2\rho_{10}^2 + \rho_{20}^2}} \frac{\rho_{10}^2}{2\rho_{10}^2 + \rho_{20}^2} = 0 \qquad (5.39)$$

$$-\rho_{20}\eta_{10}^2 + \alpha\beta\frac{\eta_{20}^2}{\rho_{20}} - \frac{2\sqrt{2}\eta_{10}^2\eta_{20}}{\sqrt{2\rho_{10}^2 + \rho_{20}^2}} \frac{\rho_{20}^2}{2\rho_{10}^2 + \rho_{20}^2} = 0 \qquad (5.40)$$

The above equations can be rewritten, in the following way, to separate η_{10}, η_{20} from ρ_{10}, ρ_{20}

$$\left(\rho_{10}^2 + 1\right)^2\left(2\rho_{10}^2 + \rho_{20}^2\right) = 2\rho_{10}^2\eta_{20}^2 \qquad (5.41a)$$

$$\left(\rho_{20}^2 + \alpha\beta\right)^2\left(2\rho_{10}^2 + \rho_{20}^2\right) = 2\rho_{10}^2\eta_{10}^4 / \eta_{20}^2 \qquad (5.41b)$$

$$\left(-\rho_{10}^2 + 1\right)\left(2\rho_{10}^2 + \rho_{20}^2\right) = 2\rho_{10}^2\left(\rho_{10}^2 + 1\right) \qquad (5.41c)$$

$$\left(-\rho_{20}^2 + \alpha\beta\right)\left(2\rho_{10}^2 + \rho_{20}^2\right) = 2\rho_{20}^2\left(\rho_{20}^2 + \alpha\beta\right) \qquad (5.41d)$$

For $\alpha\beta = 1$, a simple solution can be found i.e. $\eta_{10} = \eta_{20} = \frac{3}{5}\sqrt{6}, \rho_{10} = \rho_{20} = \frac{1}{\sqrt{5}}$ so $w = v = \frac{3\sqrt{6}}{5}\exp\left(-\frac{x^2}{5}\right)$. The original coupled equations, for this special value of $\alpha\beta$, exhibit the exact value $w = v = \frac{3}{2}\operatorname{sech}^2\left(\frac{x}{2}\right)$, however, so a brief comparison of the exact solution and the approximate solution is listed below, to give some confidence in the results.

	exact	approximate	error
Amplitude	1.5	1.47	2%
Total "mass"	3.59	3.62	0.83%

From equation (5.41c),

$$\rho_{20}^2 = \frac{4\rho_{10}^4}{1 - \rho_{10}^2} \qquad (5.42)$$

Substitution into equation (5.41d) yields, after some algebraic manipulation, a formula for ρ_{10} that is valid for general $\alpha\beta$ i.e.

$$\rho_{10}^6 + \frac{4-3\alpha\beta}{20}\rho_{10}^4 + \frac{\alpha\beta}{5}\rho_{10}^2 - \frac{\alpha\beta}{20} = 0 \tag{5.43}$$

This is a cubic equation for ρ_{10}^2 has three roots are expected, two of which may be complex. Only a real root is physically meaningful and, even then, when this real root becomes negative, the system does not support a solitary wave any more. This will happen when $\alpha\beta < 0$.

To solve the cubic equation, define

$$\rho_{10}^2 = y - \frac{4-3\alpha\beta}{60} \tag{5.44}$$

so that

$$y^3 + py + q = 0, \quad p = -3\left(\frac{4-3\alpha\beta}{60}\right)^2 + \frac{\alpha\beta}{5}, \quad q = 2\left(\frac{4-3\alpha\beta}{60}\right)^3 - \frac{(4-3\alpha\beta)\alpha\beta}{600} - \frac{\alpha\beta}{20} \tag{5.45}$$

If $T^2 = \left(\frac{q}{2}\right)^2 + \left(\frac{p}{3}\right)^3$, then the solution is

$$y = \sqrt[3]{-\frac{q}{2} + T} + \sqrt[3]{-\frac{q}{2} - T} \tag{5.46}$$

The stationary solutions are then

$$\rho_{10}^2 = y - \frac{4-3\alpha\beta}{60} \tag{5.47}$$

$$\rho_{20}^2 = 4\left(y - \frac{3-4\alpha\beta}{60}\right)^2 \bigg/ \left(1 - y + \frac{4-3\alpha\beta}{60}\right) \tag{5.48}$$

and, from equations (5.41a) and (4.51b),

$$-\eta_{10}^2 = \frac{2\rho_{10}^2 + \rho_{20}^2}{2\rho_{10}\rho_{20}}\left(\rho_{10}^2 + 1\right)\left(\rho_{20}^2 + \alpha\beta\right) \tag{5.49}$$

$$\eta_{20}^2 = \frac{2\rho_{10}^2 + \rho_{20}^2}{2\rho_{10}^2}\left(\rho_{10}^2 + 1\right)^2 \tag{5.50}$$

The ratio η_{20}/η_{10} versus $\alpha\beta$, obtained by this Lagrangian analysis, is plotted in figure 13, together with the *exact ratio* (obtained numerically). Excellent agreement is achieved so, with this confidence in the method, it now only remains to examine the stability of some of the stationary solutions.

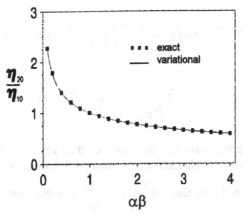

Figure 13. Ratio of stationary state second-harmonic amplitude to fundamental amplitude as a function of $\alpha\beta$. Note that the variational analysis is an excellent representation.

5.3 STABILITY ANALYSIS

As was shown earlier, the evolution of the parameters around the stationary values is determined by the full set of ten ordinary differential equations emerging from the application of the Euler-Lagrange equations. All the 10 parameters are coupled to each other, but, based upon the physical meaning they convey, these parameters can be classified into two useful groups: (1) beam position and propagation direction $[x_1, x_2, \xi_1, \xi_2]$ and (2) beam size and phase: $[\eta_1, \eta_2, \rho_1, \rho_2, \theta_1, \theta_2]$. Coupling within these groups is much stronger than coupling between the groups. Indeed, in the vicinity of the stationary point, the two groups are uncoupled to each other, up to a first-order perturbation. In the following, therefore, we will introduce perturbations in accordance with the above physical group classifications. The final verdict concerning the stability of the whole physical system is arrived at after performing a linear summation of the results, obtained from the two separate investigations [15].

- *Perturbations on the beam position and direction*

Assume that $x_1 = \delta x_1$, $x_2 = \delta x_2$, $\xi_1 = \delta\xi_1$, $\xi_2 = \delta\xi_2$ but that the other parameters take their stationary values. Clearly $|\delta x_j| \ll 1$ and $|\delta\xi_j| \ll 1$ are the perturbations so by substituting these into the ten evolution equations, and then linearising them in terms of δx_j, $\delta\xi_j$, the following results emerge

$$\frac{\eta_{10}^2}{2\rho_{10}}\frac{dx_1}{dz} = \frac{\eta_{10}^2}{2\rho_{10}}\xi_1 - \frac{\sqrt{2}}{2}\frac{\eta_{10}^2\eta_{20}}{\left(2\rho_{10}^2 + \rho_{20}^2\right)^{3/2}}(\xi_2 - \xi_1) \qquad (5.51)$$

$$\alpha \frac{\eta_{20}^2}{\rho_{20}} \frac{dx_2}{dz} = \frac{\eta_{20}^2}{\rho_{20}} \xi_2 + \frac{\sqrt{2}}{2} \frac{\eta_{10}^2 \eta_{20}}{\left(2\rho_{10}^2 + \rho_{20}^2\right)^{3/2}} (\xi_2 - \xi_1) \tag{5.52}$$

$$\frac{\eta_{10}^2}{2\rho_{10}} \frac{d\xi_1}{dz} = 4\sqrt{2} \frac{\eta_{10}^2 \eta_{20}}{\left(2\rho_{10}^2 + \rho_{20}^2\right)^{3/2}} \rho_{10}^2 \rho_{20}^2 (x_2 - x_1) \tag{5.53}$$

$$\alpha \frac{\eta_{20}^2}{\rho_{20}} \frac{d\xi_2}{dz} = -4\sqrt{2} \frac{\eta_{10}^2 \eta_{20}}{\left(2\rho_{10}^2 + \rho_{20}^2\right)^{3/2}} \rho_{10}^2 \rho_{20}^2 (x_2 - x_1) \tag{5.54}$$

All the other evolution equations are zero but this is entirely what is expected, since the two parameter groups are not coupled at the first-order perturbation level.

It can be confirmed now that $\alpha \dfrac{\eta_{20}^2}{\rho_{20}} \xi_2 + \dfrac{\eta_{10}^2}{2\rho_{10}} \xi_1 = \text{constant} = 0$, which has previously been referred to as momentum conservation. Introducing the two variables $\alpha \dfrac{\eta_{20}^2}{\rho_{20}} \xi_2 - \dfrac{\eta_{10}^2}{2\rho_{10}} \xi_1 = 2M\xi$, where $2M = \alpha \dfrac{\eta_{20}^2}{\rho_{20}} + \dfrac{\eta_{10}^2}{2\rho_{10}}$ is the total mass, and expressing the quantities ξ_1 and ξ_2, from equation (5.31), as

$$\xi_2 = \frac{\rho_{20}}{\alpha \eta_{20}^2} M\xi, \quad \xi_1 \cong -\frac{2\rho_{10}}{\eta_{10}^2} M\xi \tag{5.55}$$

the evolution equations for Δ and ξ are

$$\frac{d\Delta}{dz} = \left\{ \frac{\rho_{20}}{\alpha^2 \eta_{20}^2} + \frac{2\rho_{10}}{\eta_{10}^2} + \frac{1}{M} \left[\frac{\sqrt{2}}{2} \frac{\eta_{10}^2 \eta_{20}}{\left(2\rho_{10}^2 + \rho_{20}^2\right)^{3/2}} \right] \left(\frac{\rho_{20}}{\alpha \eta_{20}^2} + \frac{2\rho_{10}}{\eta_{10}^2} \right) \right\} M\xi = a\xi \tag{5.56}$$

$$\frac{d\xi}{dz} = -4\sqrt{2} \frac{\eta_{10}^2 \eta_{20}}{\left(2\rho_{10}^2 \rho_{20}^2\right)^{3/2}} \rho_{10}^2 \rho_{20}^2 \left[\frac{2\rho_{10}}{\eta_{10}^2} + \frac{\rho_{20}}{\alpha \eta_{20}^2} \right] \Delta = -b\Delta \tag{5.57}$$

The stability equations, in this case, become

$$\frac{d^2\Delta}{dz^2} = -ab\Delta, \quad \frac{d^2\xi}{dz^2} = -ab\xi \tag{5.58}$$

which combine to give

$$\frac{1}{2}\left(\frac{d\Delta}{dz}\right)^2 + U(\Delta) = 0, \quad \frac{1}{2}\left(\frac{d\xi}{dz}\right)^2 + U(\xi) = 0 \qquad (5.59)$$

where the "potentials" are

$$U(\Delta) = \frac{1}{2}ab\Delta^2, \quad U(\xi) = \frac{1}{2}ab\xi^2 \qquad (5.60)$$

Since $a > 0$, $b > 0$, the potential $U(\Delta)$ or $U(\xi)$ is always concave at $\Delta = 0$ or $\xi = 0$. This means that the stationary solitary solutions are always stable under this type of perturbation. Note that the current perturbation represents a mismatch in position or direction between the fundamental and harmonic wave at launching.

- *Perturbations beam size and phase*

In this kind of perturbation represents the beam size is an energy fluctuation of the solitary wave beam and ξ_1, ξ_2, x_1, x_2 are not perturbed. They will be set, initially, as zero, but it be easily confirmed, however, from their evolution equations that they *remain* at zero. Hence, ξ_j, x_j can be set to zero for the *whole evolution*. The evolution equations for the other parameters, generated from the Euler-Lagrange equations, are

$$\frac{d}{dz}\left(\frac{\eta_1^2}{2\rho_1}\right) = \frac{\sqrt{2}\eta_1^2\eta_2}{\sqrt{2\rho_1^2 + \rho_2^2}}\sin(\theta_2 - \theta_1) \qquad (5.61a)$$

$$\alpha\frac{d}{dz}\left(\frac{\eta_2^2}{\rho_2}\right) = \frac{\sqrt{2}\eta_1^2\eta_2}{\sqrt{2\rho_1^2 + \rho_2^2}}\sin(\theta_2 - \theta_1) \qquad (5.61b)$$

$$\frac{d\theta_1}{dz} = -2\rho_1^2 - 2 + 2\sqrt{2}\frac{\rho_1\eta_2}{\sqrt{2\rho_1^2 + \rho_2^2}}\cos(\theta_2 - \theta_1) \qquad (5.61c)$$

$$\alpha\frac{d\theta_2}{dz} = -\frac{\rho_2^2}{2} - \frac{\alpha\beta}{2} + \frac{\sqrt{2}}{2}\frac{\rho_1\eta_1^2/\eta_2}{\sqrt{2\rho_1^2 + \rho_2^2}}\cos(\theta_2 - \theta_1) \qquad (5.61d)$$

The Euler-Lagrange equations also generate two algebraic equations which relate to ρ_1, ρ_2 to η_1, η_2, θ_1, θ_2. These are

$$2\rho_1^2\left(2\rho_1^2 + \rho_2^2\right)^3 = \rho_2^4\eta_2^2\cos^2(\theta_2 - \theta_1) \qquad (5.62)$$

$$\rho_1\eta_1^2\left(2\rho_1^2 - \rho_2^2\right) = \rho_2^3\eta_2^2 \qquad (5.63)$$

The quantities ρ_j, η_j, θ_j are related to their stationary values by ρ_{jo}, η_{jo}, θ_{jo} through first-order perturbations i.e. $\rho_j = \rho_{10} + \delta\rho_j$, $\eta_j = \eta_{jo} + \delta\eta_j$, $\theta_j = \delta\theta_j$, where $\theta_{jo} = 0$, and $|\delta\rho_j| \ll 1$, $|\delta\eta_j| \ll 1$, $|\delta\theta_j| \ll 1$ are the actual perturbations. The introduction of the two new variables $\quad N_1 = \dfrac{\eta_1^2}{2\rho_1}$, $N_2 = \dfrac{\eta_2^2}{\rho_2} \quad$ gives $\quad \delta\eta_1 = \dfrac{\rho_{10}}{\eta_{10}}\delta N_1 + \dfrac{\eta_{10}}{2\rho_{10}}\delta\rho_1$,

$\delta\eta_2 = \dfrac{\rho_{20}}{2\eta_{20}}\delta N_2 + \dfrac{\eta_{20}}{2\rho_{20}}\delta\rho_2$. Physically, the two new variables represent the total energy (mass) in the fundamental and harmonic beams, respectively. The procedure now is the same as it was for the previous perturbation section on beam position and direction. The perturbations are substituted into equations (5.61a) to (5.61d) and only the leading orders in $\delta\rho_j$, δN_j, $\delta\theta_j$ are kept.

The calculations, in this case, are quite extensive, but very straightforward. This considerable algebraic effort is the reason why the original paper, in which this work can be found, contains *all* the relevant detail, for this part of the calculation. It need not be repeated here. The main results are

$$\frac{dN}{dz} = a\theta, \quad \frac{d\theta}{dz} = -bN \tag{5.64}$$

where $2N = \delta N_2 - \delta N_1$ and $\theta = \theta_2 - \theta_1$ and a and b are complicated forms of η_{10}, η_{20}, α, ρ_{10} and ρ_{20}.

Equations (5.64) give

$$\frac{1}{2}\left(\frac{dN}{dz}\right)^2 + U(N) = 0, \quad \frac{1}{2}\left(\frac{d\theta}{dz}\right)^2 + U(\theta) = 0 \tag{5.65}$$

where $U(N)$ and $U(\theta)$ are potentials and a "particle in a well" representation immediately comes to mind.

Decisions concerning the stability of single-peak solitary waves with respect to perturbations come down to an examination of the *shape* of the potentials $U(\Delta)$, $U(\xi)$, $U(N)$ and $U(\theta)$. Indeed, it has already been shown that, for perturbations to the beam position and direction, $U(\Delta)$ or $U(\xi)$ is always *concave* at $\Delta = 0$ or $\xi = 0$. This means that the solitary waves are always stable under this type of perturbation. The radius of curvature of the potential U is an important quantity, therefore, and will be positive when $\dfrac{d^2U}{dg^2} > 0$, for example, where $g = \Delta, \xi, N, \theta$. Hence, for $\dfrac{d^2U}{dg^2} > 0$ the U curves will be *concave upwards*, with a positive radius of curvature and when $\dfrac{d^2U}{dg^2} < 0$ i.e. the curves are *convex upwards*, with a negative radius of curvature.

The potentials in (5.65) are $U(N)$ and $U(\theta)$ for which

$$\left[\frac{d^2U(N)}{dN^2}\right]_{N=0} = \left[\frac{d^2U(\theta)}{d\theta^2}\right]_{\theta=0} = ab \qquad (5.66)$$

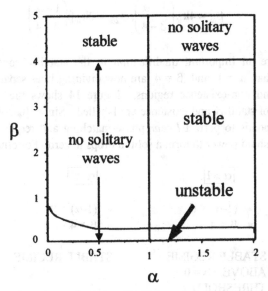

Figure 14. Stability diagram for single-peak solitary waves in quadratically nonlinear crystals.

Whenever $ab > 0$, the stationary, solitary, waves are stable but must be unstable if $ab < 0$. In the unstable regime, a small energy, or phase, fluctuation will cause all of the energy to be transferred into one of the coupled modes and, consequently, be radiated away. From the arguments developed above, the boundary $ab = 0$ is a clear division between stable and unstable regimes, for the single-peak solitary waves. In addition, it is also necessary to observe that *balanced* conditions are sustained by the solitary waves, which demand that

$$2k_\omega - k_{2\omega} - \beta_{2\omega} + 2\beta_\omega = 0 \qquad (5.67)$$

where $\beta_{2\omega} > 0$, $\beta_\omega > 0$, $\alpha\beta > 0$. From the definitions $\alpha = \dfrac{k_{2\omega}}{2k_\omega}$, $\beta = \dfrac{2\beta_{2\omega}}{\beta_\omega}$ the restraint (5.67) can be written as

$$2k_\omega - 2\alpha k_\omega - \frac{\beta\beta_\omega}{2} + 2\beta_\omega = 0 \qquad (5.68)$$

or

$$2k_\omega - 2\alpha k_\omega - \beta_{2\omega} + \frac{2(2\beta_{2\omega})}{\beta} = 0 \qquad (5.69)$$

which yield

$$\beta_\omega = 4k_\omega \left(\frac{1-\alpha}{\beta-4}\right), \ \beta_{2\omega} = 2k_\omega \beta \left(\frac{1-\alpha}{\beta-4}\right) \qquad (5.70)$$

Equations (5.69) are an important qualification to the stability condition $ab > 0$ because they show that $\alpha = 1$ and $\beta = 4$ are now dividing lines separating out stable existence regions and non-existence regions. Figure 14 shows the (β,α) plane in which the domains of stability and existence are labelled. Since $\beta_\omega \geq 0$ and $\beta_{2\omega} \geq 0$ then $\alpha = 1$ corresponds to perfect *linear* phase-matching and requires that $\beta_\omega = 0$, $\beta_{2\omega} = 0$ i.e. the threshold power to form a soliton drops to zero. For other values of α,

$\boxed{\alpha > 1}$	$\boxed{\alpha < 1}$
• $(1-\alpha) < 0$	• $(1-\alpha) > 0$
• $\beta < 4$	• $\beta > 4$
STABLE REGIME ABOVE $ab = 0$ THRESHOLD	STABLE REGIME

As $\beta \to 4$, $\beta_\omega \to \infty$ and $\beta_{2\omega} \to \infty$, a state which would require a lot of power to achieve, for $\alpha \neq 1$. Figure 15 shows some numerical verifications of these stability conclusions. The stability conditions are also qualitatively summarised in figure 16, in a sketch-plot of power against α. This illustrates that, for $\alpha < 1$, all solutions *that can exist are stable* yet, for $\alpha > 1$, the solutions are unstable, until a certain threshold power is reached.

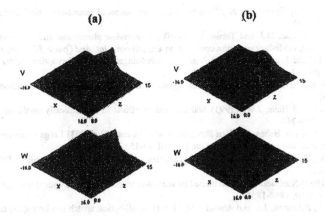

Figure 15. Numerical test of the conclusions summarised by figure 14. v and w refer to single-peak solutions of equations (5.7a) and (5.7b).

(a) $(\beta, \alpha) = (2.0, 1.01)$, (b) $(\beta, \alpha) = (0.2, 1.01)$

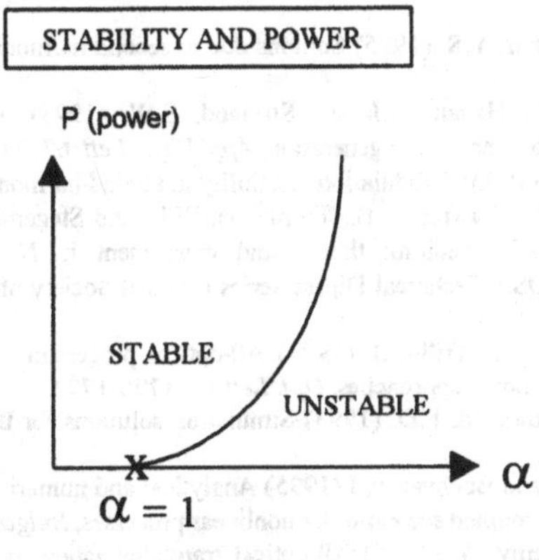

STABILITY AND POWER

Figure 16. Qualitative picture showing the threshold between stable and unstable regimes in terms of power flow and the parameter α

References

1. Boyd, R.W. (1992) *Nonlinear Optics* (Academic, Boston, Mass).
2. Butcher, P.N. and Cotter, D. (1990) *The Elements of Nonlinear Optics*, Cambridge University Press, Cambridge, England.
3. Schubert, M. and Wilhelmi, B. (1986) *Nonlinear Optics and Quantum Electronics* (Wiley-Interscience, New York).
4. Midwinter, J.E. and Warner, J. (1965) The effects of phase-matching method and of uniaxial crystal symmetry on the polar distribution of second-order nonlinear optical polarization, *Brit. J. App. Phys.* **16**, 1135-1142.
5. Dmitriev, V.G., Gurzadyan, G.G. and Nikogosyan, D.W. (1990) *Handbook of Optical Crystals*, Springer-Verlag, Berlin.

56

6. Kaplan, A.E. (1993) Eigenmodes of $\chi^{(2)}$ wave mixings: cross-induced second-order nonlinear reflections, *Opt. Lett.* **18**, 1223-1228.
7. Stegeman, G.I., Hagan, D.J. and Torner, L. (1996) $\chi^{(2)}$ cascading phenomena and their applications to all-optical processing, mode-locking, pulse compression and solitons, *Opt. And Quant. Elec.* **38**, 1691-1740.
8. Kalocsai, A.G. and Haus, J.W. (1994) Nonlinear Schrödinger equation for optical media with quadratic nonlinearity, *Phys. Rev. A* **49**, 574-585 (1994).
9. Buryak, A.V. and Kivshar, Y.S. (1994) Spatial optical solitons governed by quadratic nonlinearity, *Opt. Lett.* **19**, 1612-1614.
10. Kalocsai, A.G. and Haus, J.W. (1993) Self-modulation effects in quadratically nonlinear materials, *Opt. Commun.* **97**, 239-244.
11. Stegeman, G.I., Sheik-Bahae, M., Van Stryland, E. and Assanto, G. (1993) Large nonlinear phase shifts in second-order nonlinear optical processes, *Opt. Lett.* **18**, 13-15.
12. Menyuk, C.R., Schiek, R. and Torner, L. (1994) Solitary waves due to $\chi^{(2)}$: $\chi^{(2)}$ cascading, *J. Opt. Soc. Am. B* **11**, 2434-2443.
13. Schiek, R. (1993) Nonlinear refraction caused by second-order nonlinearity in optical waveguide structures, *J. Opt. Soc. Am. B* **10**, 1848-1855.
14. Ironside, C.N., Aitchison, J.S. and Arnold, J.M. (1993) An all-optical switch employing the cascaded second-order nonlinear effect, *IEEE Journal of Quantum Electronics* **29**, 2650-2654.
15. Boardman, A.D., Xie, K. and Sangarpaul, A. (1995) Stability of scalar spatial solitons in cascadable nonlinear media, *Physical Review A*, **52**, 4099-4105.
16. Desalvo, R., Hagan, D.J., Sheik-Bahae, M. and Stegeman, G.I. (1992) Self-focusing and self-defocusing by cascaded second-order effects in KTP, *Optics Letters* **17**, 28-30.
17. Menyuk, C.R., Schiek, R. and Torner, L. (1994) Solitary waves due to $\chi^{(2)}$: $\chi^{(2)}$ cascading, *J. Opt. Soc. Am. B* **11**, 2434-2443.
18. Buryak, A.V. and Kivshar, Y.S. (1995) Solitons due to second-harmonic generation, *Physics Letters A* **197**, 407-412.
19. Assanto, G., Wang, Z., Hagan, D.J. and Stryland, E.W. (1995) All-optical modulation via nonlinear cascading in type II second-harmonic generation, *App. Phys. Lett.* **67**, 2120-2122.
20. Trillo, S. and Ferro, P. (1995) Modulation instability in second-harmonic generation, *Opt. Lett.* **20** 438-440; Baboiu, D-M., Fuerts, R., Lawrence, B., Torruellas, W.E. and Stegeman, G.I. (1996) Spatial modulational instability in a quadratic medium: theory and experiment in *Nonlinear Guided Waves and Their Applications* Vol. 15, OSA Technical Digest Series (Optical Society of America, Washington, D.C), pp.2-4, (postdeadline paper).
21. Assanto, G., Torelli, I. And Trillo, S. (1994) All-optical processing by means of vectorial interactions in second-order cascading: novel approaches, *Opt. Lett* **19**, 1720-1722.
22. Werner,, M.J. and Drummond, P.D. (1993) Simulation solutions for the parametric amplifier, *J. Opt. Soc. Am. B.* **10**, 2390-2393.
23. Saltiel, S., Koynov, K. and Buchvarov, I (1995) Analytical and numerical investigation of opto-optical phase modulation based upon coupled second-order nonlinear processes, *Bulgarian Journal of Physics* **22**, 39-47.
24. Lefort, L. and Barthelemy, A. (1995) All-optical transistor action by polarisation rotation during type II phase-matched second-harmonic generation, *Electron Lett.* **31**, 910-911.
25. Lefort, L. and Barthelemy, (1995) Intensity-dependent polarization rotation associated with type II phase-matched second-harmonic generation: applications to self-induced transparency, *Opt. Lett.* **20**, 1749-1751.
26. Bakker, H.J., Planken, P.C.M., Kuipers, L. and Lagenddijk (1990) Phase modulation in second-order nonlinear optical processes, *Phys. Rev. A* **42** 4085-4101.
27. Assanto, G., Stegeman, G.I., Sheik-Bahae, M. and Van Stryland, E. (1995) Coherent interactions for all-optical signal processing via quadratic nonlinearities, *IEEE Journ. Quant. Elec.* **31**, 673-681.
28. Werner, M.J. and Drummond, P.D. (1994) Strongly coupled nonlinear parametric solitary waves, *Opt. Lett.* **19**, 613-615.
29. Karamzin, Y.N. and Sukorukov, A.P. (1974) Nonlinear interaction of diffracted light beams in a medium with quadratic nonlinearity: mutual focusing of beams and limitation on the efficiency of optical frequency converters, *JETPLett.* **20**, 339-342.
30. Boardman, A.D. and Xie, K. (1997) Vector spatial solitons influenced by magneto-optic effects in cascadable nonlinear media. *Physical Review E* **55**, 1899-1910.
31. Rossi, A. De., Conti, C. and Assanto, G. (1997) Mode interplay via quadratic cascading in lithium niobate waveguide for all-optical processing, *Opt. And Quant. Elec.* **29**, 53-63.
32. Kobyakov, A., Schmidt, E. and Lederer, F. (1997) Effect of group-velocity mismatch on amplitude and phase modulation of picosecond pulses in quadratically nonlinear media, *J. Opt. Soc. Am. B.* **14**, 3242-3252.

33. Kobyakov, A. and Lederer, F. (1996) Cascading of quadratic nonlinearities: An analytical study, *Physical Review A* **54**, 3455-3471.
34. Baboiu, D-M. and Stegeman, G.I. (1997) Solitary-wave interactions in quadratic media near type I phase-matching conditions, *J. Opt. Soc. Am. B* **14**, 3143-3150.
35. Leo, G. and Assanto, G. (1997) Collisional interactions of vectorial spatial solitary waves in type II frequency-doubling media, *J. Opt. Soc. Am. B* **14**, 3151-3160.
36. Torner, L., Mihalache, D., Mazilu, D. and Akhmediev, N.N. (1997) Walking vector solitons, *Opt. Comm.* **138**, 105-108.
37. Buryak, A.V., Kivshar, Y.S. and Trillo, S. (1997) Parametric spatial solitary waves due to type II second-harmonic generation, *J. Opt. Soc. Am. B.* **14**, 3110-3118.
38. Boardman, A.D., Bontemps, P. and Xie, K. (1997) Transverse modulation instability of vector optical beams in quadratic nonlinear media, *J. Opt. Soc. Am. B* **14**, 3000-3008.
39. Kobyakov, A., Peschel, U., Muschall, R., Assanto, G., Torchigin, V.P. and Lederer, F. (1995) Analytical approach to all-optical modulation by cascading, *Opt. Lett.* **20**, 1686-1688.
40. Hutchings, D.C., Aitchison, J.S. and Ironside, C.N. (1996) All-optical switching based on nondegenerate phase shifts from a cascaded second-order nonlinearity, *Opt. Lett.* **18**, 793-795.
41. Torruellas, W.E., Assanto, G., Lawrence, B.L., Fuerts, R.A. and Stegeman, G.I. (1996) All-optical switching by spatial walkoff compensation and solitary-wave locking, *Appl. Phys. Lett.* **68**, 1449-1451.
42. Kobyakov, A., Peschel, U. and Lederer, F. (1996) Vectorial type II interaction in cascaded quadratic nonlnearities – an analytical approach, *Opt. Commun.* **124**, 184-194.
43. Butcher, P.N. (1965) *Nonlinear Optical Phenomena*, **Bulletin 200**, Engineering Experiment station, Ohio State University.
44. Boardman, A.D. and Xie, K. (1994) Bright spatial soliton dynamics in a symmetric optical planar waveguide structure, *Physical Review A* **50**, 1851-1866.
45. Boardman, A.D., Xie, K. and Zharov, A.A. (1995) Polarisation interaction of spatial solitons in optical planar waveguides, *Physical Review A* **51**, 692-705.
46. Akhmanov, S.S., Sukhorukov, A.P. and Khokhlov, R.V. (1968) Self-focusing and diffraction of light in a nonlinear medium, *Sov. Phys. Usp.* **93**, 609-636.
47. Durrin, J., Miceli, J.J. and Eberly, J.H. (1987) Diffraction-free beams, *Phys. Rev. Letts.* **58**m 1499-1501.
48. Durrin, J. (1987) Exact solutions for non-diffracting beams. I. The scalar theory, *J. Opt. Soc. Am. A* **4**, 651-654.
49. Belostotsky, A.S., Leonov, A.S. and Meleshko, A.V. (1994) Nonlinear phase change in type II second-harmonic generation under exact phase-matched conditions, *Opt. Lett.* **19**, 856-856.

50. Boardman, A.D., Bontemps, P. and Xie, K. (1998) Vector solitary optical-beam control with mixed type I-type II second-harmonic generation, *Opt. And Quant. Elec.* [to appear later in 1998].
51. Hecht, E. (1975) *Theory and Problems of Optics* (McGraw-Hill, New York).
52. Petykiewicz, J. (1992) *Wave Optics* (Kluwer, Dordrecht).
53. Trillo, S. and Assanto, G. (1994) Polarisation spatial chaos in second-harmonic generation, *Opt. Lett.* **19**, 1825-1827.
54. Kaup, D.J., Malomed, B.A. and Tasgal, R.S. (1993) Internal dynamics of a vector soliton in a nonlinear optical fiber, *Phys. Rev. E* **48**, 3049-3053

PLANE AND GUIDED WAVE EFFECTS AND DEVICES VIA QUADRATIC CASCADING

Gaetano Assanto, Katia Gallo and Claudio Conti

Department of Electronic Engineering
Terza University of Rome
84, Via della Vasca Navale, Rome 00146, ITALY

Outline:

A. D. Boardman et al. (eds.),
Advanced Photonics with Second-order Optically Nonlinear Processes, 59–87.
© 1999 *Kluwer Academic Publishers. Printed in the Netherlands.*

1. Introduction

Quadratic cascading, i. e. the sequence of two second-order nonlinear processes, has assumed an important role in recent years as one of the possibilities for a low power, lossless, ultrafast approach to all-optical processing of signals.[1-5] The foundation of quadratic cascading is as old as quadratic nonlinear optics itself, but its implications were brought up later on in several theoretical [6-15] and experimental [16-22] reports outlining phase effects intrinsic to parametric interactions. More recently, however, it was the second-harmonic generation (SHG) experiment performed by DeSalvo *et al.* [1] in Potassium Tytanil Phosphate (KTP) the one which triggered a considerable effort in investigating both applications and novel effects of cascading, from switching to modulation, and from solitary waves to gap solitons [3-5, 23]. It is remarkable the number of papers published on the subject in the past five years, and the rapid progress in a field which is benefitting both from the established knowledge and understanding of cubic nonlinearities and phenomena, and from the advancements of material sciences and technology in the area of noncentrosymmetric crystals for electro-optical and parametric interactions.

In this Chapter, we will review the basic ideas of cascading through two- and three-wave interactions in bulk and in guided-wave geometries, with particular emphasis on the simple process of SHG. Trying to provide an overview which is at the same time comprehensive and concise, we will limit our considerations to the cw or quasi-cw approximations, with a few exceptions related to the formation and use of parametric gap-solitons in time. The situations involving cascading in parametric oscillators, resonators, spatial solitary waves and competing/cooperative nonlinear mechanisms will be left out and are the subject of other chapters in this Book.

The Chapter is divided in Sections: Section 2 introduces the model equations for cw three-wave interactions, discussing nonlinear phase-shift and amplitude modulation of a fundamental frequency wave in Type I and Type II SHG. Section 3 is devoted to guided-wave effects and illustrates various nonlinear integrated devices for all-optical switching and processing. Finally, Section 4 deals with parametric gap-solitons and their excitation.

2. Fundamentals and Plane-Wave effects

2.1 BASIC EQUATIONS FOR THREE-WAVE INTERACTIONS

The simplest case of quadratic cascading is an upconversion process when $\omega_3=\omega_1+\omega_2$, followed by downconversion with $\omega_1=\omega_3-\omega_2$ or $\omega_2=\omega_3-\omega_1$, the subscripts 1, 2 and 3 denoting the three waves involved in the interaction [2, 24]. The consecutive application of two nonlinear operators, namely second-order susceptibility tensors, to the electromagnetic field(s) is conveniently described by the product of susceptibilities $d^{(2)}(\omega_3, -\omega_{2,1}) : d^{(2)}(\omega_1, \omega_2)$. The most straightforward implementation is Type I SHG, with $\omega_1=\omega_2=\omega=\omega_3/2$ and the two fundamental frequency (FF) fields at ω oscillating with the same polarization in space [2, 3]. The other frequency degenerate case involves optical rectification through the product $d^{(2)}(0, \omega) : d^{(2)}(\omega, -\omega)$, with the first term representing the electro-optic effect [7, 25]. This particular situation relies on the generation of a DC field and will not be addressed hereby.

When dispersion and nonlinear tensor elements are such that a single three-wave interaction is nearly phase matched (i.e. the three waves have almost the same phase velocities), the standard slowly-varying-envelope approximation allows us to reduce Maxwell equations to a simple set of three coupled ordinary differential equations modelling the evolution of the amplitudes $E_j(z)$. For three z-copropagating plane-wave electric fields:

$$\frac{dE_1}{dz} = -i\frac{\omega_1}{n(\omega_1)c} d^{(2)}(-\omega_2, \omega_3) E_2^* E_3 e^{-i\Delta kz}$$

$$\frac{dE_2}{dz} = -i\frac{\omega_2}{n(\omega_2)c} d^{(2)}(-\omega_1, \omega_3) E_1^* E_3 e^{-i\Delta kz} \qquad (1)$$

$$\frac{dE_3}{dz} = -i\frac{\omega_3}{n(\omega_3)c} d^{(2)}(\omega_1, \omega_2) E_1 E_2 e^{+i\Delta kz}$$

with $\Delta k=[\omega_3 n(\omega_3)-\omega_2 n(\omega_2)-\omega_1 n(\omega_1)]/c$ the wavevector mismatch. In writing eqns. (1) the interaction has been reduced to a scalar one through the projections $d^{(2)}(-\omega_2, \omega_3) = e_1\, d^{(2)}(-\omega_2, \omega_3) : e_2\, e_3$ and similar ones, with e_1, e_2, e_3 the unit polarization vectors for the three fields. Assuming to operate at wavelengths far removed from material resonances such that Kleinman symmetry applies, and defining the amplitudes $\Phi_j=\sqrt{\varepsilon_0 n(\omega_j)c/2}\, E_j$ such that $|\Phi_j|^2$ (j=1,2,3) are the intensities in W/m^2, system (1) reduces to:

$$\frac{d\Phi_1}{d\zeta} = -i\omega_1\, \kappa\, \Phi_2^*\, \Phi_3\, e^{-i\Delta\zeta}$$

$$\frac{d\Phi_2}{d\zeta} = -i\omega_2\, \kappa\, \Phi_1^*\, \Phi_3\, e^{-i\Delta\zeta} \qquad (2)$$

$$\frac{d\Phi_3}{d\zeta} = -i\omega_3\, \kappa\, \Phi_1\Phi_2\, e^{+i\Delta\zeta}$$

with normalized propagation distance $\zeta=z/L$ and normalized mismatch $\Delta=\Delta kL$, and $\kappa=d^{(2)}L\sqrt{2/c^3\varepsilon_0 n(\omega_1)n(\omega_2)n(\omega_3)}$ the nonlinear constant.

It is convenient to treat first the case of Type I interactions, i.e. those with complete degeneracy between the fields with subscripts 1 and 2.

2.2   TYPE I SECOND HARMONIC GENERATION

When the fields E_1 and E_2 refer to the same fundamental frequency (FF) wave, eqns. (2) reduce to two coupled nonlinear ordinary differential equations:

$$\frac{d\Phi_1}{d\zeta} = -i\,\Gamma\,\Phi_1^*\,\Phi_3\,e^{-i\Delta\zeta}$$

$$\frac{d\Phi_3}{d\zeta} = -i\,\Gamma\,\Phi_1^2\,e^{+i\Delta\zeta} \tag{3}$$

with $\Gamma = \dfrac{\omega d^{(2)} L}{n(\omega)}\sqrt{\dfrac{2}{\varepsilon_0 c^3 n(2\omega)}}$ and $\Delta = [n(2\omega)-n(\omega)]\,2\omega L/c$.

In the undepleted pump (FF) approximation for SHG, i.e. taking $\Phi_3(0)=0$ and $\Phi_1(\zeta)\equiv\Phi_1(0)$, the second of (3) can be integrated and $\Phi_3(\zeta)$ substituted in the first of (3), to give:

$$\frac{d\Phi_1}{d\zeta} = i\,\frac{\Gamma^2}{\Delta}\left[(1-\cos\Delta\zeta)+i\,\sin\Delta\zeta\right]|\Phi_1|^2\,\Phi_1 =$$

$$= i\,k_0 L\,n_{2,\text{eff}}(\zeta)\,|\Phi_1|^2\,\Phi_1 - L\,\beta_{\text{eff}}(\zeta)\,|\Phi_1|^2\,\Phi_1 \tag{4}$$

with $n_{2,\text{eff}}(\zeta)$ a z-dependent effective Kerr coefficient (focusing or defocusing), and $\beta_{\text{eff}}(\zeta)$ an equivalent z-dependent two-photon absorption. The intensity dependence in the FF field evolution establishes a parallelism with the effects of a cubic nonlinearity characterized by a complex susceptibility, provided the z-variant feature is averaged out in the limit of large mismatches, i.e. when the undepleted pump approximation holds. In this limit, the nonlinear absorption vanishes and the Kerr coefficient takes the value:

$$\langle n_{2,\text{eff}}\rangle = \frac{4\pi}{\varepsilon_0 c\Delta}\,\frac{\left(d^{(2)}\right)^2}{n(2\omega)n^2(\omega)}\,\frac{L}{\lambda} \tag{5}$$

This Kerr-like response, obtained via second-order (SHG) processes, is linearly dependent on the usual figure of merit $\left(d^{(2)}\right)^2/n^3$, and inversely on the phase mismatch. The effective change in index (or, more appropriately, the nonlinear phase shift) changes sign according to Δ, implying that a focusing or defocusing behaviour can be obtained by acting on the field orientation with respect to the material (birefringent matching), or on the temperature, or on the period of a ferroelectric domain inversion (Quasi Phase Matching), etc. Some of these important characteristics were verified in bulk by DeSalvo et al. in a 1mm long KTP crystal at 1.064μm via Z-scan [1], and later by Danielius et al. in BBO via self-diffraction[26], by Ou in potassium niobate inside a cavity [27], by Fazio et al. in KDP [28], by Wang et al. in the organic NPP [29] and by Vidakovic et al. in periodically poled lithium niobate (PPLN) [30]. The factor L/λ in (5) can be considered as an enhancement factor, maximized in waveguide geometries.

When transverse resonances at the generated frequency are arranged, this factor can be replaced by the quality factor of the resonator [31-32].

Although (5) allows a crude estimate of the potential of cascading in mimicking the Kerr effect in a quadratic medium [1], it applies only in the limit of large mismatch and fails to underline that no measurable index change is actually induced. It is rather more instructive to examine the nonlinearly induced phase shift by a numerical integration of eqns. (3). [2-3] The evolution of phase shift and throughput versus propagation distance for various mismatches and a fixed excitation $\Gamma|\Phi_1(0)|= 20$ are shown in Fig.1. The FF intensity oscillates, and the periodicity decreases for increasing mismatch, as the coherence length correspondingly shortens. Similarly, pump (FF) depletion becomes more severe when approaching phase matching. The evolution of the FF phase exhibits a stepwise growth, with steps corresponding to a back-flow of energy from SH to FF via downconversion, and plateaus separated by nearly $\pi/2$ for large FF depletion. Clearly, the linear behaviour described by (5) is adequate only in the initial stages (small ζ) or for large Δ, when the corresponding FF pump remains nearly constant. An intuitive explanation of such behaviour can be provided in terms of energy which, partially converted from FF to SH, propagates at a different velocity and acquires a phase difference before eventually flowing back to the FF after a coherence cycle. A sketch of this intuitive mechanism is shown in Fig. 2, where the $-\pi/2$ rotation in the generated polarization at SH ("-i" factor in the second of (3)) causes a phase shift close to π-$\pi/2=\pi/2$ for conversion approaching 100%.

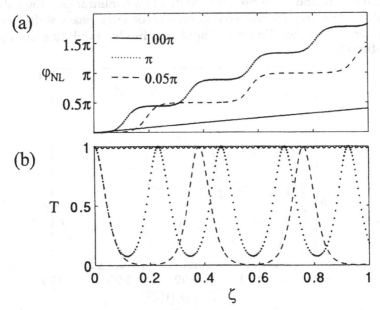

Figure 1. a) FF nonlinear phase shift φ_{NL} in units of π and b) FF throughput T vs. propagation for various mismatches (solid lines). Here $\Gamma|\Phi_1(0)|=20$.

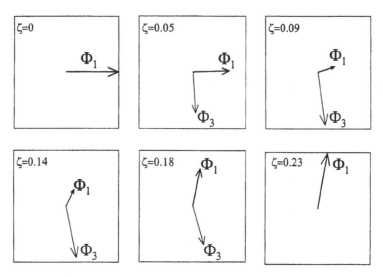

Figure 2. Phasorial representation of the cascaded FF phase rotation during the first cycle (see Fig. 1). At various propagation distances, the FF and SH components are graphed. Here $\Delta=\pi$ and $\Gamma|\Phi_1(0)|=20$.

The evolution of the FF throughput and phase versus normalized input excitation is shown in Fig. 3 for a fixed mismatch. As the nonlinear process is driven harder, the phase plateaus become progressively extended, with FF depletion more and more pronounced due to a nonlinear pull towards phase matching. For very intense excitations and/or large phase shifts, the phase dependence becomes linear in input amplitude, exhibiting an apparent saturation [2].

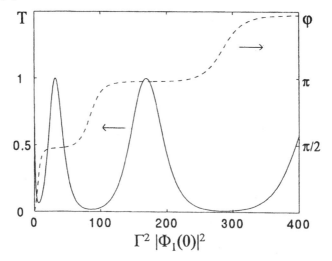

Figure 3. FF throughput (left axis) and phase (right axis) vs. excitation for $\Delta=\pi$.

2.3 TYPE I SHG ALL-OPTICAL COHERENT TRANSISTOR

The features illustrated above refer to standard SHG, i.e. to a process with no second harmonic waves injected at the crystal input. However, the studied interaction is inherently coherent, and lends itself to the possibility of *seeding* the process with an additional input at SH. This allows to control the overall field evolutions (at ω and 2ω) by acting on the relative amplitude or phase of $\Phi_1(0)$ and $\Phi_3(0)$ [3, 33]. To illustrate this concept, Fig. 4 shows FF throughput and phase versus the relative phase of a small (0.1%) coherent seed at SH. It is apparent that the phase plays a major role even though the extra input might be rather small, suggesting various possibilities for phase-to-amplitude modulators or small-signal amplifiers. If the operating Δ-point is chosen in order to maximize the phase-to-throughput modulation, a large change in transmission can result from a small variation in input phase. Conversely, keeping the phase constant, the presence of an SH input determines the response of the device, i.e. results in an intensity controlled transistor.

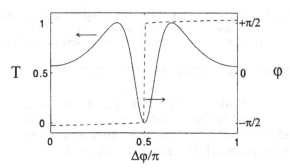

Figure 4. FF transmission and phase versus the relative phase of an SH seed 0.1% in intensity. Here the excitation is $\Gamma|\Phi_1(0)| = 6.3$ and $\Delta=0$.

Fig. 5 shows the results of the experiments performed by Hagan et al. in KTP, together with numerical simulations [34]. The input phase between the FF pump and the SH seed was controlled by letting both waves through an N_2 gas cell, the pressure of which could be varied in order to affect the differential dispersion at ω and 2ω. The FF throughput, despite the use of temporal pulses (25ps) with a Gaussian beam distribution (waist 80µm), clearly exhibits a modulation versus input relative phase (Fig. 5a), with contrast as high as 4.6:1 when seeding with SH pulses of energy 1.2% of the pump at 1.064µm. The graphs of throughput versus input intensity (Fig. 5b) show further bendings at very high excitations (above 25GW/cm^2) due to the presence of a small Kerr correction to the mismatch. In these experiments the FF field was injected at 45° with respect to ordinary and extraordinary directions in KTP, resulting in standard Type I cascading.

The drawbacks of this modulator are the use of a (phase) signal of frequency twice the output, and the need for an interferometrically stable setup in order to control the phase. While the former problem could be solved by working with an SH pump downconverting to a FF signal, the latter can be avoided by employing a Type II SHG configuration, as discussed later.

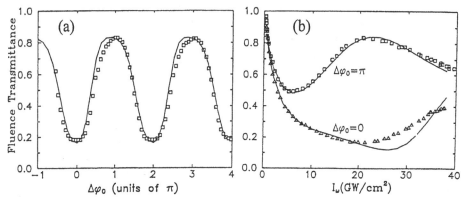

Figure 5. FF transmission vs. a) the relative phase between SH seed and FF pump for a peak FF intensity of 20GW/cm², and b) the input intensity for a relative phase of 0 or π with respect to the SH signal. In both cases the seed energy fraction is 1.2%. (After Ref. 34)

2.4   TYPE II SECOND HARMONIC GENERATION

When two input waves interact to generate a new frequency, extra degrees of freedom are available and can be exploited for cascading. This is the case of Type II phase-matched SHG and, in general, of all three-wave mixing processes with identifiable field components [24]. From Eqns. (2) we obtain:

$$\frac{d^2\Phi_1}{d\zeta^2} + i\,\Delta\,\frac{d\Phi_1}{d\zeta} - \omega_1\kappa^2\left[\omega_2\left|\Phi_3\right|^2 - \omega_3\left|\Phi_2\right|^2\right]\Phi_1 = 0$$

$$\frac{d^2\Phi_2}{d\zeta^2} + i\,\Delta\,\frac{d\Phi_2}{d\zeta} - \omega_2\kappa^2\left[\omega_1\left|\Phi_3\right|^2 - \omega_3\left|\Phi_1\right|^2\right]\Phi_2 = 0 \qquad (6)$$

$$\frac{d^2\Phi_3}{d\zeta^2} - i\,\Delta\,\frac{d\Phi_3}{d\zeta} + \omega_3\kappa^2\left[\omega_1\left|\Phi_2\right|^2 + \omega_2\left|\Phi_1\right|^2\right]\Phi_3 = 0$$

Eqns. 6 show that the evolution of each wave depends on the intensity of the others. Assuming that the third wave is entirely generated through the nonlinear interaction, i.e. $\Phi_3(0)=0$, if we let the two inputs be unbalanced with $\left|\Phi_2(0)\right| \gg \left|\Phi_1(0)\right|$, in the low depletion approximation the first of (6) reduces to:

$$\frac{d^2\Phi_1}{d\zeta^2} + i\,\Delta\,\frac{d\Phi_1}{d\zeta} + \left[\omega_1\omega_3\kappa^2\left|\Phi_2(0)\right|^2\right]\Phi_1 = 0 \qquad (7)$$

showing that the intense wave induces a phase shift on the weaker component, even for a vanishing phase mismatch $|\Delta| \to 0$. This consideration is implicit in the non-degenerate analysis[24], and was outlined in the context of SHG by Belostotsky *et al.* [35]   Experiments on cascaded phase-shifts via frequency non-degenerate interactions have been performed by Nitti *et al.* in the organic MBA-NP [36] and by Asobe *et al.* in PPLN [37]. For simplicity, in the following we restrict to the case of Type II SHG, i.e. to frequency doubling by means of two orthogonally polarized components at FF.

Calculated phase shift versus excitation and throughput of the weaker FF component versus imbalance for a Type II SHG with various mismatches and unequal FF components in orthogonal polarizations are shown in Fig. 6. A phase-shift is induced on the weak FF field even in perfect matching, with the phase of the strong wave clamped to zero in analogy to Type I. For a given imbalance (10:1 in Fig. 6a), in contrast to Type I SHG the maximum phase jump is π, with a steplike evolution which gets sharper for better phase matching and/or higher input imbalances [38]. The FF throughput versus imbalance (Fig. 6b) exhibits a sharp transition as the input ratio deviates from 1. This abrupt change can be related to a drastic shortening of the characteristic length-scale of the energy exchange between the three fields near phase matching [39-40], causing FF transmission modulation at the end of a finite length crystal (the strong component is shown in Fig. 6b, but a similar graph can be traced for the orthogonal field). For perfectly balanced inputs at FF, conversely, cascading via Type II and Type I SHG are equivalent [41].

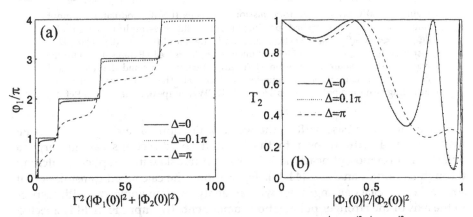

Figure 6 a) Type II SHG phase shift of Φ_1 vs. excitation with $|\Phi_1(0)|^2 / |\Phi_2(0)|^2 = 10\%$ and various mismatches. b) Throughput T_2 of the strong component Φ_2 vs. ratio $|\Phi_1(0)|^2 / |\Phi_2(0)|^2$ for various Δ.

2.5   TYPE II SHG: PHASE INSENSITIVE AMPLIFIER

Type II SHG provides more degrees of freedom (e.g. the input imbalance), with insensitivity to phase when inputting the FF fields only. This feature, on the basis of the transmission response shown in Fig. 6, stimulated the investigation of a frequency degenerate, phase insensitive, all-optical transistor [38-39, 42-44]. If the interaction is close to or at phase matching, the FF transmission in either or both polarizations will abruptly change when a small imbalance is introduced between the orthogonal FF inputs. This process is independent from the relative phase between Φ_1 and Φ_2, and can be exploited at an appropriate bias point in order to achieve small-signal amplification, as sketched in Fig. 7a and demonstrated in KTP at 1.064µm using 25ps pulses (Fig. 7b). Amplifications as high as 21.6dB were reported, although the use of gaussian beams and pulses degraded to some extent the expected performance [42-44].

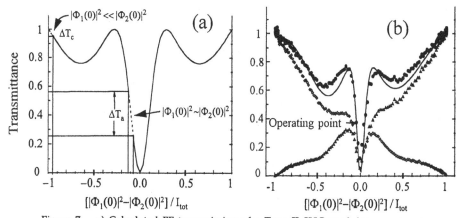

Figure 7. a) Calculated FF transmission of a Type II SHG modulator vs. input imbalance. The two dashed portions identify the pre-amplifier, or coupler $(|\Phi_1(0)|^2 << |\Phi_2(0)|^2$, see Section 2.6) and the amplifier $(|\Phi_1(0)|^2 \approx |\Phi_2(0)|^2)$, respectively. I_{tot} is the total FF input intensity. b) Measured FF transmission for the extraordinary (squares) and ordinary (triangles) components and the total (filled circles) vs imbalance through a 2mm-long KTP crystal. The solid curve is calculated including beam and pulse profiles. Here I_{tot} =12GW/cm^2 (peak value). (After Ref. 43)

Through the phase shift of the weakest FF input, a linearly polarized FF wave at $\sim\pi/4$ with respect to the "e" and "o" directions can undergo a polarization rotation of nearly $\pi/2$, which can translate in all-optical switching through an output analyzer [44-45]. Fig. 8 shows the calculated transmission of such a Type II SHG arrangement with an analyzer at an angle $-\alpha_{in}$ with respect to the axis of the stronger polarization component. The input FF field is injected at α_{in} with respect to the same axis. As α_{in} is varied, the excitation required for complete transmission shifts, with better extinction for angles closer to 45°.

2.6 DOUBLE TYPE II SHG: INCOHERENT TRANSISTOR

For best operation, the Type II SHG modulator described above should operate with comparable FF input intensities, the signal being the difference between them[38-40]. In a realistic application, then, it would require a weak signal to be

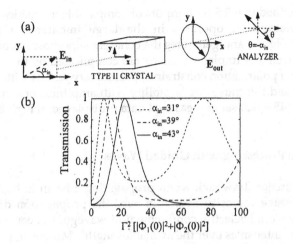

Figure 8   a) Sketch of a modulator via polarization rotation in unbalanced Type II SHG. b) Calculated transmission versus FF excitation for various input angles.

coherently added to a large background in order to satisfy the small-imbalance requirement.   This impasse can be overcome through the introduction of another nonlinear stage, the coupler, able to transfer a small signal on the appropriate large background required by the Type II amplifier. The overall two-stage transistor is sketched in Fig. 9a, and is pumped by the same intense FF wave: the first crystal couples a weak FF signal to a large FF background (one half of the total pump) in the orthogonal polarization through Type II SHG, and the second

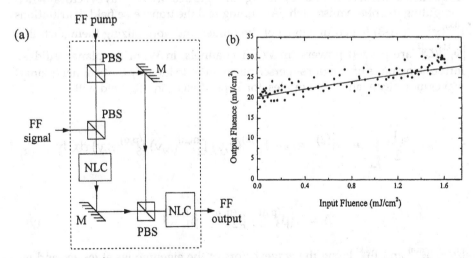

Figure 9.   a) Scheme of a two-stage Type II SHG all-optical transistor. PBS=polarizing beam-splitter, Coupler and Amplifier are crystals for Type II SHG. (After Ref. 46) b) Measured output vs. input fluences at 1.064μm, using 2 and 5mm KTP crystals for first and second stages, respectively. (After Ref. 48)

operates as described in § 2.5 with inputs of comparable intensities [46-47]. The coupler or preamplifier operates in the large imbalance region of the characteristic of Fig. 7a, and generates the coherent superposition needed by the amplifier. Experimental implementations are complicated by the use of pulsed beams and by the polarization constraints; preliminary results with KTP crystals have demonstrated the transistor feasibility with amplifications of the order of 7dB at 1.064µm [48-49].A set of measured data is displayed in Fig. 9b.

3. All-optical Processing with Guided Waves

Diffraction is a major drawback when propagating a beam in bulk, because it reduces the average intensity available over the propagation distance for a nonlinear interaction. Dielectric confinement in waveguides, conversely, allows to mantain large intensities over the available length. Moreover, guided modes provide extra degrees of freedom for phase matching and permit the implementation of integrated devices with added functionalities [49]. Nonlinear waveguides for quadratic cascading have been employed in KTP [51-52], in lithium niobate [53-58] and in DAN {4-(N, N-dimethylamino)-3-acetaminonitrobenzene} [59]. In this section, mainly focussing on cw waves in lossless channels, after a brief summary of the most important waveguide concepts relevant to quadratic interactions, we will describe the operation of several all-optical devices demonstrated or simply proposed to date.

The model introduced for plane wave effects can be extended to the guided wave scenario by a) taking into account the overlap of the modal fields at the frequencies involved, and b) using the guided-wave wavevectors when computing the phase mismatch. Assuming real the transverse field distributions $f_j^{(l,k)}(x,y)$, (j=1-3) (eigenfunctions) of the modes and normalizing them such that $\left|\Phi_j^{(l,k)}(\zeta)\right|^2$ are guided powers (in W for channels, in W/m for planar guides), introducing pairs of (single) superscripts in order to identify the channel (planar) waveguide eigenvalues and modes, for Type I SHG eqns. (2) hold with

$$\kappa = \frac{\varepsilon_0 L}{2} \int_{-\infty}^{\infty} e_3 \, d^{(2)} : e_1 \, e_2 \, f_1^{(l,k)}(x,y) \, f_2^{(m,n)}(x,y) \, f_3^{(p,q)}(x,y) \, dxdy \qquad (8)$$

$$\Delta = L\left(\beta_3^{(p,q)} - \beta_2^{(m,n)} - \beta_1^{(l,k)}\right) \qquad (9)$$

$\beta_3^{(p,q)}$, $\beta_2^{(m,n)}$ and $\beta_1^{(l,k)}$ being the wavevectors of the eigenmodes at ω_3, ω_2 and ω_1, respectively. The overlap integral in (8) reduces the bulk nonlinear coefficient, and proper parity and shape of the involved modes have to be chosen/engineered in order to maximize the effectiveness of the interaction.

A z-nonuniform distribution of the propagation wavevectors or the nonlinearity can be introduced in order to taylor the quadratic response. Through coupled mode theory, this gives rise to a modified set of equations. For Type I SHG, for example:

$$\frac{d\Phi_1}{d\zeta} + iL\, \delta\beta_1(\zeta)\, \Phi_1 = -i\,\Gamma(\zeta)\, \Phi_1^*\, \Phi_3\, e^{-i\Delta_0\zeta}$$

$$\frac{d\Phi_3}{d\zeta} + iL\, \delta\beta_3(\zeta)\, \Phi_3 = -i\,\Gamma(\zeta)\, \Phi_1^2\, e^{+i\Delta_0\zeta}$$

$$(10)$$

with $\Gamma(\zeta)$ the longitudinally varying nonlinearity, and $\Delta(\zeta) = \Delta_0 + L[\delta\beta_3(\zeta) - 2\delta\beta_1(\zeta)]$ the propagation-dependent phase mismatch. The ζ-variant nonlinearity includes quasi-phase-matching and segmentation, and can be taylored to specific bandwidth-enhanced high-efficiency applications [60-61].

Defining the phase-shifted amplitudes $a_1=\Phi_1\, e^{i\varphi_\omega(\zeta)}$, $a_3=\Phi_3\, e^{i\varphi_{2\omega}(\zeta)}$ and the phase

$$\varphi(\zeta)=\varphi_{2\omega}(\zeta) - 2\varphi_\omega(\zeta)=L\int_0^\zeta d\zeta\left[\delta\beta_3(\zeta) -2\delta\beta_1(\zeta)\right] \qquad (11)$$

eqns. (10) become:

$$\frac{da_1}{d\zeta} = -i\,\Gamma(\zeta)\, a_1^*\, a_3\, e^{-i(\Delta_0\zeta+\varphi(\zeta))}$$

$$\frac{da_3}{d\zeta} = -i\,\Gamma(\zeta)\, a_1^2\, e^{+i(\Delta_0\zeta+\varphi(\zeta))}$$

$$(12)$$

Assuming no FF depletion, the second of (12) gives:

$$a_3(1) = -ia_1(0)^2 F\{rect(\zeta-1/2)\,\Gamma(\zeta)\, e^{+i\varphi(\zeta)}\}_{\Delta_0} \qquad (13)$$

i.e. the SH amplitude $a_3(1)$ results proportional to the Fourier transform of the perturbation over the extent of the interaction. Although this result is rigorous for low conversion efficiencies, it is useful when analyzing or designing linear and/or nonlinear profiles for specific operations. A full integration of (12) has to be carried out in cases of highly effective interactions. Some of the examples discussed below show a few advantages provided by z-varying profiles.

72

3.1 SINGLE CHANNELS

In Type I SHG a small phase-mismatch is required for efficient cascading to take place. In waveguides this requirement translates into proper orientation of the crystal, polarization of the involved eigenmodes, temperature or electro-optical adjustments of the effective indices, profile of the nonlinearity along the propagation direction (QPM), etc. [62-66] Monomode waveguides at the FF are often desirable, with a limited set of modes at the higher frequencies. In the case of SHG, for instance, when several modes can propagate at the SH, competitions between nearly matched interactions take place, as reported by Treviño-Palacios *et al.* in a QPM lithium niobate channel [57].

3.1.1. *Modulators.*

Channel waveguides can operate as amplitude or phase modulators controlled by the intensity of the incoming FF wave(s) through SHG. In fact, this is the signature of quadratic cascading, as it was experimentally verified by Sundheimer *et al.* in KTP segmented channels [51-52]. The measurements were carried out by evaluating the spectral broadening of laser pulses through cascading-induced self-phase modulation, and phase shifts of the order of π were demonstrated in the near infrared. Schiek *et al.* also demonstrated cascading through SHG in thermo-optically matched y-cut Ti:indiffused lithium niobate waveguides, using a stabilized interferometer with 100ps pulses from a Mode-Locked laser at 1.3μm [53-54]. Tuning was achieved in an oven with a nonuniform profile of temperature along the channel. The latter configuration, with lower temperature at the crystal (oven) ends, allowed efficient SHG conversion internally to the sample, with cascading phase shifts as large as 1.5π but with low FF depletion at the output. Notice that a high FF throughput is a desirable feature for all-optical devices based on quadratic cascading. The phase matching temperature T_{PM} for Type I SHG between the $TM_{00}(\omega)$ and $TE_{00}(2\omega)$ modes was 336.6°C. Temperatures lower (higher) than T_{PM} produced a positive (negative) phase shift, such that a large positive phase shift at the end of the crystal could be obtained for $T<T_{PM}$ [53]. Fig. 10 shows measured and calculated responses for FF throughput and nonlinear phase shift versus oven temperature.

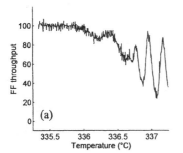 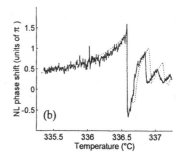

Figure 10. a) FF throughput and b) nonlinear phase shift vs. oven temperature in a 15μm-wide Ti:indiffused lithium niobate waveguide.Peak power is 300W and wavelength 1.319μm.(After Ref. 54)

3.1.2 Wavelength Shifters.

In modern communication networks based on wavelength division multiplexing, wavelength shifters able to route a given channel into another preserving its protocol are of primary interest. Particularly appealing are those solutions which can provide a wide range of tunability in the fiber communication window and can be operated with semiconductor laser sources. An elegant approach to wavelength shifting can be implemented by cascading two quadratic frequency non-degenerate processes, namely SHG and difference frequency generation (DFG). In this case, through the cascaded interaction:

$$d^{(2)}(-\omega-\delta\omega; 2\omega, -\omega+\delta\omega)\ d^{(2)}(-2\omega; \omega, \omega) \qquad (14)$$

a small frequency shift can be obtained at the expense of an FF input pump at ω. Similar effects were demonstrated in bulk BBO by Tan *et al.* [67], and proposed by Gorbounova *et al.* in a MQW waveguide with a vertical cavity at SH [68], and by Gallo *et al.* in a QPM lithium niobate channel [69]. The latter configuration offers a large 3dB bandwith and amplification at the expense of the input pump, which can operate at a frequency in the 3rd window for fiber communications. A sketch of the device and representative numerical results are shown in Fig. 11. The response of such a QPM channel in z-cut lithium niobate can be calculated from:

$$\frac{d\Phi_1}{d\zeta} = -i\Gamma_{SHG}\ \Phi_1^*\ \Psi\ e^{-i\Delta_{SHG}\zeta}$$

$$\frac{d\Phi_2}{d\zeta} = -i\left(\frac{\omega_2}{\omega}\right)\Gamma_{DFG}\ \Phi_3^*\ \Psi\ e^{-i\Delta_{DFG}\zeta}$$

$$\frac{d\Phi_3}{d\zeta} = -i\left(\frac{\omega_3}{\omega}\right)\Gamma_{DFG}\ \Phi_2^*\ \Psi\ e^{-i\Delta_{DFG}\zeta} \qquad (15)$$

$$\frac{d\Psi}{d\zeta} = -i\Gamma_{SHG}\ \Phi_1^2\ e^{+i\Delta_{SHG}\zeta} - i2\Gamma_{DFG}\ \Phi_2\ \Phi_3\ e^{+i\Delta_{DFG}\zeta}$$

with $\Phi_1(0) \neq 0$, $\Phi_2(0) \neq 0$, $\Phi_3(0) = \Psi(0) = 0$, and

$$\Delta_{SHG} = \frac{4\pi L}{\lambda}\ (n_{eff}^{2\omega} - n_{eff}^{\omega}) - \frac{2\pi L}{\Lambda} \xrightarrow{QPM} 0$$

$$\Delta_{DFG} = 2\pi L\left(\frac{n_{eff}^{2\omega}}{\lambda/2} - \frac{n_{eff}^{\omega_2}}{\lambda_2} - \frac{n_{eff}^{\omega_3}}{\lambda_3}\right) - \frac{2\pi L}{\Lambda} \qquad (16)$$

Preliminary measurements carried out with a color center and a fiber laser indicate the feasibility of this approach and a qualitative agreement with the simulations [70]. Implementations with an SH reflector at one end of the waveguide, introducing a signal which counterpropagates with respect to the pump, should allow to maximize the conversion and easily separate output and pump [71].

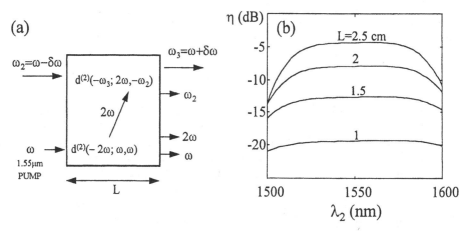

Figure 11. a) Device sketch and b) calculated conversion efficiency ($\omega{-}\delta \rightarrow \omega{+}\delta$) versus signal wavelength λ_2 for wavelength-shifters in QPM lithium niobate channels of various lengths L.

3.1.3   Isolators.

A non-reciprocal response upon excitation from different ends can be induced by profiling either the linear or the nonlinear properties of a quadratic waveguide in a non-symmetric way along the propagation direction. In the limit of complete FF depletion in one sense, and complete transmission in the opposite one, such a configuration would correspond to an all-optical isolator or diode. A demonstration of this concept has been reported by Treviño-Palacios *et al.* [58] A QPM lithium niobate channel was partially overcoated with a thin film in order to introduce a z-nonsymmetric variation of linear wavevectors and, consequently, SHG mismatch. The non-reciprocity was clearly verified with a pulsed color center laser, but the waveguide parameters could not be optimized to yield a diode-like response.

When employing QPM for efficient SHG, a convenient way to act on mismatch and obtain non-reciprocity is represented by chirping the QPM grating itself. For various such cases, Fig. 12 shows the calculated responses of QPM waveguides with a flat + various chirp profiles: the isolation effect is apparent at various excitation levels, with a large contrast when propagating cw waves in opposite directions.

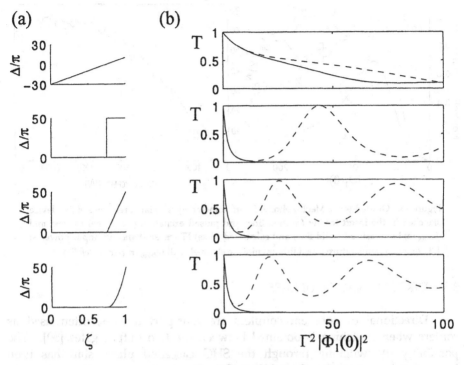

Figure 12. a) Various flat + chirped mismatch profiles versus propagation coordinate in QPM waveguides and b) calculated all-optical diode FF transmittance versus cw excitation. Forward (solid line) and backward (dashed line) throughputs show that the diode exhibits a good degree of isolation in certain power intervals.

3.2   MACH-ZEHNDER INTERFEROMETERS

An interferometer with a nonlinearly driven relative phase between distinct arms is an all-optical switch able to transform the cascading phase shift into throughput modulation [50, 72].   A simple   configuration with identical waveguides characterized by opposite phase-mismatches allows to minimize the switching power, because the opposite phases in the two arms will effectively combine to produce throughput switching, as shown in Fig. 13a. [72-73] Moreover, the plateaus of phase vs excitation (see Fig. 3) allow this device to work with good contrast even in the presence of temporal pulses with a continuous distribution of instantaneous power. A demonstration of a lithium niobate Mach-Zehnder with nonuniform temperature tuning has been reported by Baek *et al.* [55], and a typical set of experimental results is reproduced in Fig. 13b together with a numerical simulation.

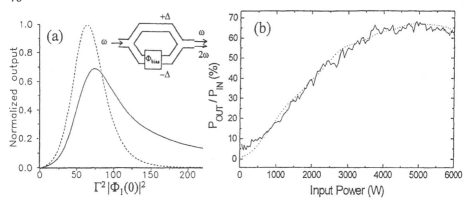

Figure 13. Guided-wave Mach-Zehnder interferometer: a) calculated response of the device sketched in the insert for cw (dashed line) and pulsed excitation (solid line) for $\Delta=\pi$ and $\Phi_{bias}=\pi$; b) measured (solid line) and calculated (dots) FF transmittance vs. input power at 1.319μm in a temperature tuned lithium niobate channel with $\Phi_{bias}=\pi$. (After Ref. 55)

3.3 DIRECTIONAL COUPLERS

Directional or coherent couplers are four-port devices, often used as routers when switching is obtained between the two output gates [50]. The possibility of switching through the SHG cascaded phase shift has been investigated by several authors [72, 74-79]. The pertinent equations, when the coupling between the two parallel adjacent channels is negligible at the SH, are:

$$\frac{d\Phi_1}{d\zeta} = i\kappa\Psi_1 e^{-i\delta\zeta} - i\,\Gamma_\Phi\,\Phi_1^*\,\Phi_3\,e^{-i\Delta_\Phi\zeta} \qquad \frac{d\Phi_3}{d\zeta} = -i\,\Gamma_\Phi\,\Phi_1^2\,e^{+i\Delta_\Phi\zeta}$$

$$\frac{d\Psi_1}{d\zeta} = i\kappa\Phi_1 e^{+i\delta\zeta} - i\,\Gamma_\Psi\,\Psi_1^*\,\Psi_3\,e^{-i\Delta_\Psi\zeta} \qquad \frac{d\Psi_3}{d\zeta} = -i\,\Gamma_\Psi\,\Psi_1^2\,e^{+i\Delta_\Psi\zeta} \tag{17}$$

being Φ_3 and Ψ_3 the guided field amplitudes at SH, κ the linear coupling strength between the modal fields Φ_1 and Ψ_1 at FF, $\delta=L(\beta_1^{(\Phi)}-\beta_1^{(\Psi)})$ the linear detuning.

Let us consider the simple case of a symmetric half-beat length coupler (HBLC), i.e. with identical arms ($\Gamma_\Phi = \Gamma_\Psi$, $\delta=0$) and of length such that at low excitations the input power is transferred to the other channel ("cross" state). Since in SHG with $\Delta \neq 0$ the FF field undergoes periodic oscillations with propagation, a "clean" switching in a uniform HBLC is expected only in the "cross" channel, which exchanges power in favour of the "bar" (or input) channel. The intensity in the "bar" channel, instead, will exhibit dips as the excitation increases, as shown in Fig. 14a. One way to avoid this problem is the introduction of a nonlinear asymmetry which, in the limiting case of a HBLC with bar and cross channels linear and nonlinear, respectively, provides ideal switching in both ports [76]. Experiments have been performed in lithium niobate HBLC's using a non-uniform phase-matching profile via temperature tuning in an oven. In contrast to the simulations carried out for uniform mismatches, these devices exhibited good switching even though both arms were nonlinear [56]. A set of experimental data is displayed in Fig. 14b. More complex operations, such as coherent modulation in the presence of a seed, pulse compression, logic gating and demultiplexing can also be exploited in directional couplers. Lee *et al.* have studied couplers in geometries involving surface emitted second-harmonic [78].

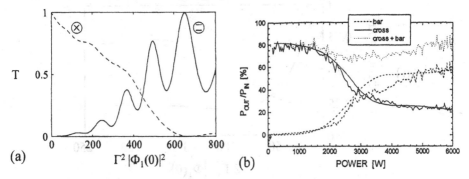

Figure 14. Half-beat length symmetric directional coupler. a) Calculated FF transmission through bar ("=") and cross ("X") arms vs. excitation ($\Delta=8\pi$); b) measured and calculated FF transmission in a non-uniformly temperature-tuned (T=343.5°C) lithium niobate device operating at 1.319µm. (Ref. 56)

3.4   MODE MIXERS

Waveguides able to support at least two modes at a given wavelength and in the same polarization can be used for spatial routing of information, provided the interference between the corresponding eigenmodes is controlled [50]. In bimodal waveguides two FF modes (fundamental and first-order, Φ_1 and Ψ_1) can change their relative phase - and therefore the distribution deriving from their spatial overlap - through cascading, i.e. depending on input power or on the presence of a coherent seed (see Fig. 15a). A two-moded guide at FF, however, will in general support several modes at the SH, with various Type I and Type II interactions potentially contributing to the overall phase shift. With reference to the situation sketched in Fig. 15a, De Rossi *et al.* considered a channel waveguide in temperature tuned z-cut lithium niobate and took into account the prevailing interactions between $TE_{00}(\omega) + TE_{00}(\omega) \rightarrow TM_{00}(2\omega)$ (i.e. Type I SHG) and $TE_{00}(\omega) + TE_{01}(\omega) \rightarrow TM_{01}(2\omega)$ (i.e. Type II SHG) [80]. When increasing the FF excitation with equal weights in the two FF modes, the output

- defined by the maximum overlap in the region where the modes add in phase
- switches from one side to the other. The calculated exchange of power between the two corresponding halves of the channel end-section is shown in Fig. 15b for the wavelength $\lambda=1.55\mu m$. Preliminary experimental results have confirmed the simulated behaviour.

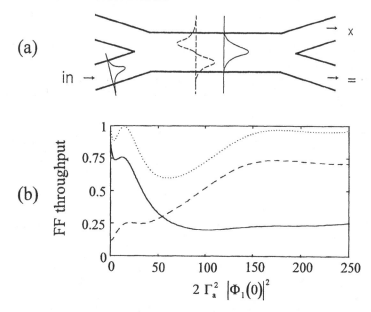

Figure 15. Mode mixer. a) Sketch of a channel with two FF modes involved; b) calculated FF throughputs in the two channel halves (Right: dashed line, Left: solid line) vs. FF excitation (equally distributed between the two input modes). The dotted line is the total FF throughput, showing no substantial depletion.

3.5 BRAGG GRATINGS

A perturbation of period $\Lambda=\pi/\beta$ on a waveguide can operate as a Bragg reflector (see Fig. 16a) by coupling forward and backward modes of wavevector β, i.e. by opening a stop-gap in the dispersion (momentum vs frequency) diagram. When employed in a quadratically nonlinear guide for SHG, the grating can induce a Bragg reflection at one or at both harmonics, depending on their dispersion, polarization, index/depth profile. This linear coupling, together with the parametric interaction, is able to deform the stop-gap at large input excitations, potentially leading to induced transparency and optical bistability through the inherent feedback. This phenomenon is more complicated when propagation-forbidden gaps are present at both frequencies, e.g. when the grating is doubly resonant, as depicted in Fig. 16b.

For the simple case of a single stop-gap at FF, Picciau et al. have analyzed the occurrence of optical bistability versus FF excitation or coherent SH seeding [81-82]. In this case, neglecting out-coupling effects and resorting to a scalar

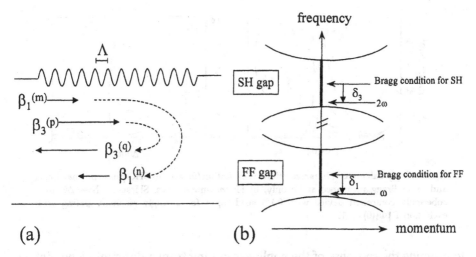

Figure 16. Bragg grating. a) Sketch of a doubly-resonant Bragg reflector operating at both FF and SH; b) the two stop-gaps created by the grating in the dispersion diagram.

approximation, for Type I SHG the SVEA model is given by the coupled equations:

$$\frac{d\Phi_1}{d\zeta} = i\gamma L\Psi_1 e^{-i\delta^\omega\zeta} - i\,\Gamma_\Phi\,\Phi_1^*\,\Phi_3\,e^{-i\Delta\zeta}$$

$$\frac{d\Psi_1}{d\zeta} = -i\gamma L\Phi_1 e^{+i\delta^\omega\zeta} + i\,\Gamma_\Psi\,\Psi_1^*\,\Psi_3\,e^{+i\Delta'\zeta}$$

$$\frac{d\Phi_3}{d\zeta} = -i\,\Gamma_\Phi\,\Phi_1^2\,e^{+i\Delta\zeta}$$

$$\frac{d\Psi_3}{d\zeta} = +i\,\Gamma_\Psi\,\Psi_1^2\,e^{-i\Delta'\zeta}$$

(18)

with $\delta = L(2\pi/\Lambda - \beta_1^{(m)} - \beta_1^{(n)})$ the Bragg detuning between the grating and counterpropagating modes of orders m and n. For identical modes ($\Gamma_\Phi = \Gamma_\Psi = \Gamma$, $\beta_1^{(m)} = \beta_1^{(n)} = \beta_1$), Fig. 17a shows how bistable loops versus FF excitation can be originated at low enough phase mismatches Δ for a grating of given linear strength $\gamma L = 1.5$ tuned at the Bragg resonance $(\delta = L(2\pi/\Lambda - 2\beta_1) = 0)$. Fig. 17b displays bistability versus the relative phase of a 0.1% seed at SH in a grating with $\delta = \Delta = 0$ (i.e. linear and SHG matching).

The ocurrence of optical bistability can be interpreted in terms of energy localization inside the perturbed structure. This approach to the problem leads

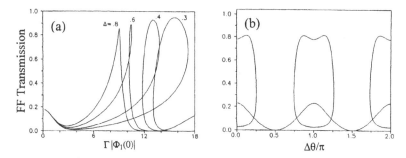

Figure 17. a) Calculated FF transmission vs. excitation for various SHG phase mismatches Δ and exact Bragg resonance at FF only; b) FF transmission vs. SH seed phase $\Delta\Theta$ in a coherently controlled device with 0.1% seed input for a singly resonant grating with excitation $\Gamma |\Phi_1(0)| = 13$

to examine the response of the nonlinear grating from a different viewpoint, i.e. the formation of gap-solitons along the propagation coordinate [83]. In the case of the quadratic nonlinearities of concern here, these particular eigensolutions can be called parametric gap simultons, and will be addressed in the following section.

4. Parametric Gap Simultons

4.1 GAP SIMULTONS IN PERIODIC QUADRATIC STRUCTURES

Including the time coordinate in eqns. (18), considering the scalar approximation for a weakly-coupled grating at both FF and SH, defining κ and v the ratios between the linear strengths γ_3 and γ_1 and group velocities v_{g3} and v_{g1} at SH and FF, respectively, neglecting the material dispersion and rescaling the field amplitudes in order to eliminate the nonlinear coefficients, the system of four coupled equations becomes:

$$-i\frac{\partial u_1^+}{\partial \xi} = i\frac{\partial u_1^+}{\partial \tau} + \delta_1 u_1^+ + u_1^- + u_3^+(u_1^+)^*$$

$$+i\frac{\partial u_1^-}{\partial \xi} = i\frac{\partial u_1^-}{\partial \tau} + \delta_1 u_1^- + u_1^+ + u_3^-(u_1^-)^*$$

$$-i\frac{\partial u_3^+}{\partial \xi} = i\frac{\partial u_3^+}{v\partial \tau} + \delta_3 u_3^+ + \kappa u_3^- + \frac{(u_1^+)^2}{2}$$

$$+i\frac{\partial u_3^-}{\partial \xi} = i\frac{\partial u_3^-}{v\partial \tau} + \delta_3 u_3^- + \kappa u_3^+ + \frac{(u_1^-)^2}{2} \qquad (19)$$

with $\xi=z\gamma_1$, $\tau=\gamma_1 t v_{g1}$, and detunings δ_1 and δ_3 defined with respect to the centers of the stop-bands at FF and SH, respectively (see Fig. 16b). Looking for localized energy states within the periodic structure, it is convenient to solve system (19) in the proximity of the band edges, using a multiple scale expansion $\xi_n=\varepsilon^n\xi$, $\tau_n=\varepsilon^n\tau$ with $\varepsilon\ll1$ and solving for the envelopes u_1 and u_3 of the fields at FF and SH, respectively. This allows to use the linear (Bloch) solutions of the fast variables ξ_0 and τ_0, and to describe the evolution of the nonlinear envelopes with two equations of the form (for positive κ):

$$i\frac{\partial u_1}{\partial\tau} - \frac{\rho_1}{2}\frac{\partial^2 u_1}{\partial\xi^2} + \frac{1}{2\sqrt{2}}(\rho_3+1)\, u_3 u_1^* \, e^{-i\alpha\tau} = 0$$

$$i\frac{\partial u_3}{\partial\tau} - \frac{v\rho_3}{2\kappa}\frac{\partial^2 u_3}{\partial\xi^2} + \frac{v}{2\sqrt{2}}(\rho_3+1)\, u_1^2 \, e^{+i\alpha\tau} = 0 \tag{20}$$

with $\rho_{1,3}=+1$ ($\rho_{1,3}=-1$) if the corresponding frequency is close to the lower (upper) bandedge, and $\alpha=2(\delta_1+\rho_1) - v(\delta_3+\rho_3\kappa)$. It is therefore apparent from (20) that the nonlinearity is effective only when 2ω is close to its lower bandedge (LB) in Fig. 16b [84-85]. Equations (20) resemble those describing dispersive SHG in uniform media (without grating) after interchanging τ and ξ, and are known to exhibit solitary-wave solutions. Otherwise stated, eqns. 20 support the existence of gap solitons in quadratic media [84, 86-87]. Such z-localization, or mutual trapping of the FF and SH fields, is given by the counterbalance of grating dispersion with parametric mixing, provided the nonlinearity is effective [84, 88]. Stable and stationary bright-bright solitons of the sech^2-type (LB at FF and SH) and twin-hole dark solitons (UB for FF and LB for SH) have been identified for the two envelopes, with several non-stationary and unstable solutions as well [84, 89]. Fig. 18 shows the FF profiles of two such stationary parametric simultons, with the SH component following the FF.

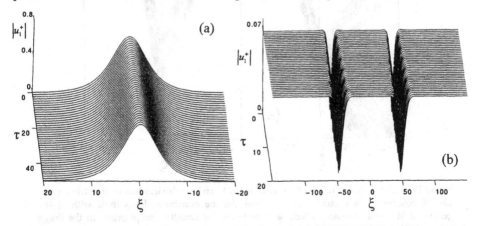

Figure 18. FF profile vs. time and propagation of a) a bright-bright (FF-LB, SH-LB) and b) a twin-hole dark (FF-UB, SH-LB) stationary gap-simulton. The localization gives origin to a steady eigenstate.

It is worth mentioning that exact bright-bright simultons can also be found for frequencies inside the stop-gaps, resorting in these cases to the solutions of the evolution equations (19).

4.2   EXCITATION OF SLOWLY-TRAVELLING GAP-SIMULTONS

Non-stationary gap-solutions of (19) are able to travel at reduced speeds inside the structure, making ideal candidates for power-dependent delay lines or optical buffers. Since typical material dispersions of uniform media are not large enough for observing temporal solitons of the quadratic nonlinearity (*simultons*), propagation in structures with Bragg gratings, including the cases of out-gap frequencies [88, 90], could open the way to their use in ultrafast all-optical devices for processing and switching. To this extent, Fig. 19 describes the excitation of a simulton: an incident sech–shaped FF pulse reaches the quadratic periodic medium (z>0), where SH is generated and, inside the grating, locks with the FF to form a simulton propagating at 30% of the FF velocity. In the case of a singly resonant Bragg grating ($\kappa=0$), Fig. 20a shows snapshots of reflected and transmlitted pulses at FF and SH, after a gap-simulton has been launched.   Finally, Fig. 20b displays the case of an incident FF pulse which, due to the dispersion provided by the periodicity, shrinks and slows down after the linear/nonlinear interface. However, in this example the FF is out-gap and only negligible reflection takes place.

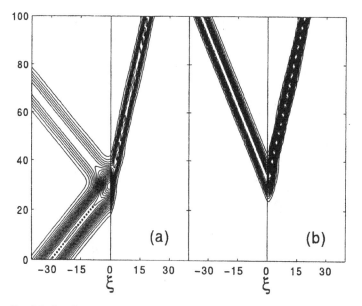

Figure 19.   Calculated excitation of a simulton due to an FF pulse incident on the structure. a) The FF component is partially reflected and, for the remaining, locks itself with b) the generated SH field in order to form a slow two-color simulton propagating in the Bragg structure. Here v=0.5, κ=1, δ_1=δ_2=-0.9.

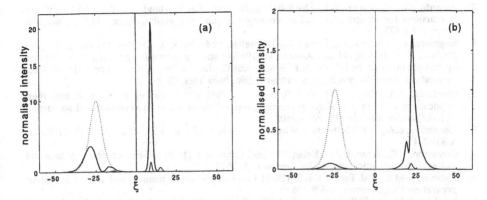

Figure 20. a) Snapshots of reflected (ξ<0) and transmitted (ξ<0) pulses at FF (thick solid lines) and SH (thin solid lines) after the nonlinear interaction. The incident pulse (dotted line) is 100ps in duration and the grating interacts with FF only. b) Similar to a) but in a doubly resonant grating and for an out-gap FF pulse: after the interface, the reflected portion is negligible while the transmitted pulse slows down and compresses.

5.   Conclusions

Quadratic cascading is a convenient approach to all-optical processing in numerous situations, ranging from wavelength multiplexing to analog/digital modulation to temporal solitons. Plane- and guided-wave results have demonstrated, experimentally and/or theoretically, that quadratic nonlinear effects can be effectively employed in novel signal processors. More efficient materials might indeed allow their implementation in applied physics and engineering. After almost four decades from Franken's experiment [91], research in quadratic nonlinear optics is still dispensing the excitement of new phenomena and effects.

Acknowledgments   We are grateful to G.I. Stegeman and E.W. Van Stryland of CREOL-UCF in Orlando (USA), and to S. Trillo of FUB in Rome (Italy). Partial support was provided by the Italian Ministry for Research (MURST 40% "Photonic Technologies...") and the National Research Council (grants 96.01844.CT11 and 96.02238.CT07).

6.   References

1. DeSalvo, R., Hagan, D.J., Sheik-Bahae, M., Stegeman, G., Van Stryland, E.W. and Vanherzeele, H. (1992) Self-focusing and self-defocusing by cascaded second-order effects in KTP, *Opt. Lett.* **17**, 28-30.
2. Stegeman, G.I., Sheik-Bahae, M., VanStryland, E. W. and Assanto, G. (1993) Large nonlinear phase shifts in second-order nonlinear optical processes, *Opt. Lett.* **18**, 13-15.

3. Assanto, G., Stegeman, G.I., Sheik-Bahae, M. and Van Stryland, E. W. (1995) Coherent interactions for all-optical signal processing via quadratic nonlinearities, *IEEE J. Quantum Electron.* **31**, 673-681.
4. Stegeman, G.I., Schiek, R., Torner, L., Torruellas, W.E., Baek, Y., Baboiu, D., Wang, Z., Van Stryland, E.W., Hagan, D.J. and Assanto, G. (1997) Cascading: a promising approach to nonlinear optical phenomena, in I.C. Khoo, F. Simoni and C. Umeton (eds.), *Novel Optical Materials and Applications*, John Wiley & Sons, Interscience Div., New York, Ch.2 .
5. Stegeman, G.I., Hagan, D.J. and Torner, L., (1996) $\chi^{(2)}$ cascading phenomena and their applications to all-optical signal processing, mode-locking, pulse compression and solitons, *J. Opt. & Quantum Electron.* **28**, 1691-1740.
6. Ostrovskii, L.A., (1967) Self-action of light in crystals, *JETP Lett.* **5**, 272-275; (1967) *ZhETF Pis'ma* **5**, 331-334.
7. Gustafson, T.K., Taran, J.-P.E., Kelley, P.L. and Chiao, R.Y. (1970) Self-modulation of picosecond pulses in electro-optic crystals, *Opt. Commun.* **2**, 17-21.
8. Thomas, J.-M.R. and Taran, J.-P.E. (1972) Pulse distortion in mismatched second harmonic generation, *Opt. Commun.* **4**, 329-334.
9. Klyshko, D.N. and Polkovnikov, B.F., (1974) Phase modulation and self-modulation of light in three-photon processes, *Sov. J. Quant. Electron.* **3**, 324-326.
10. Eckardt, R.C. and Reintjes J. (1984) Phase matching limitations of high efficiency second harmonic generation, *IEEE J. Quant. Electron.* **20**, 1178-1187.
11. Manassah, J.T. (1987) Amplitude and phase of a pulsed second harmonic signal, *J. Opt. Soc. Am.. B* **4**, 1235-1240.
12. Bakker, H.J., Planken, P.C.M.,, Kuipers, L. and Lagendijk, A. (1990) Phase modulation in second order nonlinear processes, *Phys. Rev. A* **42**, 4085-4101.
13. Andrews, J.H., Kowalski, K.L. and Singer, K.D. (1992) Pair correlations, cascading, and local-field effects in nonlinear optical susceptibilities, *Phys. Rev. A* **46**, 4172- 4184.
14. Pliszka, P. and Banerjee, P.P. (1993) Self-phase modulation in quadratically nonlinear crystals, *J. Mod. Opt.* **40**, 1909-1916.
15. Kalocsai, A.G. and Haus, J.W. (1993) Self-modulation effects in quadratically nonlinear materials, *Opt. Commun.* **97**, 239-244.
16. De Martini, F. (1967); Coherent Raman amplification, *Il Nuovo Cimento* **51 B**, 16-41; Coffinet, J.P. and De Martini, F. (1969) Coherent excitation of polaritons in gallium phosphide, *Phys. Rev. Lett.* **22**, 60-64.
17. Yablonovitch, E., Flytzanis, C. and Bloembergen, N. (1972) Anisotropic interference of three-wave and double two-wave frequency mixing in GaAs, *Phys. Rev. Lett.* **29**, 865-868.
18. Akhmanov, S.A., Dubovik, A.N., Saltiel, S.M., Tomov, I.V. and Tunkin, V.G. (1974) Nonlinear optical effects of fourth order in the field in a lithium formiate crystal, *JETP Lett.* **20**, 117-118.
19. Meredith, G.R. (1981) Cascading in optical third-harmonic generation by crystalline quartz, *Phys. Rev. B* **24**, 5522-5532; (1982) Second-order cascading in third-order nonlinear optical processes, *J. Chem. Phys.* **77**, 5863–5871.
20. Qiu, P., and Penzkofer, A. (1988) Picosecond third-harmonic light generation in β-BaB$_2$O$_4$, *Appl. Phys. B* **45**, 225 -236.
21. Belashenkov, N.R., Gagarskii, S.V., and Inochkin, M.V. (1989) Nonlinear refraction of light on second-harmonic generation, *Opt. Spectrosc.* **66**, 806-808.
22. Zgonik, M. and Gunter, P. (1992) Second and third-order optical nonlinearities in ABO$_3$ compounds measured in fast four-wave mixing experiments, *Ferroelectrics* **126**, 33-38.
23. Assanto, G., Conti, C., Leo, G., Stegeman, G. I., Torruellas, W. E. and Trillo, S., Three-wave simultons: quasi-particles in quadratic media, *J. Nonlinear Opt. Phys. and Mat.*, in press
24. Hutchings, D.C., Aitchison, J.S. and Ironside, C.N. (1993) All-optical switching based on nondegenerate phase shifts from a cascaded second-order nonlinearity, *Opt. Lett.* **18**, 10-12.
25. Bosshard, Ch., Spreiter, R., Zgonik, M. and Günter, P. (1995) Kerr nonlinearity via cascaded optical rectification and the linear electro-optic effect, *Phys. Rev. Lett.* **74**, 2816-2819.
26. Danielius, R., Di Trapani, P., Dubietis, A., Piskarskas, A., Podenas, D. and Banfi, G.P. (1993) Self-diffraction through cascaded second-order frequency-mixing effects in β-barium borate, *Opt. Lett.* **18**, 574-576.
27. Ou, Z.Y. (1995) Observation of nonlinear phase shift in CW harmonic generation, *Opt. Commun.* **124**, 430-437.
28. Fazio, E., Sibilia, C., Senesi, F. and Bertolotti, M. (1996) All-optical switching during quasi-collinear second-harmonic generation, *Opt. Commun.* **127**, 62-68.

29. Wang, Z., Hagan, D.J., Van Stryland, E.W., Zyss, J., Vidakovic, P. and Torruellas, W.E. (1997) Cascaded second-order effects in N-(4-nitrophenyl)-L-prolinol, in a molecular single crystal, *J. Opt. Soc. Am. B* **14**, 76- 86.

30. Vidakovic, P., Lovering, D.J., Levenson, J.A., Webjorn, J. and Russell, P.St.J. (1997) Large nonlinear phase shift owing to cascaded $\chi^{(2)}$ in quasi-phase-matched bulk LiNbO$_3$, *Opt. Lett.* **22**, 277-279.

31. Khurgin, J.B. and Ding, Y.J. (1994) Resonant cascaded surface-emitting second-harmonic generation: a strong third-order nonlinear process, *Opt. Lett.* **19**, 1016-1018.

32. Khurgin, J.B., Obeidat, A., Lee, S.J. and Ding, Y.J. (1997) Cascaded optical nonlinearities: Microscopic understanding as a collective effect, *J. Opt. Soc. Am. B* **14**, 1977-1983.

33. Russell, P.St.J. (1993) All-optical high gain transistor action using second-order nonlinearities, *Electron. Lett.* **29**, 1228-1229.

34. Hagan, D.J., Sheik-Bahae, M., Wang, Z., Stegeman, G.I., Van Stryland, E.W. and Assanto, G. (1994) Phase controlled transistor action by cascading of second-order nonlinearities, *Opt. Lett.* **19**, 1305-1307.

35. Belostotsky, A.L., Leonov, A.S. and Meleshko, A.V. (1994) Nonlinear phase change in type II second-harmonic generation under exact phase-matched conditions, *Opt. Lett.* **19**, 856-858.

36. Nitti, S., Tan, H.M., Banfi, G.P. and Degiorgio, V. (1994) Induced 'third-order' nonlinearity via cascaded second-order effects in organic crystals of MBA-NP, *Opt. Commun.* **106**, 263-268.

37. Asobe, M., Yokohama, I., Itoh, H. and Kaino, T. (1997) All-optical switching by use of cascading of phase-matched sum-frequency-generation and difference-frequency-generation processes in periodically poled LiNbO$_3$, *Opt. Lett.* **22**, 274-276.

38. Assanto, G. and Torelli, I. (1995) Cascading effects in type II second-harmonic generation: applications to all-optical processing, *Opt. Commun.* **119**, 143-148.

39. Kobyakov, A., Peschel, U., Muschall, R., Assanto, G.,Torchigin, V.P., and Lederer, F. (1995) An analytical approach to all-optical modulation by cascading, *Opt. Lett.* **20**, 1686-1688.

40. Kobyakov, A., Peschel, U. and Lederer, F. (1996) Vectorial interaction in cascaded quadratic nonlinearities - an analytical approach, *Opt. Commun.* **124**, 184-194.

41. Kobyakov, A. and Lederer, F. (1996) Cascading of quadratic nonlinearities: an analytical study, *Phys. Rev. A* **54**, 3455-3471.

42. Lefort, L. and Barthelemy, A. (1995) All-optical transistor action by polarisation rotation during type-II phase-matched second-harmonic generation, *Electron. Lett.* **31**, 910-911.

43. Assanto, G., Wang, Z., Hagan, D.J. and VanStryland, E.W. (1995) All-optical modulation via nonlinear cascading in type II second harmonic generation, *Appl. Phys. Lett.* **67**, 2120-2122.

44. Lefort, L. and Barthelemy, A. (1995) Intensity-dependent polarization rotation associated with type-II phase-matched second-harmonic generation: application to self-induced transparency, *Opt. Lett.* **20**, 1749-1751.

45. Buchvarov, I., Saltiel, S., Iglev, Ch. and Koynov, K. (1997) Intensity dependent change of the polarization state as a result of non-linear phase shift in type II frequency doubling crystals, *Opt. Commun.* **141**, 173-178.

46. Wang, Z., Hagan, D.J., VanStryland, E.W. and Assanto, G. (1996) Phase-insensitive, single wavelength, all-optical transistor based on second-order nonlinearities, *Electron. Lett.* **32**, 1135-1136.

47. Assanto, G., Hagan, D.J., Stegeman, G.I., Van Stryland, E.W. and Torruellas, W. E. (1996) Vectorial quadratic interactions for all-optical signal processing via second-harmonic generation, *Optica Applicata* **XXVI**, 285-291.

48. Wang, Z., Hagan, D.J., Van Stryland, E.W. and Assanto, G. (1996) A phase-insensitive all-optical transistor using second-order nonlinear media, in *Int. Quantum Electronics Conf.*, 1996 OSA 1996, Tech. Dig. Series, Washington D.C., 264-265.

49. Kim, S., Wang, Z., Hagan, D. J., Van Stryland, E. W., Kobyakov, A., Lederer, F. and Assanto, G., Phase-insensitive All-optical Transistors based on Second-order Nonlinearities, in preparation

50. Assanto, G. (1995) All-optical switching in integrated structures, in: *Nonlinear Optical Materials: Principles and Applications*, V. DeGiorgio and C. Flytzanis ed., IOS Press, Amsterdam .

51. Sundheimer, M.L., Bosshard, Ch., Van Stryland, E.W., Stegeman, G.I. and Bierlein, J.D. (1993) Large nonlinear phase modulation in quasi-phase-matched KTP waveguides as a result of cascaded second-order processes, *Opt. Lett.* **18**, 1397-1399.

52. Sundheimer, M.L., Villeneuve, A., Stegeman, G.I. and Bierlein, J.D. (1994) Cascading nonlinearities in KTP waveguides at communications wavelengths, *Electron. Lett.* **30**, 1400- 1401.

86

53. Schiek, R., Sundheimer, M.L., Kim, D.Y., Baek, Y., Stegeman, G.I., Seibert, H. and Sohler, W. (1994) Direct measurement of cascaded nonlinearity in lithium niobate channel waveguides, *Opt. Lett.* **19**, 1949-1951.

54. Baek, Y., Schiek, R. and Stegeman, G.I. (1995) All-optical switching in a hybrid Mach-Zehnder interferometer as a result of cascaded second-order nonlinearity, *Opt. Lett.* **20**, 2168-2171.

55. Baek, Y., Schiek, R., Krijnen, G., Stegeman, G.I., Baumann, I. and Sohler, W. (1996) All-optical integrated Mach-Zehnder switching in lithium niobate waveguides due to cascaded nonlinearities, *Appl. Phys. Lett.* **68**, 2055-2057.

56. Schiek, R., Baek, Y., Krijnen, G., Stegeman, G.I., Baumann, I. and Sohler W. (1997) All-optical switching in lithium niobate directional couplers with cascaded nonlinearity, *Opt. Lett.* **21**, 940-942.

57. Treviño-Palacios, C.G., Stegeman, G.I., De Micheli, M.P., Baldi, P., Nouh, S., Ostrowsky, D.B., Delacourt, D. and Papuchon, M. (1995) Intensity dependent mode competition in second harmonic generation in multimode waveguides, *Appl. Phys. Lett.* **67**, 170-172.

58. Treviño-Palacios, C.G., Stegeman, G.I. and Baldi, P. (1996) Spatial nonreciprocity in waveguide second-order processes, *Opt. Lett.* **21**, 1442-1444.

59. Kim, D.Y., Torruellas, W.E., Kang, J., Bosshard, C., Stegeman, G.I., Vidakovic, P., Zyss, J., Moerner, W.E., Twieg, R. and Bjorklund, G. (1994) Second-order cascading as the origin of large third-order effects in organic single-crystal-core fibers, *Opt. Lett.* **19**, 868-870; Torruellas, W.E., Krijnen, G., Kim, D.Y., Schiek, R., Stegeman, G.I., Vidakovic, P. and Zyss, J. (1994) Cascading nonlinearities in an organic single crystal core fiber: the Cerenkov regime, *Opt. Commun.* **112**, 122-130.

60. Bortz, M.L., Fujimura, M. and Fejer, M.M. (1994) Increased acceptance bandwidth for quasi phase-matched second harmonic generation in LiNbO3 waveguides, *Electron. Lett.* **30**, 34-35.

61. Mizuuchi, K., Yamamoto, K., Kato, M. and Sato, H. (1994) Broadening of the phase matching bandwith in quasi-phase matched second harmonic generation, *IEEE J. Quantum Electron.* **30**, 1596-1604.

62. See, for example, *Guided Wave Nonlinear Optics*, D.B. Ostrowsky and R. Reinisch eds., Kluwer Academic Publ, London (1992)

63. Stegeman, G.I. (1990) Overview of nonlinear integrated optics, in *Nonlinear Waves in Solid State Physics*, A.D. Boardman, M. Bertolotti, T. Twardowski eds., Plenum Press, New York.

64. Suhara, T. and Nishihara, H. (1990) Theoretical analysis of waveguide second-harmonic generation phase matched with uniform and chirped gratings, *IEEE J. Quantum Electron.* **26**, 1265-1276.

65. Gase, T. and Karthe, W. (1997) Quasi-phase matched cascaded second order processes in poled organic polymer waveguides, *Opt. Commun.* **133**, 549-556.

66. Kelaidis, C., Hutchings, D.C. and Arnold, J.M. (1994) Asymmetric two-step GaAlAs quantum well for cascaded second-order processes, *IEEE Trans. on Quantum Electron.* **30**, 2998-3006.

67. Tan, H., Banfi, G.P. and Tomaselli, A. (1993) Optical frequency mixing through cascaded second-order processes in β-barium borate, *Appl. Phys. Lett.* **63**, 2472-2474.

68. Gorbounova, O., Ding, Y.J., Khurgin, J.B., Lee, S.J. and Craig, A.E. (1996) Optical frequency shifters based on cascaded second-order nonlinear processes, *Opt. Lett.* **21**, 558-560.

69. Gallo, K., Assanto, G. and Stegeman, G.I. (1997) Efficient wavelength shifting over the Erbium amplifier bandwidth via cascaded second order processes in Lithium Niobate waveguides, *Appl. Phys. Lett.* **71**, 1020-1022.

70. Treviño-Palacios, C.G., *Private communication*

71. Gallo, K., Assanto, G. and Stegeman, G.I., A Lithium Niobate quadratic device for wavelength multiplexing around 1.55μm, *this book*.

72. Assanto, G., Stegeman, G.I., Sheik-Bahae, M. and Van Stryland, E. (1993) All optical switching devices based on large nonlinear phase shifts from second harmonic generation, *Appl. Phys. Lett.* **62**, 1323-1325.

73. Ironside, C.N., Aitchison, J.S. and Arnold, J.M. (1993) An all optical switch employing the cascaded second-order nonlinear effect, *IEEE J. Quantum Electron.* **29**, 2650-2654.

74. Schiek, R., (1993) Nonlinear refraction caused by cascaded second-order nonlinearity in optical waveguide structures, *J. Opt. Soc. Am. B* **10**, 1848-1855.

75. Schiek, R. (1994) All-optical switching in the directional coupler caused by nonlinear refraction due to cascaded second-order nonlinearity, *Opt. & Quantum Electron.* **26**, 415-431.

76. Assanto, G., Laureti-Palma, A., Sibilia, C. and Bertolotti, M. (1994) All-Optical Switching via Second Harmonic Generation in a Nonlinearly Asymmetric Directional Coupler, *Opt. Commun.* **110**, 599-603.

77. Karpierz, M.A. (1996) Directional coupling for solitary waves in quadratic nonlinearity, *Optica Applicata* **XXVI**, 391-397.
78. Lee, S.J., Khurgin, J.B. and Ding, Y.J. (1997) Directional couplers based on cascaded second-order nonlinearities in surface-emitting geometry, *Opt. Commun.* **139**, 63-68.
79. De Angelis, C., Laureti-Palma, A., Nalesso, G.F. and Someda, C.G. (1997) On the modelling of nonlinear guided-wave optics for all-optical signal processing, *Opt. & Quantum Electron.* **29**, 217-238.
80. De Rossi, A., Conti, C. and Assanto, G. (1997) Mode interplay via quadratic cascading in a lithium niobate waveguide for all-optical processing, *Opt. & Quantum Electron.* **29**, 53-63.
81. Picciau, M., Leo, G. and Assanto, G. (1996) A versatile bistable gate via quadratic cascading in a Bragg periodic structure, *J. Opt. Soc. Am. B* **13**, 661-670.
82. Leo, G., Picciau, M. and Assanto, G. (1995) Guided-wave optical bistability through nonlinear cascading in a phase-matched distributed reflector, *Electron. Lett.* **31**, 1661-1662.
83. Chen, W. and Mills, D.L. (1987) Gap solitons and the nonlinear optical response of superlattices, *Phys. Rev. Lett.* **58**, 160-163.
84. Conti, C., Trillo, S. and Assanto, G. (1997) Doubly resonant Bragg simultons via second-harmonic generation, *Phys. Rev. Lett.* **78**, 2341-2344.
85. Conti, C., Trillo, S. and Assanto, G. (1997) Bloch function approach for parametric gap solitons, *Opt. Lett.* **22**, 445-447.
86. Peschel, T., Peschel, U., Lederer, F. and Malomed, B.A. (1997) Solitary waves in Bragg gratings with a quadratic nonlinearity, *Phys. Rev. E* **55**, 4730-4739.
87. He H. and Drummond, P. (1997) Ideal Soliton Environment using Parametric Band Gaps, *Phys. Rev. Lett.* **78**, 4311-4314.
88. Schiek, R. (1997) Soliton-like propagation and second harmonic generation in waveguides with second-order optical nonlinearities, *AEÜ Int. J. Electron. Commun.* **2**, 77-86.
89. Conti, C., Trillo, S. and Assanto, G., Optical gap solitons via second-harmonic generation: exact solitary solutions, *Phys. Rev. E*, in press
90. Conti, C., Assanto, G. and Trillo, S. (1997) Excitation of self-transparency Bragg solitons in quadratic media, *Opt. Lett.* **22**, 1350-1352.
91. Franken, P.A., Hill, A.E., Peters C.W. and Weinreich, G. (1961) Generation of optical harmonics, *Phys. Rev. Lett.* **7**, 118-121.

CONTROL OF LASER LIGHT PARAMETERS BY
$\chi^{(2)}:\chi^{(2)}$ NONLINEAR OPTICAL DEVICES

S. SALTIEL, I. BUCHVAROV, K. KOYNOV

University of Sofia, Faculty of Physics,
Quantum Electronics Department,
5 J. Bourchier Blvd., 1164 Sofia, Bulgaria

Abstract

Nonlinear optical devices based on cascaded second order processes ($\chi^{(2)}:\chi^{(2)}$ cascading) able to control laser light parameters are reviewed. Such parameters of the laser light as phase, intensity, polarization and pulse duration are considered. Special attention is paid to cascaded $\chi^{(2)}$ nonlinear optical devices suitable for mode-locking and intracavity pulse control. Theoretical approaches used to investigate these types of nonlinear optical devices are also reviewed.

1. Introduction

The possibility to use second order nonlinear optical processes to measure and control parameters of the laser light was noticed long time ago. First typical example is the measurement of laser pulse duration. The pulse duration can be recalculated from intensity autocorrelation function obtained with autocorrelator based on second harmonic generation (SHG) [1-3]. For measuring frequency chirp of the pulse was developed interferometric autocorrelation technique [4,5]. For full characterisation of the measured pulses (duration, shape and frequency chirp) more sophisticated technique called FROG technique - frequency resolved optical gate [6-9] is now used.

Laser light polarization can also be controlled by nonlinear optical devices based on second order processes. In [10,11] is shown that crystal for second harmonic generation can be used for measurement of the polarization extinction coefficient of laser light. This method is especially suitable for measurement very small extinction coefficients (10^{-5}-10^{-8}).

In the late 1980s and during 1990s a number of different experiments demonstrated that a second order nonlinear optical device called nonlinear mirror [12] can be used as a mode-locking device [13-17] and for intracavity pulse control [18-20]. This nonlinear mirror is constructed from a SHG crystal and dichroic mirror. At this early stage for explanation of the principle of operation of the nonlinear mirror single quadratic processes were used. At present, when we have much more understanding on

89

A. D. Boardman et al. (eds.),
Advanced Photonics with Second-order Optically Nonlinear Processes, 89–112.

90

cascaded second order processes can be shown that $\chi^{(2)}:\chi^{(2)}$ cascading as we will discuss in this review can play important role in the mode-locking properties of the nonlinear mirror as well.

Rediscovering of cascaded processes in 1992 [21-25] opened new frontiers for development of $\chi^{(2)}$ nonlinear optical devices. $\chi^{(2)}:\chi^{(2)}$ cascading can be used successfully for the control of the phase of the fundamental beams involved in different types second order processes [26-31]. From the other hand generation of the nonlinear phase shift (NPS) can be the base for construction of nonlinear optical devices with intensity dependent transmission [32-34] or with possibilities to control the polarization of the transmitted light [35-36]. Pulse compression and mode-locking can also be realized with nonlinear optical devices based on cascaded second order processes [17,36].

The idea of cascaded second order processes was for the first time investigated with respect to the exploration the possibility to generate third harmonic in single quadratic crystal [37-39].

Let us consider single $\chi^{(2)}$ media with fundamental input wave $E(\omega) = a_{10} \exp(i\varphi)$ at frequency ω (fig. 1).

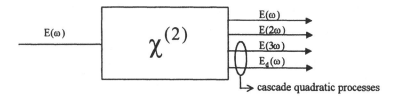

Figure 1. Cascade processes in single $\chi^{(2)}$ media.

Nonlinear polarization at frequency 2ω is a source for generation of second harmonic wave $E(2\omega)$

$$E(2\omega) \propto d_{eff}^{(2)}E(\omega)E(\omega).$$ (1)

Second harmonic wave (SHW) and the fundamental wave (FW) create also polarization at frequency 3ω, that becomes a source for generation of third harmonic wave $E(3\omega)$

$$E(3\omega) \propto d_{eff}^{(2)}E(\omega)E(2\omega)$$ (2)

and as a result

$$E(2\omega) \propto \frac{d_{eff}^{(2)}d_{eff}^{(2)}}{\Delta k}(E(\omega))^3. \qquad (3)$$

The term $\frac{d_{eff}^{(2)}d_{eff}^{(2)}}{\Delta k}$ is in fact "effective" cascade third order nonlinearity $\chi_{casc}^{(3)}$ responsible for the process of third harmonic generation in single quadratic media. The magnitude of the $\chi_{casc}^{(3)}$ can be bigger or smaller than the intrinsing cubic nonlinearity $\chi_{int}^{(3)}$, depending on the magnitude of the phase mismatch Δk for the step that is nonphase matched. Usually one of the two steps ($\omega + \omega = 2\omega$ or $2\omega + \omega = 3\omega$) is phase matched and the other is not phase matched. It is impossible using the natural birefegence of the crystal to achieve simultaneous phase matching for both second-order processes. Only recently it was shown that using quasi phase matched technique it is possible to phase match both $\chi^{(2)}$ steps simultaneously [40].

Single crystal third harmonic generation was used as a standard source for $\chi^{(3)}$ measurement [38-39]. Note that $\chi_{casc}^{(3)}$ is in most of the cases positive, so can be used for absolute measurement of the sign of natural $\chi^{(3)}$.

2. Control of laser light phase

Totally different result we have when another consequence of $\chi^{(2)}$ processes $\omega + \omega = 2\omega \rightarrow 2\omega - \omega = \omega$ is considered.

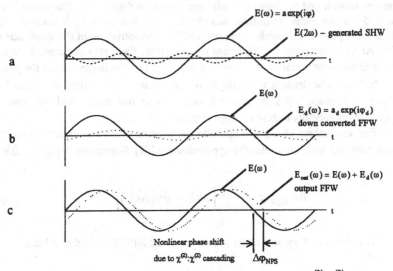

Figure 2. Simplified physical explanation of the process of phase shifting via $\chi^{(2)}:\chi^{(2)}$ cascading.

First step is the same - generation of SHW $E(2\omega)$ (fig.2,a). The second step is downconversion of generated SHW (fig. 2, b). As a result downconverted FW $E_d(\omega)$ at fundamental frequency ω is generated

$$E_d(\omega) \propto d_{eff}^{(2)}E(\omega)E(2\omega) \qquad (4)$$

or

$$E_d(\omega) \propto \frac{d_{eff}^{(2)}d_{eff}^{(2)}}{\Delta k}(E(\omega))^3. \qquad (5)$$

If the process of SHG is slightly mismatched ($\Delta k \neq 0$) then the local phase of $E_d(\omega)$ is different from the phase of the FW. The two waves $E(\omega)$ and $E_d(\omega)$ interfere at the output of the crystal and as a result the output FW obtains NPS with magnitude that depends on the input intensity (fig. 2,c). The higher is the input intensity the higher is the nonlinear phase shift. This process is in fact self phase modulation due to $\chi^{(2)}$ processes and is observed when SHG process is with Type I interaction. As we will see later the effect of cross phase modulation can be obtained with SHG Type II or sum frequency generation. Additional degree of control of the NPS can be realized when one uses seeded SHG [23] or two pass phase shifting devices [41].

2.1. PHASE SHIFTING WITH TYPE I SHG.

Typical dependencies of the NPS and the fundamental depletion as a function of the distance, mismatch and the input intensity are shown on figures 3-5. The dependence of the NPS on the input amplitude and the distance has stepwise behaviour. The middle of each plato corresponds to the point of full reconstruction of the amplitude of the FW. At the same point the SHW has its minimum. Each step corresponds to the nonlinear phase jump equal to $\pi/2$ or less depending on the magnitude of the phase mismatch ΔkL. The dependence of the NPS as a function of the mismatch ΔkL has dispersion like shape. NPS is zero for exact phase matching, ΔkL=0, and is decreasing asymptotically to zero for big values of ΔkL.

For low values of input amplitude the process of NPS is described by the formulae obtained with fixed intensity approximation [42]. Suggesting $\sigma a_{10} \ll \Delta k$

$$\Delta\varphi_{NPS} = \frac{\sigma^2 a_{10}^2}{\Delta k^2}\left[\Delta kL - \sin(\Delta kL)\right]. \qquad (6)$$

For $\Delta kL = \pi$, where is the maximum of the NPS & ΔkL dependence

$$\Delta\varphi_{NPS} = \frac{1}{\pi}\sigma^2 a_{10}^2 L^2 = \frac{1}{\pi}\frac{L^2}{L_{NL}^2}. \tag{7}$$

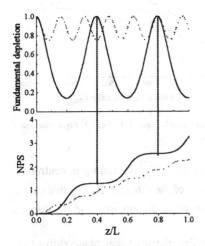

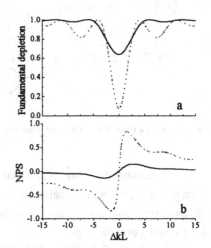

Figure 3. Nonlinear phase shift and fundamental depletion versus normalized distance z/L. Normalized input intensity is $\sigma a_{10}L = 10$. Normalized mismatch ΔkL is 3 (solid line) and 30 (dotted line)

Figure 4. Fundamental depletion (a) and nonlinear phase shift (b) versus normalized mismatch ΔkL. Normalized input intensity $\sigma a_{10}L$ is 0.7 (solid line) and 2 (dotted line).

In this limit is possible to introduce $n_{2,casc}$ but this analogy will be valid only for low input intensities ($L_{NL} \gg L$)

$$n = n_o + n_{2,caso}a_{10}^2 + n_2 a_{10}^2, \tag{8}$$

where $n_{2,casc} = \frac{\lambda\sigma^2}{2\pi^2}L$. In this limit $n_{2,casc} > n_2$ [21].

For higher input intensities is impossible to introduce $n_{2,casc}$, that is due to $\chi^{(2)}:\chi^{(2)}$ cascading. For these intensities $L_{NL} < L$

$$n = n_o + n_{1,oaso}a_{10} + n_2 a_{10}^2. \tag{9}$$

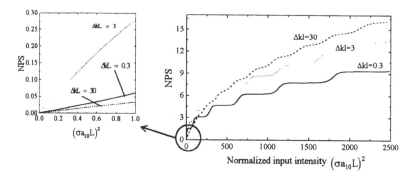

Figure 5. Nonlinear phase shift as a function of the input intensity. Left the case of $L/L_{NL} < 1$. Right - the case of $L/L_{NL} \gg 1$.

Our group [43] has shown that using $\chi^{(3)} : \chi^{(3)}$ cascading in centrosymmetric media is possible to simulate the behaviour of the inherent n_2. In this case at high input intensities $\Delta\varphi_{NLS}$ has linear dependence on the intensity and the distance after averaging over the steps.

In the case of nonzero input signal at 2ω the nonlinear phase shift of the fundamental wave depends on the input relative phase difference $\Delta\varphi_0 = 2\varphi_1 - \varphi_2$ between the phases of the FW and SHW. As it is seen from Fig. 6 the presence of seeding allows for obtaining NPS of the fundamental waves at exact phase matching condition ($\Delta k = 0$). In this case the change of the magnitude of the shift can be controlled not only by the amplitude of the pump FW but also by the input phase of the seeded SHW wave. In the approximation of negligible fundamental depletion when $\Delta k = 0$ we have for the NPS of the FW [30]:

$$\Delta\varphi_{NPS}(L) = -\sigma_1 \frac{a_{30}a_{20}}{a_{10}} L \cos(\Delta\varphi_0). \tag{10}$$

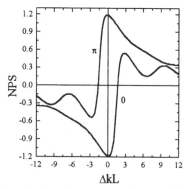

Figure 6. Nonlinear phase shift as a function of the wavevector mismatch ΔkL for two different values of the input phase difference. Normalized input amplitude is $\sigma a_{10}L=2$.

The induced phase shift in this case is due to the single second order process. The sign of the shift depends on the sign of the second order susceptibility tensor and the magnitude of $\Delta\varphi_0$. The analysis of the "seeding case" is important for understanding the phase shifting properties of two pass devices as nonlinear mirror.

2.2. CROSS PHASE MODULATION WITH TYPE II SHG.

In the case of type II SHG the input waves are respectively ordinary and extraordinary wave for the nonlinear birefringent media (fig. 7). Nonlinear interaction of these two waves under phase matched condition $\Delta k = k_{2e} - k_{1o} - k_{1e} = 0$ leads to generation of extraordinary SHW.

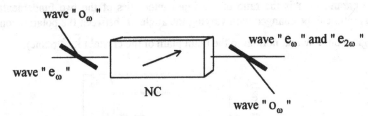

Figure 7. Type II SHG in negative uniaxial quadratic nonlinear crystal.

For maximum conversion efficiency the amplitudes of the two input waves must be equal. Type II with equal amplitudes at the input is practically equivalent to Type I SHG. From point of view to obtain large NPS more interesting is the case of nondegenerate Type II SHG ($A_{1o} \neq A_{1e}$). In this case the weaker wave has stronger NPS and the amount of the NPS of the weaker wave is defined by the amplitude of the stronger wave. This statement can be easily illustrated by simple mathematical analysis of the system of equations (11) that describes Type II SHG:

$$\frac{dA_{1o}}{dz} = -i\sigma_1 A_2 A_{1e}^* \exp(-i\Delta kz),$$

$$\frac{dA_{1e}}{dz} = -i\sigma_1 A_2 A_{1o}^* \exp(-i\Delta kz), \tag{11}$$

$$\frac{dA_2}{dz} = -i\sigma_2 A_{1o} A_{1e} \exp(i\Delta kz),$$

where σ_i are the nonlinear coupling coefficients.

Using fixed intensity approximation that is valid for low input intensities expressions for the NPS of the two FW can be obtained [30,31]

$$\Delta\varphi_{1o} = \sigma_2 \sigma_1 a_{1e}^2 L^2 \frac{(\Delta kL)}{(\Lambda L)^2}\left[1 - \text{sinc}(\Lambda L)\right], \tag{12.1}$$

$$\Delta\varphi_{1e} = \sigma_2 \sigma_1 a_{1o}^2 L^2 \frac{(\Delta kL)}{(\Lambda L)^2}\left[1 - \text{sinc}(\Lambda L)\right], \tag{12.2}$$

where $\Lambda = \sqrt{\Delta k^2 + \sigma_2 (\sigma_{1o} a_{1e}^2 + \sigma_{1e} a_{1o}^2)}$.

It is seen that the nonlinear phase shift of the "o" wave is defined by the intensity a_{1e}^2 of the "e" wave and the NPS of the "e" wave is defined by the intensity a_{1o}^2 of the "o" wave. Another important feature of this cross phase modulation process is the fact that the obtained by the fundamental waves NPS does not depend on the relative phase difference between them at the input of nonlinear media.

In fig. 8 is shown NPS of the weaker and the stronger waves as a function of normalized input amplitude σaL of the stronger fundamental wave. The amount of the phase jump between two neighbour steps is twice more than for the case of Type I SHG and degenerate Type II SHG ($r = 1$) and for small values of ΔkL is close to π. The parameter r is the ratio of the input intensities of the two fundamental waves. This ratio can be changed with varying the angle α between the polarization plane of single input FW and normal to the main plain of the crystal ($\vec{k}, \vec{z}$ plane).

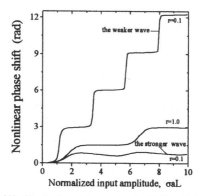

Figure 8. Nonlinear phase shift of the fundamental waves involved in Type II SHG. Normalized mismatch is $\Delta kL = 0.3$. As a parameter is indicated the ratio of the input intensities.

In fig. 9 is presented the amount of NPS collected by the both waves at the

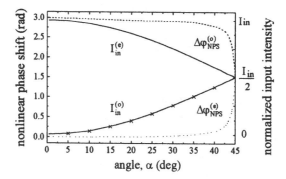

Figure 9. Input intensities (solid line) and NPS (dotted line) of the ordinary and extraordinary wave as a function of angle α (see the text). Normalized mismatch is $\Delta kL = 0.3$.

point of first full reconstruction as a function of the angle α. From the figure is seen that small disbalance of the input intensities of the fundamental waves leads to big difference of NPS obtained by them. This fact as we show later can be used for obtaining intensity dependent polarization rotation [36].

2.3. NONLINEAR FREQUENCY DOUBLING MIRROR AS A DOUBLE PASS PHASE SHIFTING ELEMENT

Nonlinear frequency doubling mirror (NFDM) shown in fig. 10 was initially known as a device with intensity dependent reflection [12]. As a rule for obtaining intensity dependent reflection with this device is necessary (I) frequency doubling crystal to be at exact phase matched condition and (ii) the mirror M to have maximum reflection at second harmonic frequency and partial reflection at fundamental frequency.

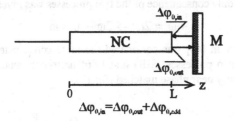

Figure 10. Nonlinear frequency doubling mirror. NC - crystal for SHG, M - mirror, $\Delta\varphi_0 = 2\varphi_1 - \varphi_2$.

Recently [41] we have shown that this device, when tuned not exactly but near exact phase matching can be efficient phase shifting element. By proper choosing the distance between the crystal and the mirror one can always obtains 100% reflection of the NFDM. However, in contrast to original "Stankov" type nonlinear mirror, the back reflector of the NFDM for phase shifting should have maximum reflection for both waves - fundamental and second harmonic. In fig. 11 is shown the dependence of the intensity reflection and the nonlinear phase shift as a function of the linear phase difference $\Delta\varphi_{add}$, that depends on the mirror - crystal distance.

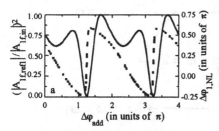 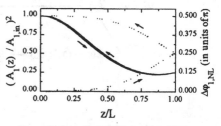

Figure 11. NPS and intensity of the reflected fundamental wave versus linear phase difference $\Delta\varphi_{add}$ for the case of Type I SHG ($\sigma a_{10}L=2$, $\Delta kL=1$).

Figure 12. Intensity (solid line) and the NPS (dotted line) of the fundamental wave inside the crystal: on the way to and back from the back reflector $\sigma a_{10}L=2$, $\Delta kL=1$, $\Delta\varphi_{add} = 1.67\pi$).

98

It is seen that by proper tuning of $\Delta\varphi_{add}$ both 100% reflection and large NPS can be obtained. In fig. 12 is shown the dynamic of the NPS collection and the reconstruction of fundamental intensity during the two pass of the fundamental wave through the crystal for proper chose $\Delta\varphi_{add}$.

NFDM is two to four times more efficient phase shifting element than single pass second harmonic crystals. It can be used as an element for all optical switching devices [41] and for mode-locking [17] of pulsed and CW lasers.

2.4. ENHANCED NPS VIA MULTISTEP $\chi^{(2)}$ CASCADING

Another way to improve phase shifting efficiency due to cascade processes is to increase the number of the second order processes in which the fundamental wave is involved. Up to now only consequence of the two processes was investigated.

$$\omega + \omega = 2\omega \rightarrow 2\omega - \omega = \omega .$$

Only recently in [44] was considered the more complicate case when in the crystal are possible two processes: SHG and third harmonic generation and both of them are simultaneously near phase matched (fig.13).

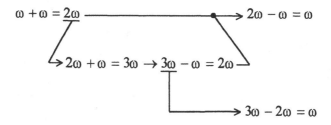

Figure 13. The idea of multistep $\chi^{(2)}$ cascading. NC - nonlinear crystal, F - harmonic stop.

The process of SHG is taken to be Type I. Then following second order processes must be considered for analysing NPS collected by the fundamental wave.

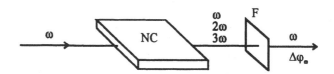

We see that two, three and four step $\chi^{(2)}$ cascading contribute to the process of reconverting depleted fundamental wave yelding this way large NPS.

In fig. 14 is compared NPS efficiency for two and multistep cascading. With multistep cascading almost four times less input intensity is necessary for obtaining NPS with amount of π. This more efficient way to generate large NPS can be used for reduction of the switching intensity of existing all optical switching devices based on $\chi^{(2)} : \chi^{(2)}$ cascading and for construction intracavity nonlinear optical devices for mode- locking and pulse duration control.

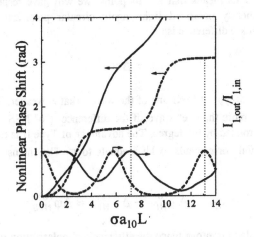

Figure 14. NPS and fundamental throughput as a function of input amplitude $\sigma a_{10}L$ for SHG Type I only (dashed line), $\Delta k_2 L = 0.3$ and simultaneous SHG and THG (solid line) $\Delta k_2 L = 4.8$, $\Delta k_3 L = 30$

$$(\Delta k_2 = k_2 - 2k_1 ; \Delta k_3 = k_3 - k_2 - k_1)$$

3. Polarization Rotation as a Result of Cascaded Processes

Polarized light entering the birefringent crystal splits into two waves: ordinary wave with polarization plane perpendicular to the main plane of the crystal and extraordinary wave with polarization plane parallel to the main plain of the crystal. If ordinary and extraordinary waves obtain different nonlinear shift than at the output of the crystal will be observed change of the polarization state of the input light. Since NPS is intensity dependent then the change of polarization state will be also intensity dependent.

    Intensity dependent change of polarization state can be realised by exploring $\chi^{(2)} : \chi^{(2)}$ cascading during Type I [34] or Type II [35,36] process of SHG (fig. 15).

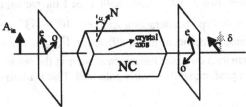

Figure 15. Main idea of polarization rotation due to quadratic cascading. NC - nonlinear crystal, N - normal to the plane formed by wavevector k and crystal axis, α - angle between input polarization and normal N, δ - angle of rotation of output polarization.

In the case of Type I SHG only the wave "o" collects NPS. If its value is π the polarization is rotated by 90 degree. By definition, for small ΔkL, we have full reconstruction of the amplitude of the fundamental wave for values for NPS equals to $\pi/2$, π, $3\pi/2$, ... This means that at this points we will have respectively, circular polarized light, linearly polarized light, again circular polarized light, ... The intensity induced phase difference is:

$$\Gamma_{NPS} = \Delta\varphi_{NPS}^{(o)} \ .$$

In the case of type II SHG one of the waves, that is weaker, for example, "o" wave, collects NPS more than "e" wave. If the difference ["o" NPS - "e" NPS] is π the polarization is rotated by 90 degree. The advantage of Type II is that the first point of full reconstruction corresponds to NPS equals to π. The intensity induced phase difference is:

$$\Gamma_{NPS} = \Delta\varphi_{NPS}^{(o)} - \Delta\varphi_{NPS}^{(e)} \approx \Delta\varphi_{NPS}^{(o)} \ .$$

It is possible to improve twice the efficiency of polarization rotation in case of Type I SHG by use of two crystal in a row (fig. 16). The axes of the two crystals are in perpendicular planes. Let us consider the input polarization as a sum of two linearly polarized wave with perpendicular polarization vectors. In the first crystal named A the wave "a" is ordinary wave and collects NPS. In the second crystal named B the wave "b" is ordinary wave and collects also NPS, but with opposite sign. This way the intensity induced phase difference between the two waves is doubled. Respectively, the intensity for 90 degree rotation is twice less [34].

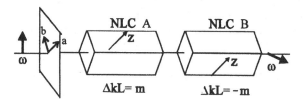

Figure 16. Two crystal scheme for polarization rotation.

Both schemes: two crystal scheme with Type I interaction and $\alpha = 45°$ and one crystal scheme with Type II interaction and $\alpha = 40° - 43°$ give similar results. Since nonlinear intensity dependent phase difference Γ_{NPS} is close to π at the first point of the full reconstruction of the fundamental wave at the output of the crystal, the output wave remains predominantly linearly polarized. This is illustrated in fig. 17.

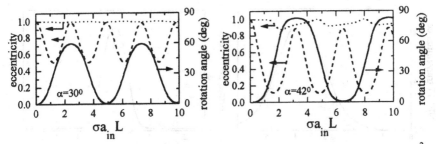

Figure 17. The eccentricity (dotted line), the rotation angle δ (solid line) and the square of the big axis, a^2, (dashed line) as a function of the normalized input intensity for two different input angles α. The normalized phase mismatch $\Delta kL = 0.3$.

4. Nonlinear Frequency Doubling Polarization Interferometer (NFDPI)

The effect of intensity induced polarization rotation can be used for construction of so called nonlinear frequency doubling polarization interferometer [34,36]. The scheme of the interferometer is shown in fig. 18.

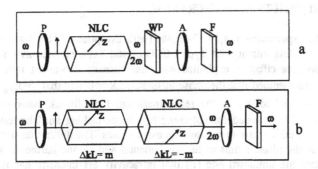

Figure 18. Nonlinear frequency doubling polarization interferometer: a) single crystal scheme (with Type II SHG); b) double crystal scheme (with Type I SHG).

It consists polarizer, frequency doubling crystal(s), phase corrector and analyzer. The phase corrector is used to compensate the different linear phase shift that the two waves obtain in the crystal. Depending on the mutual orientation of the polarizer and the analyser NFDPI exhibits self induced transparency or self induced darkening. The throughput of the NFDPI shown in fig. 18a is presented in fig.19. This device has also the capability to shorten input pulses. When the polarizer and the analyser are crossed strong shortening of the pulses can be obtained (fig. 20).

It is clear that the two capabilities of the NFDPI: pulse shortening and self induced transparency effect at relatively low power (we calculate 60 MW/cm^2 for 10 mm long KTP crystal) make this device suitable for mode locking of lasers with ring resonators. Another possible application may be all optical switching and sensor protection.

Experimentally NFDPI has been realised with bulk phase matched KTP [35] and with periodically poled LiNbO$_3$ [45].

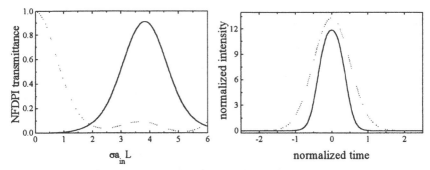

Figure 19. Fundamental throughput for parallel (dotted line) and crossed (solid line) polarizer and analyser for the single crystal NFDPI shown in fig. 18a. Input angle is $\alpha = 40°$. Normalized mismatch $\Delta kL = 0.3$.

Figure 20. Single pass shortening of Gaussian pulse with the NFDPI shown in fig. 18a. Dashed line - input pulse, solid line - transmitted pulse. Input angle is $\alpha = 40°$. Normalized mismatch $\Delta kL = 0.3$.

5. Role of $\chi^{(2)}$ Cascade Processes on the Performance of NFDM as a Mode-Locker

5.1. NFDM WITH TYPE I SHG CRYSTAL

In all of the experiments [14,15,17-20,46] (with exception of [17]) where nonlinear frequency doubling mirror with Type I SHG crystal has been used as a mode locker, the mode-locking effect is explained with the intensity dependent reflection of the NFDM. It was assumed that the phase mismatch $\Delta k = 0$ and that $\Delta\varphi_{add}$ (defined in part 2.3.) is equal to π. In the real experimental conditions, however, these two parameters had been tuned for obtaining best mode-locking and shortest generating pulses without taking care to measure its exact values. The difficulties arises from the facts (i) that the phase jumps of the fundamental and second harmonic waves at the back reflector are unknown and (ii) that intracavity SH intensity has its maximum when one achieves good mode locking and no care was taken to determine if this point corresponds to exact phase matching condition.

As we already discussed in part. 3 NFDM is an effective double pass phase shifter if the wavevector mismatch and the distance SH crystal - back reflector are properly chosen.

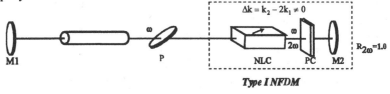

Type I NFDM

Figure 21. Typical set-up for mode - locking with nonphasematched NFDM with Type I SHG crystal

Here we investigate intensity reflection and NPS for typical experimental set up (fig. 21) for mode locking with NFDM with Type I SHG crystal. The process is considered to be with $\Delta kL = 3$ and $\Delta\varphi_{add}$ is taken to be π - the value that is recommended when NFDM is used as an amplitude mode locker [12].

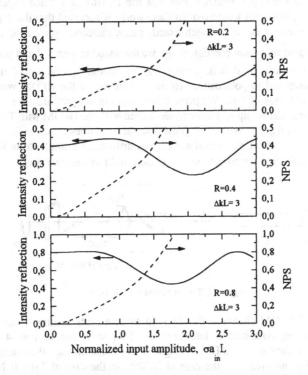

Figure 22. Intensity reflection (solid line) and nonlinear phase shift (dotted line) of the fundamental wave as a function of normalized input amplitude for three different values for reflection of the mirror M2.

The results of the computer simulations are shown in fig. 22 for three different values of the linear reflection coefficient of the mirror M2: 20, 40 and 80%. The graphs show that the effect of intensity dependent increase of the reflection of the NFDM is accompanied by strong intensity dependent phase modulation of the fundamental wave. This means that NFDM when used as intracavity laser element should be considered as a nonlinear optical device with combined amplitude and phase passive mode locking effect. From fig. 22 is clearly seen that the higher is the linear reflection of M2, the higher is the phase modulation efficiency for one and the same input intensities. The requirement of low linear reflection of M2 was one of the drawback for NFDM working at exact phase matching conditions since the small R_{M2} leads to very high threshold for the laser. The fact that NFDM can work with high reflecting back mirror makes this device suitable for CW pumped lasers [15,17]. The presence of intracavity diaphragm can lead to effective Kerr lens mode locking as it was observed in [17].

5.2. NFDM WITH TYPE II SHG CRYSTAL

In almost half of the experimental works that use NFDM as a mode locker Type II SHG crystals [13,46,48-50] were used. In these works is accepted that this type NFDM has the same behaviour as NFDM with exactly phase matched Type I SHG crystal i.e. intensity dependent reflection coefficient. As we discussed in part 2.2 and part 3 $\chi^{(2)}$ cascade processes in Type II SHG crystal lead to stronger phase modulation of the fundamental wave and to polarization rotation. Analysis of the NFDM with Type II SHG crystal is not done by now. We present here some of our preliminary results.

Experimental set up of a laser mode-locked with the NFDM with Type II SHG crystal is shown in fig. 23. We will suggest that the intensities of the ordinary "o" and extraordinary "e" waves in SH crystal are slightly different, i.e. angle α is 40-42°. M2 has maximum reflection for both second harmonic and fundamental wave.

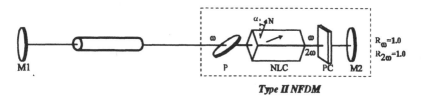

Figure 23. Type II NFDM as mode locker

Type II NFDM shown in fig. 23 is in some aspects similar to the nonlinear frequency doubling polarization interferometer [36] discussed in part 4 with this difference that in the case of NFDM the light passes twice through the crystal. Instead of self induced transparency in the case of NFDPI, in the case of Type II NFDM we should expect intensity dependent reflection coefficient. Due to the double pass geometry the intensity needed for maximum reflection will be less than intensity needed for maximum transmission of the scheme shown in fig. 18a. Additional advantage of the Type II NFDM is the possibility to change $\Delta\varphi_{add}$ by playing with mirror - crystal distance or inserting phase plates.

In fig. 24 are compared reflection coefficients of NFDM with a) Type II SHG crystal, $\Delta\varphi_{add} = \Delta\varphi_{in} - \Delta\varphi_{out} = 2\pi$, $\alpha = 40°$, $\Delta kL = 0.3$ and b) Type I SHG crystal at exact phase matched condition [12]. It is clearly seen that Type II NFDM exhibits much stronger increase of reflection with increase of fundamental intensity in comparison with the conventional "Stankov" nonlinear mirror [12]. The physical explanation of the intensity dependent reflection of Type II NFDM is different. With the help of the phase corrector PC and by changing mirror-crystal distance linear reflection of the NFDM can be made to take any value from 0 to 1. Due to intensity induced rotation of the polarization (part 3 or [36]) Type II NFDM has intensity induced change of the reflection coefficient. Type II NFDM can be either positive or negative feedback for the laser depending of the chosen linear reflection coefficient of the NFDM as a whole.

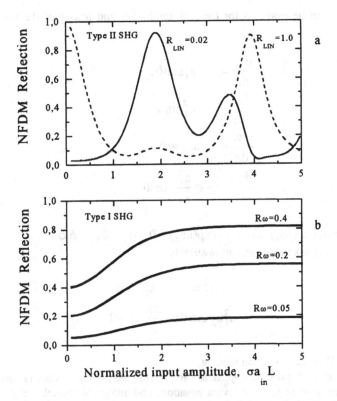

Figure 24. NFDM reflection as a function of normalized input amplitude for case of using Type II crystal for SHG (a) and Type I crystal for SHG (b) and different values of the reflection of mirror M2.

6. Approximate theoretical approaches for description of nonlinear optical devices based on $\chi^{(2)} : \chi^{(2)}$ cascading

In this part will be presented some approximate theoretical approaches used by us for investigation nonlinear optical devices based on $\chi^{(2)} : \chi^{(2)}$ cascading. Let us take as example Type I SHG process. Generalization for Type II SHG, sum frequency mixing and other second order processes is not difficult.

In the case of SHG Type I relevant equations are:

$$\frac{dA_1}{dz} = -i\sigma A_2 A_1^* \exp(-i\Delta kz) \, , \qquad (13.1)$$

$$\frac{dA_2}{dz} = -i\sigma A_1 A_1 \exp(i\Delta kz) \, . \qquad (13.2)$$

This system can be rewritten for the real amplitudes and phases of the interacting waves:

$$\frac{da_1}{dz} = \sigma a_2 a_1 \sin\Phi,$$ (14.1)

$$\frac{da_2}{dz} = -\sigma a_1^2 \sin\Phi,$$ (14.2)

$$\frac{d\varphi_1}{dz} = -\sigma a_2 \cos\Phi,$$ (14.3)

$$\frac{d\varphi_2}{dz} = -\sigma \frac{a_1^2}{a_2} \cos\Phi,$$ (14.4)

where $A_1 = a_1 \exp(i\varphi_1)$, $A_2 = a_2 \exp(i\varphi_2)$, $\Phi = \varphi_2 - 2\varphi_1 - \Delta kz$.
The system (14) has two important invariants

$$a_1^2 + a_2^2 = a_{10}^2 + a_{20}^2 = u^2,$$ (15)

$$\sigma a_1^2 a_2 \cos\Phi + \frac{\Delta k}{2} a_2^2 = \Gamma.$$ (16)

where $\Gamma = 0$ in the case of no seeding.

In general, possible approaches for solving exactly the system (2) are direct numerical integration of the relevant equations and analytical formulae expressed in Jacobi elliptic functions and integrals. Direct numerical integration is the most popular method for investigation of cascade processes. Big part of the results shown in this review are obtained this way. The use of analytical formulae expressed in Jacobi elliptic functions and integrals is rather difficult since the elliptic sinus and the elliptic integral of the third kind (used in the expressions for the NPS) have to be evaluated by complicated numerical calculations. There is a need for simple approximate approaches that allow to express and explain the phenomena based on $\chi^{(2)}$ cascading and to optimise nonlinear optical devices based on cascaded effects.

Following analytical approximations have been used by us:
a) fixed intensity approximation . Describes NPS only. Valid for low input intensities;
b) low depletion approximation (LDA). Describes fundamental depletion and NPS. Valid for $\sigma a_{in} < \frac{1}{2}\Delta k$;
c) high depletion approximation (HDA) - describes fundamental depletion and NPS at arbitrary input power intensities.

6.1. FIXED INTENSITY APPROXIMATION

Fixed intensity approximation was introduced for the first time for description NPS via Type I SHG [42]. Later fixed intensity approximation was extended for description of

Type II SHG and sum frequency generation [30,31] and for description of NFDM as mode locker [17]. In comparison with the well known fixed amplitude approximation [37] in which both real amplitude and phase of the fundamental wave are considered constant, in the case of fixed intensity approximation only the fundamental amplitude is constant but $\varphi_1 = \varphi_1(z)$.

For the case of $a_{20} = 0$ with this approximation we have:

$$\left(\frac{a_2}{a_{10}}\right)^2 = a_{10}^2 \left(\frac{\sigma}{Q}\right)^2 \sin^2(Qz), \qquad (17)$$

where $Q^2 = 2\sigma^2 a_{10}^2 + \Delta k^2 / 4$. Then defining $\cos(\Phi)$ from (16) and integrating (14.3) following expression for the NPS of the fundamental wave can be obtained

$$\varphi_1 = \varphi_{10} + \Delta kL \left(\frac{\sigma a_{10}}{2Q}\right)^2 \left[1 - \text{sinc}(2QL)\right]. \qquad (18)$$

This approach correctly describes disperse like shape for the Δk dependence of the NPS (fig. 4b) and also allows to define $n_{2,casc}$ (see (8)). However, does not describe the stepwise behaviour of NPS *versus* intensity dependence and does not describe fundamental depletion.

6.2. LOW DEPLETION APPROXIMATION - LDA

Expressing $\sin(\Phi)$ from (16) and substituting in (14.2) we obtain:

$$\frac{da_2^2}{dz} = -2\sigma u^3 \sqrt{\left(\frac{a_2}{u}\right)^6 - \frac{Q^2}{2\sigma^2 u^2}\left(\frac{a_2}{u}\right)^4 + \left(\frac{a_2}{u}\right)^2}. \qquad (19)$$

For low conversion, $(a_2 / u)^6$ term can be neglected and (19) can be integrated:

$$\left(\frac{a_2}{u}\right)^2 = 2M \sin^2(Qz); \qquad \left(\frac{a_1}{u}\right)^2 = 1 - 2M \sin^2(Qz), \qquad (20)$$

where $M = \frac{1}{2}(\sigma u/Q)^2$, $Q = \sqrt{2\sigma^2 u^2 + \Delta k^2 / 4}$.

Then in case of Type I SHG with no seeding

$$\Delta\varphi_{NPS} = -\frac{\Delta kL}{2} + \frac{\Delta k}{2}\frac{arctg\left[\sqrt{1-2M}\ tg(QL)\right]}{Q\sqrt{1-2M}} \qquad (21)$$

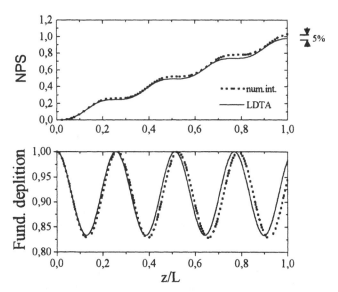

Figure 25. Comparison of exact numerical calculations (dotted line) and calculations with low depletion trigonometric approximation (solid line) for Type I SHG without seeding. $\Delta kL = 20$, $\sigma a_{10}L = 5$.

The range of validity of LDA was verified by comparison with exact numerical calculations (fig. 25). The depletion of the fundamental wave should not exceed 20%. In contrast to the fixed intensity approximation, LDA describes both the fundamental depletion and the stepwise behaviour of the NPS. LDA is suitable for description of mode locking devices based on cascaded second order processes. Similar formulae as (21) can be derived for the cases of presence of SH seeding and also for Type II SHG and sum frequency generation interactions.

6.3. HIGH DEPLETION APPROXIMATION - HDA

This approximation is valid practically for any fundamental depletion. The key idea is the replacement of the square of elliptic sinus in the expression for second harmonic intensity by:

$$sn^2\left(\frac{z}{\ae}\Big|m\right) \approx sin^2\left(\frac{\pi}{2}\frac{z}{l_{coh}}\right) + m^2F sin^2\left(\pi\frac{z}{l_{coh}}\right). \qquad (22)$$

The expressions for the intensity and the NPS of the fundamental waves are derived in [36] for the case of Type II SHG.

This approximation describes (with less than 10% deviation) the amplitude and the phase of interacting waves in quadratic processes as Type I SHG, Type II

SHG, sum frequency generation interactions and others. It explains both the disperse like shape of the NPS & ΔkL dependence (fig. 26.a) and stepwise behaviour of NPS & input amplitude (fig. 26.b) and NPS & distance dependencies.

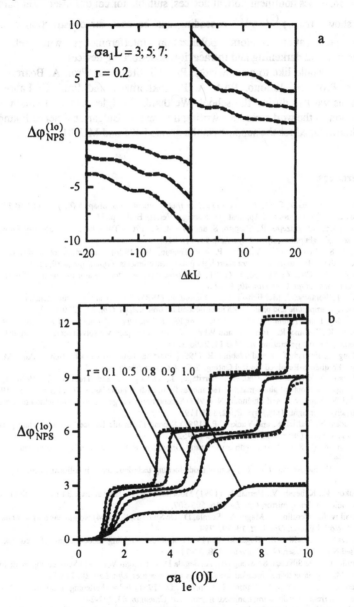

Figure 26. Comparison of exact numerical calculations (dotted line) and calculations with high depletion approximation (solid line) for Type II SHG without seeding.

110

7. Conclusion

In this article we have given a brief review of quadratic cascade processes and of based on these processes nonlinear optical devices, suitable for control laser light parameters. It was shown that $\chi^{(2)}:\chi^{(2)}$ cascading can be used for construction of intensity induced polarization rotators, polarization interferometers with self induced transparency and darkening and intracavity mode-locking devices.

We would like to acknowledge Prof. G. Stegeman, Prof. A. Boardman, Prof. M Fejer, Prof. G. Assanto, Prof. A. P. Sukhorukov and Prof. C. Fabre for the discussions we had during the school. We thank Ch. Iglev and P. Tzankov for their contribution in the works used for writing this review. Bulgarian Science Foundation is also acknowledged for the support via the grants F405 and MUF01.

8. References

1. Ippen, E. and Shank, C. (1977) Techniques for measurement, in S. Shapiro (ed.), Ultrashort Light Pulses-Picosecond Techniques and Applications, Springer-Verlag, Berlin, p. 83.
2. Giordmaine, J.,Rentzepis, P., Shapiro, S. and Wecht, K. (1967) Two-photon excitation of fluorescence by picosecond light pulses, *Appl. Phys. Lett.* **11**, 216-218.
3. Saltiel, S., Stankov, K., Yankov, P. and Telegin, V. (1986) Realization of a diffraction grating autocorrelator for single-shot ultra short light pulse measurements, *Appl. Phys. B* **40**, 25-27.
4. Kurobori, T., Cho, Y., Matsuo, Y. (1981) An intensity phase autocorrelator for the use of ultrashort pulse measurements, *Opt. Commun.* **40**, 156-60.
5. Diels, J., Fontaine, J, McMichael, I and Simoni, F. (1985) Control and measurement of ultrashort pulse shapes (in amplitude and phase) with femtosecond accuracy, *Appl. Opt.* **24**, 1270-1282.
6. Walmsley, I. and Trebino, R. (1996) Measuring Fast Pulses, *Optics & Photonics News*, No.3, 24-33.
7. Delong, K., Trebino, R., Hunter, J. and White, W. (1994) Frequency resolved optical gating with the use of second harmonic generation, *JOSA B* **11**, 2206-2215.
8. Delong K., Fittinghof, D. and Trebino, R. (1996) Practical Issues in Ultrashort-Laser-Pulse Measurement Using Frequency-Resolved Optical Gating, *IEEE J. Quant. Electr.* **32**, 1253-1264.
9. Sullivan, A., White, W., Chu, K., Herritage, J., Delong, K. and Trebino, R. (1996) Quantitative investigation of optical phase matched techniques for ultrashort pulse lasers, *JOSA B* **13**, 1965-1978.
10. Saltiel, S., Yankov, P. and Zheludev, N. (1987) Second harmonic generation as a method for polarizing and analysing laser light, *Appl.Phys. B*, **42**, 115-119.
11. Zheludev, N., Saltiel, S. and Yankov, P. (1987) Second Harmonic Devices - a new class polarizers and analysers, *Sov.J.Quantum Electron.* **17**, 948-952.
12. Stankov, K. (1988)A Mirror with an Intensity Dependent Reflection Coefficient, *Applied Physics B* **45**, 191-194.
13. Stankov K. and Jethwa, J. (1988) A new mode-locking technique using non-linear mirror, *Opt. Commun.* **66**, 41-44.
14. Stankov, K., Kubecek, V., Hamal, K. (1991) Mode locking of a laser at the 1.34 μm transition by a second harmonic non-linear mirror, *Opt. Lett.* **16**, 505-507.
15. Danailov, M., Cerullo, G., Magni, V., Segala, D., De Silvestri, S. (1994) Non-linear mirror mode locking of a CW Nd:YLF laser, *Opt. Lett.* **19**, 792-794.
16. Buchvarov, I., Saltiel, S., Gagarski, S. (1995) on-linear doubling mode-locking of feedback controlled pulsed Nd:YAG laser, *Opt.Commun.* **118**, 51-54.
17. Cerullo, G.,, De Silvestri, S., Monguzzi, A., Segala D. and Magni V. (1995) Self-starting mode locking of a CW Nd:YAG laser using cascaded second order nonlinearities, *Opt. Lett.* **20**, 746-748.
18. Buchvarov, I., Stankov, K., Saltiel, S., Georgiev, D. (1991) Pulse shortening in an actively mode locked laser with frequency doubling nonlinear mirror, *Opt. Commun.* **83**, 241-245.
19. Buchvarov, I., Saltiel, S., Stankov, K., Georgiev, D. (1991) Extremely long train of ultra short pulses from an actively mode locked pulsed Nd:YAG laser, *Opt. Commun.* **83**, 65-70.
20. Stankov, K. (1991) negative feedback uding a nonlinear mirror for the generation of a long train of short light pulses, *Appl. Phys. B* **52**, 158-162.

21. De Salvo, R., Hagan, D., Sheik Bahae, M., Stegeman, G., Van Stryland, E. (1992) Self-focusing and self-defocusing by cascaded second order effects in KTP, *Opt. Lett.* **17**, 28-30.

22. Stegeman, G., Sheik-Bahae, M., Van Stryland, E., Assanto, G. (1993) Large nonlinear phase shift in second-order nonlinear optical processes, *Opt. Lett.* **18**, 13-15.

23. Assanto, G., Stegeman, G., Sheik-Bahae, M., Van Stryland, E. (1995) Coherent interactions for all-optical signal processing via quadratic nonlinearities, *IEEE J.Quant.Electron*, **31**, 673-681.

24. Stegeman, G., Hagan, D. and Torner, L. (1996) $\chi^{(2)}$ cascading phenomena and their applications to all-optical signal processing, mode-locking, pulse compression and solitons, *J. Opt.&Quantum Electron.* **28**, 1691-1740.

25. Stegeman, G., Scheik, R., Torner, L., Torruellas, W., Baek, Y., Baboiu, D., Wang, Z., Van Stryland E., Hagan, D. and Assanto, G (199) Cascading: a promising approach to nonlinear optical phenomena, in *Novel Optical Materials and Applications* ed. by I-C. Khoo F. Simoni and C. Umeton (JOHN WILEY&SONS, INC), pp. 49-76.

26. Hutchings, D., Aitchison, J., Ironside, C. (1993) All-optical switching based on nondegenerate phase shifts from a cascaded second-order nonlinearity, *Opt.Lett.* **18**, 793-795.

27. Belostotsky, A., Leonov, A., Meleshko, A. (1994) Nonlinear phase change in type II second harmonic generation under exact phase-matched conditions, *Opt.Lett.* **19**, 856-858.

28. Assanto, G. and Torelli, I. (1995) Cascading effects in type II second harmonic generation: applications to all-optical processing, *Opt. Commun.* **119**, 143-148.

29. Assanto, G., Wang, Z., Hagan, D. and VanStryland, E. (1995) All-optical modulation via nonlinear cascading in type II second harmonic generation, *Appl. Phys. Lett.* **67**, 2120-2122.

30. Saltiel, S., Koynov, K., Buchvarov, I. (1995) Analytical and numerical investigation of opto-optical phase modulation based on coupled second order nonlinear processes, *Bulg. J. Phys.* **22** 39-47.

31. Saltiel, S., Koynov, K., Buchvarov, I. (1996), Analytical formulae for optimisation of the process of lower phase modulation in a quadratic nonlinear medium, *Appl. Phys. B* **62**, 39-42.

32. Hagan, J., Sheik-Bahae, M., Wang, Z., Stegeman, G., Van Stryland, E., Assanto, G. (1994) Phase controlled transistor action by cascading of second order nonlinearities *Opt. Lett.* **19**, 1305-1307.

33. Assanto, G., Torelli, I. and Trillo, S. (1994) All-optical processing by means of vectorial interactions in second-order cascading: novel approaches, *Opt. Lett.*, **19**, 1720-1722.

34. Saltiel, S., Koynov, K., Buchvarov, I. (1996) Self-induced transparency and self-induced darkening with nonlinear frequency doubling polarization interferometer, *Appl. Phys. B* **63**, 371-374.

35. Lefort, L. and Barthelemy, A. (1995) Intensity dependent polarization rotation associated with type II phase-matched second-harmonic generation: application to self-induced transparency, *Opt. Lett.* **20** 1749-1751.

36. Buchvarov, I., Saltiel, S., Iglev, Ch., Koynov, K. (1997) Intensity dependent change of the polarization state as a result of non-linear phase shift in type II frequency doubling crystals, *Opt. Commun.* **141**, 173-179.

37. Akhmanov, S. and Khokhlov, R. Problems of Nonlinear Optics, Moscow: Acad. Nauk SSR (1964), English ed. New York: Gordon&Breach.

38. Yablonovitch, E., Flytzanis, C. and Blombergen, N., Anisotropic (1972) Interference of Three Wave Frequency Mixing in GaAs, *Phys. Rev. Lett.* **29** 865-868.

39. Akhmanov, S., Meysner, L., Parinov, S., Saltiel, S., Tunkin, V. (1977) Third Order Nonlinear Optical Susceptibilities of Crystals; Signs and amplitudes of the Susceptibilities in Crystals with and without Inversion Center, *Sov. Phys. JETP* **46**, 898-907.

40. Pfister, O., Wells, J., Hollberg, L., Zink, L., Van Baak, D., Levenson, M., Basenberg, W. (1997) Continuous-wave frequency tripling and quadrupling by simultaneous three wave mixings in periodically poled crystals: application to a two-step 1.19-10.71 μm frequency bridge, *Opt. Lett.* **22**, 1211-1213.

41. Koynov, K., Saltiel, S. and Buchvarov, I. (1997) All-optical switching by means of interferometer with nonlinear frequency doubling mirrors", *JOSA B* **14**, 830-834.

42. Tagiev, Z., Chirkin, A. (1977) Fixed Intensity Approximation in Wave Theory, *ZETPh* **73**, 1271-1281, (in Russian).

43. Saltiel, S., Koynov, K., Tzankov, P., Boardman, A., Tanev, S. (1998) Nonlinear phase shift as a result of cascaded third-order processes, *Phys Rev A*

44. Koynov, K. and Saltiel, S. (1998) Nonlinear phase shift via multistep $\chi^{(2)}$ cascading, *Opt. Commun.*

45. Asobe, M., Yokohama, I., Itoh, H., Kaino, T. (1997) All-optical switching by use of cascading of phase-matched SFG and DFG processes in periodically poled LiNbO₃, *Opt. Lett.* **22**, 274-276.

46. Kobyakov, A., Lederer, F. (1996) Cascading of quadratic nonlinearities: an analytical study, *Phys. Rev. A*, **54**, 3455-3471.

47. Cerullo, G., Danailov, M., De Silvestri, S., Laporta, P., Monguzzi, A., Magni, V., Segala, D. and Taccheo, T. (1994) A diode pumped nonlinear mirror mode-locked Nd:YAG laser, *Appl. Phys. Lett.* **65**, 2392-2394.

48. Stankov, K. (1991) 25 ps pulses from a Nd:YAG laser lode-locked by a frequency doubling β-BaB$_2$O$_4$ crystal, *Appl. Phys. Lett.* **58**, 2203-2206.
49. Barr, J. and Hughes, D. (1989) Coupled cavity mode-locking of a ND:YAG laser using second-harmonic generation, *Appl. Phys. B* **49**, 323-325.
50. Buchvarov, I., Saltiel, S., Gagarskii, S. (1995) Nonlinear Doubling Mode-Locking of Feed-back Controlled Nd:YAG Laser, *Opt. Commun.* **118**, 51-54.
51. Wu, Q., Zhou, J., Huang, X, Li, Z., Li, Q. (1993) Mode-locking with linear and nonlinear phase shifts, *JOSA B.* **10**, 2080-2084.

ASYMMETRIC QUANTUM WELLS FOR SECOND-ORDER OPTICAL NONLINEARITIES

J. M. ARNOLD
University of Glasgow
Department of Electronics and Electrical Engineering
University of Glasgow, Glasgow G12 8LT, UK

1. Introduction

Second-order optical nonlinearities are increasingly important for a number of applications including second-harmonic generation, optical rectification, and self-phase modulation using the second-order cascade effect. Implementation of any second-order nonlinearity generally requires phase-matching, which may be quite difficult to achieve in a solid-state medium lacking linear birefringence. In such cases phase-matching must be engineered into the material in some way, either during the material growth stage, or post-growth in the form of a quasi-phase-matching grating. Semiconductor materials such as gallium arsenide are attractive for nonlinear optical applications because they have a fundamental bandgap in the region of the optical part of the electromagnetic spectrum, so that near-resonant excitation can be achieved, resulting in high nonlinear responses; in addition semiconductors support mature fabrication technologies which can be used to make devices in waveguide form, which can in principle be integrated with laser sources on the same substrate.

Unfortunately gallium arsenide and related semiconductors are linearly isotropic and therefore do not permit phase-matching by birefringence. In addition it is difficult to devise viable methods of producing phase-matched structures in bulk material during growth. Quantum wells, however, offer a post-growth technology for the modification of optical properties *via* the methods of intermixing (or disordering) which is quite widely used as a routine method of fabricating a variety of guided-wave optical devices [1]. In these lectures the basic principles of quasi-phase-matching using quantum-

113

A. D. Boardman et al. (eds.),
Advanced Photonics with Second-order Optically Nonlinear Processes, 113–132.
© 1999 *Kluwer Academic Publishers. Printed in the Netherlands.*

well intermximg (QWI) will be described, along with a summary of the theory needed to properly understand the nonlinear optical behaviour of semiconductor quantum wells.

2. Quantum wells for second-order nonlinearities

A quantum well consists essentially of a thin layer (~ 10 nm) of semiconductor, denoted as the *well*, enclosed between two media of higher bandgap energy than the well, known as *barriers*. It is possible in some material systems such as the gallium-aluminium arsenide III-V system, to have the bandgaps of the well and barrier materials aligned so that the bottom of the bulk conduction band and the top of the bulk valence band in the well both lie in the bandgaps of the barrier materials; this leads to the existence of a finite number of trapped electron and hole states in the well, which behave in some respects like large atomic states. A material consisting of many parallel quantum wells is called a *multiple quantum-well* (MQW) material, except in the case where the inter-well separations are periodic and small enough to permit the discrete well states to couple across the barrier regions by tunnelling, in which case we refer to a *superlattice*. The nonlinear optical tensors of these quantum-well materials can be calculated by adaptations of methods well-established for bulk semiconductors.

A key role is played in the determination of these tensors by the symmetry of both the underlying bulk semiconductors and the altered symmetries brought about by the quantum well. In particular, for the second-order tensors a degree of inversion asymmetry is required in order to produce non-vanishing tensors. In the bulk of a semiconductor consisting of two atom types in a cubic lattice arrangement this inversion asymmetry is brought about by the location of one of the atom types in the unit cell. A quantum well inherits the same structural inversion asymmetry from the bulk materials of which it is composed, but in addition if the geometry of the quantum well is itself inversion asymmetric then new tensor components can be produced that cannot exist in the bulk material.

Several distinct types of asymmetric quantum well configurations have been studied with a view to exploiting their second-order nonlinearities in different parts of the electromagnetic spectrum [2, 3, 4, 5, 6, 7, 8, 9, 10, 11, 12]. These are:

(i) *asymmetric stepped wells*, consisting of a well of two layers of different material, with different bandgap energies both lower than the bandgap energy of the barrier material;

(ii) *asymmetric coupled wells*, with pairs of wells of different width separated by a small inter-well separation sufficient to cause coupling between discrete states;

(iii) *electric-field-induced asymmetry*, in which a static electric field is applied across a conventional symmetric well in order to superimpose over the well potential an additional asymmetric potential on the electrons.

In addition there are two distinct ways of exploiting the nonlinear properties of asymmetric quantum wells in different wavelength regimes of the electromagnetic spectrum. These are:

(i) *interband transitions*, in which the photon energy $\hbar\omega$ is near resonance with an energy difference between discrete hole and electron subband states in the well material;

(ii) *intersubband transitions*, where the photon energy is near a resonance between two discrete subband states in the same bulk band.

Of these two interactions, the first is appropriate for optical and near-infrared excitations because semiconductors have bulk bandgap energies in the range $\sim 1 - 2$eV, corresponding to resonance with photon wavelengths in the optical range. The second is appropriate for mid- and far-infrared excitations because typical intersubband energies tend to be much smaller than the interband energies, of the order $\sim 10 - 100$ meV.

The studies of these two different wavelength regimes, corresponding to interband and intersubband transitions respectively, are quite different in character, at least from an engineering point of view. The interband transitions require optical excitation which can be provided by semiconductor laser sources which are potentially integrable with the nonlinear device itself, using technologies which are very mature and well-characterised; however, the available second-order nonlinear susceptibilities tend to be modest, of the order ~ 200 pm V^{-1} for bulk gallium arsenide, and generally much less for contributions purely from the quantum well. The intersubband transitions require long wavelength excitation, where semiconductor integration is not well-developed, and the use of an excited electron state as the 'ground state' of the transition scheme requires further device engineering to ensure the sustained population of this level. Predicted second-order susceptibilities for this kind of excitation are extremely large in comparison with interband transitions, of the order ~ 1 nm V^{-1}, and orders of magnitude larger still if multiple resonances can be arranged in a ladder of energy levels. However, the engineering of these devices is difficult, the resonances are very susceptible to absorption broadening, and by definition they cannot be used in the wavelength regime of optical fibre communication which forms a major focus for current investigations of nonlinear optics.

In these lectures we shall only be concerned with interband transitions for nonlinear interactions in the optical regime. The complementary case, of intersubband transitions, is covered in the lectures by Berger [16]

3. Nonlinear susceptibilities

Maxwell's equations in a dielectric lead to the equation

$$\nabla^2 \mathbf{E} - \nabla(\nabla \cdot \mathbf{E}) - \frac{1}{c^2}\partial_t^2 \mathbf{E} = \frac{1}{\epsilon_0 c^2}\partial_t^2 \mathbf{P} \tag{1}$$

for the electric field $\mathbf{E}$ driven by a polarisation $\mathbf{P}$ [17, 18, 19]. The polarisation density can generally be given in the form of an expansion in powers of the electric field $\mathbf{E}$ as [19]

$$\hat{P}_j(\omega) = \epsilon_0 \sum_{n=1}^{\infty} (2\pi)^{-n+1} \int \cdots \int_{\omega = \omega_1 + \cdots \omega_n} \sum_{k_1,\dots,k_n}$$

$$\chi_{jk_1\dots k_n}^{(n)}(\omega; \omega_1,\dots,\omega_n)\hat{E}_{k_1}(\omega_1)\cdots\hat{E}_{k_n}(\omega_n)\mathrm{d}\omega_2\cdots\mathrm{d}\omega_n \tag{2}$$

The quantum mechanical origins of this expansion will be discussed in Section 4. Here the Fourier transforms of the field quantities have been introduced, with the convention

$$f(t) = (2\pi)^{-1} \int_{-\infty}^{\infty} \hat{f}(\omega)e^{-i\omega t}\mathrm{d}\omega. \tag{3}$$

The reason for the factor $(2\pi)^{-n+1}$ in (2) is the position of the factor 2π in the Fourier transform convention (3), which differs to that used, for example, in the book of Butcher and Cotter [19]. Eq.(2) is essentially a statement that the Fourier component at frequency ω is a superposition of all ways in which n frequency components of the electric field can multiply together such that the sum of the frequencies $\sum_{s=1}^{n} \omega_s = \omega$, with coefficients $\chi_{jk_1\dots k_n}^{(n)}(\omega; \omega_1,\dots,\omega_n)$. These coefficients are known as *nonlinear susceptibilities*. The principal object of concern here is the second-order tensor $\chi_{jk_1 k_2}^{(2)}(\omega; \omega_1, \omega_2)$ obtained at $n = 2$.

Let us suppose that the fields are restricted to being periodic in time, and therefore expressible as a Fourier series

$$\mathbf{E} = \tfrac{1}{2} \sum_{m \in \mathbf{Z}} \mathbf{E}_m \exp(-im\Omega t) \tag{4}$$

$$\mathbf{P} = \tfrac{1}{2} \sum_{m \in \mathbf{Z}} \mathbf{P}_m \exp(-im\Omega t) \tag{5}$$

with coefficients that are dependent on z but not on t, having a fundamental frequency Ω. Taking the transforms, using the convention of (3)

$$\hat{\mathbf{E}} = \pi \sum_{m \in \mathbf{Z}} \mathbf{E}_m \delta(\omega - m\Omega) \tag{6}$$

$$\hat{\mathbf{P}} = \pi \sum_{m \in \mathbf{Z}} \mathbf{P}_m \delta(\omega - m\Omega) \tag{7}$$

and substituting in eq.(2), the result is

$$\hat{P}_{jm} = \epsilon_0 \sum_{n=1}^{\infty} \sum_{m=m_1+\cdots+m_n} \sum_{k_1,\ldots,k_n} \tag{8}$$

$$2^{-n+1} \chi^{(n)}_{jk_1\ldots k_n}(m\Omega; m_1\Omega, \ldots, m_n\Omega) \hat{E}_{k_1 m_1} \cdots \hat{E}_{k_n m_n} \tag{9}$$

This expression exhibits the phenomenon of *harmonic generation*. A major objective of the theoretical study of semiconductor nonlinear optics is the calculation of the susceptibilities from first principles. In the next sections the basic principles of carrying out such a calculation will be outlined. The steps are:

(i) calculation of the bandstructure of the bulk semiconductor using the Kane approach [20, 21];

(ii) calculation of the quantum well subband dispersion and wavefunctions using the envelope wave approximation [22];

(iii) calculation of the susceptibilities using the density-matrix expansion [18, 19]in 'A · p' representation for the electromagnetic interaction [24].

4. Theoretical calculations of bulk and quantum-well susceptibilities

4.1. BANDSTRUCTURE MODEL FOR GALLIUM ARSENIDE

The Hamiltonian for an electron moving in the periodic potential of a semiconductor lattice is

$$H = \frac{1}{2m}\Xi \cdot \Xi + V + \frac{\hbar}{4m^2c^2}(\sigma \times \nabla V)\cdot\Xi \tag{10}$$

where m is the reduced electron mass, $\Xi = -i\hbar\nabla$ is the momentum operator, V is the periodic potential seen by the electron, and σ is the electron spin operator. The first term represents the kinetic energy, the second term the potential energy due to Coulomb interaction with the lattice, and the third term the spin-orbit interaction. The Hamiltonian (10) is diagonalised by the Bloch eigenfunctions

$$\Psi_{ka}(x) = e^{ik\cdot x}\sum_{g}\tilde{\Psi}_{ka}(g)e^{ig\cdot x}$$

$$= e^{ik\cdot x}u_{ka}(x). \tag{11}$$

where g is a reciprocal lattice vector; the eigenfunctions are parameterised by two indices, the Bloch wavevector k and the band index a. The Bloch functions $u_{ka}(x)$ are periodic under translations which preserve the lattice $u_{ka}(x+a)$, where a is any vector in the direct lattice. The eigenfunctions satisfy

$$E_{ka}\Psi_{ka} = H\Psi_{ka} \tag{12}$$

where E_{ka} is the energy corresponding to the state indexed by $\{k, a\}$. The eigenfunctions are mutually orthogonal for different indices $\{k, a\}$ over the entire infinite crystal, under the orthogonality relation

$$\int_{\mathbf{R}^3} u_{ka}u_{k'b}e^{-i(k-k')\cdot x}d^3x = (2\pi)^3\delta(k-k')\delta_{ab}. \tag{13}$$

In addition the Bloch wavefunctions are orthogonal over a unit cell for $k = k'$

$$\int_{\text{unit cell}} u_{ka}u_{kb}d^3x = \delta_{ab}. \tag{14}$$

The solution of the eigenvalue problem for the Bloch waves produces eigenvalues which depend on both the band index a and the Bloch wavevector k; the graph of all bands of this relation with respect to the Bloch wavevector is called the *energy band diagram* of the solid. Stationary points (maxima and minima) of this relation are points of high symmetry in the Brillouin zone which reflect the underlying symmetry of the lattice. In gallium arsenide the fundamental bandgap is located between the maximum of the valence band and the minimum of the conduction band at the stationary point $k = 0$, which is therefore referred to as the *zone centre*.

Generally one is mostly interested in electronic states lying close to the zone centre, since they contribute most to the optical interactions, there being a high density of states in the vicinity of such symmetry points. This being the case, it is possible to approximate the energy dispersion relation for small deviations in k from the zone centre. The electronic states at the zone centre are strongly constrained by the symmetry present at this point in the Brillouin zone, and in fact some operators are almost completely determined by the symmetry except for one parameter. If the form (11) for the eigenfunction is substituted in the Schrödinger equation (12), then the Bloch wavefunctions are easily found to satisfy the equation

$$E_{ka}u_{ka} = \{\frac{1}{2m}(\Xi+\hbar k)\cdot(\Xi+\hbar k)+V+\frac{\hbar}{4m^2c^2}(\sigma \times \nabla V)\cdot(\Xi+\hbar k)\}u_{ka} \quad (15)$$

After expanding the off-centre Bloch functions in (15) as a superposition of zone-centre Bloch functions

$$u_{ka} = \sum_{a'} v_{kaa'}u_{0a'}, \quad (16)$$

multiplying throughout by the Bloch function u_{0a}, integrating over a unit cell and using the orthogonality (14) of the Bloch functions, there results the equivalent Schrödinger equation

$$E_{ka}v_{kaa'} = \sum_{b} H_{a'b}v_{kab} \quad (17)$$

for the coefficients $v_{kaa'}$ in the superposition (16), with the Hamiltonian

$$H_{ab}(k) = (E_{0a} + \frac{\hbar^2\|k\|^2}{2m})\delta_{ab} + \frac{\hbar}{m}(k \cdot \Xi_{ab}) + \frac{\hbar}{m}(k \cdot \Sigma_{ab}) \quad (18)$$

where

$$\Xi_{ab} = \int_{\text{unit cell}} u_{0a}^*(x)\Xi u_{0b}(x)d^3x \quad (19)$$

$$\Sigma_{ab} = \frac{\hbar}{4mc^2} \int_{\text{unit cell}} u_{0a}^*(x)(\sigma \times \nabla V)u_{0b}(x)d^3x. \quad (20)$$

The terms in the Hamiltonian which depend on k are treated as perturbations to the $k = 0$ Hamiltonian, and ordinary perturbation theory is then applied to obtain the eigenvalues and eigenfunctions. The leading-order parts of these perturbation expansions are

$$\begin{aligned} v_{kaa} &\sim 1 \\ v_{kaa'} &\sim \frac{\hbar}{m}\frac{k \cdot \Xi_{a'a}}{E_{0a} - E_{0a'}}, \quad a \neq a' \\ E_{ka} &\sim E_{0a} + \frac{\hbar^2\|k\|^2}{2m} + \frac{\hbar^2}{m^2}\sum_{a\neq b}\frac{\|k \cdot \Xi_{ab}\|^2}{E_{0a} - E_{0b}} \end{aligned} \quad (21)$$

Here we have neglected the spin-orbit contribution to the matrix elements in order to simplify the notation; the expressions can easily be corrected by the replacement $\Xi \to \Xi + \Sigma$. It is clear from (21) that the dominant contributions to the perturbation expansion arise from levels where the energy denominators in (21) are small.

The above discussion applies to a simple band structure where there is a single dominant band far removed in energy from all others. In practice several bands may have energies sufficiently close to cause significant coupling, or there may even be degeneracies such as occurs at the top of the valence band in GaAs, producing the light and heavy hole bands. In such cases the simple band theory above may be expected to be insufficient. It is then necessary to retain all the bands which are expected to be coupled in a single group, and allow for corrections of these bands due to coupling to the remote bands. These corrections were described by Luttinger and Kohn [25]; they produce modifications of the exact form of the Hamiltonian, but do not alter the basic structure of (18).

Because the optical interactions are dominated by terms in which the transition energy is close to the photon energy, it is permissible to neglect all bands except those whose transition frequencies lie close to the optical frequencies likely to be encountered, thereby selecting the most resonant interactions. In this way the number of bands included in the Hamiltonian can be greatly restricted, and it is possible to obtain satisfactory predictions of the optical susceptibilities by retaining no more than 1 conduction band, 2 valence bands which are degenerate at the zone centre and the spin-orbit band. Along with the spin multiplicity in each of these bands, this leads to an 8-band model for the approximate Kane theory. Solution of the eigenvalue problem (17) determines the approximate energy eigenvalues $E_{\mathbf{k}a}$ and coefficients $v_{\mathbf{k}aa'}$ in the perturbation expansion (16) applicable for small $\|\mathbf{k}\|$.

4.2. QUANTUM WELL BAND STRUCTURE

The next step is the determination of the electron states of a quantum well. A quantum well consists of a barrier material of a wide bandgap semiconductor on either side of a planar layer of semiconductor having a lower bandgap than the barrier. Thus it is necessary to obtain wavefunctions which are separately valid in each semiconductor region, to apply boundary conditions at each interface, and to determine the allowed states of electrons in the composite material. For this purpose the *envelope wave approximation* is used. The electron wave function ψ is represented in the form

$$\psi(\mathbf{x}) = \sum_a F_a(\mathbf{x})u_{0a}(\mathbf{x}). \tag{22}$$

Because the quantum well retains translation invariance in the directions parallel to the plane of the well, it follows that the dependence of the envelope function F on the in-plane coordinate must have exponential form; accordingly, F factors into

$$F_a(\mathbf{x}) = F_{\mathbf{k}_\| a}(z)e^{i\mathbf{k}_\| \cdot \mathbf{x}_\|}, \tag{23}$$

where we have introduced the decomposition of the spatial coordinate $\mathbf{x}$ into components parallel and perpendicular to the quantum-well plane by $\mathbf{x} = (\mathbf{x}_\|, z)$ and a similar decomposition for the Bloch wavevector $\mathbf{k} = (\mathbf{k}_\|, k_z)$. The envelope function $F_{\mathbf{k}_\| a}(z)$ now depends only on the normal coordinate z. The Hamiltonian for the electron in each separate region of the quantum well is taken to be the Hamiltonian appropriate to that layer considered as a bulk medium, with the semiconductor material parameters expressed as z-dependent functions, and the z-component k_z of the Bloch wavevector $\mathbf{k}$ replaced by the operator $-i\hbar\partial_z$. In this way the eigenvalue equation

$$E_{\mathbf{k}_\|}F_{\mathbf{k}_\| a} = \sum_b H_{ab}(\mathbf{k}_\|, -i\partial_z)F_{\mathbf{k}_\| b} \tag{24}$$

is a system of ordinary differential equations with respect to z for the envelope wavefunction F, in which the in-plane wavevector $\mathbf{k}_\|$ appears as a parameter. Boundary conditions must be specified for the connection of the envelope wavefunction across interfaces between different semiconductor media, as well as for the behaviour of the wavefunction at infinity in the direction normal to the plane of the well. The simplest boundary conditions at interfaces are that F and $\int_0^z H(z')F(z')dz'$ are continuous at interfaces [23], although other boundary conditions have been proposed and are generally considered to be more accurate [26]. The boundary conditions at infinity in the direction normal to the well are that the wavefunction be bounded; in addition, a certain discrete set of eigenfunctions can be found for which the eigenfunctions exhibit exponential decay to zero, corresponding to the bound states of the quantum well. The system of differential equations is solved numerically by replacing the z-derivatives by finite-differences, and solving the resulting discrete system of linear algebraic equations by a standard numerical package to determine the eigenvectors and eigenvalues. The eigenvalues are the energies of electron states in the quantum well, and the eigenvectors represent discretised approximations to the wavefunctions.

The dispersion relation for the dependence of the energy on the in-plane Bloch wavevector $\mathbf{k}_\|$ consists of a number of discrete states at each $\mathbf{k}_\|$, along with associated continua. The eigenvalues and eigenfunctions corresponding to these states must be labelled in some way. Here we introduce a further set of indices denoted by the label s, so that a quantum well eigenstate is denoted by energy $E_{\mathbf{k}_\| s}$, the corresponding eigenfunction by $\Psi_{\mathbf{k}_\| s}(\mathbf{x})$, and the coefficients of the wavefunction by $\tilde{F}_{\mathbf{k}_\| s a}$, so that

$$\Psi_{\mathbf{k}_\| s}(\mathbf{x}) = \sum_a \tilde{F}_{\mathbf{k}_\| s a}(z) e^{i\mathbf{k}_\| \cdot \mathbf{x}_\|} u_{0a}(\mathbf{x}) \tag{25}$$

The span of a single discrete level (fixed s) over all $\mathbf{k}_\|$ in the 2-dimensional Brillouin zone is known as a *subband*. These subbands exhibit strong non-parabolicity, strong mixing between bulk electron and hole wavefunctions, and pronounced anti-crossing features, all of which have a significant effect on the subsequent calculation of the nonlinear susceptibilities. In addition the spin degeneracy is broken by the asymmetry of the quantum well.

5. Density matrix calculation of nonlinear susceptibilities

5.1. ELECTROMAGNETIC INTERACTION

Electromagnetic interactions with the *bulk* semiconductor are given by the *minimal replacement* $\Xi \to \Xi + e\mathbf{A}$ of the momentum operator in the Hamiltonian (10), which induces the change $H \to H + (e/m)\Xi$. The additional perturbation to the Hamiltonian caused by the electromagnetic interaction is both spatially- and temporally-dependent. However, generally we are interested in the macroscopic variations of the field, which occur over much larger length scales then the microscopic variations of the electron wavefunction. As a first approximation it is reasonable to neglect the spatial variation of the potential, treating its dependence on position as a parameter which can be appended after the susceptibilities have been calculated for a position-independent potential. This has the added advantage that the extra interaction introduced to the Hamiltonian by the presence of a uniform electromagnetic field does not break the periodic symmetry of the lattice. Now the time-dependent Schrödinger equation

$$i\hbar \partial_t \psi = H\psi \tag{26}$$

must be solved with the minimal replacement Hamiltonian for H. This is carried out with the representation

$$\psi(\mathbf{x}, t) = (2\pi)^{-3} \sum_a \int_{BZ} \tilde{F}_{\mathbf{k}a}(t) \Psi_{\mathbf{k}a}(\mathbf{x}) d^3\mathbf{k}, \tag{27}$$

where the crystal eigenfunctions $\Psi_{ka}(x)$ are introduced as a basis, and BZ represents the Brillouin zone containing $k = 0$. Substituting (27) in (26), and using the orthogonality over the crystal of the eigenfunctions, the Schrödinger equation

$$i\hbar\partial_t \tilde{F}_{ka} = \sum_b H_{ab}\tilde{F}_{kb}(t) \tag{28}$$

is obtained, where H_{ab} are the matrix elements of the energy-representation Hamiltonian

$$H_{ab}(\mathbf{k}) = E_a(\mathbf{k})\delta_{ab} + \frac{e}{m}\mathbf{A}\cdot\Xi_{ab}(\mathbf{k}), \tag{29}$$

where a small change in the notation has been used to set $E_{ka} = E_a(\mathbf{k})$. Here the momentum interaction matrix is given by

$$\Xi_{ab}(\mathbf{k}) = \int_{\text{unit cell}} \Psi_{ka}^*(x)\Xi\Psi_{kb}(x)d^3x \tag{30}$$

The only essential difference for a quantum well from the bulk case is the replacement of $\mathbf{k} \in \text{BZ}$ by $\mathbf{k}_\| \in \text{BZ}_2$, where BZ_2 is the 2-dimensional Brillouin zone, and the replacement of the sum over bulk bands by a sum over the quantum-well eigenstates. Thus we have in place of (29) and (30)

$$H_{ss'}(\mathbf{k}_\|) = E_s(\mathbf{k}_\|)\delta_{ss'} + \frac{e}{m}\mathbf{A}\cdot\Xi_{ss'}(\mathbf{k}_\|), \tag{31}$$

and

$$\Xi_{ss'}(\mathbf{k}_\|) = \int_{-\infty}^{\infty} \int_{\text{unit cell}} \Psi_{\mathbf{k}_\| s}^*(x)\Xi\Psi_{\mathbf{k}_\| s'}(x)d^2x_\| dz, \tag{32}$$

where s and s' are the indices for quantum-well eigenstates.

5.2. NONLINEAR SUSCEPTIBILITIES

Once the electronic states of the semiconductor have been determined, the susceptibilities can be obtained, using the density matrix according to a method used by Shen [17] and Bloembergen [18] in the early development of the theory of nonlinear optics. The density matrix associated to any Hamiltonian operator H satisfies the Liouville equation

$$i\hbar\partial_t\rho = [H, \rho] \tag{33}$$

where the bracket [,] represents the commutator of two matrices. Using the density matrix, expectation values representing the values of observables X

are computed by $< X >= \text{Tr}\{\rho X\}$, where $\text{Tr}\{\ \}$ represents the trace of a matrix. The Hamiltonian H in (33) will be taken to be given by eq.(29) for the bulk case, and (31) for the quantum-well case.

The density matrix calculation proceeds by solving the differential equations (33) using Picard's method with the Hamiltonian (29) or (31). First we shall obtain the bulk susceptibility; the quantum-well case can be easily constructed from this by the replacements described at the end of the last section for the in-plane Bloch vector and the quantum state indices. The field-independent part of the Hamiltonian, H_0, is removed by means of a unitary transformation, which transforms the Hamiltonian to *interaction representation*; the unitary operator which achieves this transformation is given by

$$U = \exp((i\hbar)^{-1}H_0 t) \tag{34}$$

and the transformed Hamiltonian is

$$
\begin{aligned}
H' &= U^\dagger H U - U^\dagger \partial_t U \\
&= U^\dagger V U \\
&= \frac{e}{m}\mathbf{A} \cdot \Xi'
\end{aligned}
\tag{35}
$$

where V is the perturbing potential due to the electromagnetic field and $\Xi' = U^\dagger \Xi U$ is the unitarily transformed momentum operator in the interaction representation. The transformed momentum operator has matrix elements

$$\Xi'_{ab}(\mathbf{k}) = \Xi_{ab}(\mathbf{k})\exp(-i\Omega_{ba}(\mathbf{k})t), \tag{36}$$

where $\Omega_{ba}(\mathbf{k}) = (E_b(\mathbf{k}) - E_a(\mathbf{k}))/\hbar$. In this representation the Liouville equation (33) is

$$i\hbar\partial_t\rho' = [H',\rho'] \tag{37}$$

with $\rho' = U^\dagger \rho U$. The last equation is integrated, to give

$$\rho'(t) = \rho'(0) + (i\hbar)^{-1}\int_0^t [H'(t'),\rho'(t')]dt', \tag{38}$$

which can be solved in series by successive substitution to give

$$
\begin{aligned}
\rho'(t) &= \rho'(0) + (i\hbar)^{-1}\int_0^t [H'(t'),\rho'(0)]dt' \\
&\quad + (i\hbar)^{-2}\int_0^t\int_0^{t'} [H'(t''),[H'(t'),\rho'(0)]]dt''dt' + \cdots.
\end{aligned}
\tag{39}
$$

The important feature of this series is that its terms are in ascending powers of the field $\mathbf{A}$. The purpose of the density matrix calculation is to compute expectation values of quantum-mechanical operators, of which the momentum operator Ξ' is the most important here. This is obtained by the 'trace' prescription outlined earlier. After multiplying the density operator ρ' by the momentum operator Ξ', taking the trace of the resulting operator, and computing the integrals over the various time variables $(t', t'', \text{etc.})$, the equation (39) reduces to

$$
\begin{aligned}
< \Xi' > \; = \; & -\frac{e^2}{2(2\pi)^2 m^2 \hbar^2} \\
& \times \sum_{b,c} \sum_{k_1, k_2} \int_{\mathbf{R}^2} \int_{\mathrm{BZ}} \frac{\Xi_{j\,ab} \Xi_{k_1\,bc} \Xi_{k_2\,ca}}{(\Omega_{ba} - \omega_1 - \omega_2)(\Omega_{ca} - \omega_2)} \\
& \times \hat{A}_{k_1}(\omega_1) \hat{A}_{k_2}(\omega_2) e^{-i(\omega_1 + \omega_2)t} \mathrm{d}^3 k \mathrm{d}\omega_1 \mathrm{d}\omega_2 \\
& + \cdots
\end{aligned}
\tag{40}
$$

Computing the current density from the momentum we get

$$
\mathbf{J} \; = \; -\frac{e}{m}(< \Xi' > + e\mathbf{A})
\tag{41}
$$

from which the polarisation density follows as

$$
\mathbf{J} \; = \; \partial_t \mathbf{P}
\tag{42}
$$

The susceptibilities at various orders can be read directly from this equation by comparing it with equation (2), along with the frequency-domain relation $\hat{\mathbf{E}} = -i\omega \hat{\mathbf{A}}$ in the Coulomb gauge. The results for the second-order susceptibilities, which are the topic of concern in these lectures, are

$$
\begin{aligned}
\chi^{(2)}_{jk_1 k_2}(\omega; \omega_1, \omega_2) \; = \; & -\frac{e^3}{2\epsilon_0 m^3 \hbar^2} \frac{i}{\omega \omega_1 \omega_2} \\
& \times \sum_{b,c} \int_{\mathrm{BZ}} \frac{\Xi_{j\,ab} \Xi_{k_1\,bc} \Xi_{k_2\,ca}}{(\Omega_{ba} - \omega_1 - \omega_2)(\Omega_{ca} - \omega_2)} \mathrm{d}^3 k \\
& + \mathrm{perm}\{(j, -\omega), (k_1, \omega_1), (k_2, \omega_2)\}
\end{aligned}
\tag{43}
$$

for a bulk material, and

$$
\begin{aligned}
\chi^{(2)}_{jk_1 k_2}(\omega; \omega_1, \omega_2) \; = \; & -\frac{e^3}{2\epsilon_0 m^3 L\hbar^2} \frac{i}{\omega \omega_1 \omega_2} \\
& \times \sum_{s_1, s_2}' \int_{\mathrm{BZ}_2} \frac{\Xi_{j\,ss_1} \Xi_{k_1\,s_1 s_2} \Xi_{k_2\,s_2 s}}{(\Omega_{s_1 s} - \omega_1 - \omega_2)(\Omega_{s_2 s} - \omega_1)} \mathrm{d}^2 k_{\parallel} \\
& + \mathrm{perm}\{(j, -\omega), (k_1, \omega_1), (k_2, \omega_2)\}
\end{aligned}
\tag{44}
$$

for a quantum well. Here the notation 'perm S' means 'permute the objects in the set S and sum the results of all permutations', and represents the total permutation symmetry required by all nonlinear susceptibilities [19]. The ground state is assumed to be numbered 1, and $a = 1$ (or $s = 1$), so $\Omega_{ba} \geq 0$ ($\Omega_{s's} \geq 0$) for all levels b (s'). The parameter L is the inter-well separation of a multiple quantum well (MQW), assumed to be large enough that the bound states of the wells are not coupled, but still much smaller than the optical wavelength.

Although the expressions given above contain no relaxation effects in the model used to derive them, it is simple to include these effects phenomenologically in the usual way, by making the resonant frequencies Ω_{ba} complex.

6. Asymmetric quantum well susceptibilities

In our most recent calculations for asymmetric quantum wells [24] we have adopted the momentum interaction $(\mathbf{A} \cdot \mathbf{\Xi})$ consistently throughout. This differs from most other calculations, which have used a version of the electric dipole $(\mathbf{E} \cdot \mathbf{Q})$ interaction to specify the electromagnetic interaction. The use of the electric dipole interaction, which is well-suited to the calculation of nonlinear susceptibilities of ensembles of noninteracting atoms, appears to significantly overestimate the magnitudes of the second-order nonlinear coefficients of quantum wells, for reasons which have been recently explored in the literature [24].

The configurations proposed for asymmetric quantum wells applicable for second-order optical nonlinearities have been discussed in Section 2. The general design considerations for these structures have been to assure two well-separated subband energy levels in the conduction band of either (i) an asymmetric stepped well or (ii) an asymmetric coupled well pair. Sample calculations have been carried out on two such structures [24]; these are:

(i) *asymmetric stepped well*: 6 monolayer thickness GaAs and 25 monolayer thickness $Al_{0.4}Ga_{0.6}As$ with barriers of $Al_{0.6}Ga_{0.4}As$;

(ii) *asymmetric coupled wells*: 9 and 18 monolayer thickness GaAs wells separated by 5 monolayers of $Al_{0.4}Ga_{0.6}As$.

These structures are designed so that the second harmonic of a fundamental excitation at a wavelength of $1.55\,\mu m$ has photon energy just below resonance with the lowest energy interband transition between the valence and conductions subbands.

The nonlinear susceptibilities are computed from the density matrix formalism; the result for $\chi^{(2)}$ is essentially eq.(44). These calculations give values of second-order nonlinear susceptibilities at a wavelength of 1.55 μm for an asymmetric quantum-well which are quite small, of the order 1 pm V^{-1} compared with bulk values around 200 pm V^{-1}. The calculations of the second-order coefficients for both the bulk material and the quantum-well material are in agreement with values indicated by experiments. The largest symmetry-breaking second-order susceptibility that has been so far been observed in these calculations is $\chi^{(2)}_{131}$ for the asymmetric coupled wells, which is around 4 pmV^{-1} at a fundamental photon energy detuned by 0.1% from the resonance between the second-harmonic and the lowest subband zone-centre level.

The reasons for the small values of $\chi^{(2)}$ are several, partly because of internal cancellations in the overall summation for $\chi^{(2)}$, and partly because of the low density of discrete states in a quantum-well. Although individual resonances in the expression (44) give contributions to the second-order susceptibility comparable to that of the bulk susceptibility, there are always at least two contributing resonances from the two lowest subband states in the conduction band, and these can be shown to have opposite signs in the $\mathbf{A} \cdot \mathbf{\Xi}$ interaction scheme. Since these energy levels are separated only by the intersubband energy, which is much smaller than the interband energy, the two resonant terms tend to cancel when the excitation frequency is removed from either of the resonances, leaving a smaller residual contribution to the net value of second-order susceptibility. This effect was overlooked in earlier calculations of these effects [9], since only the principal resonant term was generally selected, and the $\mathbf{E} \cdot \mathbf{Q}$ electric dipole formalism used previously obscures the cancellation that takes place in the summations to evaluate the second-order coefficients if a truncated basis is used.

In addition, in the simplest case of a MQW structure where the wells are arranged with a period L, each well contributes one discrete electronic state per subband as described above, whereas the bulk material contributes one electronic state per lattice period, so the density of the electronic states is reduced by a factor proportional to a/L, where a is the unit cell size of the crystal lattice, further reducing the contribution from the discrete states of the well in comparison with the bulk nonlinear susceptibilities. The recipe for increasing $\chi^{(2)}$ significantly above these small values is to design the quantum wells to be narrow and deep to increase the sub-band energy separation in the conduction band, and to pack the wells closely together to minimise L; the limit of close packed quantum-wells is a superlattice, and if the well and barrier dimensions are a few monolayers then the asymmetric superlattice begins to resemble a new type of bulk material, and the values of $\chi^{(2)}$ should then become comparable with those of bulk gallium arsenide.

7. Phase-matching and quasi-phase matching

An additional feature of second-order interactions that we have so far not considered is that of phase-matching. Due to the frequency-dependence of the linear susceptibility $\chi^{(1)}$ the phase velocities of the two waves at the fundamental and the second harmonic may often be different. Expressed alternatively, the propagation coefficients at the two frequencies generally do not satisfy the relation $\kappa_2 = 2\kappa_1$, where $\kappa_m = \kappa(m\Omega)$ for $m = 1, 2$. At points along the propagation axis where the two waves are in-phase, the flow of electromagnetic energy is from the fundamental to the second harmonic; when the two waves are in antiphase, the energy flows in the opposite direction from the second harmonic to the fundamental. Consequently, when the two waves have different phase velocities the energy oscillates between the two waves as they pass from being in-phase to being in-antiphase, with a period given by $p = 2\pi/\Delta\kappa$, where $\Delta\kappa = 2\kappa_1 - \kappa_2$. This means, for instance, that second-harmonic generation from a pump wave at the fundamental does not grow indefinitely along the propagation direction unless the interaction is phase-matched, with $\Delta\kappa = 0$.

There are two basic methods of achieving phase matching in a solid-state crystal. In the first method the linear birefringence, if any, af the material is used. Since in a birefringent material the propagation coefficients are dependent on direction of propagation through the crystal, and direction of polarisation, it may be possible to find a direction of propagation and associated polarisations for the two waves so that the phase velocities are equalised. This is known as *birefringent phase matching*. This of course presupposes that the material is birefringent in the first place. Unfortunately, crystals with cubic symmetry such as gallium arsenide are linearly isotropic, and this method is therefore ruled out in this case by fundamental symmetry considerations. The second method of achieving phase matching involves the introduction of a spatially periodic variation in the second-order tensor, such that the period is equal to the period p of the phase mismatch defined above. This periodic modulation contains a Fourier component at the spatial frequency $\Delta\kappa/2\pi$, and this frequency component is changing its phase exactly in step with the change in phase differnce between the two waves; this method is known as *quasi-phase-matching*.

The implementation of the quasi-phase-matching method is clearly of critical importance for second-order interactions in semiconductors which have no linear birefringence. Fortunately there exists a mature technology for achieving this in quantum-well materials, known as *disordering* or *intermixing*. In this method, vacancies are induced in the semiconductor sample by out-diffusion of gallium atoms into a cap placed on the top surface of the sample; these vacancies are then encouraged to diffuse through the material by annealing. The availability of vacant lattice sites in the material also permits the diffusion of Ga and Al atoms to equalise the concentrations of these atoms throughout the sample, thereby destroying the differential concentrations that exist between the well and barrier regions. In this way the quantum well material is restored to bulk material. This process can be carried out in a spatially selective manner by appropriate masking, so that the quantum well is retained or removed at will at different spatial locations in the wafer, with a spatial resolution that is known to be better than 2 μm [27].

The effect of the disordering of the quantum well is to remove any contributions to the second-order tensor $\chi^{(2)}$ which are solely due to the quantum well. It can be shown from general symmetry considerations that the symmetry breaking caused by the introduction into a bulk semiconductor of the asymmetric quantum well induces new tensor components which cannot exist in the bulk material because of its symmetry restrictions. In cubic symmetry crystals such as gallium arsenide the only nonvanishing bulk tensor components of $\chi^{(2)}$ are those where all three indices are different, *i.e.* $\chi^{(2)}_{123}$ and its permutations [19]. (The notation here is for Cartesian components of the form $\mathbf{x} = (x_1, x_2, x_3)$.) The quantum well may have additional tensor components $\chi^{(2)}_{333}$, $\chi^{(2)}_{131}$ and $\chi^{(2)}_{113}$, as well as those obtained by replacing the index 1 by 2, where the coordinate x_3 is perpendicular to the well plane. These additional components are reduced to zero by completely disordering the well.

In addition to the tensor components induced by symmetry breaking, the quantum well also has a tensor component allowed by the bulk symmetry. Due to the additional effect of quantum confinement in the quantum well, this tensor component will be different for quantum well material than its value in the bulk material, and this difference also will be affected by the state of intermixing of the quantum well. Experimentally, it appears that this effect is much larger than any effect due to symmetry-breaking. It has been possible to observe a change in the bulk tensor coefficient $\chi^{(2)}_{123}$ of 17% due to disordering of a quantum well [27]; since the bulk $\chi^{(2)}_{123}$ is about 200 pmV^{-1}, this corresponds to a change of 34 pmV^{-1}, which is of the same order of magnitude as some other well-established nonlinear crystals such as lithium niobate. This effect is due to the asymmetry of the crystal structure of gallium arsenide and not to any engineered asymmetry in the geometry of the quantum well.

8. Conclusions

Quantum wells in gallium arsenide show two distinct effects on the second-order susceptibility of the material. First, the bulk tensor coefficient is significantly modified by the presence of a quantum well because the discrete subband energy levels are closer to resonance with the photon energy than the bulk bandedge of the barrier material; hence, intermixing the quantum well produces a large change in the value of the only nonzero bulk tensor coefficient $\chi_{123}^{(2)}$ as the material reverts from quantum-well to bulk. Second, asymmetric quantum wells break the symmetry of the bulk material, introducing new tensor coefficients that are not present in the bulk material due to its symmetry restrictions. When the quantum wells are intermixed these symmetry-breaking coefficients vanish altogether. The effects due to this second mechanism are, so far, smaller than the effects on the bulk susceptibility due to the bandedge shift in intermixing, ~ 1 pm V^{-1} for symmetry-breaking as compared with ~ 30 pm V^{-1} for bandgap-shifting.

Despite their smallness, the symmetry-breaking coefficients may still be extremely useful. They permit the nonlinear coupling of polarisations that cannot be accessed through the bulk coefficients, such as TM-TM *via* the coefficient $\chi_{333}^{(2)}$. Even small second-order coefficients can be usefully exploited as intracavity components in semiconductor lasers, where the recirculating power is very much higher than that coupled out to the external radiation field. Further research on materials growth and fabrication is likely to yield improvements in the symmetry-breaking coefficient values as the interwell spacing is reduced towards the superlattice limit. The magnitude of the bandedge-shifting effect, on the other hand, is already large enough to permit high-efficiency optical second-harmonic generation using quasi-phase-matching gratings produced by periodic intermixing of the quantum wells in a spatially selective manner. The ability to produce QPM gratings in this way is a key ingredient in the eventual integration of semiconductor laser sources with nonlinear components on the same semiconductor substrates.

References

1. Marsh, J. H. (1993) Quantum well intermixing, *Semiconductor Science and Tech-*

    nology, **8**, 1136-1155.

2. Khurgin, J. (1987) Second-order susceptibility of asymmetric coupled quantum-well structures, *Appl. Phys. Lett.*, **51**, 2100-2102.

3. Khurgin, J. (1988) Second-order nonlinear effects in asymmetric quantum well structures, *Phys. Rev. B*, **38**, 4056-4066.

4. Khurgin, J. (1989) Second-order intersubband nonlinear optical susceptibilities of asymmetric quantum-well structures, *J. Opt. Soc. Am. B*, **6**, 1673-1682.

5. Tsang, L. , Ahn, D. and Chuang, S. L (1988) Electric field control of optical second harmonic generation in a quantum well, *Appl. Phys. Lett.*, **52**, 697-699.

6. Tsang, L., Chuang, S. L. and Lee, S. M. (1990) Second-order nonlinear optical susceptibility of a quantum well with an applied electric field, *Phys. Rev. B*, **41**, 5942-5951 .

7. Harshman, P. J. and Wang, S. (1992) Asymmetric AlGaAs quantum wells for second harmonic generation and quasiphase matching of visible light in surface emitting waveguides, *Appl. Phys. Lett.*, **60**, 1277-1279.

8. Shimizu, A., Kuwata-Gonokami, M. and Sakaki, H. (1992) Enhanced second-order optical nonlinearity using interband and intraband transition in low-dimensional semiconductors, *Appl. Phys. Lett.*, **61**, 399-401.

9. Kelaidis, C., Hutchings, D. C. and Arnold, J. M. (1994) Asymmetric two-step GaAlAs quantum well for cascaded second-order processes, *IEEE J. Quant. Elec.*, **30**, 2998-3005.

10. Atanasov,R., Bassani, F. and Agranovich, V. M. (1994) Second-order nonlinear optical susceptibility of asymmetric quantum wells, *Phys. Rev. B*, **50**, 7809-7819.

11. Fiore, A., Rosencher, E., Vinter, B., Weill, D. and Berger, V. (1995) Second-order susceptibility of biased quantum-wells in the interband regime, *Phys. Rev. B*, **51**, 13192-13197.

12. Huang, Y. Wang,, C. and Lien, C. (1995) Electric-field enhancement and extinguishment of optical second harmonic generation in asymmetric coupled quantum-wells, *IEEE J. Quantum Electron.*, **31**, 1717-1725.

13. Janz, S., Chatenoud, F. and Normandin, R. (1994) Quasi-phase-matched second-harmonic generation from asymmetric coupled quantum-wells, *Optics Lett.*, **19**, 622-624.

14. Qu, X. H., Ruda, H., Janz, S. and Spring-Thorpe, A. J. Enhancement of second-harmonic generation at 1.06 μm using a quasi-ohase -matched AlGaAs/GaAs asymmetric quantum-well structure, *Appl. Phys. Lett.*, **65**, 3176-3178 (1994).

15. Fiore, A., Rosencher, E., Berger, V. and Nagle, J. (1995) Electric-field induced interband second harmonic generation in GaAs/AlGaAs quantum wells *Appl. Phys. Lett.*, **67**, 3765-3767.

16. Berger, V. (1997) Quantum engineering of optical nonlinearities NATO ASI *Advanced photonics with second-order optically nonlinear processes*, Sozopol, Bulgaria.

17. Shen, Y. R. (1984) *The principles of nonlinear optics*, Wiley Interscience, New York.

18. Bloembergen, N. (1965) *Nonlinear optics*, Benjamin, New York.

19. Butcher, P. and Cotter, D. (1990) *Elements of nonlinear optics*, Cambridge University Press.

20. Kane, E. O. (1955) Band structure of indium antimonide, *J. Chem. Solids*, **1**, 249-261.

21. Harrison, W. A. (1970) *Solid state theory*, McGraw-Hill.

22. Bastard, G. (1988) *Wave mechanics applied to semiconductor heterostructures*, Halsted Press.

23. Altarelli, M. (1986) Band structure, impurities and excitons in superlattices, *Proc. Les Houches Winterschool: Semiconductor superlattices and heterojunctions*, G. Allan, G. Bastard, N. Boccara, M. Lanoo and M. Voos eds., pp12-37, Springer, Berlin.

24. Hutchings D. C. and Arnold, J. M. (1997) Determination of second-order nonlinear coefficients in semiconductors using pseudospin equations for three-level systems,*Phys. Rev. B*, **56**, 4056-4067.

25. Luttinger, J. M. and Kohn, W.(1955) Motion of electrons and holes in perturbed periodic fields, *Phys. Rev.*, **97**, 869-883.

26. Burt, M. G. (1992) The justification for applying the effective mass approximation to microstructures, *Jour. Phys. Cond. Matter.*, **4**, 6651-6690.

27. Street, M. W., Whitbread, N. D., Hutchings, D. C., Arnold, J. M., Marsh, J. H., Aitchison, J. S., Kennedy, G. T. and Sibbett, W. (1997) Quantum well intermixing for the control of second-order nonlinear effects in AlGaAs multiple-quantum-well waveguides, *Optics Letters*, **22**,1600-1603.

EXPERIMENTS ON QUADRATIC SPATIAL SOLITONS

G.I. STEGEMAN, R. SCHIEK[†], R. FUERST, Y. BAEK[‡],
D. BABOIU, W. TORRUELLAS[‡], L. TORNER[*] AND B. LAWRENCE[•]
Center for Research and Education in Optics and Lasers (CREOL), University of Central Florida

† - Permanent address: Lehstruhl für Technische Electrophysik, Technische Universität, Arcisstrasse 21, D-800 Univ.
‡ - Dept. of Physics, Washington State University
* - Polytechnic University of Catalonia, Department of Signal Theory and Communications, POB 30002, 08080 Barcelona, Spain
•- MOEC, Rensselaer Waterfliet Facility, 877 25th Street, Watervliet, NY 12189

1. Introduction

Spatial solitons are beams which do not diffract on propagation in a material due to the presence of some optical nonlinearity. Their properties were first documented by John Scott Russell when he reported his observations on non-spreading water waves which consisted of a single "hump" propagating in a canal in Scotland.[1] In the very early days of nonlinear optics, interest was quickly evoked by what were then called "self-focused filaments", initiated by observations of self-focusing of powerful lasers in optical media, frequently leading to stable filaments or even material damage.[2,3] However it was not until the late 1990s that systematic experimental research into spatial solitons was initiated.[4] Since then there has been a surge of activity and many new solitons have been observed.[5-16]

Some physical mechanism for self-trapping or self-focusing of beams is required for spatial solitons. Traditionally this has been provided by third order ($\chi^{(3)}$) nonlinearities in the form of an intensity-dependent refractive index n_2 so that a local, intensity (I) induced, index change of the form $\Delta n = n_2 I$ provides the self-focusing required.[4-6] This is sufficient for spatial solitons in a slab waveguide, i.e. the 1D case. By 1D is meant that the beam can diffract in one dimension only. For bulk media, i.e. 2D, an additional contribution to the third order nonlinearity is required, either saturation of the index with increasing intensity or a negative fifth order susceptibility $\chi^{(5)}$.[15,16] A different, very recent approach has been to induce the required index changes via the photorefractive effect, a combination of photoinduced charge liberation from trap sites, charge migration and the electro-optic effect.[9,10]

In fact intensity-induced index changes are not needed for spatial solitons in optics. This was recognized first in the mid 1970s by Karamzin and Sukhorukov who predicted that self-

A. D. Boardman et al. (eds.),
Advanced Photonics with Second-order Optically Nonlinear Processes, 133–161.
© 1999 *Kluwer Academic Publishers. Printed in the Netherlands.*

trapping should also occur in second order nonlinear media during second harmonic generation.[17,18] One of the unique features of these *quadratic* solitons was that they consisted of two or three beams with different frequencies which are mutually self-trapped, and each beam is necessary for a soliton to exist. This prediction remained untested until the mid 1990s when quadratic spatial solitons were reported in both 1D and 2D geometries near phase-matched second harmonic generation.[13,14] In this chapter we will describe a series of experiments on quadratic solitons and their interactions. Questions regarding the similarity to, and difference from the better known n_2 based solitons will be addressed. Detailed theoretical discussions of their existence, stability and properties can be found in the literature and will be covered in other chapters in this book.[19-21]

2. Theoretical Overview

All of the properties of quadratic solitons can be obtained from the usual coupled wave equations for second harmonic generation when spatial diffraction is also included. Consider a slab waveguide (1D) with propagation along the z-axis in the y-z plane so that the guided wave confinement occurs along the x-axis. For the simplest 1D case of Type I phase-matching, the interacting fundamental and harmonic fields can be written respectively as:

$$\mathbf{E_1}(\mathbf{r},t) = \tfrac{1}{2}\, a_1(y,z)\exp[i(\omega t - k_1 z)] + cc$$

$$\mathbf{E_3}(\mathbf{r},t) = \tfrac{1}{2}\, a_3(y,z)\exp[i(\omega t - k_3 z)] + cc \tag{1}$$

where the subscripts 1 identify parameters at the fundamental frequency ω, and the subscript 3 refers to the second harmonic (2ω). (Later the subscript 2 will be used for the second, orthogonally polarized, input fundamental field in Type II phase-matching.) The complex field amplitudes $a_i(y,z)$ can also be written in a way to highlight the changes in phase and amplitude in the form $a_i(y,z) = |a_i(y,z)|\exp[-i\phi_i^{NL}(y,z)]$ where in the present context the modulus term describes changes in the field amplitudes. The origin of such changes can be due to spatial diffraction and/or the energy exchange due to the coupling between the fundamental and harmonic fields which occurs during harmonic generation. ϕ_i^{NL} is the "cascaded nonlinear phase shift".[22] The corresponding coupled mode equations are:

$$-2ik_1\frac{\partial}{\partial z}a_1(y,z) + \frac{\partial^2}{\partial y^2}a_1(y,z) = -\Gamma a_1^*(y,z)a_3(y,z)\exp[i\Delta kz]$$

$$-2ik_3\frac{\partial}{\partial z}a_3(y,z) + \frac{\partial^2}{\partial y^2}a_3(y,z) = -\Gamma a_1^2(y,z)\exp[-i\Delta kz] \tag{2}$$

Here $\Delta k = 2k_1 - k_3$ is the linear wavevector mismatch (and ΔkL the phase-mismatch) and Γ, the nonlinear coupling coefficient, is proportional to the effective second order susceptibility $\chi^{(2)}$ for the appropriate material symmetry class and field geometry. It includes the "overlap integral" of the fundamental and harmonic guided wave fields. Note that walk-off between the two beams is absent for propagation along the principal optical axes of a crystal, the case consider here. However, in general, walk-off does complicate the soliton forming process in many doubling crystals. It is important to note that these equations describe all of the soliton phenomena discussed in this chapter, with the exception of the evolution from input launching conditions which do not correspond to the stable steady-state solutions. Missing is the coupling to radiation fields which occurs with non-ideal launching conditions (of the type actually used in the experiments described here). As a result the total launched power does not all end up in the guided fundamental and harmonic modes propagating along the z-axis. In general, the further away the launching conditions are from the final soliton "modes", the more non-adiabatic is the field evolution. For launching of the fundamental only (at positive phase mismatch) at high input powers, an example of the evolution of the fields (at their peaks) is shown in Figure 1. Note the oscillatory behavior that dies off with distance as the fundamental and harmonic fields readjust their relative amplitudes and phases to that of a quadratic soliton.[23]

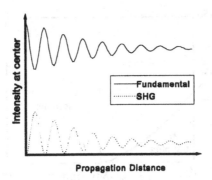

Figure 1 Simulated spatial evolution towards a stable quadratic soliton of the peak fundamental and harmonic beam intensity when only the fundamental is excited at the input.

Stationary solutions, i.e. ones in which the spatial profile and peak amplitude of the fundamental and harmonic components do not change with propagation distance, are obtained by setting the $\partial/\partial z$ terms to zero. Unfortunately there is only one specific value of the phase-mismatch ΔkL for which an analytical solution possible, and this predicts fields whose transverse profiles (along the y-axis) vary as $sech^2$.[17] They are of the form:

$$a_1(y) = \frac{3}{\sqrt{2}} sech^2(s) \qquad a_3(y) = 3sech^2(s), \tag{3}$$

and they exist for the normalized wavevector-mismatch given by $sgn(\Delta k)\ 2\pi\ell_{d1}/\ell_c = -2(\alpha+2)$, where ℓ_{d1} (ℓ_{d2}) and ℓ_c are the fundamental (second harmonic) diffraction length and the SHG coherence length respectively, and $\alpha = -\ell_{d1}/\ell_{d2} \cong -1/2$. The normalized transverse coordinate is $s = y[k_1/2l_{d1}]^{1/2}$. Note that the fundamental and harmonic field components are in phase for a stationary soliton, in contrast to the initial $\pi/2$ phase shift associated with launching only a fundamental beams for SHG. There is another limit in which the approximate field

136

distributions are known, i.e. when $\Delta k \rightarrow \infty$ the fundamental field closely resembles a Kerr soliton because the harmonic field contribution becomes vanishingly small, i.e. $a_1 \propto \text{sech}(s)$ and $a_2 \propto \text{sech}^2(s)$ with $|a_1|^2 \gg |a_2|^2$. [24] Otherwise, numerical techniques are used to estimate the soliton properties for other parameter values.

Some of the properties of 1D quadratic solitons are quite different from those of Kerr solitons. The region of existence of quadratic solitons is shown in Figure 2.[25] Stable 1D solitons exist for total peak intensities above a threshold value (which is zero for $\Delta k > 0$), i.e. there is a continuum of these solitons. Note that a vanishingly small peak intensity corresponds to very wide beams. Furthermore, the fraction of the power carried by the fundamental and harmonic along the threshold curve varies with detuning from phase-matching, see Figure 3.[25] Note that in the limit of large positive phase-mismatch, the harmonic component becomes negligible (but still must be finite). As mentioned above, these quadratic solitons asymptotically approach Kerr-law (pure electronic $\chi^{(3)}$) 1D spatial solitons. The key result is that there is a large variety of quadratic solitons which can be excited.

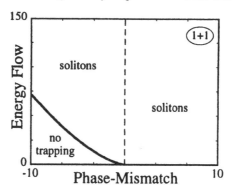

Figure 2 The range of total intensities over which 1D spatial solitons can exist for Type I phase-matching versus phase-mismatch $\Delta k L$.[25]

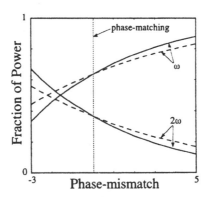

Figure 3 The fraction of power in the fundamental and harmonic waves of a 1D spatial soliton for Type I phase-matching versus the phase-mismatch.[25]

Although these equations (and the corresponding more complicated ones for Type II phase-matching and 2D) do describe the details of quadratic solitons, it is useful to discuss these solitons in terms of two simple physical mechanisms.

2.1 CASCADED PHASE SHIFT

In all of our experiments we launched just the fundamental fields and the second harmonic component required for the solitons was generated with distance into the sample. Thus the harmonic fields start of $\pi/2$ out of phase with respect to the fundamental. Away from phase-matching, the fields acquire an additional nonlinear phase-shift due to successive up and down conversion processes. The details have been recently reviewed and here we discuss only the pertinent results.[22] Neglecting for the moment spatial diffraction, assuming that the phase mismatch is large so that the power loss to the harmonic is very small, i.e. $|a_1(y,z)| = a_1(y,0)$, solving for the second harmonic in the second of equations 2 and substituting into the first gives:[22]

$$\frac{d}{dz}a_1(z) = -i\frac{\Gamma^2}{\Delta k}[(1 - \cos(\Delta kz)) - i\sin(\Delta kz)]|a_1(z)|^2 a_1(z) \qquad (4)$$

It is instructive to compare this equation with the equivalent one derived from the third order nonlinearity $\chi^{(3)}$ which leads to a local index change $\Delta n(I)$, i.e.

$$\Rightarrow \quad n_{2,eff}(z) = \frac{2\pi[d_{eff}^{(2)}]^2}{\varepsilon_0 cn^4 \lambda \Delta k}[1 - \cos(\Delta kz)] \qquad (5)$$

$$\Delta n = n_2 |a_1(z)|^2 \quad \Rightarrow \quad \frac{d}{dz}a_1(z) = -in_2 k_{vac}(\omega)|a_1(z)|^2 a_1(z) \qquad (6)$$

Note that the description via an intensity-dependent refractive index is not a good one because the resulting "n_2" depends on position, which is not physical. In fact, it is ϕ_1^{NL} that is the *key* parameter and its dependence on detuning from phase matching is shown in Figure 4. However, it is clear that for $\Delta k > 0$, an effective self-focusing mechanism is active and we know from Kerr solitons in 1D that this leads to stable solitons.

138

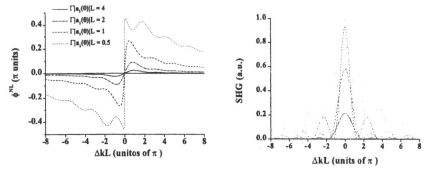

Figure 4 The cascaded nonlinear phase shift ϕ^{NL} (a) and second harmonic power (b) versus detuning from phase-match (ΔkL) for three different values of $\Gamma|a_1(0)|L$.

2.2 BEAM NARROWING VIA STRONG COUPLING BETWEEN THE FUNDAMENTAL AND HARMONIC

The physics of beam narrowing can be deduced from the SHG evolution equations 2. The first terms on the left-hand-side describe respectively the change in the complex amplitude of each beam
1. due to spatial diffraction (second left-hand-side terms with double derivatives) and
2. due to the source terms on the right-hand-side.
The key to self-trapping is the structure of the source terms which always consist of the product of two fields, both of which are of finite spatial extent.[22] Consider first the generation of the second harmonic which is driven by the term $a_1^2(y,z)$. The generated second harmonic is therefore initially narrower along the y-axis than the fundamental as shown in Figure 5. (Note that the beams are actually spatially superimposed: They are separated in Figure 5 for clarity.) The characteristic distance over which significant power transfer to the harmonic occurs is called the parametric gain length $L_{pg} = [\Gamma a_1(0,0)]^{-1}$. Some of the fundamental is regenerated via the product $a_1^*(y,z)a_2(y,z)$ and if, the second harmonic field is narrower than the fundamental, this regenerated fundamental field is also narrower than the fundamental field that was not up-converted to the harmonic. Therefore under appropriate conditions the spatial diffraction of the fundamental and the harmonic can be arrested. Both beams undergo a mutual focusing or lensing effect due to energy exchange. If this occurs over a distance comparable to or less than the diffraction length, i.e. $L_D > L_{PG}$, then the two effects can compensate resulting in a mutually locked soliton. This argument can be extended to the (detuned) phase-mismatched case, i.e. $\Delta k \neq 0$ by also including the cascaded phase shift.

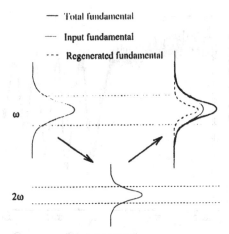

Figure 5 Schematic representation of beam narrowing for 1D Type I phase-matching with fundamental only excitation. The two beams are actually superimposed in space. The regenerated fundamental which has undergone both up- and down-conversion (dashed line) is narrower than both the input and diffracted (thin solid line) fundamentals so that the total fundamental propagates without spatial diffraction.

These two mechanisms, the cascaded phase shift and the beam narrowing can either reinforce, or interfere with one another. For example, for $\Delta kL > 0$ the two effects both lead to self-focusing effects. On the other hand, $\Delta kL < 0$ leads to self-defocusing so that the power threshold for soliton generation would be expected to be higher in this case.

3. Experiments on 1D Spatial Solitons in Lithium Niobate Planar Waveguides

The simplest type of quadratic soliton occurs in 1D for Type I phase-matching. These were investigated in y-cut Ti:in-diffused lithium niobate slab waveguides.[14] (The x,y,z-axes in the crystal reference frame correspond to the z,x,y-axes in our equations and solutions.) For a fundamental wavelength of 1319 nm, a $TM_0(\omega)$ guided mode (polarized along the crystal's y-axis) is phase-matched to a $TE_0(2\omega)$ harmonic guided mode (polarized along the crystal's z-axis) at about 335°C for propagation along the crystal x-axis (z-axis in our equations). This elevated temperature requires placing the sample inside a temperature-controlled oven. In order to couple into and out of the waveguide, 10x to 20x microscope objectives are used which have relatively short working distances. The result of these limitations is that the sample end-faces are very near to the windows of the oven and the resulting temperature is not uniform along the sample. Because the refractive index is temperature dependent, the wavevector mismatch for SHG is not uniform along the full length of the 5 cm long sample. This in turn complicates the variation in the nonlinear phase shift and fundamental depletion with propagation distance.[26] Typical results are shown in Figure 6 for the net change in ϕ_1^{NL} with temperature (as measured in the middle of the sample). Note that the slow monotonic increase in ϕ_1^{NL} with increasing temperature in the region of average $\Delta k > 0$ is commensurate with low fundamental depletion. This is the region in which the soliton experiments were performed.[14] Despite the complicated SHG evolution that occurs, quadratic solitons can still be generated although their detailed analysis requires that the temperature distribution be taken into account in detail.

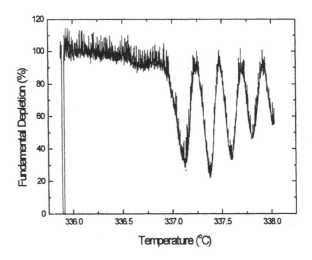

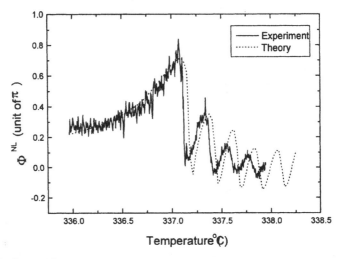

Figure 6 The fundamental throughput (upper) and nonlinear phase shift (lower) versus temperature as measured in the middle of the waveguide.[26]

3.1 EXPERIMENTAL DEMONSTRATION OF 1D QUADRATIC SOLITONS

In Figures 7 and 8 are shown the beam profiles obtained at the output end-face of the slab waveguides when the power or temperature (average phase-mismatch ΔkL) is varied.[14] For $\Delta kL \cong 10\pi$, the harmonic content is only of the order of 5% of the soliton power so that the oscillations associated with the evolution of the harmonic are a minimum, especially when the power and spatial profile launched are very close to the stable soliton parameters.[27] Then the input power was increased, the output beam narrowed until it reached the calculated soliton peak power of 1.2 KW, see Figure 7. Subsequent increase in input power led only to a small additional narrowing of the output beam, indicating that a soliton has indeed been generated. However, the experiments as described cannot rule out the possibility that these solitons could be due to the Kerr effect. This possibility was tested by changing the detuning from phase-match (i.e. average temperature by a few degrees) which changes the effective cascaded nonlinearities dramatically, but does not affect the third order nonlinearity significantly. Figure 8 shows that for a fixed input power, the beam self-focuses as the positive ΔkL is decreased (temperature increased) towards the phase-matching condition. This verified that indeed the beam narrowing and soliton formation are connected with second harmonic generation. Additional experiments were performed at different values of the phase-mismatch. For example, at $\Delta kL \cong 3\pi$, the harmonic component rises to about 40% and again quadratic solitons were observed.[28]

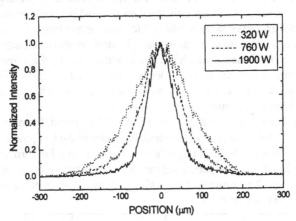

Figure 7 Power dependence of the output beam profile of the fundamental for $\Delta kL \cong 10\pi$.[14]

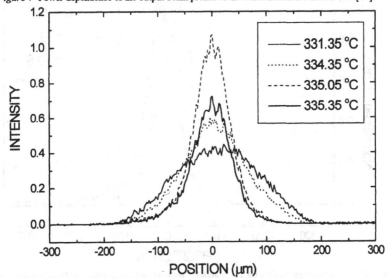

Figure 8 Temperature dependence of the output beam profile of the fundamental for $\Delta kL \cong 10\pi$.[14]

3.2 WALKING QUADRATIC SOLITONS

The strong coupling between the fundamental and harmonic fields not only leads to beam narrowing and soliton formation, but it can also change the propagation direction in space of one or both beam components.[29] In the slab waveguide experiments discussed above, the propagation wavevector and polarization directions of the fundamental and harmonic fields were essentially parallel to the principal axes of the crystal. As a result, for normal incidence onto the input end of the crystal, the fundamental and harmonics propagated in the same direction in the waveguide. When the fundamental beam is incident at an external angle α from the x-axis in the x-z plane, the fundamental remains polarized along the y-axis (o-ray) and the wavevector and Poynting vectors are collinear. But the harmonic, now polarized in the crystal's x-z plane, is now an extra-ordinary ray and the Poynting and wavevector directions are different. As a result, the fundamental and harmonic beams should separate within the crystal. However, when a soliton is formed, the two beams become locked together in space and travel in the same direction which is intermediate between the Poynting vectors of the uncoupled fundamental and harmonic beams. That is, the Poynting vector of the fundamental beam shifts towards that of the harmonic when the threshold for soliton formation is reached. These are called "walking" solitons because soliton locking causes them to "walk" away from their original propagation directions.[25]

The experimental results are shown in Figure 9 for three different incidence angles α.[30] In these graphs, the position at which the center of the low power harmonic beam (no soliton generated) exits the waveguide has been normalized to zero, i.e. the deflection due to the usual Snell's law is subtracted out. As a result the shifts shown are those solely due to the soliton locking process. The "walking" of the soliton's output position due to the locking of the two beams is clear.

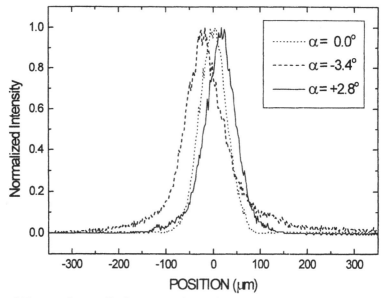

Figure 9 The output beam profiles for 1D quadratic "walking" solitons for three different incidence angles relative to the x-axis in the x-z plane. Note that the deflection due to Snell's law has been subtracted off.[30]

3.3 INTERACTIONS BETWEEN 1D QUADRATIC SOLITONS

It is well-known that closely spaced solitons in Kerr and saturable Kerr media interact with one another.[31-33] The same is true with quadratic solitons.[34-36] Two geometries were used to investigate this interaction.[28,37] In the first, the solitons were launched parallel to one another and for the second the solitons were launched so that their initial trajectories would cross in the middle of the waveguide. Experiments were performed both far from ($\Delta kL \cong 10\pi$) and near ($\Delta kL \cong 3\pi$) phase-matching for the parallel and the "crossing" launched fundamental beams.

First consider the ideal but hypothetical case in which stationary solitons are launched in parallel, i.e. they have the amplitude and phase for both the fundamental and harmonic appropriate to a stationary quadratic soliton. A simulation of the interaction is shown in Figure 10 for the case $\Delta kL \cong 40\pi$.[37] This case is near the Kerr limit of quadratic solitons so that the behavior expected should resemble that of Kerr solitons.[31] The response clearly depends on the relative phase angle ϕ between the two solitons at the input. For $\phi=0$, the solitons attract one another and then separate. This periodic motion continues, but with the separation becoming successively smaller until finally the two solitons fuse. The ultimate fusion of the solitons is a characteristic different from Kerr soliton behavior.[31] The smaller the detuning ΔkL and hence the larger the harmonic content, the shorter the distance required for the fusion to occur. For out of phase solitons, they repel each other. At other phase angles, the solitons weakly repel and the energy of one grows at the expense of the other.[38] Which one grows depends on the whether it leads or lags the second soliton in phase. Near phase-match, i.e. $\Delta kL \cong 3\pi$, the behavior closely resembles the previous case, see Figure 11. For the in-phase case, the two solitons fuse immediately without separating. Furthermore, the power exchange between the solitons occurs over much shorter distances. Also, much more radiation is emitted from the interaction than far from phase-match.

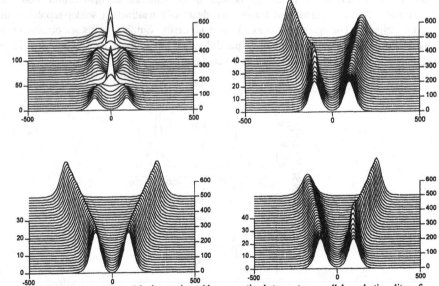

Figure 10 Numerical simulation of the interaction with propagation between two parallel quadratic solitons for four different relative phase angles and far from phase-matching at $\Delta kL \cong 40\pi$.[28,37]

144

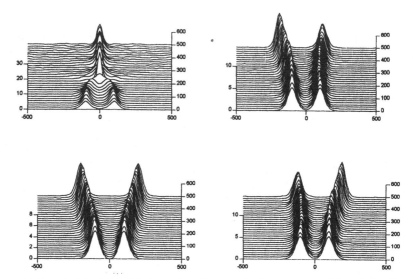

Figure 11 Numerical simulation of the interaction with propagation between two parallel quadratic solitons for four different relative phase angles and near phase-matching at $\Delta kL \cong 3\pi$.[28]

Although the available propagation distances in our samples are only of the order of three diffraction lengths, the experiments do mirror closely the behavior expected theoretically.[34-6] Again only the fundamental beam is launched and we rely on the first part of the waveguides to generated the required harmonic for soliton locking. The results are shown in Figures 12 and 13 for parallel launching ($\Delta kL \cong 10\pi$), along with detailed theoretical outputs which take into account the details of the temperature distribution along the waveguides. All of the three basic features are clear: soliton attraction, soliton repulsion, and power exchange. The agreement is excellent in all cases. Similar results are obtained near phase-match ($\Delta kL \cong 3\pi$) with the principal difference being the larger amount of radiation generated, indicating less adiabaticity in the collision.

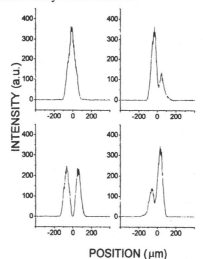

POSITION (μm)

Figure 12 Beam profiles measured at the output for two fundamental beams launched in parallel and relative phase angles (clockwise) of 0, $\pi/2$, π and 3 $\pi/2$. The detuning from phase-match is $\cong 10\pi$ and the quadratic soliton is generated during propagation into the sample.[28,37]

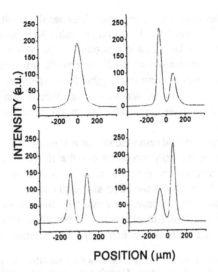

Figure 13 Calculated beam profiles at the output for two fundamental beams launched in parallel and relative phase angles (clockwise) of 0, $\pi/2$, π and $3\pi/2$. The detuning from phase-match is $\cong 10\pi$ and the quadratic soliton is generated during propagation into the sample.[28,37]

When the fundamental beams are launched at a crossing angle, for small crossing angles the response is the same as for the parallel input beams case. However, for larger enough angles of incidence, the solitons pass through each other, independent of the relative phase angle at launch. This is shown for two cases in Figure 14.[28,37] There is however a small lateral deflection in each case.

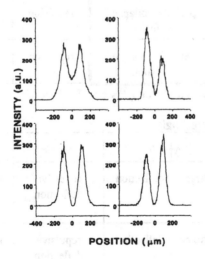

Figure 14 Measured output beam profiles for the two soliton interaction in the cross launching case for large net phase-mismatch. The relative phase difference between the two beams is (a) 0, (b) $\pi/2$, (c) π and (d) $3\pi/2$.[28,37]

It is instructive to summarize the comparison between the results obtained for quadratic 1D solitons with those for solitons in Kerr and saturable Kerr media.[31-36,37] This is shown in Table I below. The key point is that only for the near phase-matching case can quadratic solitons produce interactions which differ from the pure Kerr case. In all cases the interactions resemble those which occur in saturable third order media. These similarities in many ways are very reassuring in the sense that even though quadratic solitons are very different in composition from those in saturable media, yet the collision phenomena are very similar.

Why quadratic solitons should resemble saturable media can be argued as follows. First of all, the total electromagnetic energy is conserved so that if one beam increases power, it can only do so only at the expense of the other beam. This, in turn, weakens the effective self-trapping for at least one beam unless the beam size also decreases. However, this leads to stronger diffraction which again increases the beam size. Because of this multi-beam nature of the solitons, this self-regulating mechanism will not allow catastrophic self-focusing, independent of dimensionality, effectively behaving like a saturable nonlinearity.

TABLE 1. Comparison of the results of collisions between Kerr, saturable Kerr and quadratic solitons for different relative phases between the input solitons. The only case for which the results are different from the Kerr case is outlined in bold line.

Parallel Launching

Material	$\Delta\phi = 0$	$\Delta\phi = \pi$	$\Delta\phi = \pi/2$
$\chi^{(2)}$ ΔkL small ΔkL large	Periodic collapse $\Rightarrow$ fusion fusion	repulsion " " "	repulsion + power exchange " " "
Kerr $\chi^{(3)}$	periodic collapse no fusion	repulsion	repulsion + power exchange
Saturable $\chi^{(3)}$	Periodic collapse $\Rightarrow$ fusion	repulsion	repulsion + power exchange

Launch at Small Crossing Angle for $\chi^{(2)} \equiv$ Parallel Launch

Launch at Large Crossing Angle

Material	$\Delta\phi = 0$	$\Delta\phi = \pi$	$\Delta\phi = \pi/2$
$\chi^{(2)}$ or saturable $\chi^{(3)}$	Attractive deflection	repulsive deflection	repulsive deflection + power exchange
Kerr $\chi^{(3)}$	Attractive deflection	repulsive deflection	repulsive deflection + power exchange

4. 2D Quadratic Solitons

In contrast to pure Kerr media, but in agreement with what occurs in saturable media, quadratic solitons can exist as stable beams in bulk $\chi^{(2)}$-active media near phase-matching.[2,9,15,39,40]

4.1 2D FAMILY OF TYPE II SOLITONS

A series of experiments have been performed on the crystal KTP cut for Type II second harmonic generation with a 1064 nm input beam in the form of 30 psec pulses.[13] As shown in Figure 15 there are two orthogonally polarized fundamental beams, one of which is an o-ray and the other is an e-ray that propagates at an angle relative to the o-ray. The harmonic is also an o-ray so that at low powers all three beams walk away from each other on propagation. The diameter of the input beams was 20 μm, and they were focused onto the input face of the crystal. The beam output end-face of the crystal was imaged onto a camera.

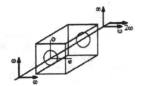

Figure 15 The crystal geometry of KTP for 2D, Type II phase-matching quadratic solitons.[13]

The equations describing the harmonic generation process in the 2D, Type II case are more complicated than for the previously discussed 1D, Type I case. Here the extra fundamental quantities are identified with the subscript 2 and

$$\frac{\partial a_1}{\partial z} - \frac{1}{2ik_1}[\frac{\partial^2 a_1}{\partial x^2} + \frac{\partial^2 a_1}{\partial y^2}] = -i\Gamma a_2^* a_3 \exp(i\Delta kz)$$

$$\frac{\partial a_2}{\partial z} + \rho_\omega \frac{\partial a_2}{\partial x} - \frac{1}{2ik_2}[\frac{\partial^2 a_2}{\partial x^2} + \frac{\partial^2 a_2}{\partial y^2}] = -i\Gamma a_1^* a_3 \exp(i\Delta kz) \qquad (7)$$

$$\frac{\partial a_3}{\partial z} + \rho_{2\omega} \frac{\partial a_3}{\partial x} - \frac{1}{2ik_3}[\frac{\partial^2 a_3}{\partial x^2} + \frac{\partial^2 a_3}{\partial y^2}] = -2i\Gamma a_1 a_2 \exp(-i\Delta kz)$$

where $\Delta k = k_1 + k_2 - k_3$ and ρ is the Poynting vector walk-off angle from the z-axis. For the Type II KTP used, $d_{eff}^{(2)} = 2.7$ pm/V, $\rho_\omega = 0.0033$ rad. and $\rho_{2\omega} = 0.0049$ rad.[13] Note that the existence of the walk-off angles implies that if solitons can be generated they will be "walking solitons".[29]

The results of the first experiment on KTP are shown in Figure 16.[13] For an input beam width of 20μm, the width of the fundamental at the output is ~90μm for low input intensity. This represents the effects of linear diffraction. At high intensities, the beam width stabilizes at about 12μm, smaller than the input width. The harmonic beam width is about 20% smaller. In the inset is the dependence of beam width on input intensity for two specific detunings from phase-matching. One can see the influence of the cascaded phase shift, i.e. for positive phase shifts (detuning) which produce self-focusing, the threshold for soliton locking is reduced and for ΔkL < 0, the self-defocusing phase shift leads to an increased threshold energy for soliton locking.

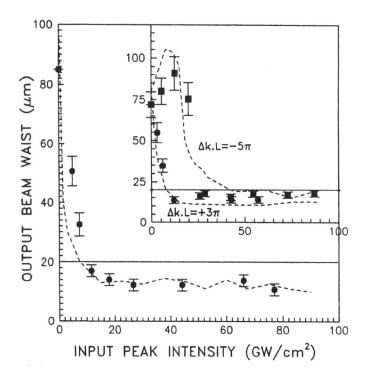

Figure 16 The output beam width of the fundamental (1064 nm) component of a 2D quadratic soliton versus peak fundamental input intensity on phase-match, and for detunings of ΔkL = 3π and -5π in the inset. The input fundamental beam width was 20μm. [13]

The measured threshold pulse energy for soliton locking versus phase-mismatch is shown in Figure 17. Again, the cascaded phase shift leads to a smaller threshold for ΔkL >0. The 2D soliton "existence" (for Type I, not the Type II KTP case) is shown for comparison in Figure 18.[41] There is an apparent discrepancy between the observed and theoretical dependence for positive detuning, i.e. the experimental result is essentially independent of detuning for ΔkL>0 where-as the existence curve minimum power rises with increasing ΔkL. However, it is important to recall that the experiments are done with just a fundamental input where-as the calculations are based on launching a stable soliton consisting of all three waves right at the input. The proper interpretation is that the experimental threshold actually lies in the continuum of soliton states, above the theoretical existence curve.

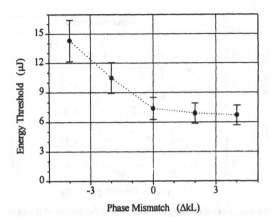

Figure 17 Threshold pulse energy (35 psec pulses) versus phase-mismatch at 1064 nm for quadratic soliton formation with a fundamental only input and $I_o = I_e$.[13]

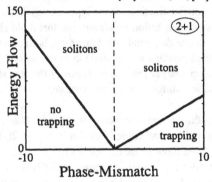

Figure 18 The range of total soliton powers over which 2D spatial solitons can exist for Type I phase-matching versus phase-mismatch.[41]

The above experiments involved excitation with equal powers along the ordinary and extraordinary axes, i.e. the two fundamental polarizations are equally excited. However, quadratic solitons should exist over a broad range of excitation conditions with respect to the ratio of powers in the two fundamental polarizations.[42,43] Although the lowest threshold energy for excitation corresponds to the 50:50 case, as shown in Figure 19, the threshold energy rises almost symmetrically on either side of this minimum.[44] It is interesting to note the absolute pulse energy required in the weaker eigenmode is approximate a constant, independent of the relative polarizations.

150

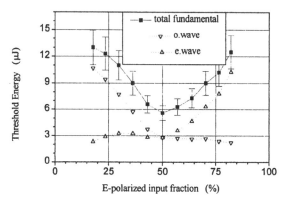

Figure 19 The threshold pulse energy for soliton formation versus the fraction of the input fundamental that was e-polarized.[44]

For progressively smaller fractions of the input energy in the weaker polarization, i.e. further away from the equal excitation condition, the threshold should increase from very simple considerations. For fundamental only launching, both polarization eigenmodes must supply an equal number of photons for conversion to the second harmonic. Therefore, the weaker eigenmode gives up a larger fraction of its power than the stronger eigenmode. This means that the threshold for soliton generation must increase with detuning from the equal excitation condition.

The composition of the generated solitons, i.e. what fraction of the pulse energy is in the two fundamental beams and the harmonic is shown in Figure 20 for three different phase-mismatches and for the fixed input pulse energy = $13\mu J$ which is above threshold for all the cases studied.[44] Consider first equal excitation of the two polarizations at the different phase mismatches. The fraction of second harmonic power decreases with increasing phase-mismatch, starting from 30% harmonic at $\Delta kL = -4\pi$ to only 10% at $\Delta kL = +4\pi$. However, as the polarization angle of the incident beam is detuned from $I_o = I_e$, the harmonic content decreases to a value as small as 5% for $\Delta kL = +4\pi$. This implies that the solitons should become more Kerr-like but to date there are no measurements available to test this question by, for example, soliton collisions. Since the fraction of the weaker fundamental eigenmode in the soliton also decreases with angular detuning of the polarization, the fraction of the power carried by the strong eigenmode can become very large. For example, for $\Delta kL = +4\pi$, the strong fundamental component carries more than 65% of the total soliton energy. This means that a strong "pump" beam of one polarization can be used with a "weak" signal beam of the orthogonal polarization to create a soliton. Additional experiments showed that the ratio of the energy in the two fundamental polarizations constituting the soliton is approximately a constant (to ±20%) for input pulse energies varying from 6 to $14\mu Js$ for $\Delta kL = +4\pi$ and $I_o/I_e = 0.36$.

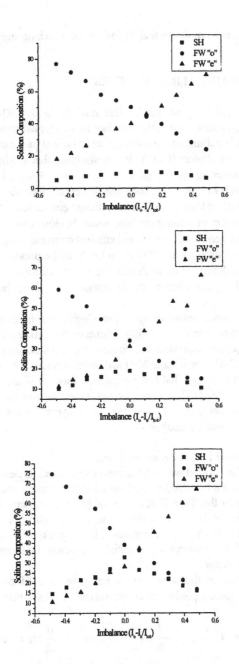

Figure 20 The fraction of the total quadratic soliton pulse energy in the two fundamental and harmonic components versus the polarization angle of the fundamental input from the e-beam axis for three different detunings (ΔkL) from phase-match.[44]

These experiments have shown that Type II phase-matching supports a very rich family of quadratic solitons.

4.2   SOLITONS AND BEAM INSTABILITIES

There are a number of well-known instabilities in nonlinear optics.[45] Here we describe experimental investigations of two of them as they relate to quadratic solitons.
(i)   The parametric instability is the cornerstone of optical parametric generators (OPG), amplifiers (OPA) and oscillators (OPO). It is well-known that a "higher" frequency photon breaks up into two "lower" frequency photons, i.e. $\omega_3 = \omega_1 + \omega_2$, subject to the wavevector matching condition $k_3 = k_1 + k_2$. This last condition determines which set of frequencies (if any) are phase-matched and hence can lead to efficient conversion. A "seed" beam at ω_1 or ω_2 can trigger this process, or it can grow from noise. In either case, exponential gain occurs, at least initially, for the generated photons and efficient conversion can occur. When $\omega_1 = \omega_2$, this is the "degenerate" case for an OPG or an OPA. In the present "degenerate" case, the injection of a very weak seed at ω_1 is shown to generate sufficient fundamental to create a soliton, which includes the ω_3 beam. In the absence of seeding, the ω_3 beam diffracts on propagation.
(ii)   It is also well-known in nonlinear optics that a high power beam incident into a $\chi^{(3)}$ (third order) nonlinear medium can break up into filaments, the process known as modulational instability.[2,45] In a pure Kerr medium, catastrophic self-focusing can occur and no stable beams are produced.[38] For a saturable third order medium, i.e. one in which there is a maximum to the index change that can be induced (not via damage), then stable spatial solitons can be generated.[15]   In the present case, an intense, spatially asymmetric, fundamental beam incident near an SHG phase-matching condition is shown to break-up into solitons via MI in a $\chi^{(2)}$-active medium.[46]

4.2.1 Quadratic Solitons Via a Parametric Instability

The same KTP doubling crystal was used in these experiments as described previously. First the 35 psec pulses from a Nd:YAG were doubled to 532 nm in a conventional doubler and then were injected into the Type II KTP crystal.[47] The pump (harmonic) beam was polarized along the e-axis. In this case, the harmonic beam diffracts as it passes through the crystal. When a seed at 1064 nm is also injected, the output beam at 532 nm collapses down to a diameter about 50% narrower than the initial input beam. Furthermore, strong conversion to the 1064 nm beam occurs.

The amplification of the seed can be easily understood from the coupled mode formalism. Consider the coupled mode equations right at phase-matching (simplest case)

$$\frac{d}{dz}a_1 = -\Gamma a_2^* a_3 \qquad \frac{d}{dz}a_2 = -i\Gamma a_1^* a_3 \qquad \frac{d}{dz}a_3 = -2i\Gamma a_1 a_2 \qquad (8)$$

Under seeding conditions, $a_1(0) \ll a_3(0)$ and $a_2(0) \ll a_3(0)$ and the depletion of the pump can be neglected. This leads to exponential growth of the seed beam, i.e.

$$a_1(z) = a_1(0)\exp[\Gamma \mid a_3(0) \mid^2 z] \quad a_2(z) = a_2(0)\exp[\Gamma \mid a_3(0) \mid^2 z] \tag{9}$$

As soon as the intensities in the various beams are above the soliton-locking threshold, a quadratic soliton is formed.

The threshold pulse energy of the pump beam required for the generation of a quadratic soliton is shown in Figure 21 for a fixed fundamental seed (20nJ) polarized along the o-axis. As seen in previous examples, the threshold is more or less constant for $\Delta kL > 0$ and rises for negative detuning. Note that only 3 μJ is needed in the pump for generating the solitons.

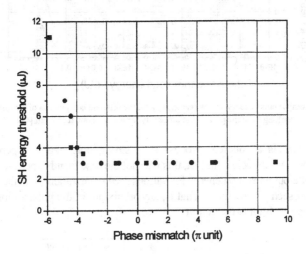

Figure 21 Threshold pulse energy at 532 nm for the parametric generation of a quadratic soliton with a weak (20 nJ) 1064 nm seed versus phase-mismatch (ΔkL).[47]

The variation in the output pulse energy at both 532 and 1064 nm is shown in Figure 22 as function of the seed pulse energy for a fixed, 4μJ, 532 nm input "pump" pulse.[47] These results show that a seed energy larger than a pJ will trigger soliton generation and lead to 0.25μJ outputs at 1064 nm. This corresponds to a net gain in excess of 10^5 (50 dbB) for the seed beam! Note also that for seed energies up to about 10 nJ, the output pulse energy in the 532 nm is "clamped" at 0.25μJ, i.e. this "device" behaves like a saturating amplifier with arbitrary gain. When the seed becomes 10% or more of the "pump", the output begins to increase with the input seed energy.

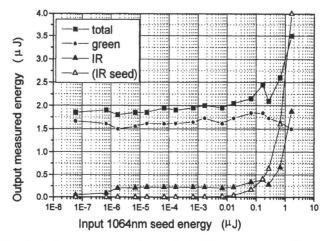

Figure 22 The output pulse energy versus seed energy for the different components of the quadratic soliton resulting from the parametric instability for a 4 µJ input at 532 nm.[47]

However, this is a coherent process so a strong polarization dependence of the fundamental would be expected. In fact, this was not the case found experimentally. Shown in Figure 23 is the output soliton energy for various seed polarizations, and the soliton output is independent of seed polarization. That is, any admixture of the polarization eigenmodes leads to the "same" soliton at the output.

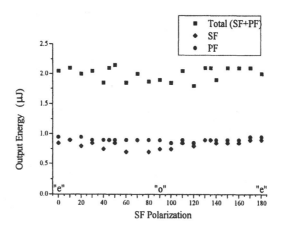

Figure 23 The dependence of the quadratic soliton energy versus the polarization of the 15 nJ 1064 nm seed beam. Here PW and SW refer to the 532 nm and 1064 nm components respectively.[47]

This initially unexpected result is a consequence of the input boundary conditions (or lack of them). The evolution of the quadratic solitons is governed by the Hamiltonian H where:

$$\mathcal{H} = -\frac{1}{2} \int \left[-\sum \frac{\omega_i}{\omega_1} |\nabla_\perp a_i|^2 + \sqrt{2}[a_1^* a_2^* a_3 + a_1 a_2 a_3^*] \right] dr_\perp \qquad (10)$$

For seeding initial conditions, only $a_3(0)$ is large and hence the $|\nabla_\perp^2 a_3|^2$ term dominates $\mathcal{H}$. Because the seed amplitude and phase-dependence only occur in the $a_1^* a_2^* a_3 + a_1 a_2 a_3^*$ terms, the evolution is essentially independent of the details of the seeding conditions, including the polarization.

4.2.2 Beam Break-Up Via Modulational Instability

Modulational instability is a well-known effect in third order nonlinear optics.[45] For example, when a high intensity plane wave has superimposed on it spatial noise or an externally applied spatial modulation, the beam can break up into filaments. A similar effect can occur in second order nonlinear optics.

The analysis is based on the well-known $\chi^{(3)}$ case and is relatively straightforward. For the second harmonic case, the "modulated" fundamental and harmonic fields are expanded around the stationary solutions to the plane wave problem to give the initial evolution due to MI as:[46,48]]

$$a_i(x,z) = a_i(0)[1 + \delta(\kappa)\cos(\kappa y)\exp(\gamma z)] \qquad (11)$$

where κ is the modulation wavevector, $\delta(\kappa)$ is the amplitude of the modulation, γ is the exponential gain coefficient and the $a_1(0)$ and $a_3(0)$ correspond to stationary solutions. For the plane wave case there are stationary solutions for the 1D case with the properties $k_1 = a_3(0)$ and $k_2 = a_1^2(0)/a_3(0)$.[49] Assuming that the perturbation is small, inserting back into the coupled mode equations, and retaining only the leading terms in $\delta(\kappa)$ gives a quadratic equation in γ^2. There are four solutions for the growth rate γ, two of which are real, equal in magnitude and with opposite signs, and two complex conjugated imaginary values. Of practical interest is the real positive solution, corresponding to an exponentially growing periodic perturbation. This root always exists so that for a range of κ there is always exponential growth, i.e. MI. A plot of γ^2 versus κ^2 is shown in Figure 24.[48] For a given intensity there is a maximum in the gain. For noise initiation of the MI, the value of κ will correspond closely to the peak of the gain curve. Note that this peak in γ moves towards larger κ as the intensity is increased, i.e. the pattern becomes "finer" in space.

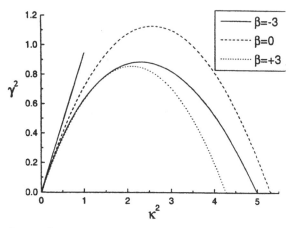

Figure 24 Plot of γ^2 versus κ^2 for 2D, Type II phase-matching for two different normalized detunings from phase-match.[48]

The case studied experimentally is for 2D, and no strictly analytical treatment is currently available for this case.[48] However, the basic features of the 1D case are expected to be same for 2D, i.e. a maximum in the gain for some value of κ. This leaves numerical Beam Propagation Method (BPM) investigations as the only route possible for investigating the properties of MI in this case. The spatial evolution with propagation distance over many diffraction lengths of a noisy 1D plane wave with a small periodic perturbation (amplitude relative to the plane wave amplitude of 10^{-3}) is shown in Figure 25. Well defined maxima are formed with distance and these eventually evolve into cylindrically symmetric beams with soliton-like properties. Note the radiation that is emitted when only the fundamental is launched.

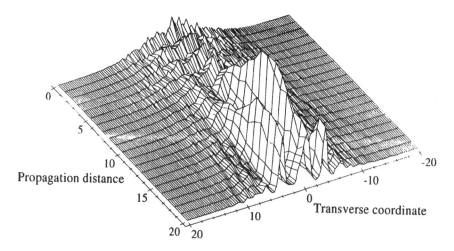

Figure 25 Evolution of a noisy, finite width, high intensity fundamental wave at Type I phase-matching for the 1D case.

In the experiments a noisy (i.e. real) elliptical beam was launched into a bulk KTP (2D) sample.[50] The ratio of the major to minor axes of the ellipse was varied from 1:1 to 12:1. The input beam, and the resultant pattern obtained at the output for various intensities is shown in Figure 26. As the intensity is increased, the spatial modulation frequency (κ) increases, in agreement with theory and progressively more maxima appear. At very high peak intensities (not shown), > 100 GW/cm^2, the pattern becomes two-dimensional. That is, a second row of maxima appears. This occurs when the spacing between the intensity maxima becomes of the order of the width of the maxima themselves, a necessary condition for the beam to break up in the second dimension.

Figure 26 The MI break up of an elliptical beam (upper trace) in Type II, 2D KTP into a series of solitons observed at the output for increasing intensity.[50]

A key question is whether these maxima are quadratic solitons. Shown in the table is a comparison between the properties of the quadratic solitons obtained with a cylindrically symmetric input beam, and the maxima obtained with an elliptical beam whose short dimension is the same as that of the cylindrical beams. It is clear that for all practical purposes the output is a series of quadratic solitons.

TABLE 2: The properties of the MI induced maxima compared to quadratic solitons.

	Circular Input	Elliptical Input
Input Beam	20µm	20µm x variable
Output Beams	12µm	9.5±1.5µm
Output Beam Shape	circular	circular to ± 8%

158

These results differ markedly from what would be expected for the 1D (waveguide) case. There MI should also occur for finite width (along y) beams. In both cases the maxima try to evolve into solitons by shedding radiation. In the waveguide case, this excess electromagnetic energy cannot leak out into the bounding media. It tends to be trapped between the local maxima and to slowly leak out along the y-axis. For elliptical beams in bulk media, the excess energy can be radiated out in the direction along the minor ellipse axis and hence stabilization into solitons should occur over smaller distances than the 1D case. This is the principal reason why in our experiments the properties of the maxima resemble strongly spatial solitons after only five diffraction lengths.

There is a range of input intensities for which only a single soliton emerges from the MI. This case corresponds to "beam clean-up" for asymmetrically shaped input beams.[51] In the present case, the beams were elliptical and the threshold intensity of forming a single cylindrically symmetric soliton was measured and is shown in Figure 27. Note that the larger the ellipticity, the smaller the peak intensity required for the formation of a single soliton. This indicates that the soliton has gathered energy from a progressively larger area as the ellipticity increases. Also, as expected, the threshold is lower for the positive detuning case due to the additive effect of the cascaded phase shift.

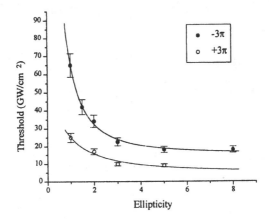

Figure 27 Peak threshold intensity needed for stable generation of a single spatial soliton from an elliptical input fundamental beam as a function of beam ellipticity.[51]

5. Summary

Interesting things happen to beams during parametric processes such as second harmonic generation. In particular, spatial solitons can be formed which rely on the strong coupling between multiple beams during SHG. For example, for Type II phase-matching, the two fundamental waves and the second harmonic all lock together in space and co-propagate, independent of any spatial walk-off which exists prior to beam locking. Discussed here were such quadratic solitons in slab waveguides (1D case) based on Type I SHG, and in bulk (2D case) KTP with Type II SHG. In all cases, excellent agreement with theory was obtained.

In slab LiNbO₃ waveguides, the interactions between two parallel and two crossing spatial solitons were investigated. As an overall conclusion, the features of the interactions resembled very closely those well-known for saturable Kerr media. In phase solitons attracted one another, eventually fusing together, and they repelled each other for all other values of the relative phase. For relative phases between 0 and π, one of the solitons takes power from the other with propagation distance. Except for some details, similar behavior was observed in crossing geometries.

An extensive family of solitons was found for Type II phase-matching in bulk SHG media. The ratio of the powers in the two fundamental polarizations can be varied over a wide range. In fact, the composition of the solitons obtained by launching the fundamental beam only had only a limited amount of SHG present. It varied over a broad range of parameters between 5% and 30%. In each case there was more fundamental than SHG power in the solitons.

We have investigated the coupling of quadratic solitons to two different well-known instabilities in nonlinear optics. The seeding of the parametric instability in which a pump photon breaks up into two lower energy photons and the subsequent generation of a quadratic soliton involving all three beams was observed. The properties of this process were found to be potentially useful for a high gain clipping amplifier. Modulational instability induced break-up of an elliptical beam into a line of quadratic solitons and the potential of this process for cleaning up asymmetric beams was demonstrated.

In summary, quadratic solitons are interesting objects. They are different in some ways from the more familiar single frequency, $\chi^{(3)}$ based spatial solitons associated with Kerr and saturating nonlinearities because they contain multiple frequencies. Being $\chi^{(2)}$ based, they should be pervasive in frequency conversion phenomena and have indeed been observed to date in second harmonic generation and degenerate optical amplification. There are also many similarities to other optical solitons. For example, quadratic soliton interactions resemble closely those associated with saturating $\chi^{(3)}$ nonlinearities, modulational instabilities exist and are closely linked to the existence of solitons etc. In fact, one of the fascinating aspects of quadratic solitons is that they allow us to identify many of the characteristics common to all solitons.

The support of this research in the U.S. by DARPA, ARO and AFOSR is gratefully acknowledged.

160

References:

1. Russell, J.S. (1984) in *"British Association Reports"*, John Murray, London.
2. for an early review see Akhmanov, S.A., Khokhlov, R.V. and Sukhorukov, A.P. chapter titled "Self-focusing, self defocusing and self-modulation of laser beams", in *Laser Handbook*, ed. F. T. Arecchi and E. O. Schulz-DuBois, (North Holland, Amsterdam, 1972), pp1171-2
3. Rasmussen, J.J. and Rypdal, K. (1986) *Phys. Scr.* 33, 481.
4. Maneuf, S. and Reynaud, F. (1988) *Opt. Commun.*, 66, 325.
5. Aitchison, J.S., Weiner, A.M., Silberberg, Y., Oliver, M.K., Jackel, J.L., Laird, D.E.A., Vogel, E.M. and Smith, P.W.E. (1990) *Opt. Lett.*, 15, 471-3.
6. Aitchison, J.S., Al-Hemyari, K., Ironside, C.N., Grant, R.S. and Sibbett, W. (1992) *Electr. Lett.*, 28, 1879; Kang, J.U., Stegeman, G.I., Villeneuve, A., Aitchison, J.S. (1996) *J. European Opt. Soc.*, Part A Pure and Applied Optics, 5, 583-94.
7. Swartzlander Jr. G.A. and Law, C.T. (1992) *Phys. Rev. Lett.*, 69, 2503.
8. Skinner, S.R., Allan, G.R., Andersen, D.R. and Smirl, A.L. (1991) *IEEE J. Quant. Electron.*, 27, 2211.
9. Duree Jr. G.C., Shultz, J.L., Salamo, G.J., Segev, M., Yariv, A., Crosignani, B., Di Porto, P., Sharp, E.J. and Neurgaonkar, R.R. (1993) *Phys. Rev. Lett.*, 71, 533-6.
10. Duree, G., Morin, M., Salamo, G., Segev, M., Crosignani, B., DiPorto, P., Sharp, E. and Yariv, A. (1995) *Phys. Rev. Lett.*, 74, 1978.
11. Tikhonenko, V., Christou, J. and Luther-Davies, B. (1995) *J. Opt. Soc. Am.*, 12, 2046.
12. Tikhonenko, V., Christou, J., Luther-Davies, B. and Kivshar, Y.S. (1996) *Opt. Lett.*, 21, 1129.
13. Torruellas, W.E., Wang, Z., Hagan, D.J., VanStryland, E.W., Stegeman, G.I., Torner, L. and Menyuk, C.R. (1995) *Phys. Rev. Lett.*, 74, 5036-9.
14. Schiek, R., Baek, Y. and Stegeman, G.I. (1996) *Phys. Rev. A*, 53, 1138-41.
15. Zakharov, V.E., Sobolev, V.V. and Synakh, V.C. (1971) *Sov. Phys. JETP* 33, 77; Bjorkholm, J.E. and Ashkin, A. (1974) *Phys. Rev. Lett.* 32, 129.
16. Torruellas, W.E., Lawrence, B. and Stegeman, G.I. (1996) *Electr. Lett.*, 32, 2092-4.
17. Karamzin, Y.N. and Sukhorukov, A.P. (1974) *JETP Lett.* 20, 339-42.
18. Karamzin, Yu. N., Sukhorukov, A.P. (1975) *Zh.Eksp.Teor.Phys* 68, 834-40 ((1976) *Sov. Phys.-JETP* 41, 414-20).
19. Kivshar, Y.S. chapter titled "Bright and Dark Solitary Waves in Non-Kerr Media", *Advanced Photonics with Second-Order Optically Nonliner Processes*, ed. by A.D. Boardman
20. Boardman, A.D. *Advanced Photonics with Second-Order Optically Nonliner Processes*, ed. by A.D. Boardman.
21. Sukhorukov, A.P. chapter titled "Wave Interactions in Quadratic Nonlinear Optics", *Advanced Photonics with Second-Order Optically Nonliner Processes*, ed. by A.D. Boardman
22. Reviewed in Stegeman, G.I., Hagan, D.J. and Torner, L. (1996) *J. Optical and Quant. Electron.*, 28, 1691-1740.
23. Torner, L., Menyuk, C.R. and Stegeman, G.I. (1994) *Opt. Lett.*, 19, 1615-7.
24. Guo, Q. (1993) *Quantum Opt.* 5, 133-139.
25. Torner, L. (1995) *Opt. Commun.*, 114, 136-40.
26. Schiek, R. Sundheimer, M.L., Kim, D.Y., Baek, Y., Stegeman, G.I., Suche, H. and Sohler, W. (1994) *Opt. Lett.*, 19, 1949-51.
27. Schiek, R. (1993) *J. Opt. Soc. Am. B* 10, 1848-1854.
28. Baek, Y. (1997) "Cascaded Second Order Nonlinearities in Lithium Niobate Waveguides", PhD thesis, Univ. Central Florida.
29. Torner, L., Mazilu, D. and Mihalache, D. (1996) *Phys. Rev. Lett.*, 77, 2455-8.
30. Schiek, R.,Baek, Y., Stegeman, G.I., Baumann, I. and Sohler, W. "One-Dimensional Quadratic Walking Solitons", to be published.
31. Aitchison, J.S., Weiner, A.M., Silberberg, Y., Leaird, D.E., Oliver, M.K., Jackel, J.L. and Smith, P.W.E. (1991) *Opt. Lett.*, 16, 15 et al, collisions.
32. Snyder, A.W., Sheppard, A.P. (1993) *Opt. Lett.* 18, 482.
33. Shih, M. and Segev, M. (1996) *Opt. Lett.*, 21, 1538.
34. Buryak, A.V., Kivshar, Y.S. and Steblina, V.V. (1995) *Phys Rev. A* 52, 1670-4.
35. Baboiu, D.-M., Stegeman, G.I. and Torner, L. (1995) Opt. *Lett.*20, 2282-4.
36. Etrich, C., Peschel, U., Lederer, F. and Malomed, B. (1995) *Phys. Rev. A* 52, 3444-3447.

37. Baek, Y., Schiek, R., Stegeman, G.I., Baumann, I. and Sohler, W. "Interactions Between One-Dimensional Quadratic Solitons", *Opt. Lett*, in press.

38. Shalaby, M., Reynaud, F. and Barthelemy, A. (1992) *Opt. Lett.*, **17**, 778.

39. Kelley, P.L. (1965) *Phys. Rev. Lett.* **15**, 1005.

40. Pelinovsky, D.E., Buryak, A.V. and Kivshar, Y.S. (1995) *Phys. Rav. Lett.* **75**, 591-595; Torner, L., Mihalache, D., Mazilu, D. and Akhmediev, N.N. (1995) *Opt.Lett.* **20**, 2183-2185.

41. Torner, L., Mihalache, D., Mazilu, D.,Wright, E.M., Torruellas, W.E. and Stegeman, G.I..(1995*) Opt. Commun.* **121**, 149-155.

42. Buryak, A.V. and Kivshar, Y.S. (1996) *Phys. Rev. Lett.*, **77**, 5210.

43. Peschel, U., Etrich, C., Lederer, F. and Malomed, B. (1997) *Phys, Rev. E*, **55**, 7704.

44. Fuerst, R.A., Canva, M.T.G., Baboiu, D. and Stegeman, G.I. (1997) *Opt. Lett.*, **22**, 1748.

45 Infeld, E. and Rowlands, R. (1990) *Nonlinear Waves, Solitons and Chaos*, (Cambridge Press, Cambridge); Newell, A.C. (1985) *Solitons in Mathematics and Physics*, (Soc. For Ind. And Appl. Math., Philadelphia).

46. Trillo, S. and Ferro, P. (1995) *Opt. Lett.*, **20**, 438-440.

47. Canva, M.T.G., Fuerst, R.A., Baboiu, D., Stegeman, G.I. and Assanto, G. (1997) "Quadratic Spatial Soliton Generation By Seeded Down Conversion of a Strong Pump Beam", *Opt. Lett.*, **22**, 1683.

48. Baboiu, D.M. and Stegeman, G.I. "Modulational Instability of a Strip Beam in a Bulk Type I Quadratic Medium", *Opt. Lett.*, in press.

49. Kaplan, A.E. (1993) *Opt.Lett.* **18**, 1223-1225.

50. Fuerst, R.A., Baboiu, D.-M., Lawrence, B., Torruellas, W.E., Stegeman, G.I. and Trillo, S. (1997) *Phys. Rev. Lett.*,**78**, 2760-3.

51. Fuerst, R.A., Lawrence, B.L., Torruellas, W.E. and Stegeman, G.I. (1997) *Opt. Lett.*,**22**, 19-21.

DIFFRACTION BEAM INTERACTION IN QUADRATIC NONLINEAR MEDIA

A. P. SUKHORUKOV

Physics Faculty, Moscow State University
Vorobjevy Gory, Moscow 119899, Russia
E-mail: aps@nls.phys.msu.su

Nonlinear dispersion of coupled cw plane waves in quadratic media is considered. The parametric 2D- and 3D-soliton properties are discussed taking into account diffraction, dispersion and walk-off effect. Mutual self-focusing and defocusing are examined in the frame of ray optics theory. The formation of half-period parametric solitons due to femtosecond pulse interaction is also presented.

1. Introduction

The parametric coupled optical solitons due to quadratic nonlinearity, which existence was predicted more than 20 years ago [1-3], attract the attention of scientists and engineers. However, till the beginning of 90's only theoretical works devoted to this topic were published (see for example [4,5]). The situation was changed dramatically in 1995 when the experimental evidence of these type of solitons were observed in nonlinear films and bulk crystals by Stegeman and co-workers [6,7]. Undoubtedly, this causes the powerful hitch to the further development of the theory and observation of this interesting phenomena in the leading scientific centers of many countries. As a result, the number of publications on the theory of parametric solitons greatly increased. Fundamental problems of the parametric solitons are discussed on special workshops, conferences, summer schools, in particular on this NATO ASI in Sozopol. This special attention to the parametric solitons is caused by the following reasons: the larger than in third-order case value of quadratic nonlinearity (especially in organic materials), that leads us to the lower threshold of soliton formation, the greater quadratic soliton stability to weak perturbations, and the capability of soliton's parameters controlling using all optical methods (direction-switching, "cleaning-up" and "reshaping" etc.). In this lecture I will discuss the nonlinear dispersion of waves in quadratic media, formation and properties of 2D and 3D parametric solitons, in particular under walk-off effect, bistability of solitons, the prosess of femtosecond soliton formation.

The investigation of nonlinear wave diffraction in Moscow State University started in 1963-65. In these years I took part in the development of the theory of laser beams using the slowly varying envelope approximation, as a PhD student under the supervision of R. V. Khokhlov. The application of this method to the beam diffraction

A. D. Boardman et al. (eds.),
Advanced Photonics with Second-order Optically Nonlinear Processes, 163–184.
© *1999 Kluwer Academic Publishers. Printed in the Netherlands.*

problems leads us naturally to the parabolic equations with imaginary transverse diffusion coefficient. This method was developed independently in many laboratories all around the world. The first objects of investigation were self-focusing and defocusing phenomena in media with Kerr-like, thermal and other nonlinearities [8]. At that time the diffraction theory of second harmonic generation under the presence of walk-off effect in undepleted pump wave approximation was developed. In the begging of 70's the numerical simulations of strong nonlinear interactions of FH and SH, fortified by original analytical theory were performed by Karamzin (Applied Mathematics Institute) and the author. This leads us to the discovery of two new physical phenomena: mutual-focusing and parametrically coupled solitons in the media with quadratic nonlinearity in 1974.

2. Nonlinear dispersion of cw plane optical waves

The wave interaction in nonlinear media results in the changing of amplitudes and phases along the direction of wave propagation. The behaviour of interacting waves depends on the type of nonlinearity and dispersion. In case of quadratic nonlinearity under strong dispersion the three-waves' interaction on the frequencies $\omega_1 + \omega_2 = \omega_3$

is very representative. Two waves $\omega_2 = 2\omega_1$ interaction is the frequency degenerated

case. The theory of propagation of this kind of waves was proposed by Khohlov [9, 10] and Bloembergen [11]. He used the slow-varying amplitude approach. Nonlinear optics becomes the most significant application for this method. In the frequency degenerated case the behavior of the fundamental and the second harmonics are described by the following equations:

$$\frac{d A_j}{d z} = -i\beta_j \frac{\partial U_2}{\partial A_j^*}, \quad U_2 = A_1^2 A_2^* \exp\left(-i\Delta_2 k z\right) + c.c., \quad (j = 1, 2), \qquad (1)$$

where A_j are complex amplitudes, $\beta_j = 2\pi\omega_1 \chi_2 / \left(c\, n_j\right)$ are coefficients of quadratic

nonlinearity, $\Delta_2 k = k_2 - 2k_1$ is the wave vector mismatch.

In publications [9-11] equations (1) were thoroughly analyzed for cw SHG, frequency mixing and parametric amplification. The same type of equations arises in plasma physics, hydrodynamics, optoelectronics. Khohlov [9] showed that the full energy transfer from the fundamental to the second harmonic is possible with phase synchronism ($\Delta_2 k = 0$, $n_1 = n_2$). In such a case the amplitudes are

$$A_1 = E_{10}\, ch^{-1}\left(z / l_{nl}\right), \quad A_2 = -i E_{10}\, th\left(z / l_{nl}\right), \qquad (2)$$

where $l_{nl} = 1 / \left(\beta_1 E_{10}\right)$ is the nonlinear interaction length. The integrals of motion,

which are the constants along the wave propagation coordinate, could be extremely helpful:

$$I_1 = n_1 B_1^2 + n_2 B_2^2, \quad I_3 = \beta_1 B_1^2 B_2 \cos\Phi + \Delta_2 k B_1^2. \qquad (3)$$

Here, we substituted the real amplitudes and phases: $A_j = B_j \exp\left(-i\varphi_j\right)$; the "generalized" phase difference $\Phi = \varphi_2 - 2\varphi_1 - \Delta_2 kz$ determines the direction of energy transfer from one wave to another. This becomes apparent after analyzing of integral curves of I_3, which family is presented in the Fig. 1 ($X = B_2 \sin\Phi$, $Y = B_2 \cos\Phi$). In general, the amplitudes suffers the spatial beatings.

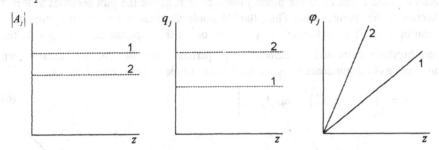

Fig. 1. The behavior of the amplitudes B_j (a), phases φ_j (b) and additions to wave numbers q_j (c) of stationary FH (1) and SH (2) waves.

The specific points, namely $X_{1,2} = 0$ and $Y_{1,2} = \pm I_3 / \sqrt{3n_2}$, are of main interest. In this points the interaction becomes pure reactive: the amplitudes of the steady waves are constant, the only varying parameter is the phase difference. This fact was mentioned in [10], but the consequences were not fully examined at that time.

It is easier to explore the parametric self-action of coupled waves, considering (see Fig. 2)

$$A_j = E_{jo} \exp(-i q_j z). \tag{4}$$

Substitution of (4) into (1) gives us the equations of nonlinear dispersion

$$q_1 = \beta_1 E_{20}, \; q_2 = \beta_2 E_{10}^2 / E_{20}, \; q_2 = 2q_1 - \Delta_2 k. \tag{5}$$

In the case of synchronism, $\Delta_2 k = 0$, we obtain $E_{10}^2 = 2E_{20}^2$ and $q_2 = 2\beta_1 E_{20}$ (the lines 1 and 2 in Fig. 3). In the region, where $E_{20} > 0$, or $\cos\Phi = 1$, nonlinear additions to wave numbers are positive, and the medium has focusing properties. The change of the sign of second harmonic amplitude, $E_{20} < 0$, or $\cos\Phi = -1$, causes defocusing. In other words, simple phase-switching of SH by π could alter the self-interaction type. In the same time the analogous phase-switching of FH does not affect the character of self-interaction. Let's stress, that in the process of coherent SHG the phase settles on the value of $\Phi = \pi/2$, providing the maximum

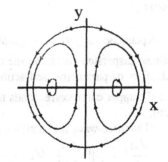

Fig. 2. Integral curves of $I_3 = \text{const}$ describe various interaction regimes of FH and SH.

rate of energy transfer, according to (2).

If $\Delta_2 k \neq 0$, and $Y = 0$, $X > 0$, then the lines q_2 shift to left or right depending on the sign of the mismatch (lines 3 and 4 in Fig. 3). In addition, if the mismatch is positive there are no solutions in the area $0 < \beta_1 E_{20} < \Delta_2 k / 2$ (there are breaks in lines 1 and 2); if the phase mismatch is negative the plot becomes a mirror reflection of the previous case. Thus, the SH amplitude threshold of focusing properties of nonlinear medium is formed if $\Delta_2 k > 0$, or, in the opposite case, $\Delta_2 k < 0$, the same threshold is formed for defocusing properties. This effect is more clearly seen from analysis of dependence of q_1 on the FH amplitude. As it follows from (5),

$$q_1 = \frac{1}{4}\left[\Delta k \pm \sqrt{(\Delta k)^2 + 8\beta_1\beta_2 E_{10}^2}\right].$$

(6)

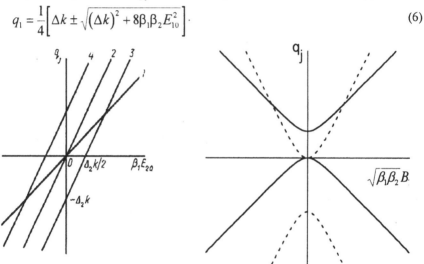

Fig. 3. The dependences of nonlinear additions to wave numbers of FH (1) and SH (2-4) on the SH amplitude with $\Delta_2 k = 0$ (2), $\Delta_2 k < 0$ (3) and $\Delta_2 k > 0$ (4).

Fig. 4. The dependences of nonlinear additions to wave vectors of FH (solid lines) and SH (dashed lines) on the first harmonic amplitude with $\Delta_2 k > 0$.

This dependence is presented graphically in the Fig. 4 for $\Delta_2 k > 0$. There are two nonlinear dispersion curves on the plot. The lower curve corresponds to the cascading mechanism of parametric self-action (the defocusing case), there $q_1 = -\beta_1\beta_2 E_{10}^2 / \Delta k$, while the upper curve corresponds to focusing. The analogous effect exists for solitons [4, 26].

The three-wave interaction can be analyzed employing the same approach:

$$\frac{d A_j}{d z} = -i\beta_j \frac{\partial U_3}{\partial A_j^*}, \quad U_3 = A_1 A_2 A_3^* \exp\left(-i\Delta_3 k z\right) + c.c., \quad (j = 1,2,3)$$

(7)

where $\beta_j = 2\pi\omega_j \chi_2 / (c n_j)$, $\Delta_3 k = k_3 - k_1 - k_2$. Now, the parametric dispersion relations for stationary waves become more complicated:

$$q_j = \beta_j E_{10} E_{20} E_{30} / E_{j0}^2, \, q_3 = q_1 + q_2 - \Delta_3 k. \tag{8}$$

According to (8), all q_j have the same sign. It's possible to derive the relationship between stationary wave amplitudes from (8):

$$\beta_3 E_{20}^2 E_{30}^2 = \beta_1 E_{10}^2 E_{30}^2 + \beta_2 E_{10}^2 E_{20}^2 - \Delta k E_{10} E_{20} E_{30} \tag{9}$$

Thus, in distinction of the previous degenerate case, q_j have two-parametric dependence on amplitudes. In other words, it's necessary to investigate the dependencies like $q_j(E_{10}, E_{20})$ and $q_j(E_{20}, E_{30})$, which are described by 2D surfaces. The similar dispersion is observed for three-wave solitons [3, 4, 26].

Let's consider the case $\Delta_3 k = 0$. First, one can see from (9) that the intensity of the third wave can't exceed the ones of the waves at lower frequencies, $n_3 E_{30}^2 \le n_1 E_{10}^2, n_2 E_{20}^2$. Second, the specific nonlinear dispersion arises only if one of the low-frequency waves has the bigger amplitude than others. Then $q_1 \approx 0$, $q_2 \approx q_3 \approx (\beta_2 \beta_3)^{1/2} E_{10}$ in case of $n_2 E_{20}^2 = n_3 E_{30}^2$. Third, if $\Delta_3 k$ is large, the cascading mechanism of self-action can be observed [12]. Here $E_{30} << E_{10}, E_{20}$, $q_1 = -\beta_1 E_{20}^2 / \Delta k$, $q_2 = -\beta_2 E_{10}^2 / \Delta k$, $q_3 \approx \Delta_3 k$ $E_{30} \approx -\beta_3 E_{10} E_{20} / \Delta k$. The significant property of this interaction is that SH changes dispersion of FH and vice versa. The character of three-wave interaction could be easily changed by shifting the phase of any wave by π because the phase difference is $\Phi = \varphi_3 - \varphi_1 - \varphi_2 - \Delta_3 k z$.

3. The parametric interaction of modulated waves

In theoretical analysis of parametric interaction of waves, which envelopes are slow varying in time and space, it is necessary to introduce additional differential operators in (1) and (7) to describe the beam diffraction, dispersive spreading of optical pulses and walk-off effect:

$$\frac{d A_{1,2}}{d z} + L_{1,2}^{(s-t)} A_{1,2} = -i \beta_{1,2} A_{2,1}^* A_3 \exp(-i \Delta k z),$$

$$\frac{d A_3}{d z} + L_{1,2}^{(s-t)} A_3 = -i \beta_3 A_1 A_2 \exp(i \Delta k z) \tag{10}$$

Here, $L_j^{(s-t)} = L_j^{(s)} + L_j^{(t)}$, and

$$L_j^{(s)} = \rho_{jx} \frac{\partial}{\partial x} + \rho_{jy} \frac{\partial}{\partial y} + i D_{jx} \frac{\partial^2}{\partial x^2} + i D_{jy} \frac{\partial^2}{\partial y^2}, \quad L_j^{(t)} = \upsilon_{jm} \frac{\partial}{\partial \eta_m} - i D_{jm} \frac{\partial^2}{\partial \eta_m^2}.$$

We introduced the following denominations: ρ_{jx} are walk-off angles; $\upsilon_{jm} = 1/u_j - 1/u_m$ is group velocity mismatch, $\eta_m = t - z/u_m$ is retarded time, $D_{jx} \cong D_{jy} \cong 1/2k_j$ are transverse diffusion coefficients of envelopes in the process of diffraction, D_{jm} are coefficients of linear second-order dispersion.

Let us discuss the nonlinear dispersion of parametrically coupled waves with the linear phase modulation. The envelopes in this case read as:

$$A_j = B_j \exp(i\Omega\eta_m - iq_j z - iq_{jx}x - iq_{jy}y) \,, \tag{11}$$

where Ω is the frequency shift, $q_{jx} = k_j\Theta_{jx}$ and $q_{jy} = k_j\Theta_{jy}$ - the wave vectors projections on transverse axis, $\Theta_{jx,jy}$ are angles of wave vector inclination. The modulation coefficients in (10) are bound together by the following relations:

$$\Omega_3 = \Omega_1 + \Omega_2, \qquad\qquad q_{3x} = q_{1x} + q_{2x}, \qquad\qquad q_{3y} = q_{1y} + q_{2y}.$$

Substitution of (11) into (10) lets us to describe the nonlinear 3D dispersion of plane cw waves in case of frequency and direction varying.

On the base of comparision of space and time operators in (10) one can formulate space-time analogy for beam diffraction and pulse dispersion, beam walk-off effect and group velocity effect [13].

4. Spatial parametric solitons

In contradictinction to the case of third-order nonlinearity [8], we don't have to search for a medium with the necessary sign of nonlinear constant. The proper selection of the amplitudes and phases of the waves provides the self-focusing properties of the quadratic medium. So, in this type of media the spatial solitons could be formed (see Fig. 5a). The parametric coupled solitons were studied for the first time during the analysis of beams interactions under the presence of diffraction and quadratic nonlinearity [1,2]. As it follows from (10), the nonlinear difraction of beams could be described by (see also [14-18]):

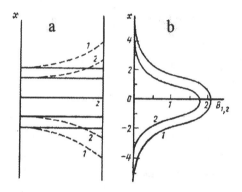

Fig. 5. Soliton propagation of beams in quadratic medium: a-schematic representation of coupled solitons at frequencies ω and 2ω. Dashed lines show the diffraction divergence of Gaussian beams in a linear medium; b-eigen modes of stationary two-dimensional waveguide.

$$\frac{\partial A_j}{\partial z} + iD_j\Delta_\perp A_j = -i\beta_j\frac{\partial U_2}{\partial A_j^*}, \tag{12}$$

where $\Delta_\perp$ is the Laplas operator in transverse coordinates x, y; $D_j = 1/(2k_j)$.

We can obtain from (12) the law of energy conservation $I_1 = \iint\left(n_1|A_1|^2 + n_2|A_2|^2\right)dx\,dy$ and Hamiltonian (integral of motion)

$$I_3 = \iint\left[2D_1(\beta_2/\beta_1)|\nabla_\perp A_1|^2 + D_2|\nabla_\perp A_2|^2 - \Delta_2 k|A_2|^2 - \beta_2|A_1^2 A_2^*|\cos\Phi\right]dx\,dy. \tag{13}$$

This Hamiltonian characterizes the interaction type of two harmonics. In case of beam diffraction in linear medium we obtain $I_3 > 0$. In the quadratic nonlinear medium it is possible to obtain $I_3 < 0$ due to strong self-action with $\cos \Phi = 1$. It means that the wave decay to non-interacting beams becomes impossible, the beams are self-trapped in parametrical coupled solitons. These solitons usually oscillate.

Let the solitons have stationary amplitude profile and nonlinear addition to the wave number:

$$A_j = B_j(x, y)\exp(-i q_j z), \quad q_2 = 2q_1 - \Delta_2 k. \tag{14}$$

The substitution of (14) into (12) gives the equation for eigen modes:

$$D_j \Delta_\perp B_j = q_j B_j - \beta_j B_1^2 B_2 / B_j. \tag{15}$$

Numerical solutions of (15) in the case of $\Delta_2 k = 0$ are presented in Fig. 5b [1]. The harmonic powers of 2D-soliton with width a are equal to $P_1 = 4,8 P_{s1}$ and $P_2 = 2,76 P_{s1}$, where $P_{s1} = D_1^2 \beta_1^{-2} a^{-3}$. The exact analytical solution of (15) for 1D soliton, that supervenes from the envelope similarity demand: $B_1(x) \propto B_2(x)$, was found in the paper [1] at the first time:

$$B_j = E_{j0} \operatorname{sech}^2(x/a) \tag{16}$$

with $E_{20} = 6a^{-2} D_1 / \beta_1$ and $E_{10} = \pm 6a^{-2}\left[D_1 D_2 / (\beta_1 \beta_2)\right]^{1/2}$, and the coefficients of diffusion are coupled by the relation $D_2 = 2D_1 - \Delta_2 k a^2 / 4$, according to (15). The solution (16) was rediscovered in some novel works (see, for example, [14]). The nonlinear additions to the wave numbers $q_1 = \frac{2}{3}\beta_1 E_{20}$, $q_2 = \frac{2}{3}\beta_2 E_{10}^2 / E_{20}$ are 1,5 times less than these additions for plane waves (5). The decrease of nonlinear dispersion is due to the existence of diffraction effects. For the same type of amplitude profile devolution the mode of cylindrical beams was found by means of numerical simulations in [1]. Here, $P_1 = 1,125 P_{s2}$ and $P_2 = 2,25 P_{s2}$, where $P_{s2} = D_1^2 \beta_1^{-2} a^{-2}$. The numerical calculations of one-dimensional soliton profiles were performed for a broad range of diffusion coefficients and mismatches [3, 19], see Fig. 6.

The numerical estimations for the typical optical crystals with $\chi_2 \approx 1$ pM/V and beam width $a \approx 10^{-2}$ cm show that relatively small power $(P_{s2} \approx c \lambda^4 \chi_2^{-2} a^{-2}$ is approximately equals to several hundred watts) are

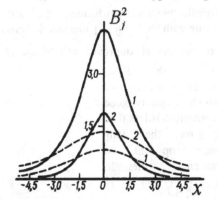

Fig. 6. Results of numerical calculations of 2D parametric solitons envelopes under group-velocity matching and $q_1 / D_1 = 1$, $q_2 / D_2 = 9$.

necessary in order to excite a spatial soliton. Nevertheless, almost 20 years passed from the first theoretical publication [1] before the first experiments were performed by Stegeman and co-workers in 1995 [6, 7]. They observed one- and two-dimentional spatial solitons in quadratic crystals.

Good analytical results can be obtained by means of variational approach [19, 20], based on the minimizing of the Hamiltonian of parametric interaction:

$$L = \iint \left[D_1 (\nabla_\perp B_1)^2 + \frac{1}{2} D_2 (\nabla_\perp B_2)^2 + q_1 B_1^2 + \frac{1}{2} q_2 B_2^2 - \beta B_1^2 B_2 \right] dx dy \cdot$$

Gaussian profiles $B_j = E_{j0} \exp\left(-r^2/a_j^2\right)$ are usually taken as trial functions. Applying this method with $\Delta_2 k = 0$, one can find the algebraic equation for beam width ratio $\xi = a_1^2/a_2^2$:

$$\xi^3 + 4\xi - 8 = 0.$$

Solving this equation, one can obtain relations between amplitudes and widths could be found: $a_2 = 0{,}856\, a_1$, $E_{20} = 8{,}3 D_1 / (\beta a_1^2)$, $E_{10} = 8{,}6 D_1 / (\beta a_1^2)$. However, the question about the difference between this solution and exact one is still open. The numerical experiments show that the amplitudes of the beams with parameters found above are varying along the propagation in the limits of 10%. It is necessary to increase the input signal power in order to hold the soliton existence with high mismatch $|\Delta_2 k|$.

The optimal soliton parameters can also be calculated using the gradient method. One could suggest, for example, the value

$$I = |A_1(0,z)|^2_{max} - |A_1(0,z)|^2_{min} + |A_2(0,z)|^2_{max} - |A_2(0,z)|^2_{min}$$

as a quality parameter in forming of almost non-oscillating solitons [21]. Using this procedure with $D_1 = 1$, $\beta_1 = 1$ one can decrease oscillations down to 83% of initial ones, and parameters of improved soliton are $E_{10} = 8{,}45, E_{20} = 8{,}15, a_2 = 0{,}84 a_1$. The soliton with listed parameters holds it's initial shape pretty good (the corruption is less than 10%) on the long distances up to 80 diffraction lengths $(l_d = k a^2/2)$, Fig 7. In linear medium at this distance the extension of beam with the same initial profile increases in 100 times.

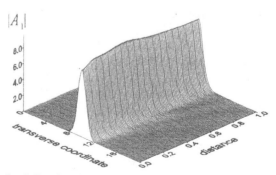

Fig. 7. Trapping of Gaussian beams in 3D spatial parametric soliton.

The spatial solitons in quadratic medium have a very important to the practice feature of stability. The results of numerical simulation of an optimal soliton propagation with the weak amplitude modulation of the FH

$$A_1(r,0) = E_{10}\exp\left(-r^2/a_1^2\right)\left(1 + 0{,}2\cos 30r\right), \quad A_2 = E_{20}\exp\left(-r^2/a_2^2\right),$$

as a small perturbation are shown in the Fig. 8. [21]. The comparison of initial and output profiles shows that the modulation disappears and profile shape becomes smooth. This self-cleaning up of amplitude profiles in quadratic media was observed experimentally [22]. As a matter of fact, I_3 in the presented example is positive, nevertheless, after the self-cleaning the propagation of parametric soliton is observed.

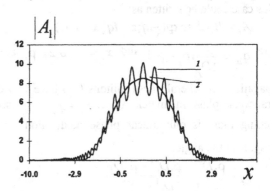

Fig. 8. Cleaning-up of FH amplitude profile due to parametric soliton formation: 1 - input profile; 2 - profile at the distance $z = 1{,}2l_d$.

5. Coupled solitons with walk-off effect

The equations (10) with $L_j^{(t)}$ can be used to describe wave packets interaction in the second-order approximation of disperse theory. Now, the local time is $\eta_j = t - z/u_j$, where $u_j = \left(\partial k_j/\partial\omega\right)^{-1}$ is a group velocity. The diffusion coefficients $D_j = -(1/2)\partial^2 k_j/\partial\omega^2$ characterize the dispersion of group velocities. The collinear geometry of beams corresponds to the case of group velocity matching, $u_1 = u_2$.

However, group synchronism couldn't always be accomplished. The presence of the group velocity mismatching increases the thresholds of soliton formation, but on the other hand the solitons acquire principally new properties, that could be used in creation of optical-switched devices [22]. In particular, phase modulation and self-tuning of solitons to some "average" velocity (direction). The interaction between wave packets is more diversified, because the coefficients of dispersion could have different values and signs. The wave beams have pre-determined values of diffraction coefficients (10). The problem of formation and the properties of the parametric solitons under the presence of walk-off effect were considered in [3] for the first time and in [23-25] recently. The properties of vector solitons was considered by [3, 4, 26, 27].

Now we take into account walk-off term in equations (10). In case of three-wave interaction:

$$\frac{\partial A_j}{\partial z} + \rho_j \frac{\partial A_j}{\partial x} + i D_j \frac{\partial^2 A_j}{\partial x^2} = -i\beta_j \frac{\partial U_3}{\partial A_j^*}, \tag{17}$$

where ρ_j is a walk-off angle, the right part is similar to (7). The envelopes of solitons in this case could be written as:

$$A_j = B_j(x_s)\exp(-iq_j z - iq_{jx} x - i\varphi_j), \tag{18}$$

where $q_{jx} = \dfrac{1}{2D_j}(\rho_j - \rho_s)$ and $x_s = x - \rho_s z$, ρ_s are the directions of the soliton

propagation. In general, these solitons (17) have very complex phase modulation due to transverse phase mismatch $\Delta q_x = q_{1x} + q_{2x} - q_{3x}$ (see Fig. 9). In case of transverse matching there is only linear phase modulation, as $\varphi_j = 0$. The coupled beams propagate at the angle:

$$\rho_s = \frac{\rho_3 D_3^{-1} - \rho_1 D_1^{-1} - \rho_2 D_2^{-1}}{D_3^{-1} - D_1^{-1} - D_2^{-1}}.$$

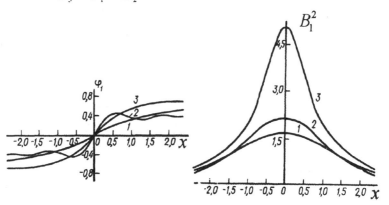

Fig. 9. Results of numerical calculations for FH soliton phase and amplitude profiles with $q_j = D_j$ and $\Delta q_x a = 1$ (1), 2 (2), 3 (3).

If the denominator is equal to zero then the transverse matching cannot be fulfilled and phase modulation is irremovable.

The evolution of beams with initial gaussian profiles $A_j(x,0) = E_{j0}\exp(-x^2/a^2)$ in quadratic media in presence of walk-off effect is shown in Fig. 10. The comparision of three pulse envelopes shapes arrived to the plane z=1.2 of the diffraction length points straight on the parametric soliton formation. By the way, these solitons have phase modulation. The properties of solitons in the media with group velocity mismatch are described in [3, 4, 23-25].

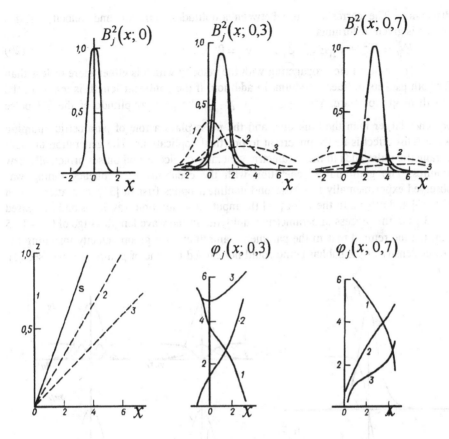

Fig. 10. Self-trapping of three Gaussian 2D beams into parametric soliton with $\beta_1 = 15$, $\beta_2 = 20$, $\beta_3 = 35$ and $\rho_1 = 0$, $\rho_2 = 6$, $\rho_3 = 10$. Dashed lines correspond to propagation in linear medium. Soliton propagates at the angle $\rho_s \approx 4$.

6. Giant soliton and kink interaction in parametric amplifier

Parametric solitons also arise in the first-order approximation of dispersion theory, when the terms with second-order differential operators aren't taken into account in the equations (10). In the frequency degenerated case steady envelopes of two harmonics [4, 28-30]

$$B_1 = E_{10} \sec h\left(\eta_s/\tau\right), \quad B_2 = E_{20} th\left(\eta_s/\tau\right), \tag{19}$$

From (19) one can easily see that localized wave - soliton is propagating at the foundamental frequency and SH responds with a kink (see Fig. 11). However, the FH intensity registration, which is common in optics, gives us $B_2^2 = E_{20}^2 th^2\left(\eta_s/\tau\right) = E_{20}^2 - E_{20}^2 \sec h^2\left(\eta_s/\tau\right)$; it can be interpreted as a dark soliton. Thus, the coupled dark soliton-kink is formed in quadratic media with first-order dispersion. The correspondence between amplitudes, duration and velocity of this couple is given by formulas

$$E_{10}^2 = E_{20}^2 + v_{21}E_{20}/\beta_1\tau, \quad v_{1s} = \beta_1 E_{20}\tau.$$ (20)

The solitons are propagating with the velocity which is either more or less than for both packets in linear medium. In addition, if the nonlinear length is less than the length of group retardation $(\tau < v_{21}/\beta_1 E_{20})$, the peak amplitude of the SH pulse becomes larger than the kink one, and the pulse plays a role of parametric pumping wave. This effect is really important to practical applications. The generation of giant parametric pulse adds to the parametric generator of picosecond pulses principally new features. The giant pulse generation with the capability of frequency tuning was observed experimentally in picosecond nonlinear optics first in [31], and after that in [32, 33] and others. In the work [31] the input pulse duration was 30 ps and decreased to 0.3 ps in the process of parametric amplifying in the wave length range of 0.8 - 1.35 μm. Let me remind that in the parametric amplifier with group velocity matching the power density of the subharmonic could not exceed the one of pumping wave, see (2), (3).

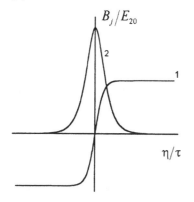

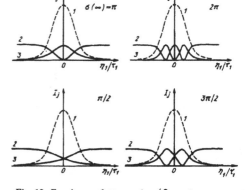

Fig. 11. Interaction of FH kink – dark soliton (1) and SH giant bright soliton (2) in the parametric traveling wave amplifier with group velocity mismatch.

Fig. 12. Envelopes of π – and $\pi/2$– pulses at sum (3) and difference (2) frequencies in the field of low-frequency pump pulse (1) with dispersion $v_{21}v_{31} > 0$.

A number of specific nonstationary effects arise in sum frequency generator with a low-frequency high-intensity short undepleted pump pulse $A_1 = E_{10}(\eta_1)$ [4,34]. In this case the solution the envelope equation, taking into account walk-off effect, at the long distance become stationary

$$|A_3|^2 = (\beta_3/\beta_2)E_{20}^2 \sin^2[n\pi\sigma(\eta_1)/\sigma(\infty)],$$ (21)

$$|A_2|^2 = E_{20}^2 \cos^2\left[n\pi\sigma(\eta_1)/\sigma(\infty)\right], \tag{22}$$

where $\sigma(\eta_1) = (\beta_2\beta_3)^{1/2} v_{31}^{-1} E_{20} \int\limits_{-\infty}^{\eta_1} A_1(\xi)d\xi$. If $\sigma(\eta_1) = n\pi$, $n = 1,2,3...$, one

can observe bright π - soliton at sum frequency (21) and dark π - soliton at idler frequency (22), see Fig. 12.

The process of π - and $\pi/2$ - soliton formation was investigated by numerical simulation, Fig. 13.

The theory of parametric solitons due to quadratic nonlinearity based on the solving of equation (19) with $D_j = 0$, was greatly developed after applying of the reverse scattering method [29].

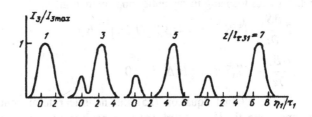

Fig. 13. Stages of π - pulse formation at sum frequency ω_3 under interaction of cw signal and Gaussian pulse of low-frequency pump wave; $v_{31} = 2v_{21}$; $\sigma(\infty) = \pi$.

7. Parametric self-focusing of coupled beams

It was shown that the medium with quadratic nonlinearity can acquire the ability of self-focusing in case of coherent three waves interaction. The self-focusing nonlinearity compensates the diffraction of beams or dispersion of pulses. This allows the soliton-forming process. If we increase the power density of coupled waves, however, we can assume the changing of soliton propagaiton by the focusing (as for third-order nonlinearity [8]). The parametric self-focusing, or mutual focusing was discovered in our first computer simulations of quadratic nonlinear propagation [2]. These phenomena are discussed in details in monograph [3]. I will review the three wave focusing in the approximation of nonlinear ray optics theory [35].

The parabolic equations (8) which describe real amplitudes and phases read as:

$$\frac{\partial\varphi_j}{\partial z} + D_j\left(\nabla\varphi_j\right)^2 = \beta_j B_1 B_2 B_3 \cos\Phi/B_j^2 + D_j\Delta_\perp B_j/B_j, \tag{23}$$

$$\frac{\partial B_j}{\partial z} + 2D_j\nabla\varphi_j\nabla B_j + D_j B_j\Delta_\perp\varphi_j = \beta_j B_1 B_2 B_3 \sin\Phi/B_j. \tag{24}$$

If we suppose $\varphi_j = q_j z$ and $\cos\Phi = 1$, then the equations of energy transfer (24) give $B_j = B_j(x,y)$, and the eikonal equations (23) will describe the localized steady profiles of parametric coupled solitons. The last terms of eikonal equations (23) disappear in ray optics approximation. However, even in this approximation (23) are still too complicated. The great simplification can be obtained in case of pure reactive interactions i.e. $\sin \Phi = 0$, $\cos \Phi = \pm 1$. The analysis of (23) gives us the similar behavior of all three waves if the diffraction coefficients $D_3^{-1} = D_1^{-1} + D_2^{-1}$, wave fronts $\varphi_j = D_3 \varphi_3 / D_j$ and amplitude profiles $B_j = B_3 \left(D_j \beta_j / D_3 \beta_3 \right)^{1/2}$ are matched. Finally, the substitution of the phase $\varphi_j = \varphi_j / 2 D_j$ gives us the equation of matched waves focusing in ray optics approximation:

$$2 \frac{\partial S_3}{\partial z} + \left(\nabla S_3 \right)^2 = \pm 4 \left(\beta_1 \beta_2 D_1 D_2 \right)^{1/2} B_3 , \tag{25}$$

$$\frac{\partial B_3}{\partial z} + \nabla S_3 \nabla B_3 + \tfrac{1}{2} B_3 \Delta_\perp S_3 = 0 . \tag{26}$$

In the eikonal equation (25) the nonlinear term contains the amplitude of electromagnetic field (for third order of nonlinearity the similar term contains the squared amplitude). That is why the parametric focusing of a beam is not as strong as for the cubic nonlinearity. In particular, the collapse of the beam doesn't occur (see Fig. 14). The quadratic solitons are extremely stable due to the same reason.

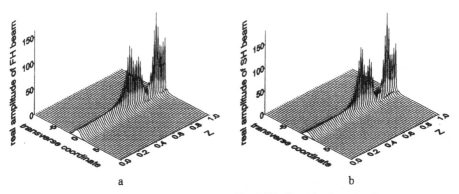

Fig. 14. Evolution of self-matched mutual focusing, $E_1 = 1414$ $E_2 = 10$. $D_1 = 0.1$, $\beta = 1$. (a) - FH beam, (b) - SH beam.

The equations (25, 26) could be solved in the aberration-free approximation [8], suggesting

$$\varphi_3 = \frac{\left(x^2 + y^2 \right)}{2 f} \frac{d f}{d z} + \varphi_0 (z) , \quad B_3 = E_{30} f^{-m/2} \exp \left[- \frac{\left(x^2 + y^2 \right)}{a^2 f^2} \right],$$

there $f(z)$ is a normolized width of the 3D beam (m=2) or 2D – beam (m=1). The ordinary methods lead us to the equation with the characteristic length of focusing (plus) or defocusing (minus) $l_s = a\left[\beta\, D E_{30}\right]^{-1/2}$

$$\frac{d^2 f}{d z^2} = \pm \frac{1}{l_s^2\, f^{(m+2)/2}},\qquad(27)$$

This length has an extremely simple form for the beams: $l_s \approx a\left[\chi_2\, E_{30}\right]^{-1/2}$. The integration of (27) is trivial, but the final formula is complicated and isn't shown in this review. It is possible to estimate the effect of compression in the point of nonlinear focus by comparing the nonlinear refraction force to the diffraction one $1/l_d^2\, f^3$, $l_d = a^2/2D$. For defocusing case it is easy to calculate the nonlinear divergence in the far field and the nonlinear limitation of the field in converging beams.

The cascading self-focusing and defocusing under SHG with large phase mismatching were observed in experiments [36, 37].

In parametric amplifier with plane pump wave one can observe anomalous diffraction (see Fig. 15).

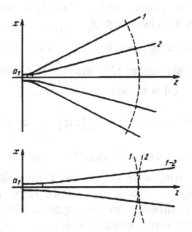

Fig. 15. The diffraction of wave beams at signal and idler frequencies in linear medium (upper) and in parametric amplifier with plane pump wave (lower). Dashed lines represent the phase fronts.

8. Interaction of waves at multiple frequencies

The synchronic interaction with multiplied frequencies ω, 2ω, 3ω allows to conduct two types of parametric interaction: the sum frequency generation ($\omega + 2\omega = 3\omega$) and SHG ($\omega + \omega = 2\omega$). As it was shown for the first time in [3], the synchronic contribution of two channels leading to the full energy transfer from the pumping to signal 2ω wave is possible. This conclusion has a great practical consequence in acheiving the maximum efficiency of parametric generators and traveling wave amplifiers. Let me remind, that in the absence of the SHG channel the energy of 2ω wave would not be greater than 2/3 of pumping beam energy [10, 11]. In another work by our group [38] it was shown that the interference of two interaction channels leads to bistability of coupled solitons. Let's describe the main properties of these novel effects.

The energy of waves' interaction with multiple frequencies consists of two terms (compare with (1) and (6)):

$$U_{23} = A_1^2 A_2^* \exp\left(-i\Delta_2 k z\right) + A_1 A_2 A_3^* \exp\left(-i\Delta_3 k z\right) + c.c. \tag{28}$$

First, we'll describe the interaction of plane monochromatic waves. Obviously, the strongest effect is observed in case of simultaneous phase synchronism for two channels: $\Delta_2 k = \Delta_3 k = 0$. If the wave vectors are parallel (collinear interaction), this case is equivalent to the phase velocities equality (refraction indexes) for all three frequencies i.e. $n_1 = n_2 = n_3$. This condition can be fulfilled in QPM crystals.

In full phase matching case it is possible to derive the equations for real amplitudes by substituting $B_1 = A_1$, $B_2 = -iA_2$, $B_3 = -A_3$. Considering the obvious relations $\beta_2 = 2\beta_1$ and $\beta_3 = 3\beta_1$, it is possible to find from equations like (6) with energy of interactions (20) the energy conservation law $I_1 = B_1^2 + B_2^2 + B_3^2$ and the second integral of motion

$$I_3 = \left(3B_1^2 + B_1 B_3 + B_3^2\right)\exp\left[\frac{-2}{\sqrt{11}}\,arctg\left(\frac{6B_1 + B_3 B}{\sqrt{11}B_3}\right)\right].$$

Using these two integrals of motion I_1 and I_3, it is easy to investigate the waves' behavior in the processes of parametric amplifying and frequency multiplying. By the way, the theory of the third harmonic cascaded generation in media with quadratic nonlinearity was developed in 1964 [10] by means of numerical analysis of (1) with interaction given by (28).. However, the parametric amplification with the waves at multiple frequencies was not discussed. Therewith, its analysis gives unexpected results.

The is a motion along the sphere. Observation of phase portrait of three waves interaction leads us to the fact of full energy transfer to the signal wave 2ω after several aperiodically oscillations. In the absence of SHG channel only 2/3 of pumping wave can be transferred to the signal one according to Manely-Rowe relations [10, 11].

The strong interaction of beams or pulses at the multiple frequencies could cause the coupled soliton formation analogous to the previous cases. The interference of two channels of interaction in the process of soliton formation is of great interest. The expected bistability effect could be demonstrated on the simple example [38], when the envelopes of all three waves have the same profile, but different amplitudes and nonlinear additions to the wave numbers:

$$A_j = E_{j0}\,sech^2\left(\eta_1/\tau\right)\exp\left(-iq_j z\right). \tag{29}$$

Substitution of (29) into (17) with $v_j = 0$ and U taken from (28) gives the relation between soliton parameters. The detunings of wave vectors obey the following condition: $\left(D_3 - D_1 - D_2\right)\Delta_2 k = \left(D_2 - 2D_1\right)\Delta_3 k$. It's clear that the full synchronism is not a necessary condition of soliton formation. In opposite, the detuning's selection allows to match the second-order dispersion coefficients. The key parameter of bistable solitons is a value of

$$C_\pm = -1 \pm \left(1 + 12 n_1 D_1 / n_3 D_3\right)^{1/2} \tag{30}$$

The peak amplitudes of all three harmonics depend on that parameter as follows:

$$E_{10} = \pm \frac{\left(6 n_2 D_2 n_3 D_3 C_\pm\right)^{1/2}}{\tau^2 \beta_1 n_1 \left(1 + C_\pm\right)^{1/2}}, \quad E_{20} = \frac{n_3 D_3 C_\pm}{\beta_1 \tau^2}, \quad E_{30} = \tfrac{1}{2} C_\pm E_{10}. \tag{31}$$

The second harmonic amplitude could be either positive or negative, according to (30), (31). It determines two types of solitons. The soliton duration of the same energy also has two different values (see Fig. 16). The first type with C_+ corresponds to synphase contribution of two channels. All channels contribute to the nonlinear medium focusing features, because $E_{20} > 0$ and $E_{10} E_{30} > 0$. As a result, the pulses become shorter. The second type of coupled solitons has C_- (i.e. $E_{20} < 0$), which causes the decrease of efficient nonlinearity and increase of duration (curve 1); here $E_{10} E_{30} < 0$.

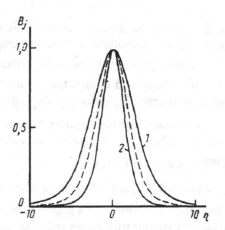

Fig. 16. Parametric soliton envelopes under one (dashed) and two (solid) interaction channels of waves with opposite (1) and same phases (2).

The analogous analysis was performed in [38] for three waves interaction on the multiple frequencies in media with cubic nonlinearity, when the the second harmonic played a role of pumping wave ($2\omega + 2\omega = \omega + 3\omega$) and the additional channel was the interaction between first and third harmonics ($\omega + \omega + \omega = 3\omega$).

9. Self-interaction of extremely short pulses

The optical radiation of extremely short duration has a lot of applications in the various fields of science and techniques. In optics this is closely related with the development of femtosecond pulse generation methods [39]. The analysis of ultra-short pulse propagation through nonlinear dispersive medium, when the spectrum width and the value of average frequency are comparable, the parabolic approximation method is unusable. That's why the methods of high-order approaches of linear and nonlinear dispersion are being developed [40], the coupled equations of Korteveg-de Vries is applied [41–43]. If the pulse contains two or three oscillations of optical field, then one should solve Maxwell's equations directly for ordinary and extraordinary waves [41,42]

$$\frac{\partial H_{o,e}}{\partial z} = -\frac{1}{c}\frac{\partial}{\partial t}\left(E_{o,e} + 4\pi P_{o,e}\right), \qquad \frac{\partial E_{o,e}}{\partial z} = -\frac{1}{c}\frac{\partial H_{o,e}}{\partial t} \qquad (32)$$

with the equations for two components of polarization vector P (anharmonic oscillators in the presence of quadratic nonlinearity model) [10]:

$$\frac{\partial^2 P_o}{\partial t^2} + \Omega_o^2 P_o + 2\beta_{oe}^o P_o P_e = \gamma_o E_o, \qquad \frac{\partial^2 P_e}{\partial t^2} + \Omega_e^2 P_e + \beta_{oo}^e P_o^2 = \gamma_e E_e. \qquad (33)$$

there z is the longitudinal coordinate, t is the time, c is the velocity of light in vacuum, Ω is the oscillator's resonance frequency, γ is the value proportional to dipole moment, β is a component of quadratic nonlineary tensor; o is the index pointing on ordinary and e on extraordinary waves. The symmetry of tensor elements of quadratic susceptibility gives: $\beta_{oo}^e / \gamma_e = \beta_{oe}^o / \gamma_o$.

The linear dispersion of every ordinary wave far from resonance is described according to (32), (33) by well known relation between the wave vector and the frequency: $k c = \omega \left[1 + 4\pi\gamma / \left(\Omega^2 - \omega^2\right)\right]^{1/2}$. It is interesting to note that in case of $4\Omega_o^2 = \Omega_e^2$ and $4\gamma_o = \gamma_e$ the global phase match appears: $2k_o(\omega) = k_e(2\omega)$ for any frequency ω. However, it's highly probable that this case could be observed only in model experiments with chains and long lines.

The following integral of motion, that could be written for coupled KdV equations derived from (32), (33) assuming the weak dispersion, is very important to analyze the character of interactions of ultra-short pulses:

$$I_3 = \int_{-\infty}^{\infty}\left[D_o n_o\left(\frac{\partial E_o}{\partial\tau}\right)^2 + D_e n_e\left(\frac{\partial E_e}{\partial\tau}\right)^2 + v n_o E_o^2 - 2\beta_{oo}^e D_o n_o \gamma_o \Omega_e^{-2} E_o^2 E_e\right]d\tau \qquad (34)$$

Here, the velocity mismatch $v = 1/v_o - 1/v_e$ is introduced. The functional (34) can be respected as a quasi-invariant of Maxwell's equations (32), (33). Indeed, the value of I_3 was a constant with a good precision during our numerical experiments. For finite signals in linear medium the value of I_3 is always positive. The sign of I_3 in nonlinear medium may become negative for strong electromagnetic fields, which indicates the parametric coupled soliton formation.

Let's consider the stationary travelling wave $H(\tau_s)$, $E(\tau_s)$, $P(\tau_s)$, where $\tau_s = t - z/v_s$. The relation between the electrical and magnetic fields is: $E = \alpha P$ and $H = v_s E/c$, the wave propagation velocity is $v_s = c\left(1 + 4\pi/\alpha\right)^{-1/2}$. The parameter of propagation α depends on profile of $E(\tau_s)$ and vice versa. The nonlinear solitary wave, however, propagates faster than in linear media $(v_s > v_o, v_e)$.

The equations (29), (30) have simple exact solution, similar to quadratic envelope soliton solution:

$$E_o = E_{om} \operatorname{sech}^2\left(\tau_s/T_s\right), \quad E_e = E_{em} \operatorname{sech}^2\left(\tau_s/T_s\right). \tag{35}$$

The parameter of propagation and duration of these solitons are equal to: $\alpha = \left(\Omega_o^2 - \Omega_e^2\right)/\left(\gamma_o - \gamma_e\right)$, $T_s^2 = 4\left(\gamma_o - \gamma_e\right)/\left(\gamma_e \Omega_o^2 - \gamma_o \Omega_e^2\right)$. The amplitude of electrical field are determined by

$$E_{em} = 3\alpha/\left(\beta_{oe}^o T_s^2\right), \quad E_{om} = \pm 6\alpha/\left(2\beta_{oe}^o \beta_{oo}^e T_s^4\right)^{1/2}. \tag{37}$$

Generally, the duration of two parts of parametric soliton with ordinary and extraordinary polarization are not equal; moreover, the profiles of solitary waves have more complicated form than (35). Since the simple exact solution was not found, we can apply to the approximate method based on minimization of Hamiltonian of interactions in order to obtain soliton parameters. Setting the form of solution as Gaussian functions

$$E_o = E_{om} \exp\left(-\tau_s^2/T_o^2\right), \quad E_e = E_{em} \exp\left(-\tau_s^2/T_e^2\right), \tag{38}$$

it is possible to obtain the following formula:

$$E_{om} = \alpha\left(2+\xi\right)^{3/2}/\left[\left(\beta_{oe}^o \beta_{oo}^e\right)^{1/2} \sqrt[4]{\xi}\left(2-\xi\right)^{1/2} T_o^2\right] \tag{39}$$

Here $\xi = T_e^2/T_o^2$ is the ratio of squared pulse durations, $\alpha = \left(\Omega_o^2 \alpha_1 - \Omega_e^2\right)/\left(\gamma_o \alpha_1 - \gamma_e\right)$ and $\alpha_1 = \xi^2\left(2+3\xi\right)/\left[\left(4+\xi\right)\left(2-\xi\right)\right]$. If the durations are the same $T_o = T_e$, $\xi = 1$ we obtain $\alpha_1 = 1$ and (38) reduces to (35) with correction due to difference in soliton profiles. Notice, that α_1 is similar to the analogous parameter introduced in [19] for envelope solitons in the second-order approximation of dispersion theory. However, unlike the envelope solitons, the velocity of parametric solitons, described by Maxwell's equations, depends on the value of electrical field.

In order to solve (32), (33) numerically, we have developed the conservative spectrum-differential method [42]. The equations in question contains the dimensionless normalized time $\omega_o t$ and length $z\omega_o/c$. The reference frequency ω_o in the example below is 9 times less than the resonant one of ordinary wave Ω_o.

The results of computing of two ultra-short pulses propagation are shown in the Fig. 17. The initial forms of the pulses are:

$$E_o = E(0)\exp\left(-t^2/T^2\right)\sin\left(t/2\right), \quad E_e = -E(0)\exp\left(-t^2/T^2\right)\sin t \tag{39}$$

with $T = 6,28$ and $E(0) = 20$ (see Fig 17, curves 1). The quasi-Hamiltonian (34) is negative in this case. The quadratic medium has the following values of parameters: $\Omega_o^2 = 81$, $\Omega_e^2 = 77.44$, $\gamma_o = 8$, $\gamma_e = 7.65$, $\beta_{oe}^o = 1$. In linear case the ordinary and extraordinary wave velocities on low frequencies are equal to $v_o = 0,6679257\,c$ and $v_e = 0,6679468\,c$.

The numerical simulation showed that at the beginning the pulses acquire an irregular form. Next, a number of solitary waves is formed and the fast-oscillating "tail" is detached due to its low velocity. At the distance of $Z = 5600$, the profiles of

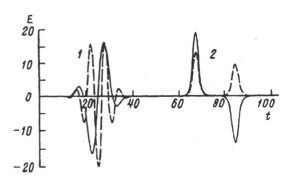

Fig. 17. Parametrically coupled soliton formation due to interaction of ordinary (solid) and extraordinary (dashed) waves. Input (1) and output (2, at the distance $z=5600$) pulse profiles are shown.

pulses becomes stable (see Fig 17, curves 2), as the couple of parametric solitons with different amplitudes and velocities. Faster soliton with $E_{om} = 19,2198$, $E_{em} = 13,4898$ has the propagation parameter $\alpha = E_{om}/P_{om} = 10,3401$, so its normalized velocity is $v_s/c = 0.67186$. The other, slower one has smaller amplitudes: $E_{om} = -13.6914$, $E_{em} = 9.6336$, propagation parameter $\alpha = E_{om}/P_{om} = 10,27914$ and velocity $v_s/c = 0.67077$. This data is in a good agreement with the results obtained using mentioned earlier variation method.

It is interesting to note that the parametric coupled soliton with the smaller amplitude was formed at the front of initial pulses, but the other soliton outran it and took the leading position during the propagation. It is obvious that this position switching was accompanied with their collision. Notice, that the first soliton has the same signs of all its components. Two parts of the second one has opposite signs. We mentioned above that the ordinary wave can have either positive or negative sign. This feature of coupled solitons takes place during the evolution of ultra-short wave packets (39).

10. Conclusion

In the nearest future we anticipate new experimental achievements in the creation of all-optical switching logical devices, employing the unique properties of quadratic solitons. This will strongly encourage further theoretical investigations. However, some interesting problems could be enumerated right now.

It is necessary to develop various kind of approximate analytical methods in the theory of parametric solitons, which is essential to arrange and perform the experiments. The development of the theory of parametric soliton collisions will continue. The high-order approximations in the theory of linear and nonlinear dispersion could be very helpful for the analysis of short temporal and narrow spatial parametric solitons.

The solutions of coupled KdV equations let us investigate the capability of parametric coupled brithers formation in quadratic media. The direct studying of Maxwell's equations allows us to investigate a lot of novel problems such as parametric amplifying and decay instability of ultra-short pulses in addition to researching of soliton properties of various duration. Finally, the phenomena of self-focusing and defocusing of beams in quadratic media are also of great interest.

It is important to notice that the parametric soliton theory could be applied to mechanics, not only to problems of nonlinear optics. The parametric self-action of two harmonics are of practical interest to nonlinear acoustics and hydrodynamics because the quadratic nonlinearity is the strongest for acoustic waves. This list of possible applications of the quadratic nonlinearity will surely grow up in the future.

Author would like to thank Yu. N. Karamzin, M. V. Komissarova, Xin Lu, A. S. Potashnikov and S. V. Polaykov for their help in preparing of this paper.

This work was partially supported by the Russian Foundation for Basic Research.

References

1. Karamzin, Ju.N. and Sukhorukov, A.P. (1974) Nonlinear interaction of diffracted light beams in a medium with quadratic nonlinearity: mutual focusing of beams and limitation on the efficiency of optical frequency converters, *JETP Lett.* **20**, 339–342.
2. Karamzin, Ju.N. and Sukhorukov, A.P. (1976) Mutual focusing of high-power light beams in media with quadratic nonlinearity, *Sov. Phys. JETP* **42**, 414–420.
3. Karamzin, Ju.N., Sukhorukov, A.P., and Filipchuk, T.S. (1978) On a new class of coupled solitons in a dispersive medium with quadratic nonlinearity, *Moscow Univ. Phys. Bull.* **33** (3), 73–79.
4. Sukhorukov, A.P. (1988) *Nonlinear Wave Interactions in Optics and Radiophysics*, Nauka Publishers, Moscow (in Russian).
5. Kanashov, A.A. and Rubenchik, A.M. (1981) On diffraction and dispersion effects on three wave interaction *Physica* 4D, 123–134.
6. Torruellas, W.E., Wang, Z., Hagan, D.J., Van Stryland, E.W., Stegeman, G.I., Torner, L., and Menyuk, C.R. (1995) Observation of two-dimensional solitary waves in a quadratic medium, *Phys. Rev. Lett.* **74**, 5036–5039.
7. Schiek, R., Baek, Y., and Stegeman, G.I. (1996) One-dimensional spatial solitons due to cascaded second-order nonlinearities in planar waveguides, *Phys. Rev. E* **53**, 1138–114.
8. Akhmanov, S.A., Sukhorukov, A.P., and Khokhlov, R.V. (1964) Self-focusing and diffraction of light in a nonlinear medium, *Sov. Phys. Usp.* **93**, 609–636.
9. Khokhlov, R.V. (1961) On propagation of waves in nonlinear dispersive media, *Radiotekh. Elektron.* **6**, 917-921 (in Russian).
10. Akhmanov, S.A. and Khokhlov, R.V. (1972) *Problems of Nonlinear Optics*, Gordon and Breach Publishers, New York (1964, Russian edition).
11. Bloembergen, N. (1965) *Nonlinear Optics*, Benjamin Publishers, New York.
12. Ostrovskii, L.A. (1967) Self-action of light in crystals, *JETP Lett.* **5**, 272–275.
13. Akhmanov, S.A., Sukhorukov, A.P., and Chirkin, A.S. (1969) Nonstationary phenomena and space-time analogy, *Sov. Phys. JETP* **28**, 748–755.
14. Hayata, K. and Koshiba, M. (1993) Multidimensional solitons in quadratic nonlinear media, *Phys Rev. Lett.* **71**, 3275–3278; (1994), **72**, 178 (E).
15. Werner, Y.J. and Drummond, P.D. (1993) Simulton solitons for the parametric amplifier, *J. Opt. Soc. Am.* B **10**, 2390–2393.
16. Werner, Y.J. and Drummond, P.D. (1994) Strongly coupled nonlinear parametric solitary waves, *Opt. Comm.* **19**, 613–615.

184

17. Kalocsai, A.G. and Haus, J.W. (1994) Nonlinear Schrodinger equation for optical media with quadratic nonlinearity, *Phys Rev. A* **49**, 574–585.
18. Buryak, A.V. and Kivshar, Yu.S. (1994) Spatial optical solitons governed by quadratic nonlinearity, *Opt. Lett.* **19**, 1612–1614.
19. Steblina, V.V., Kivshar, Y.S., Lisak, M., and Malomed, B.A. (1995) Self-guided beams in a diffractive $\chi^{(2)}$ medium: variational approach, *Opt. Comm.* **118**, 345–352.
20. Agranovich, V.M., Darmanyan, S.A., Kamchatov, A.M., Leskova, T.A., and Boardman, A.D. (1997) Variational approach to solitons in systems with cascaded $\chi^{(2)}$ nonlinearity, *Phys. Rev. E* **55** (in press).
21. Lu, X. and Sukhorukov, A.P. (1996) Three-dimensional spatial solitons in a quadratically nonlinear medium, *Bull. RAS, phys.* **60** (12), 64–69.
22. Stegeman, G.I., Hagan, D.J., and Torner, L. $\chi^{(2)}$ cascading phenomena and their applications to all-optical signal processing, mode-locking, pulse compression and solitons (1996) *J. Opt. and Quantum Electron.* **28**, 1691.
23. Torner, L., Menyuk, C.R., Torruellas, W.E., and Stegeman, G.I (1995) Two-dimensional solitons with second-order nonlinearities, *Opt. Lett.* **20**, 13–15.
24. Torner, L., Menyuk, C.R., and Stegeman, G.I (1995) Bright solitons with second-order nonlinearities, *J. Opt. Soc. Am. B* **12**, 889–897.
25. Malomed, B.A., Andersen, D., and Lisak, M. (1996) Three-wave interaction solitons in a dispersive medium with quadratic nonlinearity, *Opt. Comm.* **126**, 251–254.
26. Buryak, A.V. and Kivshar, Yu.S. (1997) Multistability of three-wave parametric self-trapping. *Phys. Rev. Lett.* **78**, 3286–3289.
27. Boardman, A.D. and Xie K. (1997) Vector spatial solitons influenced by magneto-optic effects in cascadable nonlinear media, *Phys Rev. E* **55**, 1–11.
28. Akhmanov, S.A., Chirkin, A.S., Drabovich, K.N., Khokhlov, R.V., and Sukhorukov A. P. (1968) Nonstationary nonlinear optical effects and ultra-short light pulses formation, *IEEE. Quant. Electron.* QE-**4**, 598–605.
29. Zakharov, V.E. and Shabat, A.B. (1971) *Sov. Phys. JETP* **34**, 62.
30. Kobyakov, A. and Lederer, F. (1996) Cascading of quadratic nonlinearities: An analytical study, *Phys. Rev. A* **54**, 3455–3471.
31. Danielus, R., Piskarskas, A., Sirutkaitis, V., Stabinis, A., and Jasevichute, J. (1983) *Optical parametric oscillators and picosecond spectroscopy*, Mokslas Publishers, Vilnius (in Russian).
32. Danielus, R., Dubietis, A., Valiulis, G., and Piskarskas, A. (1995) Femtosecond high-contrast pulses from a parametric generator by the self compressed second harmonic of a Nd laser, *Opt. Lett.* **20**, 2225–2227.
33. Khaydarov, J. D. V., Andrews, J. H., and Singer, K. D. (1994) Pulse compression in a synchronously pumpud optical parametric oscillator from group-velocity mismatch, *Opt. Lett.* **19**, 831–834.
34. Azimov, B.S., Karamzin, Ju.N., Sukhorukov, A.P., and Sukhorukova, A.K (1980) Interaction of weak pulses with a low-frequency high-intensity wave in a dispersive medium, *Sov. Phys. JETP*. **51**, 40–46.
35. Lu, X. and Sukhorukov, A.P. (1997) Mutual focusing in media with quadratic nonlinearity, *Bull. RAS, phys* **61** (in press).
36. Balashenkov, N.R., Gagarskii, S.V., and Inochkin, M.V. (1989) Nonlinear refraction of light on second harmonic generation, *Opt. Spectrosc.* **66**, 806–808.
37. DeSalvo, R., Hagan, D.J, Shiek-Bahae, M., Stegeman, G.I., and Vanherzeale, H. (1992) Self-focusing and self-defocusing by cascaded second-order effects in KTP, *Opt. Lett.* **17**, 28–30
38. Komissarova, M.V. and Sukhorukov, A.P. (1981) On parametric processes in the interference of several channels of wave interaction, *Laser Physics* **6**, 1036–1041.
39. Akhmanov, S.A., Vysloukh, V.A., and Chirkin, A.S. (1988) *Optics of femtosecond laser pulses*, Nauka Publishers, Moscow (in Russian).
40. Sukhorukov, A.A. On the theory of parametrically coupled solitons with high-order dispersion, *Bull. RAS, phys.* **61** (12) (in press).
41. Dubrovskaya, O.B. and Sukhorukov, A.P. (1992) Interaction of optical pulses with a number of ocsillations in media with quadratic nonlinearity, *Bull. RAS, phys.* **56** (12), 184–188.
42. Karamzin, Ju.N., Potashnikov, A.S., and Sukhorukov, A.P. (1996) Interaction extremely short electromagnetic pulses in media with quadratic nonlinearity, *Bull. RAS, phys.* **60** (12), 29–38.
43. Gottwald, G., Grimshaw, R., Malomed, B. (1997) Parametric envelope solitons in coupled Korteweg -de Vries equations, *Phys. Lett. A* **227**, 47-54.

A LITHIUM NIOBATE QUADRATIC DEVICE FOR WAVELENGTH MULTIPLEXING AROUND 1.55μm

Katia Gallo and Gaetano Assanto
Dept. of Electronic Engineering, Terza University of Rome
Via della Vasca Navale 84, 00146 Rome, Italy

George I. Stegeman
CREOL University of Central Florida
4000 Central Florida Blvd., Orlando, Florida 32816-2700 USA

A widespread interest in all-optical networks for telecommunications has generated a great deal of activity towards approaches for efficient channel shifting in systems based on wavelength-division-multiplexing (WDM). A key issue, namely the possibility of transferring an incoming stream of data from a given channel or wavelength to another, has been addressed using several techniques ranging from gain saturation in semiconductor amplifiers [1] to four-wave-mixing [2] to parametric generation [3]. Since available bandwidth, transparence to the modulation format, possibility of gain and amount of crosstalk are important characteristics of such a wavelength shifter, parametric processes are considered rather appealing. With the advances in periodically poled crystals for efficient second-harmonic-generation (SHG) and difference frequency generation (DFG) [4], a quadratic approach in guided-wave configurations appears quite affordable in terms of required powers.

Here we investigate numerically the performance of a lithium niobate channel waveguide where an SHG process is followed by DFG in order to shift an input signal at frequency ω_2 to an output at ω_3, with a pump at ω_P such that $\omega_3=2\omega_P-\omega_2$. The device, sketched in Fig. 1a, encompasses a two-pass geometry by means of a dichroic mirror of reflectivity 100% at the second-harmonic $2\omega_P$ generated in the first pass, and completely transparent in a wide interval including the pump frequency ω_P and the counterinjected signal at ω_2. This geometry is the evolution of the simple configuration discussed in [5] and demonstrated in bulk by Tan *et al.* [6]. An alternative scheme has been also proposed in Ref. 7 for a multiple quantum well waveguide. We assumed all the input wavelengths (pump and signal) to belong to the useful bandwidth of Erbium-doped fiber amplifiers, and chose $\lambda_P=1.55\mu m$ for the pump and $\lambda_2=\lambda_P-\delta\lambda$ for the signal, with $|\delta\lambda|\leq 50nm$. The output is then at $\lambda_3=(2/\lambda_P-1/\lambda_2)^{-1}\sim\lambda_P+\delta\lambda$. We considered an annealed proton exchanged (APE) channel realized in a z-cut LiNbO$_3$ crystal, with propagation along the x-axis and a transverse extraordinary refractive index profile of the form:

$$n(y, z)=n_e+\Delta n_e \exp(-z/d) \exp(-|y|/w)$$

185

A. D. Boardman et al. (eds.),
Advanced Photonics with Second-order Optically Nonlinear Processes, 185–188.
© *1999 Kluwer Academic Publishers. Printed in the Netherlands.*

We took Δn_e=0.025, T=300K, d=2μm, w=3μm, with n_e given by standard Sellmeier equations [8]. The modal structure of the waveguide was calculated with a finite difference algorithm.

A ferroelectric domain inversion grating with a period Λ=16.7 μm, achievable by electric field poling,[9] can be employed to Quasi-Phase-Match (QPM) the $TM_{00}(\lambda_P) + TM_{00}(\lambda_P) \leftrightarrow TM_{00}(\lambda_P/2)$ SHG interaction. In this case the $TM_{00}(\lambda_P/2) \leftrightarrow TM_{00}(\lambda_2) + TM_{00}(\lambda_3)$ DFG interaction is slightly mismatched as λ_2 is tuned around the degeneracy point $\lambda_2 = \lambda_P = 1.55$μm, with all the other interactions between guided modes being negligible. SHG from the ω_P pump occurs in forward propagation and the generated second harmonic is then reflected by the mirror. Then DFG is driven by the reflected SH and the ω_2 signal injected through the reflector, in order to generate the output signal propagating backward.

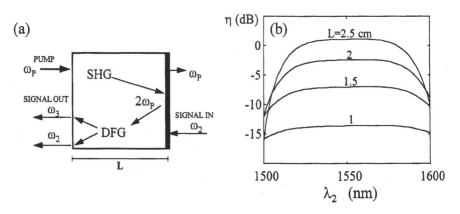

Figure 1 : a) Sketch of the $\omega_2 \rightarrow \omega_3$ converter with counter-propagating pump and signal 1; *b)* cw conversion efficiency η in dB, vs input signal wavelength (λ_2), for various waveguide lengths L. Input powers: P_{pump}=300mW, P_{signal}=1mW.

Fig. 1b shows the calculated efficiency vs signal wavelength λ_2 in the hypothesis of no losses and Kleinman symmetry. First we integrated the coupled equations for Type I SHG from z=0 to z=L, with input power P_{pump} at λ_P, in order to evaluate the SH power P_{SH} generated in the first pass. We then used P_{SH} and the input signal power P_{signal} at λ_2 to integrate the three coupled DFG equations describing the backward-propagating interaction between λ_2, λ_3 and $\lambda_P/2$ over the channel length L. Calculated effective areas in the waveguide, $S_{DFG} \sim S_{SHG}$=49.5μm² yielded, for an ideal first order QPM grating and d_{33}=27pm/V, the nonlinear strengths

$$\frac{\Gamma_{DFG}}{L} = \omega_P \left(\frac{2}{\pi} d_{33}\right) \sqrt{\frac{2\mu_0/c}{S_{DFG} \, n_{eff}^{2\omega_P} \, n_{eff}^{\omega_2} \, n_{eff}^{\omega_3}}} \cong \frac{\Gamma_{SHG}}{L} = \omega_P \left(\frac{2}{\pi} d_{33}\right) \sqrt{\frac{2\mu_0/c}{S_{SHG} \, n_{eff}^{2\omega_P} \, (n_{eff}^{\omega_P})^2}} = 0.86[cm\sqrt{W}]^{-1}$$

The corresponding normalized SHG conversion efficiency from 1.55 μm is 74% [/Wcm²] , which compares well with the reported experimental value of 43% that included propagation losses [10].

The broad 3dB conversion bandwidth, typical of DFG processes with QPM LiNbO$_3$ in the near IR,[11] is only slighly reduced with respect to the unidirectional geometry (see Ref. 5), but amplification can be achieved at substantially lower P$_{pump}$ and L. Figs. 2a and 2b show the iso-efficiency curves for the same input conditions, in the two configurations.

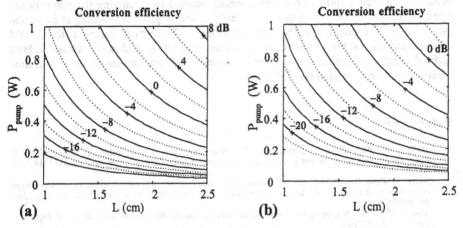

Figure 2: contour lines of the conversion efficiency (given in dB for the marked solid lines) vs waveguide length L and input pump power P$_{pump}$ for the cw case with: *a)* counter-propagating inputs (configuration of Fig.1a); *b)* co-propagating inputs. The signal input is 1mW at λ$_2$=1.57μm.

A reduction in efficiency with respect to the cw case is predicted for pulsed signal and pump, as in the case of a Time Division Multiplexed (TDM) protocol. The graphs in Fig. 3 are calculated for gaussian shaped input pulses of picosecond durations (which allow the use of a quasi-cw approximation). Here P$_{pump}$ and P$_{signal}$ refer to peak powers.

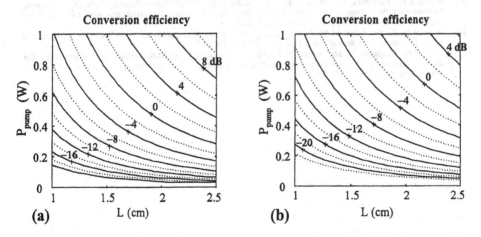

Figure 3: contour lines of the conversion efficiency (given in dB for the marked solid lines) vs waveguide length L and input pump peak power P$_{pump}$ for the pulsed case, with: *a)* counter-propagating inputs (configuration of Fig.1a), *b)* co-propagating inputs. The input signal peak power is 1mW and λ$_2$=1.57μm.

188

In conclusion, a double-pass geometry for a broadband integrated wavelength shifter for WDM in a Lithium Niobate quasi-phase-matched channel waveguide has been studied numerically. The λ-converter operates in the 1.55μm spectral window for optical fiber communications and requires reasonably low pump-powers and propagation lengths to provide lossless, or even amplified wavelength shifting. The contradirectional geometry for signal and pump offers a simple method for removing the pump beam, while the cascading of two processes is ideally suited to prevent coupling in undesired modes at the highest (pump SH) frequency. This novel element for WDM communication systems, based on quadratic cascading and well within today's technological capabilities, might indeed become a practical device in all-optical time and/or wavelength division networks.

References

1. Durhuus, T., Mikkelsen, B., Joergensen, C., Lykke Danielsen, S. and Stubkjaer, K.E. (1996) All-optical wavelength conversion by semiconductor optical amplifiers, *J. Lightwave Technol.* **14**, 942-953.
2. Zhou, J., Park, N., Vahala, K.J., Newkirk, M.A. and Miller, B.J. (1994) Broadband wavelength conversion with amplification by four-wave mixing in semiconductor optical amplifers, *Electron. Lett.* **30**, 859-860.
3. Yoo, S.J.B. (1996) Wavelength conversion technologies for WDM network applications, *J. Lightwave Technol.* **14**, 955-966.
4. Fejer M.M. (1997) Microstructured media for nonlinear optics : materials, devices and applications, *this book.*
5. Assanto, G., Gallo, K. and Conti, C. (1997) Plane and guided wave effects and devices via quadratic cascading, *this book.*
6. Tan, H., Banfi, G.P. and Tomaselli A. (1993) Optical frequency mixing through cascaded second-order processes in β-barium borate, *Appl. Phys. Lett.* **63**, 2472-2474.
7. Gorbounova, O., Ding, Y.J., Khurgin, J.B., Lee, S.J. and Craig, A.E. (1996) Optical frequency shifters based on cascaded second-order nonlinear processes, *Opt. Lett.* **21**, 558-560.
8. Nelson, D.F. and Mikulyak, R. (1974) Refractive indices of congruently melted lithium niobate, *J. Appl. Phys.* **45**, 3698-3700.
9. Kintaka K., Fujimura, M., Suhara, T. and Nishihara, H. (1996) High-efficiency LiNbO$_3$ waveguide second-harmonic generation devices with ferroelectric-domain-inverted gratings fabricated by applying voltage, *J. of Lightwave Technol.* **14**, 462-468.
10. Arbore M.A. and Fejer M.M. (1997) Singly resonant optical parametric oscillation in periodically poled lithium niobate waveguides, *Opt. Lett.* **22**, 151-153.
11. Xu, C.Q., Okayama, H. and Kamijoh, T. (1995) Broadband multichannel wavelength conversions for optical communication systems using quasiphase matched difference frequency generation, *Jpn. J. Appl. Phys.* **34**, L1543-L1545.

FULL VECTOR THEORY OF FUNDAMENTAL AND SECOND-HARMONIC CW WAVES

P.BONTEMPS AND A.D.BOARDMAN
Photonics and Nonlinear Science Group, Joule Laboratory
University of Salford
Salford, M5 4WT, United Kingdom

1. Introduction

Second-harmonic generation (SHG) is explained in terms of linear phase matching and deploys in non-centrosymmetric crystals. The classification of the SHG processes into type I and type II are well known but the more general vector case has not been studied. Type II SHG is a vector process involving two electric field components at the fundamental frequency and one component at the second-harmonic frequency. It is assumed in the literature that only one component of the second-harmonic field is phase matched but this argument is based upon the magnitude of the linear phases only. For a situation in which linear phase matching is absent the nonlinear phases are important and will balance the linear phases. The full vector theory needs to be investigated therefore.

2. Theory

Consider two vector electric fields E^ω and $E^{2\omega}$ propagating along the optic axis. Each of these fields is a combination of x and y components, as represented in figure 1. If c is the velocity of light in the vacuum and n^ω, $n^{2\omega}$ are, respectively, the refractive indices at frequencies ω and 2ω, the general wave equations are given by [1]

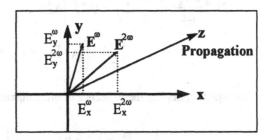

Figure 1. Vector electric fields E^ω, $E^{2\omega}$

$$\nabla^2 E^\omega - \nabla\left(\nabla.E^\omega\right) + \frac{\omega^2}{c^2}\left(n^\omega\right)^2 E^\omega + \frac{\omega^2}{c^2}P_\omega^{NL} = 0 \qquad (1a)$$

189

A. D. Boardman et al. (eds.),
Advanced Photonics with Second-order Optically Nonlinear Processes, 189–192.
© 1999 *Kluwer Academic Publishers. Printed in the Netherlands.*

$$\nabla^2 E^{2\omega} - \nabla(\nabla . E^{2\omega}) + \frac{4\omega^2}{c^2}(n^{2\omega})^2 E^{2\omega} + \frac{4\omega^2}{c^2} P_{2\omega}^{NL} = 0 \qquad (1b)$$

where P_ω^{NL} and P_ω^{NL} are the nonlinear polarisation for respectively the fundamental frequency and the second-harmonic. For a trigonal or hexagonal $(\bar{6}, \bar{6}m2, 32 \text{ and } 3m \text{ classes})$ crystal, the x-component of the polarisation at frequency ω is

$$\left(P_\omega^{NL}\right)_x = \frac{1}{2}\left[2\kappa_1\left(E_x^\omega\right)^* E_x^{2\omega} - 2\kappa_2\left(E_x^\omega\right)^* E_y^{2\omega} - 2\kappa_2\left(E_y^\omega\right)^* E_x^{2\omega} - 2\kappa_1\left(E_y^\omega\right)^* E_y^{2\omega}\right] \quad (2)$$

where the $\kappa_{1,2}$ are

$$2\kappa_1 = \chi_{xxx}^{(2)}(\omega;-\omega,2\omega) = -\chi_{xyy}^{(2)}(\omega;\omega,-2\omega) = \chi_{xxx}^{(2)}(2\omega;\omega,\omega) = -2\chi_{xyy}^{(2)}(2\omega;\omega,\omega) \quad (3a)$$

$$2\kappa_2 = \chi_{xyx}^{(2)}(\omega;-\omega,2\omega) = -\chi_{xxy}^{(2)}(\omega;\omega,-2\omega) = -\chi_{xxy}^{(2)}(2\omega;\omega,\omega) \quad (3b)$$

where $\chi_{ijk}^{(2)}$ is the second-order susceptibility tensor.

Let concentrate now on a trigonal (3m class) crystal such as lithium niobate (LiNbO$_3$), in this case, due to crystal symmetry we have $\kappa_2 = 0, \kappa_1 = \kappa$. Factoring out the average linear wave number [1], for example $E_x^\omega = A_x^\omega \exp(ik^\omega z)$, the general wave equations for the x-component at frequency ω becomes

$$2ik^\omega \frac{dA_x^\omega}{dz} + \left[\left(\frac{\omega}{c}n_x^\omega\right)^2 - \left(k^\omega\right)^2\right]A_x^\omega$$

$$- \left(\frac{\omega}{c}\right)^2 \kappa\left[\left(A_x^\omega\right)^* A_y^{2\omega} + \left(A_y^\omega\right)^* A_x^{2\omega}\right]\exp i\left(k^{2\omega} - 2k^\omega\right)z = 0 \qquad (4)$$

This equation implies the existence of birefringence parameters, defined as

$$\nu_1 = \frac{1}{2k^\omega}\left[\left(\frac{\omega}{c}n_x^\omega\right)^2 - \left(k^\omega\right)^2\right]; \nu_2 = \frac{1}{2k^\omega}\left[\left(\frac{2\omega}{c}n_x^{2\omega}\right)^2 - \left(k^{2\omega}\right)^2\right] \qquad (5)$$

The linear wavenumber ratio is $\alpha = \dfrac{k^{2\omega}}{2k^\omega}$ and $\qquad (6)$

The normalisation constant $\Gamma = \left(\dfrac{\omega}{c}\right)^2 \dfrac{\kappa}{2k^\omega}\sqrt{\dfrac{I_0}{2}}\sqrt{\dfrac{\mu_0}{\varepsilon_0}}$ can be introduced, $\qquad (7)$

where ε_0 is the permeability of free space, μ_0 is the susceptibility of free space and I_0 is the normalising intensity.

Using the transformation $A_{x,y}^{\omega,2\omega} \rightarrow \sqrt{\dfrac{2n_{x,y}^{\omega,2\omega}}{I_0}}\sqrt{\dfrac{\varepsilon_0}{\mu_0}}\,A_{x,y}^{\omega,2\omega}$ and assuming the z-axis points along the optic axis (i.e. $v_1 = v_2 = 0$), reduces the wave equations considerably. For example, the x-component is

$$i\frac{dA_x^{\omega}}{dz} - \Gamma\left[\frac{\left(A_x^{\omega}\right)^* A_y^{2\omega}}{\sqrt{n_y^{2\omega}}} + \sqrt{\frac{n_x^{\omega}}{n_y^{\omega}n_x^{2\omega}}}\left(A_y^{\omega}\right)^* A_x^{2\omega}\right] = 0 \qquad (8a)$$

$$i\alpha\frac{dA_x^{2\omega}}{dz} - 8\Gamma\left[\sqrt{\frac{n_x^{2\omega}}{n_x^{\omega}n_y^{\omega}}}A_x^{\omega}A_y^{\omega}\right] = 0 \qquad (8b)$$

where the (complex) variables $A_{x,y}^{\omega,2\omega}$ include the (real) amplitudes $U_{x,y}^{\omega,2\omega}$ and the nonlinear phases $\beta_{x,y}^{\omega,2\omega}$ because $A_{x,y}^{\omega,2\omega} = U_{x,y}^{\omega,2\omega}\exp\!\left(i\beta_{x,y}^{\omega,2\omega}z\right)$.

3. Numerical Application to Lithium Niobate

The following data for LiNbO₃ at 25°C, $\lambda = 1.064\mu m$: $n_0^{\omega} = 2.2340$, $n_e^{\omega} = 2.1554$,

$n_0^{2\omega} = 23251$, $n_e^{2\omega} = 2.2330$ is used. The other parameters are $k^{\omega} = 4.1014*10^6 \pi m^{-1}$, $k^{\omega} = 8.6267*10^6 \pi m^{-1}$, $I_0 = 1GW/cm^2$ and $\alpha = 2.10$. Figure 2 illustrates the evolution of the fields in the crystal. The input intensities for the fundamental waves are $U_x^{\omega} = 10, U_y^{\omega} = 2.7GW/cm^2$, no second-harmonic is input to the crystal. It is interesting to note that the second-harmonic generated is elliptically polarised and has, therefore, two

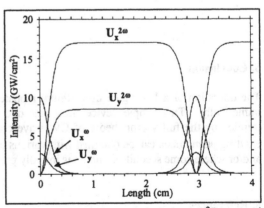

Figure 2. Generation of the two components $U_{x,y}^{2\omega}$ of the 2ⁿᵈ harmonic. In this case, $\Delta\beta = 0$.

components $E_x^{2\omega}$ and $E_y^{2\omega}$. Compared to Type II SHG, this configuration, thanks to its new degree of freedom, opens new possibilities for the design of optical devices.

4. Simple Application: Polarisation Switching at 2ω

Since the early work of P.St.J.Russell [2], G.Assanto, G.I.Stegeman and collaborators, a lot of interesting devices based on $\chi^{(2)}$ media have been suggested and the principles tested in the laboratory. The devices can be controlled with the phase and/or the amplitude of a seed at 2ω [3].

In the application shown here, the idea is to vary the polarisation of the second-harmonic (output) by changing the phase of the fundamental (signal) with an electro-optic modulator. Figure 3 gives a schematic view of the experimental setup. Figure 4 shows the variation of $U_{x,y}^{\omega}$ with the initial nonlinear phase difference $\Delta\beta = \left| \beta_x^{\omega}(z=0) - \beta_y^{\omega}(z=0) \right|$. The value of $\Delta\beta$ necessary to pass from one state to another, $\pi/4$ in our example, is fixed by the initial amplitude difference $\left| U_x^{\omega}(z=0) - U_y^{\omega}(z=0) \right|$.

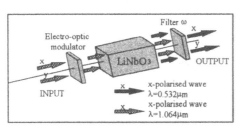

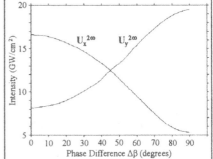

Figure 3. Proposed experimental setup

Figure 4. Evolution of $U_{x,y}^{2\omega}$ with $\Delta\beta$

5. Conclusion

The operation of a logic gate in a lithium niobate crystal has been briefly described numerically. This simple device illustrates one possible application, among many implied by the full vector theory of CW wave coupling in a quadratic media. Other exciting phenomena can be obtained with various crystal classes ($\kappa_1 \neq \kappa_2 \neq 0$) in mind and/or where some second-harmonic is initially present.

6. References

1. Boardman, A.D. and Xie, K. (1997) Vector spatial solitons influenced by magneto-optic effects in cascadable nonlinear media, *Phys. Rev.* E55, **1**, 1.
2. Russell, P.St. (1991) Theoretical study of parametric frequency and wavefront conversion in nonlinear holograms, *IEEE J.Quant.Elect.*, **27**, 830.
3. Assanto, G., Torelli, I. and Trillo, S. (1994) All optical processing by means of vectorial interactions in second order cascading: novel approaches, *Opt. Lett.*, **19**, 1720.

NONLINEAR PHASE SHIFTS IN A COUNTERPROPAGATING QUASI-PHASE-MATCHED CONFIGURATION

GARY D. LANDRY AND THERESA A. MALDONADO
The University of Texas at Arlington
Box 19016, Arlington, TX 76019 USA
maldonado@uta.edu

1. Introduction

Recently, there has been a large interest in cascaded second order processes for many applications including optical transistors,[1] and all optical switching.[2] One important aspect of these processes is the ability to produce large nonlinear phase shifts at relatively low input intensities. This poster presents a novel quasi-phase-matching scheme to obtain nonlinear phase shifts via cascaded second order processes in a counterpropagating configuration.

The device under consideration is shown in the inset of Figure 1. This configuration was first proposed for second harmonic generation (SHG) by Ding and Khurgin.[3] Like a standard quasi-phase-matched (QPM) device, the second order nonlinear coefficient $d_0 = \chi^{(2)}/2$ sign is periodically reversed. Unlike a standard QPM device, the domain inversion period Λ is equal to the wavelength of the second harmonic (SH) wave in the material. This difference leads to radically different operational characteristics. In this poster's analysis, the propagation coordinate z is normalized to the device length. The input and output interface is located at $z = 0$. There is a high reflectivity mirror located at $z = 1$.

There are four plane waves involved in the analysis, two at the fundamental frequency (FF) and two at the second harmonic (SH). The forward and reverse propagating FF waves are denoted by A(z) and B(z), respectively. The corresponding SH waves are denoted by C(z) and D(z). All fields are normalized such that the magnitude squared is equal to the power in the field.

By definition, the nonlinear grating vector is given by $K = 2\pi/\Lambda$. Retaining only nearly phase matched processes, the four coupled differential equations that describe the interaction are given by[4]

$$\frac{d}{dz}A(z) = -j\Gamma B^*(z)[C(z) + D(z)] , \qquad (1)$$

$$\frac{d}{dz}B(z) = +j\Gamma A^*(z)[C(z) + D(z)] ,$$

$$\frac{d}{dz}C(z) = -j2\Gamma A(z)B(z) - j\Delta\kappa C(z) ,$$

$$\frac{d}{dz}D(z) = +j2\Gamma A(z)B(z) + j\Delta k D(z) ,$$

where the nonlinear coupling coefficient Γ and the phase mismatch $\Delta\kappa$ are given by

A. D. Boardman et al. (eds.),
Advanced Photonics with Second-order Optically Nonlinear Processes, 193–196.
© 1999 *Kluwer Academic Publishers. Printed in the Netherlands.*

$$\Gamma = \frac{4d_0 L}{\lambda_{0\omega} n_\omega} \sqrt{\frac{2\eta_0}{n_{2\omega}}} \quad , \tag{2}$$

$$\Delta\kappa = (k_{2\omega} - K)L \quad .$$

The most important property of this device is manifest in Eqs. (1). There are six separate interactions occurring simultaneously as opposed to two in the standard QPM design. This behavior greatly enhances the energy coupling process.

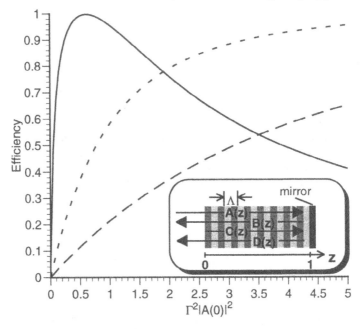

Figure 1. SHG conversion efficiency for the device under consideration (shown in inset) as a function of normalized input power. Also shown is the conversion efficiency for a standard forward QPM devices of length 1 and 2. The mirror reflectivity is taken to be 99% at both wavelengths.

2. Second Harmonic Generation

Assuming perfect phase matching $\Delta\kappa = 0$, solutions to Eqs. (1) are sought. Ding and Kurgin[3] solved this set of equations algebraically for SHG efficiency producing a transcendental solution. All solutions presented in this poster were found numerically using the Shooting method.[5]

The calculated SHG conversion efficiency is shown in Figure 1 as a function of normalized input power. For comparison, the efficiencies are also shown for forward QPM devices of normalized length of 1 and length of 2. The near 100% conversion peak for an input power of approximately $(\pi/4)^2 \cong 0.6$ is the most important aspect of this configuration.

Figure 2 shows the field magnitudes for two different input powers. As shown in Figure 2 (b), above the optimum input power level, the reverse propagating FF field reverses sign near 0.4 forcing a maximum in both SH fields.

Although all field interactions are very high in this region, a significant amount of SH power is converted back to the FF before exiting at z = 0. As the input power is increased, the SH peak increases and shifts towards z = 0.5. This behavior explains the single peak nature of the SHG efficiency plot in Figure 1.

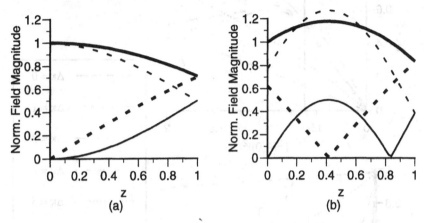

Figure 2. Normalized field profiles for input powers of (a) $\Gamma^2|A(0)|^2 = (\pi/4)^2$ and (b) $\Gamma^2|A(0)|^2 = 3$. The fields are |A(z)| (solid thick), |B(z)| (dashed thick), |C(z)| (solid thin), |D(z)| (dashed thin).

3. Cascaded Nonlinear Phase Shift

It is important to note that the field interactions are complicated and bidirectional. The energy exchange occurs through six distinct processes effectively increasing the input power and device length. Therefore, this device is promising for producing a nonlinear phase shift (NLPS) for a small amount of input power. To use the NLPS for all optical switching, a $\pm\pi/2$ phase shift is necessary for complete switching in a push-pull configuration. Consequently, the switching power is defined as the smallest input power necessary to produce a $\pm\pi/2$ NLPS. Since any power left in the SH is considered a loss, fundamental throughput is also a very important criterion.

Figure 3 shows the NLPS and fundamental throughput as a function of input power for selected values of phase mismatch. Although both positive and negative phase shifts are obtainable, only positive values are shown for clarity. The mismatch that is most important for switching applications is $\Delta\kappa = \pi$. At this value, the required switching power is approximately 0.6 and corresponds to a fundamental throughput of 96%.

It has been shown that the optimized switching power for a forward QPM device is 25 using the same normalization of power.[4] Therefore, the counterpropagating QPM configuration can be used in an all optical switching element with a switching power 42 times lower than the standard QPM design while occurring a negligible loss to the SH.

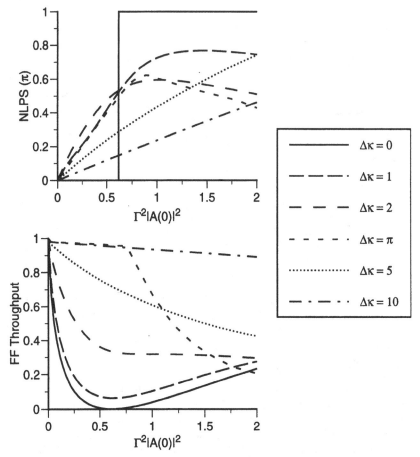

Figure 3. The NLPS (in units of π) and fundamental power throughput versus normalized input power for various values of phase mismatch.

4. References

1. G. Assanto (1995) Transistor action through nonlinear cascading in type II interactions, Optics Letters **20**, 1595-7.
2. Y. Baek, R. Schiek and G. I. Stegeman (1995) All-optical switching in a hybrid Mach-Zehnder interferometer as a result of cascaded second-order nonlinearity, Optics Letters **20**, 2168-70.
3. Y. J. Ding and J. B. Khurgin (1996) Second-harmonic generation based on quasi-phase matching: a novel configuration, Optics Letters **21**, 1445-7.
4. G. D. Landry and T. A. Maldonado (1997) Efficient nonlinear phase shifts due to cascaded second-order processes in a counterpropagating quasi-phase-matched configuration, Optics Letters **22**, 1400-2.
5. D. Zwillinger (1992) Handbook of Differential Equations, Academic Press, San Diego.

GENERATION OF HIGH POWER PICOSECOND PULSES BY PASSIVELY MODE-LOCKED Nd:YAG LASER USING FREQUENCY DOUBLING MIRROR

I. CH. BUCHVAROV, P.N. TZANKOV, V. STOEV, K. DEMERDJIEV
University of Sofia, Faculty of Physics,
Quantum Electronics Department,
5 J. Bourchier Blvd., 1164 Sofia, Bulgaria
D. SHUMOV
Institute of Applied Mineralogy, Bulgarian Academy of Sciences, Sofia,
Bulgaria

1. Introduction

In the last few years frequency doubling nonlinear mirror (FDNLM) consisting of a crystal for second harmonic generation and a dielectric mirror, placed at specific distance, was used as a means of self-mode-locking [1-3] or of pulse shortening [4-6] in pulse and CW pumped lasers. The results obtained so far for pulse pumped lasers are for the case when FDNLM operates only as an amplitude modulator (so called nonlinear mirror technique [1,2,5]). In this case the doubling in the nonlinear crystal is made with exact phase matching conditions and the dielectric mirror needs to be a dichroic mirror, i.e. to have comparatively low reflection coefficient at the fundamental frequency. These lasers suffer from two main drawbacks. Firstly the duration of the dichroic mirror output pulse is several times longer than the duration of the pulse traveling through the laser cavity [7] and secondly the laser suffers pulse parameter fluctuations [1,2,5]. Here we described a self mode-locked pulsed laser which has the aforementioned drawbacks overcame. The synchronization of modes in it is made by phase shift FDNLM, the latter operating by cascaded second order processes.

2. Laser Operation Basics

The main idea for generating short laser pulses is based upon the intensity dependent reflectivity of the FDNLM. In general the process of reflection by FDNLM represents a two-way passing of the beam through frequency doubling nonlinear crystal. A basic characteristic, forming the FDNLM application, is the inherent alteration of the phase interdependence $\Delta\varphi_{in} = \varphi_2 - 2\varphi_1$ between the first and second harmonics on the entrance of the second passing through the nonlinear crystal compared to the same at the exit of the crystal on the first passing $\Delta\varphi_{out}$. This change is physically determined by the different phase shifts, acquired by the first and second harmonics during the light propagation from the exit of the nonlinear crystal to the mirror and back to the crystal, made by the dispersion of the system air - dielectric mirror (DM). With a type I laser radiation frequency doubling, the process of energy transfer, i.e. the presence of

A. D. Boardman et al. (eds.),
Advanced Photonics with Second-order Optically Nonlinear Processes, 197–200.
© *1999 Kluwer Academic Publishers. Printed in the Netherlands.*

downconversion or upconversion, is determined by the sign of the phase interdependence $\Phi = \varphi_2 - 2\varphi_1 + \Delta kL$, where $\varphi_{1,2}$ are the phases of the fundamental and second harmonic waves, $\Delta k = k_2 - 2k_1$ the wave vector mismatch, and "L" is the distance from the entry point of the nonlinear crystal. At $\Phi \in (0, \pi/2]$ the energy is transferred from the fundamental to the second harmonic- upconversion,, when at $\Phi \in [-\pi/2, 0)$ we have downconversion, i.e. the energy is transferred back, from the second to the first harmonic.

By the change of $\Delta\varphi_{in}$ together with ΔkL_{cr} we can influence the magnitude and the sign of the FDNLM reflection coefficient change. Therefore the FDNLM reflection coefficient change will depend upon the magnitude of the input intensity and the optical characteristics of the nonlinear medium as well as on two additional construction FDNLM's parameters $\Delta\varphi_{in}$ and ΔkL_{cr}.

Figure 1.a) shows the dependence of the ratio between the intensity reflection coefficient

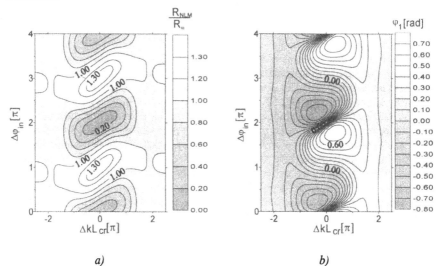

a)   *b)*

Figure 1. Normalized amplitude reflection coefficient R_{NLM}/R_ω "a)" and nonlinear phase shift φ_1 "b)" at fundamental wave versus ΔkL_{cr} and $\Delta\varphi_{in}$. The calculation is made with values of $R_\omega=50\%$ and the normalized input amplitude $U_{01}=1$ ($U_{01}=\sigma a_{10}L_{cr}$, where σ is a nonlinear coefficient, a_{10} is input amplitude and L_{cr} is the length of the nonlinear crystal).

R_{NLM} of FDNLM and the reflection coefficient of the dielectric mirror R_ω as function of $\Delta\varphi_{in}$ and ΔkL_{cr}. For very week input signals and when the influence of the nonlinear process on the reflection of the nonlinear mirror can be ignored, the value of R_{NLM} goes to R_ω. Therefore, for regions of the value of $\Delta\varphi_{in}$ and ΔkL_{cr} where FDNLM's normalized reflection coefficient is greater than 1, the reflection ability of the nonlinear mirror increases with input intensity increase. In this case FDNLM has distinct pulse shortening capability. The magnitude of the laser pulse shortening τ_{out}/τ_{in} as well as the actual existence of shortening depend primarily on $\Delta\varphi_{in}$ with values of the phase mismatch parameter ΔkL_{cr} smaller than π. For values of $\Delta kL_{cr}=0.3\pi$ the value of the pulse shortening is maximal when $\Delta\varphi_{in}$ has values between π and 2π, where the exact value in this range depends on the input peak intensity. In particular, for normalized input peak intensity value of 1, $\Delta\varphi_{in}=1.56\pi$ (Figure 2).

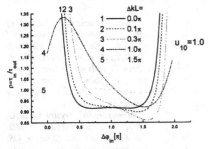

Figure 2. *Pulse shortening ratio* $\rho=\tau_{out}/\tau_{in}$ *versus* $\Delta\varphi_{in}$ *for different values of phase mismatch parameters* ΔkL_{cr} . *The calculations are made for a Gaussian temporal pulse shape and for normalized peak amplitude* $U_{01}=\sigma a_{10}L_{cr}=1$.

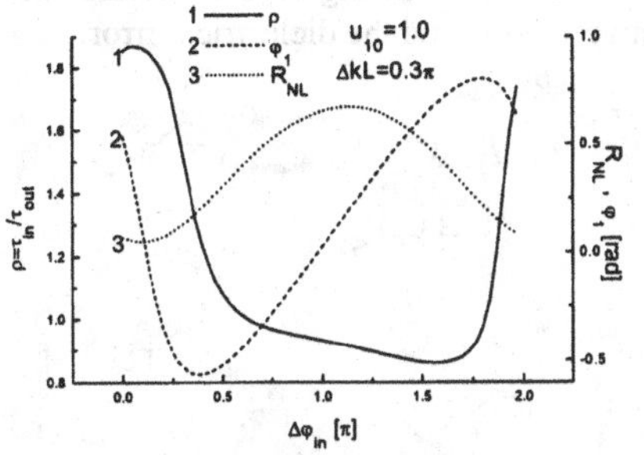

Figure 3. *Calculated pulse shortening ratio* $\rho=\tau_{out}/\tau_{in}$ *(1), nonlinear reflection coefficient* R_{NLM} *(2) and a nonlinear phase shift* φ_1 *(3) at the fundamental wave as a function of* $\Delta\varphi_{in}$

Together with the nonlinear amplitude reflection, FDNLM, depending on the $\Delta\varphi_{in}$ and ΔkL_{cr}, can incur significant nonlinear phase shift of the fundamental wave due to the so called cascaded second order processes via second harmonic generation (Fig. 1b). The value of the phase mismatch parameter (ΔkL_{cr}) for which the acquired via the reflection by FDNLM nonlinear phase shift has a maximum, has a weak dependence of the input intensity value and is about 0.3π.

By changing one of the mirrors of the laser cavity with a phase shifting FDNLM, the latter will act on the laser pulse traveling through the laser cavity as both an amplitude and a phase modulator having a periodic transmission with a period of $2L_{cav}/c$. Consequently, by the use of a phase shifting FDNLM a self-mode-locking and a generation of ultra short pulses is possible, due to the nonlinear amplitude reflection and phase modulation. Figure 3 shows the pulse shortening ratio and nonlinear phase shift as a function of the normalised input pulse peak intensity acquired by a single reflection by the FDNLM. The optimal regime of laser operation should be for $\Delta\varphi_{in}$ of about 1.5π where the three factors, namely pulse shortening, high cavity Q-factor (i.e. high R_{NLM}) and maximal phase modulation effect, are combined.

3. Experimental Setup

We have built a Nd:YAG pulse laser in which one of the cavity mirrors is replaced by FDNLM (Fig. 4). The FDNLM consists of a 15 mm long LBO crystal and a dielectric mirror (DM) with reflection coefficient at the fundamental wave of $R\omega=50\%$. Pockel's sell PC1 together with polarization mirrors P1 and P3 are used for single pulse cavity dumping. For the adjustment of the generation dynamic, an inertial electro-optical feedback and Q-switching is applied by the use of Pockel's sell PC2. By the use of a suitable negative feedback a stabilization of the laser pulses after several hundred roundtrips is achieved (quasi-steady state), allowing for independence from the initial noise distribution of the field in the resonator [8]. After achieving such quasi-steady state pulses we quickly reduce the losses in the resonator to obtain high power output

200

pulses. In this case for adjusted FDNLM the laser generates single picosecond pulses with energy of 1 mJ and duration of 22 ps with an instability by energy of no more than 3%. The choice of $\Delta\varphi_{in}$ and $\Delta kLcr$ for mode-locking of a laser is made by adjustment of the nonlinear doubling crystal around the angle of synchronism and of the distance between NLC and the dielectric mirror.

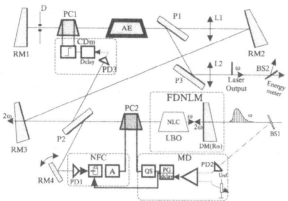

Figure 4. A schematic diagram of the passively mode-locked Nd:YAG laser by FDNLM: RM1,2,3,4: rear mirrors, PC1,2: Pockel's cells,AE: active element, P1,2,3: polar. mirrors, L1,2: lenses, FDNLM: frequency doubling nonlinear mirror, DM:mirror, NC: crystal for SHG, PD1,2,3: photodiodes, MD: modulator of cavity Q-factor, QS: electronic unit for Q modulation, PG: pulse generator, NFC: negative feedback control, A: HV large bandwidth amplifier, CDm: cavity dumper.

4. Conclusion

A high performance Nd:YAG pumped pulsed passively mode-locked laser is described. Our analysis indicates that if FDNLM works in intensity dependent phase shift mode additional to the AM phase.

Partial funding support from the Bulgarian Science Foundation under contract F601 is gratefully acknowledged.

5. References

1. Stankov, K.S. and Jethwa, J. (1988) A new mode locking technique using a nonlinear mirror, Opt. Comm. **66**, 41-46.
2. Barr. J.R.M. and Hughes, D.W. (1989) Coupled cavity modelocking of a Nd:YAG laser using second-harmonic generation, Appl. Phys **B 49**, 323-325.
3. Cerullo, G., De Silvestri, S., Monguzzi, A., Segala, D. and Magni, V. (1994) Self-starting mode locking of a cw Nd:YAG laser using cascaded second-order nonlinearities, Optics Lett. **20**, 746-748.
4. Wu, Q., Zhou, J.Y., Huang, X.G., Li, Z.X. and Li, Q.X. (1993), Mode locking with linear and nonlinear phase shifts, J. Opt. Soc. Am **B 10**, 2080-2084.
5. Buchvarov, I.Ch., Stankov, K.A., Saltiel, S.M. (1991), Pulse shortening in an actively mode-locked laser with a frequency-doubling nonlinear mirror, Opt. Comm. **83**, 241-245.
6. Zhao, X.M., McGraw, D.J. (1992), Parametric mode locking, IEEE J. Quant. Elec. **28**, 930-939.
7. Buchvarov, I., Christov, G., Saltiel, S. (1994), Transient behaviour of frequency doubling mode-locker. Numerical analysis, Opt. Comm. **107**, 281-286.
8. Buchvarov, I., Saltiel, S., Gagarskii, S. (1995), Nonlinear doubling mode-locking of feedback controlled pulsed Nd:YAG laser, Opt. Comm. **118**, 51-54.

COLLISION, FUSION, AND SPIRALLING OF INTERACTING SOLITONS IN A BULK QUADRATIC MEDIUM

V.V. STEBLINA, Y.S. KIVSHAR
Australian Photonics Cooperative Research Centre,
Optical Sciences Centre, Research School of Physical
Sciences and Engineering, Australian National University,
Canberra ACT 0200, Australia

AND

A.V. BURYAK
School of Mathematics and Statistics,
Australian Defence Force Academy,
Canberra ACT 2600, Australia

Abstract. We analyze interactions of (2+1)-dimensional parametric solitons and demonstrate non-planar beam switching in a bulk $\chi^{(2)}$ medium. This includes controllable soliton repulsion, spiralling, and fusion. Analytical model predicting results of the soliton scattering is also derived and verified by direct numerics.

Self-guided optical beams (or spatial solitons) have attracted substantial research interest because they hold a promise of ultra-fast all-optical switching and controlling light by light. Soliton interactions have been analyzed theoretically and experimentally mostly for planar geometries [1], and only recent experimental discoveries of stable (2+1)-dimensional solitons in different nonlinear media [2] initiated the experimental study of fully three-dimensional interactions between solitary beams. However, since earlier papers on the soliton interactions [3], no systematic analysis of non-planar soliton interactions in a bulk medium has been developed so far.

In this Letter we present, for the first time to our knowledge, a theory of three-dimensional collisions of optical beams taking the case of (2+1)-dimensional $\chi^{(2)}$ solitons as an important and practical example. In partic-

A. D. Boardman et al. (eds.),
Advanced Photonics with Second-order Optically Nonlinear Processes, 201–204.
© 1999 *Kluwer Academic Publishers. Printed in the Netherlands.*

ular, we derive an analytical model describing the soliton scattering, spiralling and fusion, which provides a physical insight and allows to simplify significantly the analysis of beam interactions.

Spatial solitary waves in a quadratic [or $\chi^{(2)}$] nonlinear medium are known to exist due to parametric wave mixing and nonlinear phase shift produced by cascaded nonlinearities [4]. Unlike three-dimensional solitons of an ideal Kerr medium, such solitons are stable in higher dimensions [5, 6] and have been recently observed experimentally by Tòrruellas et al. [2].

Soliton interactions in a planar geometry have been analysed theoretically (see, e.g., Ref. [7]), and have also been recently observed experimentally in a slab $\chi^{(2)}$ waveguide [8]. Interaction of (2+1)-dimensional solitons in bulk $\chi^{(2)}$ media has been only analysed for special cases when the trajectories of the interacting beams always remain in the same plane [5, 9]. Here we present a theory of non-planar collisions of $\chi^{(2)}$ solitary waves.

Equations describing spatial solitons in a bulk $\chi^{(2)}$ medium are known to possess a fundamental family of stationary (i.e., z−independent) radially symmetric solitons [5]. 'Moving' solitons can also be constructed using a simple gauge transformation. Therefore, three-dimensional soliton collisions can be readily studied numerically.

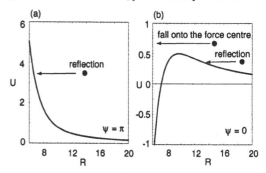

Figure 1. Qualitative sketch of the effective interaction potential $U(R, \psi)$ for (a) out-of-phase interaction ($\psi = \pi$) (b) in-phase interaction ($\psi = 0$).

For the purpose of physical insight, we analyse soliton interactions deriving a simplified dynamical model based on *effective particle approach*. Following the earlier work by Gorshkov and Ostrovsky [11], we consider two *coherently interacting* $\chi^{(2)}$ solitary waves of equal peak intensities and radial symmetry as effective particles with internal degrees of freedom (soliton phases), moving in the three-dimensional space. Then, generalizing the approach suggested in Ref. [11], we derive an effective mechanical system for the soliton collective coordinates. This mechanical model can be shown to correspond to a conservative system with the effective Hamiltonian,

$$E = \frac{1}{2}M_R\dot{R}^2 + \frac{1}{2}M_\psi\dot{\psi}^2 + U(R, \psi), \qquad (1)$$

where $R \equiv \sqrt{X^2 + Y^2}$ is the relative distance between the interacting beams, X and Y being separations between solitons in x and y directions; and ψ is the relative phase between the solitons. Elements of the effective mass matrix M_R and M_ψ can be calculated explicitly on the family of radially symmetric one-soliton solutions known numerically [5]. Therefore, effective potential energy is $U(R, \psi) = \frac{M_R s^2 C^2}{4R^2} + U_1(R) \cos(\psi) + U_2(R) \cos(2\psi)$, where the functions U_1 and U_2 are expressed in terms of the soliton overlap integrals [11]. The impact parameter s defines the distance between the trajectories of non-interacting solitons [see Figs. 2(a)-(c) below], and $C \equiv \dot{R}_0$ is the relative velocity between the solitons prior to the interaction. The effective mechanical model defined by the energy (1) can be used for predicting the outcome of soliton collisions.

In the case of out-of-phase collisions ($\psi = \pi$), the 'centrifugal force', defined by the first term of the effective potential U, and the direct interaction force, defined by the second term $U_1(R) \cos(\psi)$, are both *repulsive*. Therefore, the effective particle can not fall onto the force centre, i.e. solitons can not fuse [see Fig. 1(a)]. The interaction scenario is very different for in-phase soliton collisions ($\psi = 0$). An interplay between repulsive 'centrifugal' force and attractive interaction force leads to two qualitatively different regimes schematically shown in Fig. 1(b). For low relative velocities (or sufficiently large s) solitons can not overcome the centrifugal potential barrier, and they spiral about each other. At higher velocities (or smaller s) the soliton fusion can occur.

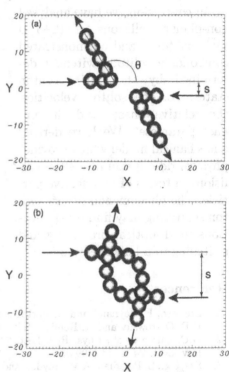

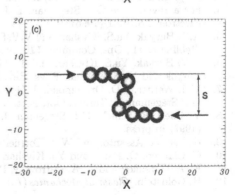

Figure 2. Snapshots of the soliton positions in (X,Y) plane at different propagation distances for (a) soliton scattering ($\psi = \pi$, $s = 4.0$) (b) soliton reflection via spiralling ($\psi = 0$, $s = 11.6$), and (c) soliton fusion ($\psi = 0$, $s = 11.0$). All results are obtained for $\beta = 0.5$, $\Delta = 1.0$ and $C = 0.1$.

204

Our direct numerical modelling confirmed the predictions of the approximate model described by Hamiltonian (1). Figures 2(a)-(c) present the characteristic examples of the soliton interaction including soliton reflection (a), soliton spiralling (b), and soliton fusion (c). To make quantitative comparison, we integrated the dynamical model for the relative coordinate and phase following from Eq. (1), applying the techniques of the well-known scattering theory of classical mechanics (see, e.g.,[12]). Figure 3 compares the results of the analytical model with direct numerical experiment showing the dependence of the scattering angle θ vs. the soliton impact parameter s. A very good agreement is found.

In conclusion, we have analysed non-planar collisions of (2+1)D $\chi^{(2)}$ solitons and demonstrated controllable soliton switching determined by the initial soliton states, i.e. the soliton velocities, the relative phase, and the impact parameter. We have derived a mechanical model which provides physical description of soliton collisions in terms of an effective particle. The phenomenon of $\chi^{(2)}$ soliton scattering may find its applications in all-optical processing and switching in a bulk medium.

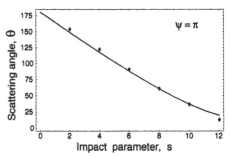

Figure 3. Soliton scattering angle θ vs. impact parameter s for out-of-phase collisions. Solid line - analytical results, filled diamonds - results of numerical simulations. Results are for $\beta = 0.5$, $\Delta = 1.0$, $C = 0.1$, and $\psi = \pi$.

References

1. See, e.g., F. Reynaud and A. Barthelemy, In: *Guided Wave Nonlinear Optics*, Eds. D.B. Ostrowsky and R. Reinisch (Kluwer, Dordrecht, 1992), p. 319.
2. G.C. Duree et al., Phys. Rev. Lett. **71**, 553 (1993); W.E. Torruellas et al. Phys. Rev. Lett. **74**, 5036 (1995).
3. See, e.g., L. Poladian, A.W. Snyder, and D.J. Mitchell, Opt. Commun. **85**, 59 (1991).
4. For a review, see G.I. Stegeman, D.J. Hagan, and L. Torner, J. Opt. Quantum Electron. **28**, 1691 (1996).
5. A.V. Buryak, Yu.S. Kivshar, and V.V. Steblina, Phys. Rev. A **52**, 1670 (1995); V.V. Steblina et al., Opt. Commun. **118**, 345 (1995).
6. A.V. Buryak, Yu.S. Kivshar, and S. Trillo, Phys. Rev. Lett. **77**, 5210 (1996); A.V. Buryak and Yu.S. Kivshar, Phys. Rev. Lett. **78**, 3286 (1997).
7. M.J. Werner, P.D. Drummond, J. Opt. Soc. Am. **B10**, 2390, (1993); D.M. Baboiu, G.I. Stegeman, L. Torner, Opt. Lett. **20**, 2282 (1995).
8. Y. Baek, R. Schiek, G.I. Stegeman, I. Baumann, and W. Sohler, Opt. Lett. **22** (1997) in press.
9. G. Leo, G. Assanto, and W.E. Torruellas, Opt. Lett. **22**, 7 (1997).
10. C. Clausen, O. Bang, and Yu. Kivshar, Phys. Rev. Lett. **78**, 4749 (1997).
11. K. A. Gorshkov and L. A. Ostrovsky, Physica D **3**, 428 (1981).
12. H. Goldstein, *Classical Mechanics* (Addison-Wesley, Massachusetts, 1980), Chap. 3.

QUADRATIC RING-SHAPED SOLITARY WAVES

D. NESHEV, M. GEORGIEV, A. DREISCHUH AND S. DINEV

Sofia University Dept. Quantum Electronics
5, James Bourchier Blvd., 1164 Sofia, Bulgaria

1. Introduction

Cascaded $\chi^{(2)} : \chi^{(2)}$ parametric interactions of intense signals in materials with quadratic nonlinearity offer a rich variety of interesting phenomena one of which is the formation of solitons [1-9]. Investigation of the evolution of spatial solitons in $\chi^{(2)}$ nonlinear media (NLM) is much more preferable because of the possibility for easier experimental observation. Two basic types of solitons exist in this case: *bright-bright* [1, 2] and *dark-dark* [5, 6]. An interesting feature of the spatial case of beam propagation is that it can be extended into two transverse dimensions (2D) [7-9].

In this work we investigate the dynamics of two special classes of 2D spatial solitary waves: bright ring-shaped solitons, first obtained in Ref.[9] (they are also known in Kerr NLM [10]) and dark ring-shaped solitons (also known to exist in Kerr NLM [11]).

2. Basic equations and initial conditions

In the slowly varying envelop approximation the propagation of the two harmonics in $\chi^{(2)}$ media is governed by the following normalized equations:

$$i\frac{\partial A}{\partial \zeta} + \frac{1}{2}\nabla_\perp^2 A + A^*B = 0, \ i\frac{\partial B}{\partial \zeta} + \frac{\alpha}{2}\nabla_\perp^2 B + A^2 = \beta B, \qquad (1)$$

where A and B are the field envelops for the fundamental and the second harmonic respectively. β is the normalized linear wave mismatch. $\alpha = k_1/k_2$ and for near phase matching $\alpha \simeq 0.5$. k_1 and k_2 are the wave vectors of the fundamental and the first harmonics respectively.

When one of the transverse coordinates is suppressed the Eqs. 1 posses exact analytical solutions like coupled waves:

A. D. Boardman et al. (eds.),
Advanced Photonics with Second-order Optically Nonlinear Processes, 205–208.

a) *bright-bright soliton solution:*[1, 2]

$$A(\zeta,\xi) = A_0 \, sech^2(\xi/a) \, e^{i\Gamma\zeta}, \ B(\zeta,\xi) = B_0 \, sech^2(\xi/b) \, e^{i\,2\Gamma\zeta}, \qquad (2)$$

where $A_0 = \pm\beta/\sqrt{2}$, $B_0 = -\beta$, $a = b = \sqrt{-3/\beta}$ and $\Gamma = -2\beta/3$ is the nonlinear shift to the wave vector. This solution is only possible for $\beta < 0$.
b) *dark-dark soliton solution:*[5, 6]

$$A(\zeta,\xi) = A_0(1 - \frac{3}{2} sech^2(\frac{\xi}{a})) \, e^{i\Gamma\zeta}, \ B(\zeta,\xi) = B_0(1 - \frac{3}{2} sech^2(\frac{\xi}{b})) \, e^{i\,2\Gamma\zeta}, \quad (3)$$

where $A_0 = \pm\frac{\sqrt{2}}{3}\beta$, $B_0 = -\frac{2}{3}\beta$, $a = b = \sqrt{3/\beta}$, $\Gamma = -2\beta/3$. This solution is only possible for $\beta > 0$.

In our investigation we extended these solutions in 2D space by "bending" the soliton stripe with transverse section equal to Eqs. 2 or 3 into ring-type formation. These new formations are called ring-shaped solitary waves. When their radius limits to infinity they are exact solution of the type of Eqs. 2, 3. It is also found that for certain size of their radius they becomes equal to the exact ring-type soliton solution found in Ref.[9].

3. Results and discussion

3.1. BRIGHT-RING SOLITARY WAVES

In this case of propagation we investigate the evolution of two bright formations (for the fundamental and the second harmonics) in the form:

$$A(\zeta,r,\varphi) = A_0 \, sech^2(r'/a) \, e^{i\Phi}, \ B(\zeta,r,\varphi) = B_0 \, sech^2(r'/b) \, e^{2i\Phi}, \qquad (4)$$

where $r' = r - R$; r, φ are the radial and azimuthal coordinate respectively. $\Phi = m\varphi + \Omega r + \Gamma\zeta$ and m is an integer number called topological charge (TC), Ω is the transverse velocity of the ring, and Γ is the nonlinear phase shift. R is the ring radius. In all our simulations we set $\Omega(\zeta = 0) = 0$.

In the transverse section this shape is very similar to the pair of interacting bright solitons [4] and one can expect similar dynamics. However in 2D case much stronger dynamics is observed. In Fig.1a,b is presented comparison between these two dynamics. It is clearly seen that while the interaction between the two 1D solitons is negligible (see also Ref.[4]) the bright ring fused into 2D soliton peak well known from Refs.[7, 8]. This behaviour is the same for rings with different radii. It is interesting to pointed out that this dynamics does not depend on the parameter β in Eqs. 1.

Similar to 1D case this dynamics is very phase sensitive (especially to the existence of TC). When m is nonzero the observed above fusion does not exist. The dynamics can goes to the oscillating behaviour (Fig.2a). The

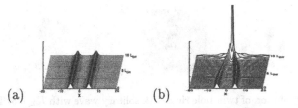

Figure 1. (a) The evolution of the pair of in-phase solitons compared to (b) the evolution of the central transverse slice of the bright-ring solitary wave with $m = 0$ (fundamental wave). The second harmonic behaves in a similar way.

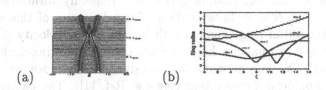

Figure 2. (a) The evolution of the central slice of the bright-ring with $R = 5$ and with $m = 1$ (fundamental wave). (b) The dependence of the ring radius versus propagation path-length for rings with different initial radii and for two values of m.

dependence on the ring radius versus propagation path-length for different ring radii and two different values of m is presented on Fig.2b. One partial case of this investigation is the exact soliton solution obtained in Ref.[9] for $m = 3$ and $R \simeq 4.8$. In this case the ring radius remains constant. However because of the presence of small perturbations (coming from the finite computational window) at $\zeta \simeq 16$ the modulational instability (MI) of these formations becomes evident (Fig.3). In addition the central vortex is decayed into more vortices which are positioned around the soliton peaks.

3.2. DARK-RING SOLITARY WAVES

The dark-ring solitary wave has the the form (with $r' = r - R$):

$$A(\zeta, r, \varphi) = A_0(1 - \frac{3}{2}sech^2\frac{r'}{a}) e^{i\Phi}, \; B(\zeta, r, \varphi) = B_0(1 - \frac{3}{2}sech^2\frac{r'}{b}) e^{2i\Phi}. \quad (5)$$

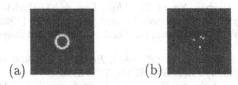

Figure 3. The initial (left) and the pattern after $\zeta = 20$ (right) of the evolution of the bright ring with $R(\zeta = 0) = 4$ after the modulational instability occurs.

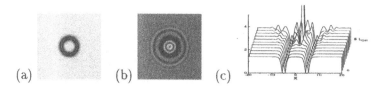

Figure 4. The evolution of twin-hole ring-dark solitary wave with $R_0 = 5$. The expanding of the ring is slightly evident. (a) Initial pattern, (b) the pattern after $5L_{Diff}$. (c) the evolution of the central slice. The appearance of the modulational instability is evident.

It should be pointed out that in this case a ring-dark formation can be obtained even for $R = 0$. Investigation of the dynamics of this structure shows that the initial ring expands with big transverse velocity ($\Omega \simeq 0.9$).

When $R \neq 0$ the pattern has the form of twin-hole ring-dark solitary wave (Fig.4a). The ring expands but the dynamics is much weaker than in the Kerr nonlinear propagation (see e.g. Ref.[11]). The pattern of the evoluted formation after propagation of $5L_{Diff}$ is shown on Fig.4b. In the both cases the MI breaks the possibility for detail investigation of the ring dynamics. The evolution of the central slice of the ring is shown in Fig.4c.

4. Conclusion

The investigation of the dynamics of ring solitary waves (RSWs) for ring radius out of the equilibrium value is made. Two different cases are viewed: *bright RSWs* and *dark RSWs*. For the first case a strong dependence of the ring dynamics on the presence of optical vortex in the beam center is observed. The existence of the MI leads to formation of soliton peaks. In the second case much weaker dynamics is observed. However the MI of the background breaks the possibility of detail investigation of the dynamics.

References

1. Karamzin, A. and Sukhorukov, A. (1974) *JETP Lett.* **11**, 339.
2. Hayata, K. and Koshiba, M. (1993) *Phys. Rev. Lett.* **71**, 3275-3278.
3. Buryak, A. and Kivshar, Yu. (1995) *Phys. Lett.*, **A197**, 407-412.
4. Clausen, C.,B., Christiansen, P. and Torner, L. (1997) *Opt. Commun.* **136**, 185-192.
5. Hayata, K. and Koshiba, M. (1994) *Phys. Rev.* **A 50**, 675-679;
6. Buryak, A. and Kivshar, Yu. (1995) *Phys. Rev.* **A 51**, 1280-1283.
7. Torner, L., Menyuk, C., Torruellas, W. and Stegeman, G. (1995) *Opt. Lett.* **20**, 13-15.
8. Torruellas, W., Wang, Z., Hagan, D., VanStryland, E. and Stegeman, G. (1995) *Phys. Rev. Lett.* **74**, 5036-5039.
9. Firth, W. and Skryabin, D. (1997) *Phys. Rev. Lett.* **79** 2450-2453.
10. Afanasjev, V. (1995) *Phys. Rev.* **E 52**, 3153-3158.
11. Neshev, D., Dreischuh, A., Kamenov, V., Stefanov, I. and Dinev, S. (1997) *Appl. Phys.* **B 64**, 429-433.

SOLITARY AND PERIODIC PULSES FOR $\chi^{(2)}$: EXPLICIT SOLUTIONS IN ABUNDANCE

D.F. PARKER AND E.N. TSOY
Dept. of Mathematics and Statistics
University of Edinburgh
Edinburgh, EH9 3JZ, U.K.

1. Introduction

Current interest in cascaded quadratic nonlinearity has led a number of authors [1] - [5] to find temporal and spatial coupled solitary waves of many classes other than the basic *bright-bright* solution due to Karamzin and Sukhorukov [6]. Like that solution, all known explicit solutions are expressible in terms of hyperbolic functions. Since hyperbolic functions are limiting cases of Jacobian elliptic functions, it is natural to adapt the search procedure used in [5] to identify classes of *dark-bright*, *dark-dark* and *brighter-brighter* profiles to seek possibilities for periodic coupled travelling waves.

2. Travelling Wave Solutions for $\chi^{(2)}$

When complex amplitudes $A(\boldsymbol{x},t)$, $B(\boldsymbol{x},t)$ of the fundamental and second-harmonic signal are sought in the form

$$A = e^{i\theta}u(\zeta) \quad , \qquad B = e^{2i\theta}v(\zeta),$$

where $\theta \equiv \boldsymbol{\beta} \cdot \boldsymbol{x} - \sigma t$ denotes an adjustment to the carrier phase while $\zeta \equiv \boldsymbol{K} \cdot \boldsymbol{x} - \Omega t$ is a travelling wave coordinate, the evolution equations for the cascaded quadratic interaction yield the ordinary differential equations

$$pu''(\zeta) + i\Pi u'(\zeta) + \Sigma u(\zeta) = 2u^*(\zeta)v(\zeta) \tag{1}$$

$$\tilde{p}v''(\zeta) + i\tilde{\Pi}v'(\zeta) + \tilde{\Sigma}v(\zeta) = u^2(\zeta) \tag{2}$$

in which primes denote differentiation. The parameters $p,\Pi,\Sigma,\tilde{p},\tilde{\Pi},\tilde{\Sigma}$ depend on the material parameters, the dispersion and diffraction coefficients for the relevant guided modes and on $\boldsymbol{K},\Omega,\boldsymbol{\beta}$ and σ.

The method used in [5] for finding new classes of solitary waves is to seek solutions to (1) and (2) depending on $S \equiv \operatorname{sech} r\zeta$, $T \equiv \tanh r\zeta$ in the

A. D. Boardman et al. (eds.),
Advanced Photonics with Second-order Optically Nonlinear Processes, 209–214.
© 1999 *Kluwer Academic Publishers. Printed in the Netherlands.*

form

$$u = aS^2 + bST + cS + dT + e,$$
$$v = \tilde{a}S^2 + \tilde{b}ST + \tilde{c}S + \tilde{d}T + \tilde{e}. \tag{3}$$

Use of the identities

$$S' = -rST, \qquad T' = rS^2, \qquad S^2 + T^2 = 1,$$
$$S'' = r^2 S(1 - 2S^2), \qquad T'' = -2r^2 S^2 T,$$

generates 18 (complex) algebraic equations for the ten coefficients a, b, ..., $\tilde{e}$. (The equations show also why the *ansatz* (3) should be only of degree 2 in $\{S, T\}$ - both two differentiations and quadratic terms yield expressions of equal degree, namely four).

In [5], two classes of solution are found, in addition to those known earlier [1] - [4]. The analogous procedure for seeking periodic explicit solutions in terms of the Jacobian elliptic functions $C \equiv \operatorname{cn} r\zeta$, $D \equiv \operatorname{dn} r\zeta$ and $N \equiv \operatorname{sn} r\zeta$ having modulus k is to use the *ansatz*:

$$u = a + bN + cC + dD + eNC + fND + gCD + hN^2,$$
$$v = \tilde{a} + \tilde{b}N + \tilde{c}C + \tilde{d}D + \tilde{e}NC + \tilde{f}ND + \tilde{g}CD + \tilde{h}N^2. \tag{4}$$

In simplifying the fourth degree expressions arising from inserting the ansatz (4) into equations (1) and (2), the relevant identities are

$$C^2 + N^2 = 1, \qquad D^2 + k^2 N^2 = 1,$$
$$C' = -rND, \qquad D' = -rk^2 NC, \qquad N' = rCD,$$
$$C'' = r^2 C(2k^2 N^2 - 1), \quad D'' = r^2 k^2 D(2N^2 - 1), \quad N'' = r^2 N(2k^2 N^2 - 1 - k^2)$$

3. Some Simple Classes of Periodic Solution

Using the above identities to eliminate C^2 and D^2 in favour of N^2 allows both sides of equations (1) and (2) to be written as linear combinations of

$$1, N, C, D; NC, ND, CD, N^2; N^2 C, N^2 D, NCD, N^3; N^3 C, N^3 D, N^2 CD, N^4$$

Comparing coefficients then yields two sets of 16 equations for the sixteen coefficients a, b, ..., $\tilde{g}$, $\tilde{h}$. This is a (grossly) overdetermined system. However, we know that it has many solutions in the limit $k \to 1$ (solitary waves) and for $\tilde{p} = \tilde{\Pi} = 0$ (which reduces (1) and (2) to a NLS equation).

Analysis commences by judiciously combining the eight equations arising from terms of degree four so that they may be solved completely to yield four types of solution for $e, \ldots, h$; $\tilde{e}, \ldots, \tilde{h}$. However, 24 equations remain to be analysed. An exhaustive search procedure would next solve the eight equations arising from terms of degree three in C, D and N as a linear system for b, c, d, $\tilde{b}$, $\tilde{c}$ and $\tilde{d}$. Finally, the remaining sixteen equations must be analysed for compatible possibilities. In this report, attention is confined to solutions which generalize some special classes known to exist as $k \to 1$:

Class I. Karamzin and Sukhorukov [6], Hayata and Koshiba [1], Werner and Drummond [2].

$$u = 6r^2\sqrt{p\tilde{p}}e^{i\alpha}\mathrm{sech}^2 r\zeta, \quad v = -6r^2 p e^{2i\alpha}\mathrm{sech}^2 r\zeta, \quad (p\tilde{p} > 0)$$

with *propagation conditions* $\Pi = \tilde{\Pi} = 0$, $\Sigma/r^2 p = \tilde{\Sigma}/r^2\tilde{p} = -4$.

Class II. Werner and Drummond [2], Menyuk, Shiek and Torner [3].

$$u = i6r^2\sqrt{-p\tilde{p}}e^{i\alpha}\mathrm{sech}\, r\zeta \tanh r\zeta, \quad v = 6r^2 p e^{2i\alpha}\mathrm{sech}^2 r\zeta, \quad (p\tilde{p} < 0)$$

with *propagation conditions* $\Pi = \tilde{\Pi} = 0$, $\Sigma/r^2 p = -1$, $\tilde{\Sigma}/r^2\tilde{p} = 2$.

Motivated by the limiting behaviour (C, $D \to \mathrm{sech}r\zeta$ and $N \to \tanh r\zeta$ as $k \to 1$, for $r\zeta$ fixed), we generalize the solitary wave solutions I by retaining in (4) only the terms in 1, CD and N^2. This yields 16 equations for a, $\tilde{a}$, g, $\tilde{g}$, h and $\tilde{h}$. These allow just the possibilities

I.1 (Using $h = \pm kg$, $\tilde{h} = \pm k\tilde{g}$)

$$u = 3r^2\sqrt{p\tilde{p}}\,e^{i\alpha}\left\{k^2\mathrm{sn}^2 r\zeta \pm k\,\mathrm{cn}\,r\zeta\,\mathrm{dn}\,r\zeta - \tfrac{1}{6}\left[1 + k^2 + L\right]\right\}$$
$$v = 3r^2 p\,e^{2i\alpha}\left\{k^2\mathrm{sn}^2 r\zeta \pm k\,\mathrm{cn}\,r\zeta\,\mathrm{dn}\,r\zeta - \tfrac{1}{6}\left[1 + k^2 + L\right]\right\}$$

where $L \equiv \pm\sqrt{(1+k^2)^2 + 12k^2}$. The corresponding *propagation conditions* are $\Pi = \tilde{\Pi} = 0$, $\Sigma/r^2 p = \tilde{\Sigma}/r^2\tilde{p} = -L$.

I.2 (Using $g = \tilde{g} = 0$; $h = 6k^2 r^2\sqrt{p\tilde{p}}e^{i\alpha}$, $\tilde{h} = 6k^2 r^2 p e^{2i\alpha}$, α arbitrary)

$$u = 6r^2\sqrt{p\tilde{p}}\,e^{i\alpha}\left\{k^2\mathrm{sn}^2 r\zeta - \tfrac{1}{3}\left[1 + k^2 + M\right]\right\}$$
$$v = 6r^2 p\,e^{2i\alpha}\left\{k^2\mathrm{sn}^2 r\zeta - \tfrac{1}{3}\left[1 + k^2 + M\right]\right\}$$

with $\Pi = \tilde{\Pi} = 0$, $\Sigma/r^2 p = \tilde{\Sigma}/r^2\tilde{p} = -M \equiv \mp 4\sqrt{1 - k^2 + k^4}$.

I.3 (Using $h = \tilde{g} = 0$; $g = i6k^2 r^2\sqrt{p\tilde{p}}\,e^{i\alpha}$, $\tilde{h} = -6k^2 r^2 p\,e^{2i\alpha}$, α arbitrary)

$$u = i6kr^2\sqrt{p\tilde{p}}\,e^{i\alpha}\mathrm{cn}\,r\zeta\,\mathrm{dn}\,r\zeta,$$
$$v = 6k^2 r^2 p\,e^{2i\alpha}\left(2(1 + k^2)^{-1} - \mathrm{sn}^2 r\zeta\right)$$

with $\Pi = \tilde{\Pi} = 0$, $\Sigma/r^2 p = 1 + k^2 - 12k^2(1 + k^2)^{-1}$, $\tilde{\Sigma}/r^2\tilde{p} = -2(1 + k^2)$.

It is readily verified that the profiles obtained from each of I.1, I.2 and I.3 as $k \to 1$ include the *bright-bright* solitary wave of Class I. Also it is clear that change of sign before k corresponds only to a translation along the $r\zeta$ axis. Observe that, in I.1 and I.2, $u/(\sqrt{p\tilde{p}}\,e^{i\alpha}) = v/(p\,e^{2i\alpha})$ while in I.3 the profiles differ. Figure 1(a), (c), (e) and (f) illustrates profiles arising for $k = 0.8$, $k = 0.9$ and $k = 0.99$ (when L, M are taken as positive). These show, as $k \to 1$, how the profile separates into a sequence of $\mathrm{sech}^2 r\zeta$ pulses. In I.1 and I.2 the phases of adjacent pulses are equal, while in I.3 the phases of u differ by π.

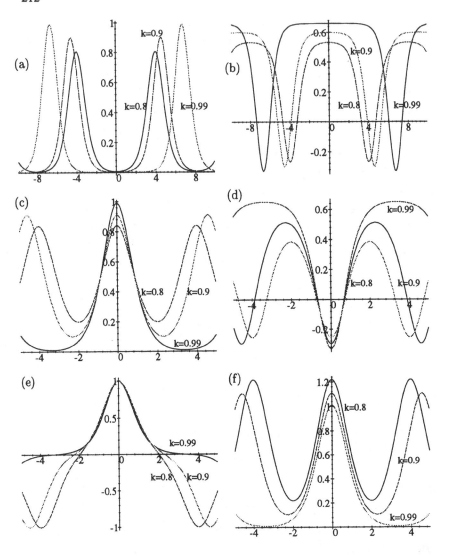

Figure 1. Profiles of $\bar{u} \equiv u/(6r^2\sqrt{p\tilde{p}}\,e^{i\alpha})$ and $\bar{v} \equiv v/(6r^2 p\,e^{2i\alpha})$, for $k = 0.8$, 0.9 and 0.99. Case I.1 (a) $-\bar{u} = -\bar{v}$ for $L > 0$, (b) $\bar{u} = \bar{v}$ for $L < 0$; Case I.2 (c) $-\bar{u} = -\bar{v}$ for $M > 0$, (d) $\bar{u} = \bar{v}$ for $M < 0$; Case I.3 (e) $-i\bar{u}$, (f) $\bar{v}$.

Figure 1(b) and (d) illustrates profiles with L, M taken as negative square roots. Both cases give a sequence of *twin-hole* dark pulses, generalizing that in Buryak and Kivshar [7] (and included in case VI of [5]).

Solutions generalizing class II arise from setting $a = \cdots = d = 0 = g = h$ and $\tilde{b} = \cdots = \tilde{f} = 0$. The resulting possibilities (requiring $p\tilde{p} < 0$) are

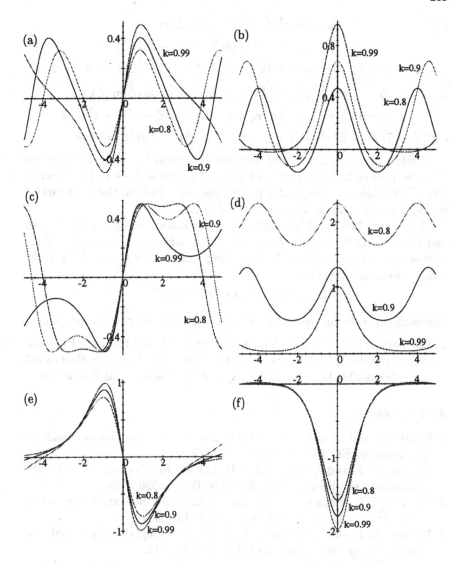

Figure 2. Profiles of $\bar{u}$ and $\bar{v}$ as in *Figure 1*, for $k = 0.8$, 0.9 and 0.99. Case II.1 (a) $\bar{u}$, (b) $\bar{v}$; Case II.2 (c) $\bar{u}$, (d) $\bar{v}$; Case II.3 (e) $\bar{u}$, (f) $\bar{v}$.

II.1
$$u = i6r^2\sqrt{-p\bar{p}}\,e^{i\alpha}k^2\mathrm{sn}\,r\zeta\,\mathrm{cn}\,r\zeta\,,$$
$$v = 6r^2p\,e^{2i\alpha}k^2\left\{(2-k^2)^{-1} - \mathrm{sn}^2\,r\zeta\right\}$$
with $\Pi = \tilde{\Pi} = 0$; $\Sigma/r^2p = 4 + k^2 - 6k^2(2-k^2)^{-1}$, $\tilde{\Sigma}/r^2\tilde{p} = 4 - 2k^2$.

II.2
$$u = -6r^2\sqrt{-p\tilde{p}}\,e^{i\alpha}k\,\mathrm{sn}\,r\zeta\,\mathrm{dn}\,r\zeta\,,$$
$$v = 6r^2p\,e^{2i\alpha}k^2\left\{(2k^2-1)^{-1}-\mathrm{sn}^2\,r\zeta\right\}$$

with $\Pi = \tilde{\Pi} = 0$; $\Sigma/r^2p = 1 + 4k^2 - 6k^2(2k^2-1)^{-1}$, $\tilde{\Sigma}/r^2\tilde{p} = 4k^2 - 2$.

II.3
$$u = -6r^2\sqrt{-p\tilde{p}}\,e^{i\alpha}\left\{\tfrac{1}{2}k^2\mathrm{sn}\,r\zeta\,\mathrm{cn}\,r\zeta \pm \tfrac{1}{2}k\,\mathrm{sn}\,r\zeta\,\mathrm{dn}\,r\zeta\right\},$$
$$v = 6r^2p\,e^{2i\alpha}\left\{\tfrac{1}{2}k^2\mathrm{sn}^2\,r\zeta \mp \tfrac{1}{2}k\,\mathrm{cn}\,r\zeta\,\mathrm{dn}\,r\zeta - k^2(1+k^2)^{-1}\right\}$$

with $\Pi = \tilde{\Pi} = 0$; $\Sigma/r^2p = 1 + k^2 - 6k^2(1+k^2)^{-1}$, $\tilde{\Sigma}/r^2\tilde{p} = 1 + k^2$.

Profiles for $k = 0.8$, 0.9 and 0.99 are illustrated in Figure 2. Although all cases behave like a sequence of Class II pulses as $k \to 1$, for chosen k Class II.3 exhibits much greater pulse separation. Indeed, the fundamental period is double that in Classes II.1 and II.2.

A further class, readily obtainable, generalizes those given by Werner and Drummond [4] in the case $\tilde{\Pi} = \tilde{p} = 0$ in which equations (1) and (2) reduce to the equation for NLS travelling waves (with $v = \tilde{\Sigma}^{-1}u^2$). Relevant periodic solutions are

$$u = (-p\tilde{\Sigma}/2)^{1/2}e^{i\alpha}\{k\,\mathrm{cn}\,r\zeta \pm \mathrm{dn}\,r\zeta\}.$$

Solutions generalizing classes IV and VI of [5] will appear in a future paper.

Acknowledgement. Dr E.N. Tsoy's visit to Edinburgh from the Physical-Technical Institute, Uzbek Academy of Sciences, Tashkent, Uzbekistan has been made possible by a Royal Society/NATO Postdoctoral Fellowship.

4. References

1. Hayata, K. and Koshiba, M. (1993) Multidimensional solitons in quadratic nonlinear media, *Phys. Rev. Lett.* **71**, 3275-3278.
2. Werner, M.J. and Drummond, P.D. (1993) Simulton solutions for the parametric amplifier, *J. Opt. Soc. Am.* B **10**, 2390-2393.
3. Menyuk, C.R., Schiek, R., and Torner, L. (1994) Solitary waves due to $\chi^{(2)} : \chi^{(2)}$ cascading, *J. Opt. Soc. Am.* B **11**, 2434-2443.
4. Werner, M.J. and Drummond, P.D. (1994) Strongly coupled nonlinear parametric solitary waves, *Opt. Lett.* **19**, 613-615.
5. Parker, D.F., (1998) Exact representations for coupled bright and dark solitary waves of quadratically nonlinear systems, to appear in *J. Opt. Soc. Am. B*.
6. Karamzin, Yu.N. and Sukhorukov, A.P. (1974) Nonlinear interaction of diffracted light beams in a medium with quadratic nonlinearity: mutual focusing of beams and limitation on the efficiency of optical frequency converters, *JETP Letters* **20**, 339-342.
7. Buryak, A.V. and Kivshar, Yu.S. (1996) Twin-hole dark solitons, *Phys. Rev. A* **51**, R41-R44.

PROPAGATION OF RING DARK SOLITARY WAVES IN SATURABLE SELF-DEFOCUSING MEDIA

E. D. EUGENIEVA and A. A. DREISCHUH * ⵌ

Institute of Electronics - BAS,
72 Tzarigradsko Chaussee,
Sofia 1784, Bulgaria
* Max-Planck-Institute for Quantum Optics,
Hans-Kopfermann-Str. 1, 85748 Garching, Germany
ⵌ Permanent address: Sofia University
Department of Physics, 5. J. Bourchier Blvd.
1164 Sofia, Bulgaria

Dark spatial solitons are self-supported dips imposed on uniform background beams. The physical mechanism of their formation is based on the balanced counteraction of beam diffraction and self-defocusing. In this work we study the spatial evolution of ring dark solitary waves (RDSWs) in nonlinear media possessing saturable self-defocusing nonlinearity. Such type of nonlinear response was observed in photovoltaic media [1].

Optical beam propagation in a defocusing non-Kerr meterial is described by the generalized nonlinear Schrödinger equation (GNSE) :

$$i\frac{\partial u}{\partial z} + \frac{1}{2}\nabla_\perp^2 u - \frac{(1+a)|u|^2}{1+a|u|^2}u = 0 \quad , \tag{1}$$

where $u = E / \sqrt{I_m}$ is the dimensionless envelope of the electric field, E is the slowly varying amplitude, I_m is the maximum background intensity, $a = I_m / I_{sat}$ is the saturation parameter, and I_{sat} is the saturation intensity. The nonlinear contribution to the refractive index has the form $\Delta n = n_\infty a / (1+a)$ with $n_\infty = \Delta n(I_m \to \infty)$. In the case of defocusing nonlinearity $n_\infty < 0$. The longitudinal scale is introduced in the following way:

$$L_{nl} = n_0(1+a) / (-kn_\infty a) \quad . \tag{2}$$

The transverse coordinates are scaled to:

$$x_0 = \left[n_0(1+a) / (-k^2 n_\infty a)\right]^{1/2} \tag{3}$$

In these units $L_{nl} \to \infty$ as $a \to 0$. In this case Eq.1 corresponds to the 'true' Kerr-type

215

A. D. Boardman et al. (eds.),
Advanced Photonics with Second-order Optically Nonlinear Processes, 215–218.
© 1999 *Kluwer Academic Publishers. Printed in the Netherlands.*

nonlinearity. At high saturation (a>>1) $\mathbf{L_{nl}} \to \mathbf{n_0}/(\mathbf{kn_\infty})$. The normalization introduced by Eq.2 provides a natural way to compare the propagation distances in the two physically different nonlinear regimes. At all values of the saturation parameter at $z = \mathbf{L_{nl}}$ the maximum nonlinear phaseshift at the beam wings is -1.

Eq. (1) was solved numerically by using the split-step Fourier method. At z=0 the RDSW is described by:

$$\mathbf{u} = \mathbf{B(r)V(r - R_0)}\exp(i\Phi) , \qquad (4a)$$

where

$$\mathbf{B(r)} = \exp\{ -(\mathbf{r}/40)^{10} \} , \qquad (4b)$$

$$\mathbf{r} = (\mathbf{x}^2 + \mathbf{y}^2)^{1/2} / \mathbf{r_0} , \qquad (4c)$$

$$\mathbf{V(r - R_0)} = |\tanh(\mathbf{r - R_0})|^{\mathbf{b}}, \qquad (4d)$$

x and y being the Cartesian coordinates. The initial phase distribution Φ contains a pair of phase jumps of π in each diametrical cross-section, thus ensuring odd initial conditions. The values of $\mathbf{r_0}$ and $\mathbf{b}$ were determined by using the quasi-stationary solution ($\partial \mathbf{u} / \partial \mathbf{z} = 0$) of the NLSE for the case of one transverse coordinate (1D case) at certain saturation. These solutions were found numerically by using the Runge-Kutta method and were further approximated with a function of the form $|\tanh(\mathbf{x} / \mathbf{r_0})|^{\mathbf{b}}$. The dependence of $\mathbf{b}$ and $\mathbf{r_0}$ on the saturation parameter is shown in Fig. 1. We found that values of $\mathbf{b}$ other than 1 give a good approximation of the amplitude profile found numerically. Physically this means that at higher saturations the wings of the exact 1D solution become flatter and finally at $\mathbf{a} \to \infty$ no dark soliton solution could be expected since the transverse variations of the refractive index disappear.

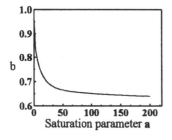
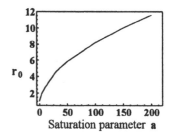

Fig.1. Plots of **b** and the dark ring width r_0 vs. the saturation parameter a (quasi-stationary solution).

We found that the transverse velocity of the RDSW depends on the saturation parameter. At low saturation levels (e.g. $\mathbf{a} \approx 2$ or less) the ring radius and the width of the dark stripe always increase as the RDSW propagates along the medium. This results in decrease of the soliton contrast. At a certain distance ($z = 30 L_{nl}$ for $a = 2$) the dark wave disappears. Higher saturation levels lead to reduction of the ring expansion rate as shown in Fig. 2.

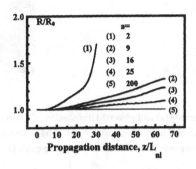

Fig.2. Evolution of the ring radius uder propagation $R_e = 10$

In nonsaturated Kerr media significant evolution of the RDSW radius occurs at propagation distances well below $10 L_{nl}$, however, L_{nl} as defined by Eq. (2) tends to infinity. In presence of saturation, as in the Kerr case [2], the rate at which the ring expands was found to depend on the initial ring radius. This tendency is shown in Fig.3.

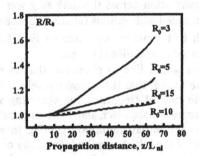

Fig.3. Longitudinal evolution of the ring radius at a=25 and different values of the initial ring radius. The dashed curve corresponds to $R_e = 15$

At values of $a \sim 10$ or higher and at initial ring radius exceeding certain critical value R_{cr} the RDSW decays into pairs of Optical Vortices (OV). This behaviour is illustrated in Fig. 4. The appearance of OV pairs can be regarded as a manifestation of the instability of the dark solitons against transverse perturbations [3,4,5]. As expected [3], the critical value R_{cr} of the RDSW radius was found to depend on the saturation parameter. We found numerically that R_{cr} decreases with increasing the saturation parameter. For the case $a = 25$ $R_{cr} \sim 5$. At $R < R_{cr}$ the ring radius and width increase monotonically under propagation and the contrast decreases, but the ring symmetry remains preserved.

218

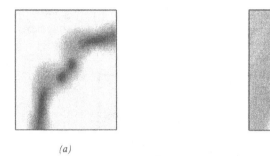

(a) (b)

Fig. 4. Gray-scale images of intensity (a) and phase (b) distribution of a part of the RDSW at $a = 25$, $R_0 = 10$ and $z = 100\ L_{nl}$. Formation of pairs of optical vortices of opposite topological charges due to the transverse modulational instability could be easily identified.

Similar behavior of the RDSWs was observed for saturation parameters **a** ranging from 15 to 30. At lower saturation OVs does not appear when R_0 increases (up to 20 in our simualtions). At high saturation levels ($a \sim 200$) no OVS-pairs are clearly observed up to $z = 100L_{nl}$, however the initial stage of their formation (i.e. the development of 'snake instability') was evident at all ring radii.

The exact quasi-stationary solution of Eq. (1) presenting the radial (one-dimensional) intensity distribution across the dark ring was found numerically. This solution could be approximated with various functions of the type $V(r - R_0)$. When a tanh-approximation with **b=1** function was used in Eq. (4d), at **a=25** we observed gradualy weaker pronounced modulational instability.

In conclusion we would like to point out that as a result of the extended numerical simulations we found, that the transverse velocity of RDSWs in saturable nonlinear media and the critical frequency for observing snake-like instability do depend on both the saturation parameter and dark ring radius. The stability of the RDSW depends also on the initial amplitude profile, thus making important the experimental clarification of the initial amplitude/intensity conditions. Splitting of the RDSW into optical vortices could be used for branching of an optical signals guided in into several channels. This waveguide structure could be written in a photorefractive crystal, thus producing a photonic coupler.

Acknowledgement

A. D. is grateful to the Alexander-von-Humboldt-Foundation for the fellowship awarded.

References:

1. Valley, G., Segev, M., Crosignani, B., Yariv, A., Fejer, M., Bashav, M., (1994), Phys. Rev., A50, pp.R4457-R4460.
2. Kivshar, Yu. S., Yang, X., (1994) Phys. Rev. , E50, pp. R40-R43.
3. Kivshar, Yu. S., Luther-Davies, B., (1997), *Dark Optical Solitons*, Phys. Reports.
4. Mamaev, A. V., Saffman, M. , Zozulya, A. A. , (1996), Phys. Rev. Lett., 76, pp. 2262-226.
5. Tikhonenko, V., Cristou, J., Luther-Davies, B. , and Kivshar, Yu. S. , (1996) Opt. Lett., 21, pp. 1129-1131.

THE N-SOLITON INTERACTIONS, COMPLEX TODA CHAIN AND STABLE PROPAGATION OF NLS SOLITON TRAINS

V. S. GERDJIKOV, E. G. EVSTATIEV
Institute for Nuclear Energy and Nuclear Research,
boul. Tzarigradsko shosse 72, 1784 Sofia, Bulgaria

D. J. KAUP
Institute for Nonlinear Studies, Clarkson University, Potsdam,
NY, 13699-5815 USA

G. L. DIANKOV
Institute of Solid State Physics, boul. Tzarigradsko shosse 72,
1784 Sofia, Bulgaria.

AND

I. M. UZUNOV
Friedrich-Schiller University Jena, D-07743 Jena, Germany
and Institute of Electronics, boul. Tzarigradsko shosse 72,
1784 Sofia, Bulgaria

1. Introduction

One of the important problems in optical fiber soliton communication is to achieve as high of a bit rate as possible. In order to do this, one needs to be able to pack the solitons into as short of a space as possible. However, if the solitons are too close together, then their mutual interactions can cause them to collide and/or separate, thereby corrupting the signal. The current solution of this problem is simply to require each soliton to be sufficiently far apart from all others (usually 6 or so soliton widths) so that such interactions can be totally neglected. However, at the same time, it was predicted [1, 2] and experimentally confirmed [3] that for certain values of relative soliton parameters, this separation can be reduced, and at the same time, still maintain signal integrity. Our purpose is to analytically and numerically detail the soliton parameter regime, inside of which, signal

A. D. Boardman et al. (eds.),
Advanced Photonics with Second-order Optically Nonlinear Processes, 219–226.
© *1999 Kluwer Academic Publishers. Printed in the Netherlands.*

integrity can be maintained. In particular, we are interested to determine how the inter-soliton interaction can be used for *stabilizing* a soliton train.

Any communication signal will be composed of "random" combinations of 0's and 1's. It can also be viewed as being composed of a random collection of N-soliton trains, with varying widths of 0's between them. Thus it is then adequate for us to simply analyze the stability of individual N-soliton trains, for finite N. This we will do and will study a nonperiodic and finite train of soliton pulses by both analytical and numerical methods developing the ideas in [4, 5].

The basic model describing N-soliton trains in optical fibers is provided by the nonlinear Schrödinger (NLS) equation and its perturbed version:

$$iu_t + \frac{1}{2}u_{xx} + |u|^2u(x,t) = iR[u]. \tag{1}$$

which describes a variety of wave interactions, including ones in nonlinear fiber optics [1, 6]-[11] and nonlinear refractive media [9].

The inverse scattering method [6] allows one to solve exactly Eq. (1) when $R[u] = 0$ and to calculate explicitly its N-soliton solutions. However for our purposes, this method is impractical for two reasons. First, there are important perturbations of this system which have no explicit solutions. Second, an approximate method can serve better than an exact approach since the N-soliton trains here can only be approximated by N-soliton solutions. Such trains are sums of 1-soliton pulses, which are spaced almost equally apart and have almost equal amplitudes and velocities. More specifically, they are the solutions to Eq. (1) with $R[u] = 0$ satisfying the initial conditions:

$$u(x,0) = \sum_{k=1}^{N} \frac{2\nu_k e^{i\phi_{k,0}(x)}}{\cosh(2\nu_k(x - \xi_{k,0}))}, \tag{2}$$

where $\phi_{k,0}(x) = 2\mu_0(x - \xi_{k,0}) + \delta_{k,0}$, and $\delta_{k,0}$, $\xi_{k,0}$, $2\nu_{k,0}$, and $2\mu_{k,0}$ are the initial phase, position, amplitude and velocity respectively of the k-th soliton; $2\nu_0$ and $2\mu_0$ are the averages of the initial amplitudes and velocities.

An effective method for studying the interaction of such trains of soliton pulses was first proposed by Karpman and Solov'ev (KS), for the 2-soliton interaction [12], see also [7, 13, 14]. This method is based on the adiabatic approximation and is valid for any collection of well separated solitons, such that their mutual interactions lead only to a slow deformation in the soliton parameters, so one derives a dynamical system for ξ_k, ν_k, μ_k and δ_k.

An infinite train of out-of-phase soliton pulses, with equal amplitudes and velocities is described by the real Toda chain [15, 16]:

$$\frac{d^2q_k}{d\tau^2} = e^{q_{k+1}-q_k} - e^{q_k-q_{k-1}}, \tag{3}$$

where $\tau = 4\nu_0 t$, $k = 0, \pm 1, \pm 2, \ldots$ and q_k are real functions related to the soliton positions. This system will be referred to as the real Toda chain (RTC). Numerically this description for finite number of solitons was verified in [17].

Recently in Refs. [4, 5, 17], the KS method was extended to N-soliton pulses, and after additional approximations, was reduced to the complex Toda chain equations (CTC) (3) with N sites and $e^{-q_0} \equiv e^{q_{N+1}} \equiv 0$. The complex valued functions $q_k(t)$ are related to the soliton parameters by:

$$q_{k+1} - q_k = 2i(\mu_0 + i\nu_0)(\xi_{k+1} - \xi_k) + \ln 4\nu_0^2 + i\,(\pi - \delta_{k+1} + \delta_k). \quad (4)$$

Thus the problem of determining the evolution of an NLS N-soliton train has been reduced to the problem of solving the CTC for N sites. The fact that CTC, like RTC [18, 19, 20] is integrable will be extensively used.

Our main results are the following:

(i) we classify the asymptotic regimes of CTC. Besides the asymptotically free motion (which is the only one possible for the RTC [19, 20]), CTC allows also for: (a) bound state regime when all the N particles may move quasi-equidistantly (QED); (b) intermediate regimes when one (or several) group(s) of particles form bound state(s) and the rest of them go into free motion asymptotics. In addition to these relatively stable regimes of motion there are also less stable regions in the space of soliton parameters, where one regime switches into another one. There one can find (c) singular solutions [23], and (d) various types of degenerate solutions.

(ii) we find that the predictions from the CTC model match very well with the numerical solutions of the NLS equation. Our analytic approach allows us to describe the set of initial parameters, for which each of these asymptotic regimes takes place. We put special stress on the bound state and QED regimes which are desirable in fiber optics communications.

After this research was completed we became aware that the CTC model has been derived earlier by Arnold in [21].

2. Asymptotic Regimes of the CTC

As in [4, 5], one can generalize the RTC [18, 19, 20] to the complex CTC case. We list the four most important points concerning this below:

S1) The CTC Lax representation is the same as for the RTC: $\dot{L} = [B, L]$:

$$L = \sum_{k=1}^{N} (b_k E_{kk} + a_k(E_{k,k+1} + E_{k+1,k})), \quad (5)$$

Here $a_k = \frac{1}{2}e^{(q_{k+1}-q_k)/2}$ and $b_k = \frac{1}{2}(\mu_k + i\nu_k)$. The matrices $(E_{kn})_{pq} = \delta_{kp}\delta_{nq}$, and $(E_{kn})_{pq} = 0$ whenever p or q becomes 0 or $N + 1$.

S2) The integrals of motion in involution are provided by the eigenvalues, ζ_k, of $L_0 = L(\tau = 0)$.

S3) The solutions of both the CTC and the RTC are determined by the scattering data S_{L_0} of L_0. If $\zeta_k \neq \zeta_j$ for $k \neq j$, then $S_{L_0} = \{\zeta_k, r_k\}_{k=1}^N$, where r_k are the first components of the properly normalized eigenvectors $\xi^{(k)}$ of L_0, [see [19, 20]].

S4) Lastly, ζ_k uniquely determine the asymptotic behavior of the solutions of the CTC and can be calculated from the initial conditions [4].

For CTC not only the dynamical variables q_k, but also $\zeta_k = \kappa_k + i\eta_k$ and r_k take complex values. The eigenvalue ζ_k, and in particular, κ_k determines the asymptotic velocity of the k-th soliton. For simplicity and without loss of generality we assume that: tr $L_0 = 0$; $\zeta_k \neq \zeta_j$ for $k \neq j$ and that the κ_k's are ordered as: $\kappa_1 \leq \kappa_2 \leq \ldots \leq \kappa_N$. Then in any train of solitons, there are three possible general configurations:

D1) $\kappa_k \neq \kappa_j$ for $k \neq j$, i.e. the asymptotic velocities are all different. Then we have asymptotically separating, free solitons, like for the RTC, see [4, 5]

D2) $\kappa_1 = \kappa_2 = \ldots = \kappa_N$, i.e. all N solitons will move with the same mean asymptotic velocity, and therefore will form a "bound state". We find the sets of initial soliton parameters, for which

$$A_k = \left(\max(\xi_{k+1} - \xi_k) - \min(\xi_{k+1} - \xi_k)\right)/r_0 \ll 1, \qquad (6)$$

for $k = 1, \ldots, N-1$, i.e., any two adjacent solitons move QED. If (6) holds then the solitons will not separate asymptotically, but instead the distance between them will oscillate with some small amplitude.

D3) One may have also a variety of intermediate situations when only one group (or several groups) of particles move with the same mean asymptotic velocity; then they would form one (or several) bound state(s) and the rest of the particles will have free asymptotic motion.

Obviously the cases D2) and D3) have no analogies in the RTC and physically are qualitatively different from D1). The same is also true for the degenerate and singular solutions of the CTC, see e.g. [23].

3. The Soliton Bound States

The case $N = 2$ solitons is analytically solved in [12]. For the sake of brevity we present here some of the analytical results only for $N = 3$; it is possible to extend these results to any N. We assume that $\sum_{k=1}^3 q_k(\tau) = 0$, i.e. the center of mass is fixed at the origin. This is compatible with tr $L_0 = 0$.

Here and below we study initially equidistant soliton trains $\xi_{k+1,0} - \xi_{k,0} = r_0$ with $\mu_{k,0} = 0$ and with two sets of initial amplitudes: $\mathbf{Am_1} \equiv \{\nu_{k+1,0} - \nu_{k,0} = \Delta\nu_0\}$ and $\mathbf{Am_2} \equiv \{\nu_{k,0} - \nu_{k+1,0} = (-1)^k \Delta\nu_0\}$. The symmetric solution to the $N = 3$ CTC, i.e. the one for which $q_2 \equiv 0$, $q_1 = -q_3$ and $\zeta_k = (2 - k)i\eta_1$ is periodic with period $T_3^\pm = \pi/(2\nu_0\eta_1^\pm)$ where

$$2\eta_1^\pm = \sqrt{(\Delta\nu_0)^2 \pm 8\epsilon_0^2} \qquad (7)$$

Here $\epsilon_0 = \nu_0 e^{-\nu_0 r_0}$ and the sign in (7) is determined by the set of initial phases $\mathbf{Ph} \equiv \{0, \delta_{2,0}, 0\}$: plus for $\delta_{2,0} = 0$ and minus for $\delta_{2,0} = \pi$. For this solution we find

$$A_1^\pm = A_2^\pm = \pm\frac{1}{2\nu_0} \ln \frac{z_0^2 \pm 1}{z_0^2}, \qquad (8)$$

where $z_0 = |\Delta\nu_0|/(2\sqrt{2}\epsilon_0)$. Note that similar results for $A_1^\pm$ hold also for the $N = 2$ case with the only difference being that z_0 should be replaced by $z_0 = |\Delta\nu_0|/(4\epsilon_0)$. The motion will be quasi-equidistant if $A_1^\pm \ll 1$. The formulas (8) show that the increase of $|\Delta\nu_0|$ diminishes $A_1^\pm$. Another way to diminish A_1 for fixed $\Delta\nu_0$ and r_0, is to increase the average amplitude ν_0.

The singular behavior of A_1^+ for $\Delta\nu_0 \to 0$ corresponds to a singular solution of the CTC; the numeric solution of the NLS shows that the solitons do not collide, but do come rather close to each other at the values of t where the CTC solution develop singularities.

On the other hand the singularity of A_1^- for $\Delta\nu_0 = \Delta\nu_{cr,3}$ corresponds to the fact that at this critical point the quasi-equidistant regime switches over into the free motion regime. For $\Delta\nu_0 < \Delta\nu_{cr,3}$ we have free asymptotic regime and the distance between the solitons grows infinitely, while for $\Delta\nu_0 > \Delta\nu_{cr,3}$ we get bound state regime and a possible QED behavior.

Let us now describe the particular sets of soliton parameters, which lead to the bound state regime. To do this we analyze the roots of the characteristic equation for L_0, which for $N = 3$ with $\text{tr}\, L_0 = 0$ is:

$$\zeta^3 + \zeta p + q = 0, \qquad (9)$$

$$p = \frac{1}{32}\left(d_1^2 + d_2^2 + d_3^2\right) - a_1^2 - a_2^2, \qquad (10)$$

$$q = \frac{i}{4}\left(a_1^2 d_3 + a_2^2 d_1\right) + \frac{i}{64}d_1 d_2 d_3, \qquad (11)$$

where $d_k = 2(\nu_{k,0} - \nu_0)$, and $a_k^2 = -\epsilon_0^2 e^{i(\delta_{k+1,0} - \delta_{k,0})}$. Now we can use Cardano formulas to evaluate the roots ζ_k which determine the asymptotic regime predicted by CTC; the results are collected in the Table, where $P_0(z) = 0$ for $\cos z > 0$ and $P_0(z) = \sqrt{-\cos z}$ for $\cos z < 0$.

We also compared the analytical results obtained above with the numerical solutions of the unperturbed NLS ($R[u] = 0$) (1) with initial conditions (2). We find an excelent agreement between the CTC and NLS in each of

TABLE 1. List of sets of soliton parameters, for which three-soliton bound state occur.

Ph	Am$_1$	Am$_2$
$\{0,0,0\}$	$|\Delta\nu_0| \geq 0$	$|\Delta\nu_0| \geq 0$
$\{0, \pm\pi, 0\}$	$|\Delta\nu_0| > 2\sqrt{2}\epsilon_0$	$|\Delta\nu_0| > 4\sqrt{2}\epsilon_0$
$\{0, \pm\pi, \pm\pi\}$	$|\Delta\nu_0| > 2 \cdot 3^{3/4}\epsilon_0$	—
$\{0, \delta_{2,0}, 0\}$	—	$|\Delta\nu_0| > 4\sqrt{2}\epsilon_0 P_0(\delta_{2,0})$

the regimes. In particular we find that the soliton propagation is a QED one, i.e. the values of $\max A_k \leq 10\%$ for the configurations described in the first column of table with $2\Delta\nu_0 > 0.1$. The agreement improves with the increasing of $\Delta\nu_0$, ν_0 and r_0 and is worse for the cases when: a) CTC has singular or degenerate solutions and b) transition from one regime to another takes place. The choice of **Am$_1$** with $\Delta\nu_0 = 0$ and $\delta_{k,0} = 0$ leads to singular solution of CTC for any N. On the other hand the NLS solutions are always analytic and never have singularities. Our numerical checks show that such sets of initial conditions correspond either to soliton collisions or soliton coalescence depending on the value of r_0. For $N = 3$ and $r_0 = 8$ the first coalescence takes place at $t_1 \simeq (T_3^+/2)|_{\Delta\nu_0=0}$ and the next ones tend to repeat periodically with period very close to $T_3^+|_{\Delta\nu_0=0}$ (7). The QED propagation of the solitons is maintained for $t \leq 0.9t_1$. Additional analytical details and numerical results can be found in [24].

4. Discussion

We end by two short remarks concerning recent results obtained by two of us (V.S.G. and E.G.E.). The first one is that the CTC describes the interactions between the solitons of all higher NLS equations, some of which are also relevant to fiber optics, see the review paper [25]. The main idea for proving this is in the fact that $|\lambda_0 - \lambda_k| \simeq \epsilon_0$, where $\lambda_k = \mu_k + i\nu_k$. This allows us to use the linearized dispersion law. Indeed, if $f(\lambda)$ is the dispersion law of the higher NLS ($f_{NLS}(\lambda) = -2\lambda^2$), then the soliton evolution is approximated by $f(\lambda_k) \simeq f'(\lambda_0)(\lambda_k - \lambda_0)$ which, up to insignificant constants, is the linear dispersion law for the CTC.

The second remark is that the CTC describes also the soliton interactions for the vector NLS solitons - Manakov model. The important difference between the scalar and the vector cases is in formula (4) in which we have to replace $\delta_k - \delta_{k+1}$ by the scalar product (n_k, n_{k+1}) of the "polarization" vectors of the corresponding solitons. We derived this starting from the Lax representation, but it can be derived also through the variational approach. Important fact in deriving this result is that all components of the vector soliton have the same functional dependence on x and t.

Acknowledgments

One of us (V. S .G.) is grateful to Prof. A. Boardman and Prof. L. Pavlov for making his participation in the ASI possible and to Prof. J. Arnold for usefull discussions. This research has been supported in part by a grant by Deutsche Forschungsgemeinschaft, Bonn, Germany, in the framework of the Innovationskolleg "Optische Informationstechnik" and the project Le–755/4. This research has also been supported in part by U.S. Air Force Office of Scientific Research and the ONR.

References

1. Desem, C. , Chu, P. L. (1992) Soliton-soliton Interactions, *In Optical Solitons – Theory and Experiment:* Taylor, J. R. Ed., chapter 5, Cambridge University Press, Cambridge, p. 127.
2. Uzunov, I. M. Stoev, V. D. , Tzoleva, T. I. (1992) N-soliton Interaction in Trains of Unequal Soliton Pulses in Optical Fibers, *Optics Lett.* **Vol. 20**, pp. 1417–1419.
3. Suzuki, M., Edagawa, N., Taga, H., Tanaka, H., Yamamoto, Feasibility Demonstration of 20 Gbit/s Single Channel Soliton Transmission over 11500 km Using Alternating-Amplitude Solitons, S., Akiba, S. (1994) *Electron. Lett.* **Vol. 30 no. 13**, pp. 1083–1084.
4. Gerdjikov, V. S., Kaup, D. J., Uzunov, I. M., Evstatiev, E. G. (1996) The asymptotic behavior of N–soliton trains of the Nonlinear Schrödinger Equation, *Phys. Rev. Lett.* **Vol. 77 no.**, pp. 3943–3947.
5. Gerdjikov, V. S., Uzunov, I. M., Evstatiev, E. G., Diankov, G. L. (1997) The nonlinear Schrödinger equation and N–soliton Interactions. Generalized Karpman–Solov'ev Approach and the Complex Toda Chain, *Phys. Rev. E* **Vol. 55 no. 5**, pp. 6039–6060.
6. Takhtadjan, L. A., Faddeev, L. D. (1986) *Hamiltonian Approach to Soliton Theory* Springer Verlag, Berlin.
7. Kivshar, Yu. S., Malomed, B. A. (1989) Dynamics of Soltons in Nearly Integrable Systems. *Rev. Mod. Phys.* **Vol. 61 no. 4**, pp. 763–915.
8. Kaup, D. J. (1990) Perturbation Theory for Solitons in Optical Fibers *Phys. Rev. A* **Vol. 42 no. 9**, pp. 5689–5694; Kaup, D. J. (1991) Second Order Perturbations for Solitons in Optical Fibers, *Phys. Rev. A* **Vol. 44 no. 7**, pp. 4582–4590.
9. Hasegawa, A., Kodama, Y. (1995) *Solitons in Optical Communications*, Oxford University Press, Oxford, UK.
10. Agrawal, G. P. (1995) *Nonlinear Fiber Optics* Academic, San Diego, (2-nd edition).
11. Wabnitz, S., Kodama, Y., Aceves, A. B. (1995) Control of Optical Soliton Interactions. *Optical Fiber Technology,* **Vol. 1 no. 1**, pp. 187–217.
12. Karpman, V. I., Solov'ev, V. V. (1981) A Perturbational Approach to the Two-Soliton Systems. *Physica D* **Vol. 3D no. 1-2**, pp. 487–502.
13. Okamawari. T., Hasegawa A., Kodama, Y. (1995) Analysis of Soliton Interactions by means of a Perturbed Inverse Scattering Transform, *Phys. Rev. A* **Vol. 51 no. 4**, pp. 3203–3225.
14. Aceves, A. B., De Angelis, C., Nalesso G., Santagiustina, M. (1994) Higher Order Effects in Bandwidth-limited Soliton Propagation in Optical Fibers, *Optics Lett.* **Vol. 19**, pp. 2104–2106.

15. Gorshkov, K. A. (1981) Asymptotic Soliton Theory: Interaction, Generation, Bound States. PhD Thesis, Inst. Appl. Phys., Gorky, USSR.
    Gorshkov, K. A., Ostrovsky, L. A. (1981) Interactions of Solitons in Nonintegrable Systems: Direct Perturbation Method and Applications, *Physica D* **Vol. 3D no. 1-2**, pp. 428–438.
16. Arnold, J. M. (1993) Soliton Pulse Position Modulation, *IEE Proc. J.* **Vol. 140 no. 6**, pp. 359–366.
17. Uzunov, I. M., Gerdjikov, V. S., Gölles, M., Lederer, F. (1996) On the description of N-soliton interaction in optical fibers, *Optics Commun.* **Vol. 125 no. 1**, pp. 237–242.
18. Manakov, S. V. (1974) On Complete Integrability and Stochastization in Discrete Dynamical Systems, *JETPh* **Vol. 67 no. 2**, pp. 543–555;
    Flashka, H. (1974) On the Toda Lattice. II. Inverse Transform Solution, *Prog. Theor. Phys.* **Vol.51 no. 3**, pp. 703–716.
19. Moser, J. (1975) Three Integrable Hamiltonian Systems Connected with Isospectral Deformations, *Adv. Math.* **Vol. 16 no. 2**, pp. 197–220.
20. Toda, M. (1989) *Theory of Nonlinear Lattices* Springer Verlag, Berlin, (2nd enlarged ed.).
21. Arnold, J. M. (1995) *Proceedings URSI Electromagnetic Theory Symposium, St. Petersburg* pp. 553–555 and private communication.
22. Uzunov, I. M., Gölles, M., Lederer, F. (1995) Soliton Interaction Near rhe Zero-Dispersion Wavelength *Phys. Rev. E* **Vol. 52 no. 3**, pp. 1059–1071.
23. Kodama, Y. (1996) Toda Hierarchy with Indefinite Metric *Physica D* **Vol. 91 no. 2**, pp. 321–339; Khastgir, S. P., Sasaki, R. (1996) Instability of Solitons in Imaginary Coupling Affine Toda Field Theory *Prog. Theor. Phys.* **Vol. 95 no. 3**, pp. 485–501.
24. Gerdjikov, V. S., Evstatiev, E. G., Kaup, D. J., Diankov, G. L., Uzunov, I. M. (1997) *Criterion and Regions of Stability for Quasi-Equidistant Soliton Trains.* **E-print solv-int/9708004**.
25. Kodama, Y., Maruta, A., Hasegawa, A. (1994) Long Distance Communications with Solitons *Quantum Opt.* **Vol. 6**, pp. 463–516.

RAY OPTICS THEORY OF SELF-MATCHED BEAMS MUTUAL FOCUSING IN QUADRATIC NONLINEAR MEDIA

A. P. SUKHORUKOV, LU XIN

Radiophysics Department, Physics Faculty, Lomonosov Moscow State
University, Vorob'evy Gory, Moscow, 119899, Russia.
aps@nls.phys.msu.su; lu@nls.phys.msu.su

1. Introduction

Self-action of waves in quadratic nonlinear media leads to the phase velocity changes, which are proportional to the amplitudes. This phenomena take place when wave energy exchange is absence. Due to parametric self-action, spatial $\chi^{(2)}$ solitons may be formed [1- 4]. In strong fields when the nonlinear refraction prevails, the parametric mutual focusing of beams with quadratic nonlinearity should occur [1, 2].

In this work we studied the reactive interaction between beams of first and second harmonics on quadratic nonlinearity. We found self-matched beams of these harmonics, which have the same wave front and the same amplitude profile with equal width, but the intensity of FH beam is two times higher than SH beam. The mutual focusing of such beams is more effective. In this work we studied the properties of self-matched beams with help of ray optics approximation and numerical simulation.

2. Parametric self-action of self-matched beams

Diffraction of two harmonics in quadratic nonlinear media with phase matching is described by the following envelope equations:

$$\frac{\partial A_1}{\partial z} + iD_1 \Delta_\perp A_1 + i\gamma A_1^* A_2 = 0,$$

$$\frac{\partial A_2}{\partial z} + iD_2 \Delta_\perp A_2 + i\gamma A_1^2 = 0 \tag{1}$$

where A_1, A_2 are the envelopes of first and second harmonics, D_1, D_2 the diffraction coefficients, γ the nonlinear coefficient in normalised values.

227

A. D. Boardman et al. (eds.),
Advanced Photonics with Second-order Optically Nonlinear Processes, 227–230.

228

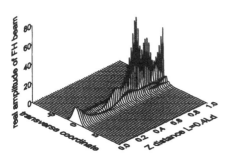
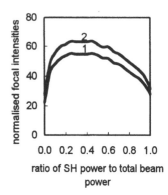

Figure 1. Evolution of Mutual focusing without SH input beam. $E_1 = 17.32$ $E_2 = 0 \cdot D_1 = 0.1$, $\gamma = 1 \cdot$

Figure 2. FH(1) and SH(2) intensities at the first focus vs SH power portion.

The equation (1) has been solved numerically. Figure 1 shows mutual focusing when the second harmonic generation occurs. To investigate the self-matched beam we have to introduce real amplitudes $B_j(r,z)$ and phases $\Psi_j(r,z)$ in the following forms:

$$A_j(r,z) = B_j(r,z)\exp\left[-i\,\Psi_j(r,z)/2\,D_j\right]. \qquad (2)$$

After substitution (2) into basic equation (1) we obtain two equations with real variables. To obtain the ray optics approximation we ignored all diffraction members. The analyses showed if the phases and amplitudes of two beams are consistent the following relations:

$$\Psi_1(r,z) = \Psi_2(r,z) \pm \pi\,N, \qquad (3)$$

$$B_1(r,z) = \sqrt{2}\,B_2(r,z), \qquad (4)$$

then the two beams will propagate without energy exchange, and the wave fronts of both beams (fundamental and SH) coincide. In other words, the amplitude and phase profiles of two beams are matched. Finally, we can obtain the equation for such beams:

$$2\frac{\partial\psi_j}{\partial z} + \left(\frac{\partial\psi_j}{\partial r}\right)^2 = \pm 4\,\gamma\,D_j\,B_1^2 B_2/B_j^2, \qquad (5.a)$$

$$\frac{\partial B_j}{\partial z} + \frac{\partial B_j}{\partial r}\frac{\partial\psi_j}{\partial r} + \frac{1}{2}B_j\Delta_\perp\psi_j = 0 \qquad (5.b)$$

We have solved the basic equation (1) by numerical simulations for gaussian beams with same width and different peak amplitudes: $A_j(0,r) = E_j\exp(-r^2)$, j =1,2, but the sum power of two beams is constant, i.e. $E_1^2 + E_2^2 = const$, in numerical experiments we choose $E_1^2 + E_2^2 = 300$. From (4) we can get the amplitudes of self-matched beams: $E_1 = 14.14$, $E_2 = 10$. Figure 2 illustrates the intensity dependencies in nonlinear focus on ratio of second harmonic power to total beam power, in Figure 2 we can see that the mutual focusing of self-matched beams is most effective. Figure 3 shows the process of self-matched mutual focusing. To pass from focusing to defocusing we can just change the phase of FH beam to π or SH beam to $\pi/2$. In Figure 4 is shown self-matched defocusing.

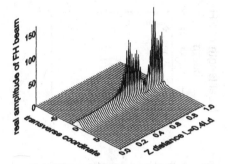
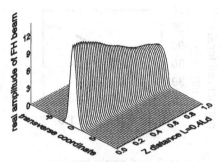

Figure 3. Evolution of self-matched mutual focusing, $E_1 = 14.14$ $E_2 = 10 \cdot$ $D_1 = 0,1$, $\gamma = 1 \cdot$

Figure 4. Evolution of self-matched mutual defocusing,. $E_1 = 14.14$ $E_2 = -10$ $D_1 = 0,1$, $\gamma = 1 \cdot$

3. Aberration free approximation

Equations (5) can be solved by the aberration free approximation, introducing spherical wave front of gaussian beam with width $f(z)$, as

$$\Psi_2 = \frac{r^2}{2f} \frac{df}{dz} + \varphi_0 (z) \tag{6}$$

$$B_2 (r,z) = E_2 f^{-1}(z)\left(1 - \frac{r^2}{f^2(z)}\right). \tag{7}$$

After substitution (6), (7) into equation (5), we can obtain

$$\frac{d^2 f}{dz^2} = \pm \frac{1}{l_s^2 f^2}, \tag{8}$$

where $l_s = \left(8\gamma\, D_2\, E_2\right)^{-1/2}$. In case of mutual focusing (minus), the result of equation (8) can be founded in the implicit form: $f(z) = \cos^2\left[\sqrt{f(1-f)} - z/l_s\right]$. And we can obtain the focal length $z_f = \pi\, l_s/2$.

If adding to right hand of equation (8) a diffraction factor which proportional to $D_2^2 f^{-3}$, we can also find the beam compression $f_f \propto D_2 /\left(\gamma\, E_2\right)$ and focal amplitude

$E_{2f} \propto \gamma E_2^2 / D_2$ of self-matched mutual focusing. In numerical experiment we also got the dependencies of focal intensity and focal length on the coefficient of nonlinear interaction γ for self-matched focusing (Figure 5, Figure 6).

In case of mutual defocusing we have $f(z) = ch^2\left[\sqrt{f(f-1)} - z/l_s\right]$. In large distance the divergence angle of beams is proportional to $1/l_s$,

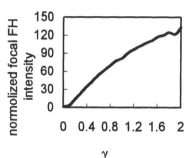
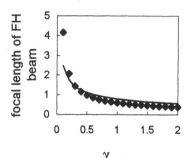

Figure 5. Focal intensity dependence vs nonlinear coefficient

Figure 6. Focal length dependence vs nonlinear coefficient. Solid line - analytical results, filled diamonds - results of numerical simulations.

4. Conclusion

We have developed the ray optics theory of chi-2 mutual focusing and defocusing for the first time. The properties of self-matched beams were analysed, we have showed the parametric focusing and defocusing of self-matched beams is most effective. The computer simulations conform the ray optics theory.

Acknowledgement

This work was supported by the Russian Foundation for Basic Research (project No. 96–02–18592) and the «Fundamental Problems of Natural Sciences» program (grant No.9–5–0–2.2–79).

References

1. Karamzin, Yu. N., Sukhorukov, A. P.(1974) Nonlinear interaction of diffracted light beams in a medium with quadratic nonlinearity: mutual focusing of beams and limitation on the efficiency of optical frequency converters, Sov Phys. *JETP Lett.* **20**, 734

2. Karamzin, Yu. N., Sukhorukov, A. P.(1975). Mutual focusing of high-power light beams in media with quadratic nonlinearity, *Sov. Phys. JETP* **41**, 414.

3. Torruellas, W. E., Wang, Z., Hagan, D. J., Van Stryland, E. W., Stegeman, G. I., Torner, L., Menyuk, C. R.(1995) Observation of two-dimensional spatial solitary waves in quadratic medium, *Phys. Rev. Lett.* **74**, 5036.

4. Torruellas, W. E., Wang, Z., Torner, L., Stegeman, G. I. (1995) Observation of mutual trapping and dragging of two-dimensional spatial solitary waves in a quadratic medium, *Opt. Lett.* **20**. 1949.

RESONANT PROPERTIES OF χ^2 IN TWO-PARTICLE FREQUENCY REGIONS

K. T. STOYCHEV
Institute of Solid State Physics,
Bulgarian Academy of Sciences, 1784 Sofia, Bulgaria

The nonlinear optical properties of crystals depend strongly on the excited quasiparticles in them and their interactions. These interactions modify the many-particle spectra and govern the resonances of the susceptibilities. Nonlinear susceptibilities have been calculated for systems of interacting phonons [1,2], excitons and phonons [3-5] and phonon-polaritons [6,7]. In all these cases a specific common feature of the nonlinear two-particle spectra can be traced, which distinguishes them from the corresponding linear optical spectra. In the present communication we analyze this feature on the basis of a simple microscopic model.

Let us consider a crystal with two types of polar excitations with Bose operators A_n^+, A_n and B_n^+, B_n interacting with each other. They can be optical phonons and/or low-density Frenkel excitons. The Hamiltonian of the system can be cast in the following form:

$$H = \sum_k \left(\varepsilon_k A_k^+ A_k + \eta_k B_k^+ B_k + \Gamma T_k^+ T_k \right) T_k = N^{-1/2} \sum_{k_1} A_{k_1} B_{k-k_1} \tag{1}$$

where ε_k and η_k are the energies of the two modes, $T_k^+ (T_k)$ are two-particle operators corresponding to the simultaneous creation (annihilation) of two quasiparticles - one from each type - with a summary wave-vector k and Γ is the nonlinear interaction constant. We shall consider local (on-site) interaction between the quasiparticles, in which case Γ does not depend on the wave-vectors.

If both single-particle states and the mixed two-particle states are polar, then the dipole moment operator of the crystal can be written as follows:

$$P(r) = \sum_k e^{ikr} \left[d_A \left(A_k + A_{-k}^+ \right) + d_B \left(B_k + B_{-k}^+ \right) + d_T \left(T_k + T_{-k}^+ \right) \right] \tag{2}$$

A. D. Boardman et al. (eds.),
Advanced Photonics with Second-order Optically Nonlinear Processes, 231–234.
© 1999 *Kluwer Academic Publishers. Printed in the Netherlands.*

where d_A, d_B and d_T are the corresponding matrix elements. The last term in (2) describes the so-called electrooptical anharmonicity and it plays a crucial role for the resonances of χ^2 in the two-particle region.

The macroscopic nonlinear polarization induced in the crystal when two external electric fields are applied is:

$$\left\langle P_i^{(2)}(\omega_1+\omega_2,k_1+k_2)\right\rangle = \chi_{ijl}^{(2)}(\omega_1,k_1,\omega_2,k_2)E_j(\omega_1,k_1)E_l(\omega_2,k_2) \tag{3}$$

where $\chi_{ijl}^{(2)}$ is the quadratic nonlinear susceptibility of the medium. It is expressed by means of the Fourier-transform φ_{ijl} of the triple-time retarded Green function

$$-\Theta(t-t')\Theta(t'-t'')\left\langle\left[\left[P_i(r,t),P_j(r',t')\right],P_l(r'',t'')\right]\right\rangle \equiv \left\langle\left\langle P_i;P_j;P_l\right\rangle\right\rangle \tag{4}$$

through [8]:

$$\chi_{ijl}^{(2)}(\omega_1,k_1,\omega_2,k_2) = \\ \pi V^{-1}\left[\phi_{ijl}(\omega_1+\omega_2,k_1+k_2;\omega_2,k_2)+\phi_{ilj}(\omega_1+\omega_2,k_1+k_2;\omega_1,k_1)\right] \tag{5}$$

where $P_i(r,t)$ is the i-th component of the dipole moment operator (2) in the Heisenberg representation and   is the crystal volume. Eq.(5) accounts for the proper permutation symmetry of the nonlinear susceptibility [9].

The calculation of φ and χ^2 leads to some specific features of the nonlinear optical spectra in the two-particle region, which will be discussed below. In order to simplify the analysis we shall neglect nonlocal effects by putting $k_1=k_2=0$. The quasiparticles considered here have energies well above the thermal ones and the statistical average in (4) can be taken over the vacuum state, in which case only terms involving one operator from each type are nonzero i.e.

$$\left\langle\left\langle T_0;A_0^+;B_{0+}\right\rangle\right\rangle,\left\langle\left\langle A_0^+;T_0;B_0^+\right\rangle\right\rangle,\left\langle\left\langle A_0;B_0;T_0^+\right\rangle\right\rangle,\dots \tag{6}$$

For the Hamiltonian (1) the integration over one of the time variables $t-t'$ or $t'-t''$ is trivial, and the problem reduces to calculations of double-time Green functions of the type:

$$G^L=\left\langle\left\langle T_0;A_0^+B_0^+\right\rangle\right\rangle=\left\langle\left\langle\sum_k A_k B_{-k};A_0^+B_0^+\right\rangle\right\rangle \tag{7}$$

This nonlinear response function describes the correlation between three microscopic polarization fields associated with the A_0, B_0 and T_0 excitations. There is an important difference between (7) and the Green function which governs the linear response in the two-particle region:

$$G^L = \left\langle\left\langle T_0; T_0^+ \right\rangle\right\rangle = \sum_{k_1 k_2}\left\langle\left\langle A_{k_1} B_{-k_1}; A_{k_2}^+ B_{-k_2}^+ \right\rangle\right\rangle \tag{8}$$

which correlates two T_0 polarization fields.

The calculation of G^L gives:

$$G^L = \frac{N^{-1}\sum_k (\omega - \varepsilon_k - \eta_k)^{-1}}{1 - \Gamma N^{-1}\sum_k (\omega - \varepsilon_k - \eta_k)} = \frac{G_0^L}{1 - G_0^L} \tag{9}$$

where

$$G_0^L = N^{-1}\sum_k (\omega - \varepsilon_k - \eta_k)^{-1} \tag{10}$$

is the harmonic linear response function corresponding to noninteracting quasiparticles $\Gamma = 0$. The poles of G_0 form a quasicontinuous band of two-particle states with a summary wave-vector $k = 0$. The presence of the nonlinear interaction Γ yields additional resonances in G^L for frequencies satisfying $\Gamma G_0^L = 1$ which correspond to bound or quasibound two-particle states.

The calculation of the nonlinear response function G^{NL} yields:

$$G^{NL} = \frac{(\omega - \varepsilon_0 - \eta_0)}{1 - \Gamma N^{-1}\sum_k (\omega - \varepsilon_k - \eta_{-k})^{-1}} \tag{11}$$

It has resonances at the sum frequency of two noninteracting quasiparticles $\varepsilon_0 + \eta_0$ and at the bound-state frequencies.

Solving the equations of motion for the nonzero terms in (4), after some algebra an analytic expression for the quadratic susceptibility (5) is derived. Without going into details, we shall focus on the resonant parts of χ^2 in the two-particle region. They have the form :

$$\Delta\chi^{(2)} \sim (\omega - \varepsilon_0 - \eta_0)G^{NL}(\omega) \tag{12}$$

where ω is any of the frequencies ω_1 , ω_2 or $\omega_1 + \omega_2$. Together with (11), (9) and (10) this gives:

$$\Delta\chi^{(2)} \sim \frac{1}{1 - \Gamma N^{-1} \sum_k (\omega - \varepsilon_k - \eta_{-k})^{-1}} = \frac{G^L(\omega)}{G_0^L(\omega)} \tag{13}$$

It is remarkable that due to the symmetry condition (5) the resonance at the sum frequency of the noninteracting quasiparticles is canceled in the nonlinear susceptibility. Its frequency dependence in the two-particle region is determined from the corresponding linear response function scaled by the harmonic linear response function. Thus the spectrum of noninteracting qusiparticles is projected out and only features relevent to the interaction remain. The quadratic nonlinear susceptibility has strong resonances at the bound-state frequencies corresponding to $\Gamma G_0^L = 1$, while the resonances in the two-particle band are weakened by a factor of Γ/ε. This explains why nonlinear two-particle spectra are usually narrow and are associated with the presence of strong anharmonicity. For weak anharmonicity $G^L \approx G_0^L, \Delta\chi^{(2)} \approx const$ and the two-particle region will not show in quadratic nonlinear spectra, while it can be clearly expressed in the linear ones. Higher-order susceptibilities [2] also exhibit a similar feature.

Acknowledgments

The author is grateful to Prof. A. Boardman and Prof. L. Pavlov for making his participation at the NATO ASI possible, and to Prof. V. Agranovich and Prof. I. Lalov for useful discussions. This work is supported in part by the National Foundation for Scientific Investigations under Grant No F 514.

References

1. Agranovich, V.M., N.A. Efremov, N.A., and Kaminskaya, E.P., (1971) *Opt. Commun.* **3**, 3887.
2. Efremov, N.A. and Kaminskaya, E.P., (1972) *Fiz. Tverd. Tela* **14**, 1185.
3. Lalov, I.J., Stoychev, K.T. and Penchev, S.P. (1983) *Phys. Stat. Sol. (b)* **119**, 419.
4. Lalov, I.J., Stoychev, K.T. and Penchev, S.P. (1983) *Phys. Stat. Sol. (b)* **120**, 671.
5. Stoychev, K.T., (1986) *Phys. Stat. Sol. (b)* **133**, 557.
6. Lalov, I.J. and Stoychev, K.T., (1985) *J. Phys. C* **18**, 6691.
7. Stoychev, K.T., (1992), *ICTP Trieste Preprint* IC/92/251.
8. Ghenkin, G.M. and Fein, V.M. (1965) *ZhETF* **49**, 1118.
9. Armstrong, J.A., Bloembergen, N., Ducing, J. and Pershan, P.S. (1962) *Phys. Rev.* **127**, 1918.

ON PARAMETRIC COUPLED SOLITONS WITH HIGH-ORDER DISPERSION

A.A. SUKHORUKOV
Physics Faculty, Moscow State University
Vorobjevy Gory, Moscow 119899, Russia
e-mail: as@nls.phys.msu.su

Short pulses propagation in quadratic media is analyzed taking into account high-order dispersion. Exact analytical solutions are obtained for bright and dark parametric solitons with symmetric envelops. The soliton propagation velocity is found to depend on its amplitude.

1. Introduction

In recent years the great interest [1] is attracted to parametric solitons, predicted in [2,3]. Coupled spatial solitons have been observed in planar structures [4] and bulk quadratic nonlinear media [5]. In the case of two short pulse propagation in such medium 1-D temporal $\chi^{(2)}$ – soliton formation can also be observed [6-12].

Different methods are used to describe soliton propagation in quadratic media. General soliton theory includes walk-off effect [6-8], second-order linear dispersion [9-14], higher-order linear dispersion [15,16], dispersion of nonlinearity [12,17]. To calculate evolution of extremely short wave packets with several oscillations of electric field, Maxwell's equations with nonlinear polarization [18,19] and coupled KDV equations [20] are solved.

In many papers describing parametric solitons in media with group velocity dispersion the nonlinearity dispersion effect is often neglected. But for short pulses it is essential to consider higher-order dispersion effects. In this paper the properties of parametrically coupled solitons are studied taking into account linear 3-rd order dispersion and nonlinear interaction dispersion.

2. Amplitude equations under high-order dispersion

We consider interaction between 1-st and 2-nd harmonics at frequencies $\omega_2 = 2\omega_1$: $E_j = A_j(\tau, z)\exp(i\omega_j t - ik_j z)$, $j = 1,2$, with phase mismatching $\Delta k = k_2 - 2k_1$. Higher-order dispersion equations for envelops are:

$$\frac{\partial A_j}{\partial z} + v_j \frac{\partial A_j}{\partial \tau} = iD_j \frac{\partial^2 A_j}{\partial \tau^2} + G_j \frac{\partial^3 A_j}{\partial \tau^3} - i\beta_j W_j - \gamma_j \frac{\partial W_j}{\partial \tau}, \tag{1}$$

where $W_1 = A_2 A_1^* \exp(-i\Delta k z)$, $W_2 = A_1^2 \exp(i\Delta k z)$ are nonlinear terms, z is

235

A. D. Boardman et al. (eds.),
Advanced Photonics with Second-order Optically Nonlinear Processes, 235–238.

propagation coordinate, $\tau = t - z/u_1$ is retarded time, $v_j = u_j^{-1} - u_1^{-1}$ is group velocity mismatch, D_j and G_j are 2-nd and 3-rd order linear dispersion coefficients respectively, β_j are nonlinear coefficients, γ_j are nonlinear dispersion coefficients.

Terms with higher-order time derivatives in equations (1) correspond to higher order approximation of linear dispersion. Parameters G_j characterize dispersive curves divergence from parabolic approximation (Fig. 1):

$$k_j(\omega) = k_j(\omega_j) + \frac{1}{u_j}(\omega - \omega_j) + \frac{1}{2}D_j(\omega - \omega_j)^2 + \frac{1}{6}G_j(\omega - \omega_j)^3.$$

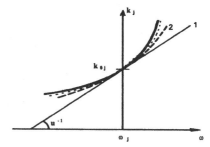

Figure 1
The plot represents dispersion curves. The solid line shows wave number on frequency dependence $k_j(\omega)$. Curves 1, 2, 3 illustrate how accurate this dependence can be described employing first, second, and third order linear dispersion approximations.

Terms with coefficients γ_j in (1) account for dispersion of nonlinear interaction, which leads to propagation velocity dependence on amplitude. If one would change reference frequencies, the nonlinear coefficients values will vary according to the following formula: $\beta_j(\omega) = \beta_j(\omega_j) + \gamma_j(\omega - \omega_j)$.

In limiting cases (1) transforms into well-known equations. Second-order nonlinear dispersion theory ($G_j = \gamma_j = 0$) is commonly used for coupled solitons analysis, frequency doubling, etc. [1-14]. The case $D_j = \beta_j = 0$ leads to set of coupled KDV equations [20].

3. Conservation laws

Equation (1) has conservation laws, if there are some specific relations between medium parameters. The total energy flux $J_1 = \int(|A_1|^2/\beta_1 + |A_2|^2/\beta_2)d\tau$ is conserved for propagating wave packets, if $2\beta_1/\beta_2 = \gamma_1/\gamma_2$. This relation is often fulfilled, as usually $\gamma_j \propto \beta_j/\omega_j$. If dispersion and nonlinear coefficients are related by the following ratio, $\beta_j/D_j = \gamma_j/G_j$, group velocities are equal and there is phase matching, the following integral remains constant along distance:

$$J_3 = \int\left(\frac{2G_1}{\gamma_1}\left|\frac{\partial A_1}{\partial \tau}\right|^2 + \frac{G_2}{\gamma_2}\left|\frac{\partial A_2}{\partial \tau}\right|^2 + \left(A_2 A_1^{*2} + A_2^* A_1^2\right)\right)d\tau.$$

4. Bright and dark parametrical solitons

To study parametric soliton existence, we search for analytical solutions of equation (1) describing propagation of coupled localized waves with equal velocities:

$$A_j = B_j(\tau + \alpha\, z)\exp(i\Omega_j\tau - iq_j z), \qquad (2)$$

where B_j is real function, α accounts for propagation velocity change, Ω_j and q_j are constants. For effective interaction between waves on fundamental and double frequencies they should have equal phase velocities: $\Omega_2 = 2\Omega_1 = 2\Omega$ and $q_2 - 2q_1 + \Delta k = 0$.

Substituting expressions (2) into equations (1), we obtain system of four equations for two real functions B_j. This system has non-trivial solutions if some relations between medium and soliton parameters are fulfilled. In this case the set of four equations reduces to two independent ones:

$$\left(3G_1\Omega + D_1\right)\frac{\partial^2 B_1}{\partial\tau^2} = \left(G_1\Omega^3 + D_1\Omega^2 + v_1\Omega - q_1\right)B_1 + \left(\gamma_1\Omega + \beta_1\right)B_1 B_2,$$

$$\left(6G_2\Omega + D_2\right)\frac{\partial^2 B_2}{\partial\tau^2} = \left(8G_2\Omega^3 + 4D_2\Omega^2 + 2v_2\Omega - q_2\right)B_2 + \left(\gamma_2\Omega + \beta_2\right)B_1^2.$$

This system has exact analytical solutions, describing parametrically coupled solitons [1,2,13]. Depending on medium characteristics, either bright or dark solitons can exist. In the absence of linear phase modulation, $\Omega_2 = \Omega_1 = 0$, simple expressions for soliton profiles can be obtained:

$$B_j(\tau + \alpha\, z) = B_{0j}\operatorname{sech}^2(\tau + \alpha\, z/T), \qquad (3a)$$

$$B_j(\tau + \alpha\, z) = E_{0j}\left(\operatorname{sech}^2(\tau + \alpha\, z/T) - 2/3\right). \qquad (3b)$$

Corresponding plots of wave packets envelopes in these two cases are shown on Fig. 2. Bright (3a) or dark (3b) solitons can exist only if some specific relations between media properties and soliton parameters are fulfilled.

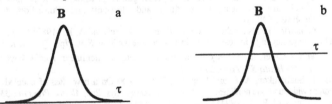

Figure 2. These plots illustrate bright (a) and dark (b) soliton profiles given by (3). The soliton envelops are symmetric, due to mutual compensation of the dispersion of nonlinearity and third-order linear dispersion effects.

It is interesting to analyze soliton characteristics. The peak amplitude values are backward proportional to soliton duration squared: $B_{02} = -6D_1/\beta_1 T^2$ and $B_{01} = \pm(6/T^2)\sqrt{D_1 D_2/\beta_1\beta_2}$; soliton duration is fixed: $T^2 = \left|4(G_2 - G_1)/v\right|$. The soliton can exist, if $G_j = D_j\gamma_j/\beta_j$. In the case of equal group velocities, $v_j = 0$, solitons of any duration can propagate in the medium (third order dispersion coefficients should be equal, $G_1 = G_2$). Due to nonlinearity dispersion the soliton propagation velocity depends on its amplitude: $\alpha = 2\gamma_1 B_{02}/3$.

5. Conclusion

In this paper propagation of two coupled waves was analyzed in $\chi^{(2)}$ medium. High-order dispersion effects was taken into account, which is essential for extremely short pulses with broad spectrum. Conservation laws were found for wave packets. Exact analytical solutions for parametric coupled bright and dark solitons were obtained. The propagation velocity change of such coupled soliton is proportional to its amplitude. It was shown that the soliton envelope form is symmetric, due to mutual compensation of the dispersion of nonlinearity and third-order linear dispersion effects.

Acknowledgement

This work was supported by RFBR and CRDF funds.

References

1. Stegeman, G. I., Hagan, D. J. and Torner, L. (1996) $\chi^{(2)}$ cascading phenomena and their applications to all-optical signal processing, mode-locking, pulse compression and solitons, *Optical and Quantum Electronics* **28**, 1691-1740.
2. Karamzin, Yu. N. and Sukhorukov, A. P. (1974) Nonlinear interaction of diffracted light beams in a medium with quadratic nonlinearity: mutual focusing of beams and limitation on the efficiency of optical frequency converters, *Sov. Phys. JETP Lett.* **20(11)**, 339-342.
3. Karamzin, Yu. N. and Sukhorukov, A. P. (1975) Mutual focusing of high-power light beams in media with quadratic nonlinearity, *Sov. Phys. JETP* **41(3)**, 414-420.
4. Schiek, R., Baek, Y. and Stegeman, G. L. (1996) One-dimensional spatial solitary waves due to cascaded second-order nonlinearities in planar waveguides *Phys. Rev. A* **53**, 1138-1141.
5. Torruelas, W. E., Wang, Z., Hagan, D. J., Van Stryland, E. W., Stegeman, G. I., Torner, L. and Menyuk, C. R. (1995) Observation of two-dimensional spatial solitary waves in quadratic medium, *Phys. Rev. Lett.* **74**, 5036-5039.
6. Akhmanov, S. A., Chirkin, A. S., Drabovich, K. N., Kovrigin, A. I., Khokhlov, R. V. and Sukhorukov, A. P. (1968) Nonstationary Nonlinear Optical Effects and Ultrashort Light Pulse Formation, *IEEE J. Quantum Electron.* **4(10)**, 598-605.
7. Azimov, B. S., Karamzin, Yu. N., Sukhorukov, A. P. and Sukhorukova, A. K. (1980) Interaction of weak pulses with a low-frequency high-intensity wave in a dispersive medium, *Sov. Phys. JETP Lett.* **51(1)**, 40-46.
8. Kobyakov, A. and Lederer, F. (1996) Photonics in quadratic nonlinear media beyond the $\chi^{(3)}$ perspective, *Optica Applicata* **XXVI(4)**, 275-284.
9. Karamzin, Yu. N., Sukhorukov, A. P. and Filipchuk, T. S. (1978) On a new class of coupled solitons in a dispersive medium with quadratic nonlinearity, *Moscow University Phys. Bull, Alerton Press Inc.* **33(4)**, 73-79.
10. Werner, M. J., Drummond, P. D. (1994) Strongly coupled nonlinear parametric solitary waves, *Opt. Lett.* **19(9)**, 613-615.
11. Torner, L. (1995) Stationary solitary waves with second-order nonlinearities, *Opt. Commun.* **114**, 136-140.
12. Sukhorukov, A. P. (1988) *Nonlinear wave interactions in optics and radiophysics*, Moscow, Nauka (in Russian).
13. Buryak, A. V. and Kivshar, Yu. S. (1995) Solitons due to second harmonic generation, *Phys. Rev. A* **197**, 407-412.
14. Boardman, A. D. and Xie, K. (1997) Vector spatial solitons influenced by magneto-optic effects in cascadable nonlinear media, *Phys. Rev. E* **55**, 1-11.
15. Hasegava, A. (1989) *Optical solitons in Fibers*, Springer-Verlag, Berlin.
16. Gromov, E. M., Talanov, V. I. (1996) Higher order approximations of the dispersion theory for nonlinear waves in homogeneous and inhomogeneous media. *Bulletin of the Russian Academy of Sciences, Physics* **60(12)**, 1836-1849.
17. Liu, Shan-liang and Liu, Xi-quiang. (1997) Mutual compensation of the higher-order nonlinearity and the third-order dispersion, *Phys. Lett. A* **225**, 67-72.
18. Karamzin, Yu. N., Sukhorukov, A. P. and Potashnikov, A. S. (1996) *Izv. RAN. Ser. Phys.* **60(12)**, 29.
19. Brabec, T. and Krausz, F. (1997) Nonlinear Optical Pulse Propagation in the Single-Cycle Regime, *Phys. Rev. A* **78**, 3282-3285.
20. Gottwald, G., Grimshaw, R., Malomed, B. (1997) Parametric envelope solitons in coupled Korteweg -de Vries equations, *Phys. Lett. A* **227**, 47-54.

LARGE SELF-PHASE MODULATION VIA SIMULTANEOUS SECOND HARMONIC GENERATION AND SUM FREQUENCY MIXING

K. KOYNOV, S. SALTIEL
University of Sofia, Faculty of Physics,
Quantum Electronics Department,
5 J. Bourchier Blvd., 1164 Sofia, Bulgaria

The effect of obtaining large self-phase modulation of the fundamental wave participating in second order nonlinear optical processes such as second harmonic generation (SHG) and sum frequency generation (SFG) via cascading was investigated extensively in last few years[1-5].

This investigations show that the process of SFG or Type II nondegenerate SHG is much more efficient if one would like to obtain bigger nonlinear phase shift (NPS) at lower input intensities [6-9]. In this case however the difficulty is that two light sources are needed and more over only the weaker input wave obtains big NPS.

In this article we propose a method for obtaining large NPS with only one fundamental wave that enter quadratic nonlinear media. Let us consider the fundamental beam with frequency ω entering second order nonlinear media. As a first step via process of type I SHG the wave with frequency 2ω is generated and as a second step via process of SFG ($\omega + 2\omega = 3\omega$) third harmonic is generated. Both processes (SHG and SFG) are supposed to be near phase matched. Generated second and third harmonic wave are downconverted to the fundamental wave ω via processes ($2\omega - \omega$); ($3\omega - 2\omega$) and ($3\omega - 2\omega$, $2\omega - \omega$) contributing to the nonlinear phase shift that the fundamental wave collects.

In the past three and four multistep $\chi^{(2)}$ processes have been considered only in connection with the possibility to generate fourth harmonic in single noncentrosymmetric crystals [10,11].

In order to find the influence of multistep processes on the nonlinear phase shift of the fundamental wave we investigate reduced amplitude equations in the slowly-varying envelope approximation, with assumption of zero absorption for all interacting waves, that have the following form:

$$\frac{dA_1}{dz} = -i\sigma_1 A_3 A_2^* \exp(-i\Delta k_3 z) - i\sigma_2 A_2 A_1^* \exp(-i\Delta k_2 z)$$

$$\frac{dA_2}{dz} = -i\sigma_4 A_3 A_1^* \exp(-i\Delta k_3 z) - i\sigma_5 A_1 A_1 \exp(i\Delta k_2 z) \qquad (1)$$

$$\frac{dA_3}{dz} = -i\sigma_3 A_1 A_2 \exp(i\Delta k z)$$

A. D. Boardman et al. (eds.),
Advanced Photonics with Second-order Optically Nonlinear Processes, 239–242.
© 1999 *Kluwer Academic Publishers. Printed in the Netherlands.*

240

where the subscripts "1" denote the fundamental waves, subscripts "2" denote the second harmonic and the subscript "3" denotes the third harmonic wave. The wave vector mismatch for the SHG process is $\Delta k_2 = 2k_2 - k_1$ and for SFM process is $\Delta k_3 = k_3 - k_2 - k_1$. We used the following relations between nonlinear coupling coefficients $\sigma_1 = \sigma_2 = \sigma_5$; $\sigma_3 = 3\sigma_1$; $\sigma_4 = 2\sigma_1$.

The results of the numerical solution of system (1) for various combination of the wave vector mismatches Δk_2 and Δk_3 are presented on figure 1, where is shown the dependence of the collected by the fundamental wave nonlinear phase shift versus normalized coordinate in the media for the normalized input amplitude of the fundamental wave $\sigma a_{in} L = 10$. With doted line is presented the contribution of only two step cascading and with solid line the full NPS. As one may see the contribution of the multistep cascading leads to increasing of the self-phase modulation of the fundamental wave.

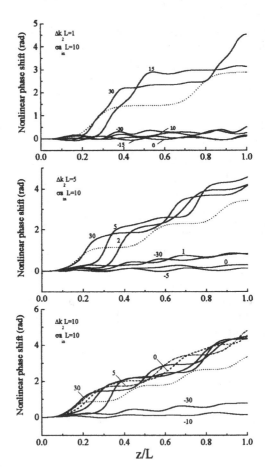

Figure 1. Nonlinear phase shift of the fundamental wave as a function of normalised distance inside media. With dotted line is presented the contribution of two step cascading only. The parameter is $\Delta k_3 L$.

This large nonlinear phase shift make the multistep cascading suitable for low power all-optical switching and signal processing. On figure 2 there are presented the NPS and the depletion of the fundamental wave versus its normalised input amplitude. The phase mismatches are respectively $\Delta k_2 L = 4.8$ and $\Delta k_3 L = 30$. These values are optimal for obtaining simultaneously value of NPS $\sim \pi$ and full reconstruction of the fundamental intensity. At it seen it occur at normalized input amplitude $\sigma a_1 L = 7$. For comparison in the same figure is presented similar optimization for Type I SHG ($\Delta k_2 L = 0.3$). It is seen that in this case the respective normalized input amplitude $\sigma a_1 L$ exceeds 13.

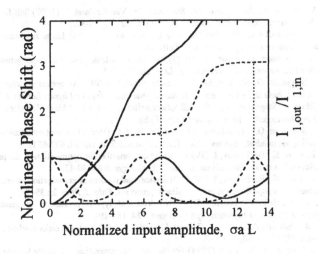

Figure 2. NPS and the depletion of the fundamental wave as a function of its normalised input amplitude for the case of multistep cascading with $\Delta k_2 L = 4.8$ and $\Delta k_3 L = 30$ (solid line) and for the case of Type I SHG with $\Delta k_2 L = 0.3$ (dashed line).

The dependencies shown on figures 2 and 3 are for relatively small values of the normalized mismatches $\Delta k_2 L$ and $\Delta k_3 L$. When $\Delta k_2 \gg \Delta k_3$ or $\Delta k_3 \gg \Delta k_2$ the resulted NPS is equal to the NPS due to the process of SHG only. This means that substantial increase of the self-phase modulation due to multistep $\chi^{(2)}$ cascading is possible when both processes SHG and SFG are near phase matched. One of the possible way to achieve simultaneous phase matching for two different nonlinear processes is the quasi phase matched technique [12]. Simultaneous phase matching of the processes of SHG and SFG have been achieved experimentally in periodically poled crystals [13].

In conclusion, here we present a method for obtaining a large self-phase modulation by the fundamental wave that pass through quadratic nonlinear media. The presence of multi step $\chi^{(2)}$ cascading leads to increase of the value of the NPS in comparison with two step cascading. This more efficient way to generate large nonlinear phase shift can be used for reduction of the switching intensity of existing all optical switching devices based on $\chi^{(2)}$ cascading [3,6] and for construction intracavity nonlinear optical devices for mode locking [14,15].

We would like to acknowledge Prof. M Fejer, Prof. G. Assanto and Prof. A. P. Sukhorukov for the discussions we had during the school. Bulgarian Science Foundation is also acknowledged for the support via the grant MUF01.

References:

1. De Salvo, R., Hagan, D., Sheik Bahae, M., Stegeman, G., Van Stryland, E. (1992) Self-focusing and self-defocusing by cascaded second order effects in KTP, *Opt. Lett.* **17**, 28-30.

2. Stegeman, G., Sheik-Bahae, M., Van Stryland, E., Assanto, G. (1993) Large nonlinear phase shift in second-order nonlinear optical processes, *Opt. Lett.*, **18**, 13-15.

3. Hutchings, D., Aitchison, J., Ironside, C. (1993) All-optical switching based on nondegenerate phase shifts from a cascaded second-order nonlinearity, *Opt.Lett.*, **18**, 793-795.

4. Assanto, G., Stegeman, G., Sheik Bahae, M., Van Stryland, E. (1995) All-optical switching devices based on large nonlinear phase shifts from second harmonic generation, *Applied Physics Letters* **62**, 1323-1326.

5. Baek, Y., Schek, R., Stegeman, G. (1995) All optical switching in a hybrid Mach Zehnder interferometer as a result of cascaded second order nonlinearity, *Opt. Letters* **20**, 2168-2170.

6. Assanto, G., Stegeman, G., Sheik-Bahae, M., Van Stryland, E. (1995) Coherent interactions for all-optical signal processing via quadratic nonlinearities, *IEEE J.Quant.Electron.*, **31**, 673-681.

7. Saltiel, S., Koynov, K., Buchvarov, I. (1996), Analytical formulae for optimisation of the process of lower phase modulation in a quadratic nonlinear medium, *Appl. Phys. B* **62**, 39-42.

8. Saltiel, S., Koynov, K., Buchvarov, I. (1995) Analytical and numerical investigation of opto-optical phase modulation based on coupled second order nonlinear processes, *Bulg. J. Phys.* **22** 39-47.

9. Kobyakov, A., Peschel, V., Lederer, F. (1996) Vectorial type-II interaction in cascaded quadratic nonlinearities - an analytical approach, *Opt. Commun.* **124**, 184-191.

10. Akhmanov, S., Dubovik, A., Saltiel, S., Tomov, I., Tunkin, V. (1974) Forth order nonlinear optical effects in LFM crystal, *JETP Lett.* **20**, 117-118.

11. Hooper, B., Gauthier, D., Madey, J. (1994) Fourth-harmonic generation in a single lithium niobate-crystal with cascaded second-harmonic generation, *Appl. Opt.* **33**, 6980-6987.

12. Fejer, M., Magel, G., Jundt, D., and Byer, R. (1992) Quasi-Phase-Matched Second Harmonic Generation: Tuning and Tolerances, *IEEE J. of Quant. Electronics.* **28**, 2631-2639.

13. Pfister, O., Wells, J., Hollberg, L., Zink, L., Van Baak, D., Levenson, M., Basenberg, W. (1997) Continuous-wave frequency tripling and quadrupling by simultaneous three wave mixings in periodically poled crystals: application to a two-step 1.19-10.71 μm frequency bridge *Opt. Lett.*, **22**, 1211-13.

14. Cerullo, G., De Silvestri, S., Monduzzi, A., Segala, D., Magni, V. (1995) Self-starting mode locking of a CW Nd:YAG laser using cascaded second order nonlinearities, *Opt. Lett.* **20**, 746-748.

15. Stegeman, G., Hagan, D. and Torner, L. (1996) $\chi^{(2)}$ cascading phenomena and their applications to all-optical signal processing, mode-locking, pulse compression and solitons, *J. Opt.&Quantum Electron.*, **28**, 1691-1740.

PULSED BEAM SELF-FOCUSING

A.V. CHURILOVA, A.P. SUKHORUKOV
Radiophysics Department, Physics Faculty,
Moscow State University
Vorobjevy Gory, Moscow, 119899, Russia
alina@nls.phys.msu.su, aps@nls.phys.msu.su

Nonlinear optical radiation propagation bounded in time (pulse) and in space (beam) is considered. We investigate temporal-spatial self-action, when pulse decompression influences on the beam self-focusing. An exact analytical solution of coupled nonlinear differential equations for pulse and beam widths is derived. Influence of dispersion spreading of the pulse on the formation of spatial soliton is considered.

1. Introduction

The nonlinear phenomena of self-action are of great interest in various fields of physics [1-4]. Earlier we have considered effects of self-localization (delocalization) of oscillations propagating in two-dimensional gratings with a cubic nonlinearity near critical frequencies of the transparency band. There is the analogy of these effects with self-focusing (defocusing) of wave beams [5].

In a present paper we consider nonlinear propagation of pulsed optical beam. Effects of self-compression (decompression) and self-focusing (defocusing) can develop in the dependence of the signs of the group velocity dispersion coefficients and the refractive index nonlinearity [6]. Obviously, pulse self-action influences on the beam self-action and on the contrary. In this work, from possible influences of interference on enumerated effects, we consider a variant of competition of opposite in action processes, namely: self-focusing (defocusing) and decompression (self-compression) in the medium with a cubic nonlinearity. It is supposed that the beam has planar structure.

2. Propagation of Waves Bounded in Time and Space

In quasi-optics approximation the equation for the complex amplitude of space and time modulated wave, propagating in dispersive medium with cubic nonlinearity, is given by

$$i\frac{\partial A}{\partial z} + D_1 \frac{\partial^2 A}{\partial x^2} - D_2 \frac{\partial^2 A}{\partial \tau^2} + \gamma |A|^2 A = 0, \tag{1}$$

where $D_1 = 1/(2k)$ is the coefficient of amplitude diffusion in transverse cross-section of

243

A. D. Boardman et al. (eds.),
Advanced Photonics with Second-order Optically Nonlinear Processes, 243–246.

the beam, $D_2 = -(1/2)(\partial^2 k/\partial \omega^2)_{\omega_0}$ is the dispersion coefficient of group velocity $v = (\partial k/\partial \omega)_{\omega_0}^{-1}$, $\tau = t - z/v$, γ is the cubic nonlinearity coefficient.

It is interesting to note that (1) is identical to the evolution equation, what describes oscillations in two-dimensional anharmonic grating near the saddle point on the dispersion surface [5].

Let us consider the case of the beam self-action and dispersion decompression of the pulse when $\gamma > 0$ and $D_2 > 0$. To solve (1) we use the aberrationless approximation [6] introducing the complex amplitude in the form of a Gaussian shape with quadratic phase modulation in space and time:

$$A = \left(E_0 / \sqrt{f_x f_\tau}\right) \exp\left[-x^2/\left(a^2 f_x^2\right) - \tau^2/\left(T^2 f_\tau^2\right) - i\psi\right], \qquad (2)$$

where phase is $\psi = f_x' x^2/\left(4D_1 f_x\right) - f_\tau' \tau^2/\left(4D_2 f_\tau\right) + \varphi(z)$; f_τ and f_x are the dimensionless magnitudes of pulse and beam widths; a, T are initial pulse and beam widths, respectively.

Substituting (2) into (1) and applying method [6], we can derive from (1) a system of two coupled equations for f_τ and f_x:

$$f_x'' = 1/\left(l_x^2 f_x^3\right) - \alpha/\left(l_x f_\tau f_x^2\right), \quad f_\tau'' = 1/\left(l_\tau^2 f_\tau^3\right) + \alpha/\left(l_\tau f_x f_\tau^2\right), \qquad (3)$$

where differentiation is performed with respect to the coordinate z, $l_x = a^2/\left(4D_1\right)$, $l_\tau = T^2/\left(4D_2\right)$ are diffraction and dispersion spreading lengths, respectively, $\alpha = 2\gamma E_0^2$ is the nonlinearity parameter.

System (3) has the following integrals

$$H = l_x \left(f_x'\right)^2 - l_\tau \left(f_\tau'\right)^2 + 1/\left(l_x f_x^2\right) - 1/\left(l_\tau f_\tau^2\right) - 2\alpha/\left(f_x f_\tau\right), \qquad (4)$$

$$I = l_x f_x^2 - l_\tau f_\tau^2 - Hz^2 + 2Cz, \qquad (5)$$

$$C = Hz - l_x f_x f_x' + l_\tau f_\tau f_\tau'. \qquad (6)$$

Under boundary conditions $f_x\big|_{z=0} = f_\tau\big|_{z=0} = 1$, $f_x'\big|_{z=0} = f_\tau'\big|_{z=0} = 0$ the Hamiltonian of the system is $H = 1/l_x - 1/l_\tau - 2\alpha$, integrals $I = l_x - l_\tau$ and $C = 0$. If rates of diffraction and dispersion wave spreading are equal, $l_x = l_\tau$, so $I = 0$.

Results of numerical solving of equations (3) with $\alpha = 10$ are presented in Fig. 1. Beam width decreases because of self-focusing and pulse width increases owing to nonlinear decompression. Then after passing the focus widths increase simultaneously.

3. The Exact Analytical Solution of Equations for Pulse and Beam Widths

If we introduce auxiliary function $g(z)$ and $f_x = (2\alpha)^{1/2} z \, \mathrm{sh}\, g$, $f_\tau = (2\alpha)^{1/2} z \, \mathrm{ch}\, g$, and when diffraction and dispersion spreading lengths are equal, $l_x = l_\tau$, and $C = 0$, $I = 0$, we get one of the exact analytical solutions of (3) in implicit form:

$$\left(\sqrt[4]{1+\alpha^2} + 1/\sqrt[4]{1+\alpha^2}\right) F(\beta, r) - 2\sqrt[4]{1+\alpha^2}\, E(\beta, r) +$$

$$+ \left(2\alpha \, \mathrm{ch}(2g)\sqrt{\alpha \, \mathrm{sh}(2g) - 1}\right) \Big/ \left(\sqrt{1+\alpha^2} + \alpha \, \mathrm{sh}(2g) - 1\right) = 2/z \,, \tag{7}$$

where $F(\beta, r)$ and $E(\beta, r)$ are the elliptical integrals of first and second type, respectively, $\beta = \arccos\left[\left(\sqrt{1+\alpha^2} + 1 - \alpha \, sh(2g)\right) \Big/ \left(\sqrt{1+\alpha^2} - 1 + \alpha \, sh(2g)\right)\right]$, $r = \sqrt{\left(\sqrt{1+\alpha^2} - 1\right) \Big/ \left(2\sqrt{1+\alpha^2}\right)}$.

The analysis of solution (7) with $\alpha \gg 1$, shows that the beam width achieves minimum value $f_{x\,min} = 0{,}83/\alpha$ at the distance $z_{min} = 1{,}18/\sqrt{\alpha}$; in this case $f_\tau(z_{min}) = 1{,}668$. Dependences of values $f_{x\,min}$ and $f_\tau(z_{min})$ on nonlinearity parameter α have been calculated by numerical methods and shown in Fig. 2.

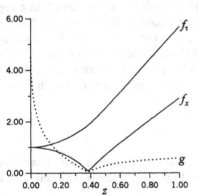

Figure 1. Beam width f_x and pulse width f_τ as functions of the distance z. Auxiliary function $g(z)$ is showed by dotted line.

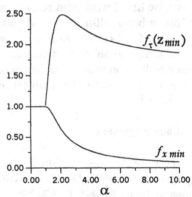

Figure 2. Minimum beam width $f_{x\,min}$ and pulse width $f_\tau(z_{min})$ as functions of the parameter of nonlinearity α.

4. Influence of Dispersion Spreading of Pulse on the Forming of the Spatial Soliton

Let us consider the problem of waveguide pulse propagation for the system (3). It follows from the first equation (3) that spatial soliton (nonlinear planar waveguide) of pulse radiation with $f_x = 1$ and $f_\tau = 1$ corresponds to a nonlinearity parameter $\alpha = 1/l_x$.

246

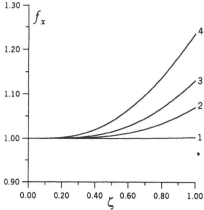

Figure 3. Beam width f_x as a function of normalized coordinate ζ ($b = 20$ (1), $b = 1$ (2), $b = 0.5$ (3), $b = 0.2$ (4)).

Introducing ratio of dispersion length to diffraction one $b = l_\tau/l_x$, we can rewrite (3) in the following form:

$$f_x'' = 1/f_x^3 - 1/(f_\tau f_x^2),$$
$$f_\tau'' = b^2/f_\tau^3 + b/(f_x f_\tau^2), \qquad (8)$$

where differentiation is performed with respect to the dimensionless spatial coordinate $\zeta = z/l_x$. Results of numerical solution of (8) are presented in Fig. 3.

The analysis of the obtained data shows that because of dispersion spreading the pulse wave amplitude decreases and self-action becomes weak, so the spatial soliton doesn't preserve its width. The stronger the dispersion spreading of the pulse (smaller parameter b) the quicker the spatial soliton is destroyed.

5. Conclusion

Thus, we have investigated self-action of optical radiation in the case of the competition of planar beam self-focusing and pulse decompression. A similar situation takes place in the case of self-compression of a short pulse with beam defocusing. We have found the exact analytical solution of two coupled differential equations for pulse duration and beam width. Formulas for minimum pulse width and distance corresponding to this width have been obtained. The problem of waveguide pulse propagation has also been considered.

Acknowledgement

This work was supported by the Russian Foundation for Basic Research (project No. 96–02–18592) and the «Fundamental Problems of Natural Sciences» program (grant No.9–5–0–2.2–79).

References

1. Rainer Scharf and Bishop, A. R. (1991) Properties of the nonlinear Schrodinger equation on a lattice, *Physical review* **43**, 6535-6544.
2. Pouget, J. (1993) Energy self-localization and gap local pulses in a two-dimensional nonlinear lattice, *Physical review* **47**, 14866-14874.
3. Bridges, E.R., Boyd, R.W., Agrawal, G. P. (1996) Multidimensional coupling owing to optical nonlinearities. I. General formulation, *J. Opt. Soc. Am. B* **13**, 553-559.
4. Bridges, E.R., Boyd, R.W., Agrawal, G. P. (1996) Multidimensional coupling owing to optical nonlinearities. II. Results, *J. Opt. Soc. Am. B* **13**, 560-569.
5. Sukhorukov, A.P., Churilova, A.V. (1996) Dynamics of vibrations in two-dimensional nonlinear lattices, *Bulletin of the Russian Academy of Sciences,Physics* **60**, 1889-1895.
6. Vinogradova, M.B., Rudenko, O.V., and Sukhorukov, A.P. (1990) *Theory of Waves,* Nauka, Moscow [in Russian].

SLOW AND IMMOBILE SOLITONS IN QUADRATIC MEDIA

S.V. POLYAKOV and A.P. SUKHORUKOV
Radiophysics Department, Physics Faculty,
Moscow State University
Vorobjovi Gory, Moscow, Russia, 119899
e-mail: polyakov@nls.phys.msu.su

1. Introduction

It is well known what solitons can be excited in wide range of nonlinear media due to the balance of dispersive and nonlinear effects [1]. The properties of solitons in the media with third-order nonlinearity were studied very intensive. In particular, the phenomenon of slow and immobile solitons forming near the bounds of non-transmission bands in the medium with complex dispersion was discovered and studied theoretically and numerically in [2-4]. This phenomenon is based on the bound frequency shift due to self-action effects. The input signal on forbidden frequencies splits into slow non-dumping solitons. This phenomenon was named "nonlinear tunneling" [2]. Solitons that propagate in forbidden bands are known as "gap solitons".

In the recent years parametric solitons engage the attention of many scientists all around the world. These solitons were predicted theoretically in 1974 [5]. In the year 1995 the scientific group directed by Stegeman G. I. proved the existence of parametric solitons [6, 7]. Many works and publications were devoted to the problems and the properties of optical parametric solitons ([8–12] and the references within). The theoretical and numerical investigation of immobile and slow solitons, or simultons, is being performed by Kivshar Yu., Assanto G. and others [10-14]. Nevertheless, there are a lot of questions on the excitation and the properties of slow parametric solitons in periodical structures. In order to excite this type of solitons one can use the media with two dispersion curves with a non-transmission gap between them (Fig. 1).

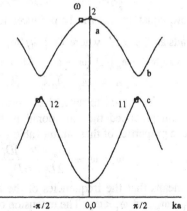

Fig. 1. The dispersion curve of a medium with periodical inhomogenity. Two fundamental waves 11 and 12 excite a second harmonic 2. Circles demonstrate the excitation of an immobile soliton, squares demonstrate the excitation of a slow one.

A. D. Boardman et al. (eds.),
Advanced Photonics with Second-order Optically Nonlinear Processes, 247–250.
© 1999 *Kluwer Academic Publishers. Printed in the Netherlands.*

2. The basic equations

Let us consider a periodically inhomogenious medium with a dispersion characteristic, shown in Fig. 1. In order to excite slow or immobile soliton the second harmonic should have the frequency $\omega_2 \approx \omega_a$ near the top of the upper curve and the fundamental one should consist of two waves with the frequency $\omega_1 = \omega_2/2$ and contrary-oriented wavevectors $k_{11} \approx -k_{12}$ (Fig. 1). The envelopes of these waves are found to obey the three equations [13, 14]:

$$\frac{\partial B_2}{\partial t} = iD_2 \frac{\partial^2 B_2}{\partial z^2} - i(\omega_2 - \omega_a)B_2 + i\beta_2 B_{11}B_{12},$$

$$\frac{\partial B_{11,12}}{\partial t} = iD_1 \frac{\partial^2 B_{11,12}}{\partial z^2} - i(\omega_1 - \omega_c)B_{11,12} + i\beta_1 B^*_{12,11}B_2, \qquad (1)$$

where t is the time, z the propagation coordinate, $D_{1,2} > 0$ the dispersion coefficients, $\beta_{1,2}$ the coefficients of quadratic nonlinearity. The equations similar to (1) were investigated in many studies devoted to three-wave interactions in the presence of second order nonlinearity [9 – 12]. In this case, however, we are dealing with nonlinear tunneling of slow solitons and excitation of immobile ones.

3. Immobile solitons

To consider the excitation of the immobile parametric solitons (Fig. 1), we substitute the envelopes into (1) as $B = A(z)$ where A are real. This leads us to the following equations:

$$D_2 A_2'' = (\omega_2 - \omega_a)A_2 - \beta_2 A_{11}A_{12}, \quad D_1 A_{11,12}'' = (\omega_1 - \omega_c)A_{11,12} - \beta_1 A_{12,11}A_2. \quad (2)$$

Now, equations (2) will be produced to normalized form. The dimensionless coordinate reads as $\xi = z/l$, where $l = [D_1/(\omega_1 - \omega_c)]^{1/2}$. We also use the dimensionless constant $\alpha = D_1(\omega_2 - \omega_a)/D_2(\omega_1 - \omega_c)$. After all substitutions we will obtain:

$$\tilde{A}_2'' = \alpha\tilde{A}_2 - \tilde{A}_{11}\tilde{A}_{12}, \quad \tilde{A}_{11,12}'' = \tilde{A}_{11,12} - \tilde{A}_{12,11}\tilde{A}_2. \qquad (3)$$

It is necessary to note, that Kivshar et. al. calculated the envelopes of parametric soliton practiced the variational method and numerically [10, 11]. However, some basic properties of these immovable solitons can be determined analytically. First:

$$\omega_1 = \omega_c + \frac{(\omega_a - 2\omega_c)D_1}{2D_1 - \alpha D_2}, \quad \omega_2 = \omega_a + \frac{(\omega_a - 2\omega_c)\alpha D_2}{2D_1 - \alpha D_2}. \qquad (4)$$

It means that the frequencies of the harmonics are detuned from the boundary ones, $\omega_1 > \omega_c$ and $\omega_1 < \omega_c$. The extension of soliton is proportional to l given by:

$$l^2 = (2D_1 - \alpha D_2)/(\omega_a - 2\omega_c). \qquad (5)$$

Thus, the immobile soliton with fixed extension exists for every constant α. The extension (5) might be substituted into (4). Then, frequencies of the fundamental and the second harmonics reads as:

$$\omega_2 = \omega_a + \alpha D_2/l^2, \quad \omega_1 = \omega_c + D_1/l^2. \qquad (6)$$

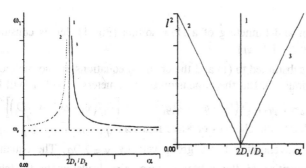

Fig. 2 The frequency of a fundamental harmonic via normalized parameter α. Three thinkable cases are marked by numbers 1, 2 and 3.

Fig. 3 The extension of immobile soliton square via normalized parameter of the soliton α.

The fig. 2 and 3 represent the frequency of the fundamental harmonic and the extension of immobile soliton via normalized parameter α. The numbers 1, 2 and 3 marks each thinkable case.

$\alpha = 2D_1/D_2$. This case is possible only when $\omega_a = 2\omega_c$. (5) is correct without reference to extension l. Thus, soliton extension depends on the harmonic's frequencies (6).

In the second case, $\alpha < 2D_1/D_2$. And $\omega_a > 2\omega_c$. The formulas (4, 5) describe the dependence of extension l and excitation frequencies ω_1 и ω_2 on the parameter α.

Finally, if $\alpha > 2D_1/D_2$, the situation is similar to previous. The only difference is the demanding $\omega_a < 2\omega_c$.

Let us remark, that $\omega_1 > \omega_c$ in all listed cases. (i. e. for all possible values of α)

The exact soliton solution in partial case of $\alpha = 1$ is:

$$A = a \operatorname{sech}^2(z/2l).$$ (7)

The amplitudes of the exact soliton a reads as:

$$a_{11} = a_{11} = \pm(3/2l^2)\sqrt{D_1 D_2/\beta_1 \beta_2}, \quad a_2 = \pm(3D_1/2l^2\beta_1).$$ (8)

Let's examine all three cases considering $\alpha = 1$.

1) $\omega_a = 2\omega_c$. Then (5) shows, that the solitons with the envelopes (7) can be excited only if $D_2 = 2D_1$. In this kind of structure the value of the soliton extension l is determined by their amplitudes (8). Every value of l is allowed.

2) If $\omega_a > 2\omega_c$, then $D_2 < 2D_1$. The soliton excitation frequencies with the envelopes (7) supervene from (6) and condition $\omega_{2i} = 2\omega_{1i}$. This frequencies are exclusive when the boundary frequencies and dispersion coefficients are fixed. The index 'i' indicates this fact. The correlations between boundary and soliton's frequencies are: $\omega_{1i} > \omega_a/2 > \omega_c$, $\omega_{2i} > \omega_a$.

3) $\omega_a < 2\omega_c$ and $D_2 > 2D_1$. This case is similar to previous, but the correlations between boundary and soliton frequencies are different, namely: $\omega_2 > 2\omega_c > \omega_a$, $\omega_1 > \omega_c$.

The extension and the amplitude of the soliton l in two bottom cases have the determined values (5, 8).

4. Slow Solitons

To examine the excitation and tunneling of a slow soliton (Fig. 1), let us consider $B = A(\xi)\exp(-iqz)$, where $\xi = (z - \mathrm{v}t)$.

These expressions are substituted to (1) and the obtained equations are normalized. This leads us to the equations (3). But the dimensionless parameters l and α will be different: $\alpha = D_1(\omega_2 - \omega_a + D_2 q_2^2)/D_2(\omega_1 - \omega_c + D_1 q_1^2)$, $l = [D_1/(\omega_1 - \omega_c + \mathrm{v}^2/D_1)]^{1/2}$. The term Dq^2 is added due to the changes of wave numbers $q_{11} = q_{12} = q_1$, $2q_1 = q_2$. The wave with corrected wavenumber has a group velocity $\mathrm{v} = 2Dq$. The equality demand of the group velocities of the soliton's harmonics leads us to the stiff condition: $D_1/D_2 = 2$.

The velocity of slow soliton can be determined as: $\mathrm{v}^2 = 4D_1[(2\omega_a - \alpha\omega_c)/(4 - \alpha) - \omega_1]$. In the limiting case ($\mathrm{v} \to 0$) the frequencies of soliton excitation and dimensionless parameters l and α converge to the values of immobile soliton (4), (5).

Acknowledgement

The work was supported by Russian Fund of Basic Research (project No. 96–02–18592) and the program "Fundamental Natural Science" (grant No. 95–0–2.2–79).

References

1. Vinogradova, M. V., Rudenko, O. V. and Sukhorukov, A. P. (1990) *Theory of waves*, Nauka, Moscow. (in Russian)
2. Newell, A. C. (1978) Nonlinear tunneling, *J. Math. Phys.* **19** (5), 1126-1133.
3. Chen, W. and Mills, D. L. (1987) Gap solitons and the nonlinear optical response of superlattices, *Phys. Rev. Lett.* **58**, 160-163.
4. Karamzin, Yu. N., Polyakov, S. V. and Sukhorukov A. P. (1996) Slow solitons excitation during the nonlinear tunneling of signals, *Moscow University Phys. Bull.* **5**.
5. Karamzin, Yu. N. and Sukhorukov, A. P. (1974) Nonlinear interaction of defracted light beams in a medium with quadratic nonlinearity: mutual focusing of beams and limitation of the efficiency of optical frequency converters, *JETP Lett.* **20**, 339-342.
6. Schiek R., Baek, Y. and Stegeman, G. I. (1996) One-dimensional spacial solitary waves due to cascaded second-order nonlinearities in planar waveguides, *Phys. Rev. A* **53**, 1138-1141.
7. Torruellas, W. E., Wang, Z., Hagan, D. J., VanStryland, E. W., Stegeman, G. I., Torner, L. and Menyuk, C. R. (1995) Observation of two-dimensional spatial solitary waves in quadratic medium, *Phys. Rev. Lett.* **74**, 5036-5039.
8. Stegeman, G. I., Hagan, D. J. and Torner, L. (1996) $\chi^{(2)}$ cascading phenomena and their applications to all-optical signal processing, mode locking, pulse compression and solitons, *J. of Opt. and Quantum Electronics* **28** (12), 1691-1740.
9. Sukhorukov, A. P. (1988) *Nonlinear wave interactions in optics and radiophysics*, Nauka, Moscow. (in Russian)
10. Kivshar, Yu. S. (1995) Gap solitons due to cascading, *Phys. Rev. E.* **51** (2), 1613-1615.
11. Steblina, V. V., Kivshar, Yu. S., Lisak, M. and Malomed, B. A. (1995) Self-guided beams in diffractive quadratic medium: variational approach, *Opt. Commun.* **118**, 345-352.
12. Conti, C., Trillo, S. and Assanto, G. (1997) Bloch function approach for parametric gap solitons, *Opt. Lett.* **22** (7), 445-447.
13. Polyakov, S. V. and Sukhorukov, A. P. (1997) The immobile parametric solitons in media with periodical inhomogeneity and quadratic nonlinearity, *Moscow University Phys. Bull.* **3**, 65-67.
14. Polyakov, S. V. and Sukhorukov, A. P. (1997) Slow and Immobile Parametric Solitons in the Media with Periodic Inhomogenity, *Izv. RAN* **12**, 2353-2358.

FERMI RESONANCE NONLINEAR WAVES AND SOLITONS IN ORGANIC SUPERLATTICES

V.M. Agranovich and A.M. Kamchatnov
Institute of Spectroscopy, Russian Academy of Sciences
Troitsk, Moscow Region, 142092 Russia

1. Introduction

1.1. WHAT IS FERMI RESONANCE?

Fermi resonance is a phenomenon which takes place in vibrational or electronic spectra of molecules. For example, let a molecule have two vibrational modes with frequencies ω_a and ω_b. If the second order resonance condition $2\omega_a \simeq \omega_b$ is fulfilled, then the $\hbar\omega_b$ transition in infrared spectrum can be split into two lines of comparable intensity and the second line cannot be explained as a result of interaction of light with the vibrational a mode because the transitions with excitation of two $\hbar\omega_a$ quanta are forbidden due to well–known $n \to n \pm 1$ selection rule for harmonic oscillator. E. Fermi explained [1]-[2] this experimental observation as a result of nonlinear resonance interaction of two vibrational modes with each other. Since that time the notion of Fermi resonance has been generalized to the processes with participation of different types of quanta (e.g., $\omega_1 + \omega_2 \simeq \omega_3$, $\omega_1 + \omega_2 \simeq \omega_3 - \omega_4$, and so on) and to electronic types of excitations as well. Further generalizations were suggested for Fermi resonance interactions of collective modes in molecular crystals and other macroscopic systems, so that Fermi resonance phenomenon became a part of not only molecular physics but solid state physics also (see, e.g., review articles [3]-[5]). Recent progress in molecular beam deposition method permitted one to obtain molecular multilayer structures [6] analogous to inorganic superlattice and quantum well structures. In case of multilayer crystalline organic structures their spectrum is created by the "overlapping" of the spectra of different crystalline compounds and new Fermi resonances arise due to the unharmonicity across the interface, which may be quite interesting from various point of view. For example, one may suppose that Fermi resonance interaction between different molecules composing the layered structure can play an important role in its optical properties [7], and so the theoretical investigation of corresponding phenomena became topical. It is worth to mention already now that there exist deep interconnections between Fermi resonance theory and another "hot spot"—cascaded $\chi^{(2)}$ nonlinearity phenomenon in optics [8]-[19] which is the main topic of this School. In these lectures we are going to discuss some recent developments in application of

A. D. Boardman et al. (eds.),
Advanced Photonics with Second-order Optically Nonlinear Processes, 251–275.
© *1999 Kluwer Academic Publishers. Printed in the Netherlands.*

Fermi resonance theory to the organic superlattices and other layered structures. We assume that this audience is more experienced in optical nonlinearities than in molecular Fermi resonance interactions and therefore we shall start from simple introduction to the Fermi resonance theory.

At first, we shall consider this problem from classical point of view. Let q_a and q_b be two normal coordinates corresponding to the vibrational degrees of freedom of the molecule with eigenfrequencies ω_a and ω_b (we neglect all other degrees of freedom of the molecule). Then the Lagrangian of the vibrational motion can be written as follows,

$$L = \frac{M_a \dot{q}_a^2}{2} + \frac{M_b \dot{q}_b^2}{2} - U(q_a, q_b), \qquad (1)$$

where the overdot stands for the derivative with respect to time t, M_a and M_b are "mass" coefficients, and the potential energy $U(q_a, q_b)$ can be expanded into series with respect to powers of variables q_a and q_b,

$$U(q_a, q_b) = U_2(q_a, q_b) + U_3(q_a, q_b) + \dots \qquad (2)$$

Since q_a and q_b are normal coordinates, in harmonic approximation we have a diagonal form of the potential energy:

$$U_2(q_a, q_b) = \frac{M_a \omega_a^2}{2} q_a^2 + \frac{M_b \omega_b^2}{2} q_b^2. \qquad (3)$$

The next nonlinear term in (2) corresponds to the third degrees of q's:

$$U_3(q_a, q_b) = \sum_{m+n=3} \alpha_{mn} q_a^m q_b^n \qquad (4)$$

(the sum over cubic terms $\alpha_{03} q_b^3 + \alpha_{12} q_a q_b^2 + \dots$). It is clear that not all terms in the sum (4) are equally important under the Fermi resonance condition

$$2\omega_a \simeq \omega_b. \qquad (5)$$

Indeed, in the equations of motion

$$\begin{aligned} M_a \left(\ddot{q}_a + \omega_a^2 q_a \right) &= -\partial U_3/\partial q_a = -3\alpha_{30} q_a^2 - 2\alpha_{21} q_a q_b - \alpha_{12} q_b^2, \\ M_b \left(\ddot{q}_b + \omega_b^2 q_b \right) &= -\partial U_3/\partial q_b = -\alpha_{21} q_a^2 - 2\alpha_{12} q_a q_b - 3\alpha_{03} q_b^2, \end{aligned} \qquad (6)$$

the different terms on the right–hand side have different physical sence. For example, the terms

$$-3\alpha_{30} q_a^2 \quad \text{and} \quad -3\alpha_{03} q_b^2$$

describe a weak nonlinearity of separate eigenmodes and are not responsible for their interaction at all; hence we can omit them from the sum (4). The other terms differ from each other by their time dependence. In harmonic approximation we have

$$q_a = q_a^0 \cos(\omega_a t + \phi_a), \quad q_b = q_b^0 \cos(\omega_b t + \phi_b), \qquad (7)$$

and the potential (4) is a small perturbation leading to slow variation of q_a^0 and q_b^0. Substitution of (7) into the right–hand sides of (6) leads to resonant and non–resonant terms. For example, in the first equation (6) the left–hand side oscillates with approximately harmonic frequency ω_a whereas the term

$$q_b^2 \propto \cos^2 \omega_b t = (1 + \cos 2\omega_b t)/2$$

does not contain such a harmonic and describes a non–resonant interaction of modes q_a and q_b. Hence we must hold in this equation only the term

$$q_a q_b \propto \cos \omega_a \cos \omega_b = [\cos(\omega_b + \omega_a)t + \cos(\omega_b - \omega_a)t]/2$$

which according to the condition (5) contains the resonant "force" oscillating with frequency $\omega_b - \omega_a \simeq \omega_a$. Analogously, in the second equation (6) we must hold only the term $\alpha_{21} q_a^2$ which contains the resonant "force" oscillating approximately with the frequency $2\omega_a \simeq \omega_b$ of the mode q_b. Both these terms arise from the potential energy $\alpha_{21} q_a^2 q_b$, which includes in addition some non–resonant terms which must be removed in the so called *rotating wave approximation* we use here. To this end, it is convenient to introduce into the classical equations of motion

$$M_a \left(\ddot{q}_a + \omega_a^2 q_a\right) + 2\alpha_{21} q_a q_b = 0, \qquad M_b \left(\ddot{q}_b + \omega_b^2 q_b\right) + \alpha_{21} q_a^2 = 0, \qquad (8)$$

where we took into account the relevant $\alpha_{21} q_a^2 q_b$ interaction term only, the complex variables

$$A = \sqrt{\frac{M_a}{2\omega_a}} \left(\omega_a q_a + i\dot{q}_a\right), \quad B = \sqrt{\frac{M_b}{2\omega_b}} \left(\omega_b q_b + i\dot{q}_b\right), \qquad (9)$$

and their complex conjugate A^* and B^*, so that

$$q_a = \frac{1}{\sqrt{2M_a\omega_a}} \left(A + A^*\right), \quad q_b = \frac{1}{\sqrt{2M_b\omega_b}} \left(B + B^*\right), \qquad (10)$$

$$\dot{q}_a = i\sqrt{\frac{\omega_a}{2M_a}} \left(A^* - A\right), \quad \dot{q}_b = i\sqrt{\frac{\omega_b}{2M_b}} \left(B^* - B\right). \qquad (11)$$

Simple transformation yields the following equations for complex amplitudes A and B:

$$\dot{A} = -i\omega_a A - 2 \cdot \frac{\alpha_{21}i}{2M_a\omega_a\sqrt{2M_b\omega_b}} \left(A + A^*\right)\left(B + B^*\right), \qquad (12)$$

$$\dot{B} = -i\omega_b B - \frac{\alpha_{21}i}{2M_a\omega_a\sqrt{2M_b\omega_b}} \left(A + A^*\right)^2. \qquad (13)$$

In harmonic approximation the last terms in these equations can be omitted, so in this case we get

$$A(t) = A(0) \exp(-i\omega_a t), \quad B(t) = B(0) \exp(-i\omega_b t). \qquad (14)$$

As we stressed above, the rotating wave approximation consists of taking into account only the terms varying with time with approximately the same frequencies, that is, according to the resonance condition (5) we neglect in eq. (12) all interaction terms except $A^*B \propto \exp\left[-i\left(\omega_b - \omega_a\right)t\right]$ and in eq. (13) all interaction terms except $A^2 \propto \exp\left(-2i\omega_a t\right)$. As a result, we arrive at the following simple system of equations

$$i\dot{A} = \omega_a A + 2\Gamma A^* B, \qquad i\dot{B} = \omega_b B + \Gamma A^2, \tag{15}$$

where the notation for the interaction constant

$$\Gamma = \frac{\alpha_{21}}{2M_a\omega_a\sqrt{2M_b\omega_b}} \tag{16}$$

is introduced.

Equations (15) arise in various physical contexts. In particular, they describe the process of second harmonic generation in nonlinear optics (see, e.g., [20]). In fact, they can be applied to any process in which two classical oscillating modes (waves) transform one into the other under the resonance condition $2\omega_a \simeq \omega_b$.

It is important that equations (15) can be derived from the Lagrangian

$$L = \frac{i}{2}\left(\dot{A}^*A - A^*\dot{A} + \dot{B}^*B - B^*\dot{B}\right) + \omega_a A^*A + \omega_b B^*B + \Gamma\left(A^{*2}B + A^2B^*\right), \tag{17}$$

where A, A^*, B, B^* are considered as independent variables. Then the Lagrange equations

$$\frac{d}{dt}\frac{\partial L}{\partial \dot{A}^*} - \frac{\partial L}{\partial A^*} = 0, \quad \frac{d}{dt}\frac{\partial L}{\partial \dot{B}^*} - \frac{\partial L}{\partial B^*} = 0$$

reproduce the equations (15). The classical Hamiltonian has the form

$$H = \omega_a A^*A + \omega_b B^*B + \Gamma\left(A^2B^* + A^{*2}B\right), \tag{18}$$

where pairs of canonically conjugate variables are A, A^* and B, B^*, respectively. Then the Hamiltonian form of the equations of motion is as follows:

$$\begin{aligned} i\dot{A} = \partial H/\partial A^*, \quad i\dot{A}^* = -\partial H/\partial A, \\ i\dot{B} = \partial H/\partial B^*, \quad i\dot{B}^* = -\partial H/\partial B. \end{aligned} \tag{19}$$

Having the Hamiltonian treatment, it is easy to proceed to quantum mechanical formulation of the Fermi resonance problem. As explained in almost any textbook on *Quantum Mechanics*, the classical variables (9) and their complex conjugates correspond to the "annihilation" and "creation" operators of quantum oscillator:

$$A \rightarrow \hbar^{1/2}\hat{a}, \quad A^* \rightarrow \hbar^{1/2}\hat{a}^\dagger, \quad B \rightarrow \hbar^{1/2}\hat{b}, \quad B^* \rightarrow \hbar^{1/2}\hat{b}^\dagger, \tag{20}$$

where "dagger" denotes the Hermitian conjugation and the operators $\hat{a}$, $\hat{a}^\dagger$, and $\hat{b}$, $\hat{b}^\dagger$ obey the usual commutation relations

$$\left[\hat{a},\ \hat{a}^\dagger\right] = 1, \quad \left[\hat{b},\ \hat{b}^\dagger\right] = 1, \quad \left[\hat{a},\ \hat{b}\right] = \left[\hat{a}^\dagger,\ \hat{b}^\dagger\right] = 0. \tag{21}$$

The classical Hamiltonian (18) converts into quantum mechanical one in the following way

$$\hat{H} = \hbar\omega_a\hat{a}^\dagger\hat{a} + \hbar\omega_b\hat{b}^\dagger\hat{b} + \hbar^{3/2}\Gamma\left(\hat{a}^2\hat{b}^\dagger + \hat{a}^{\dagger 2}\hat{b}\right), \tag{22}$$

where the product of non–commuting variables is replaced by their mean value:

$$A^*A \rightarrow \frac{\hbar}{2}\left(\hat{a}^\dagger\hat{a} + \hat{a}\hat{a}^\dagger\right) = \hbar\left(\hat{a}^\dagger\hat{a} + \frac{1}{2}\right),$$

and zero oscillation terms are dropped.

Now we are ready to discuss the energy splitting of molecular Fermi resonance states. From classical point of view, the stationary states correspond to a purely periodic dependence of the amplitudes A and B on time:

$$A = A_0 \exp\left(-\frac{i\Omega}{2}t\right), \quad B = B_0 \exp\left(-i\Omega t\right). \tag{23}$$

So we seek the solution of eqs. (15) in the form (23) which leads to the system

$$\begin{aligned} (\omega_a - \Omega/2)\, A_0 + 2\Gamma A_0^* B_0 &= 0, \\ (\omega_b - \Omega)\, B_0 + \Gamma A_0^2 &= 0. \end{aligned} \tag{24}$$

We multiply the first equation by A_0, introduce the "intensity" of vibration

$$I = |A_0|^2, \tag{25}$$

analogous to the number of quanta, and eliminate A_0^2 and B_0 from this system, which gives the equation

$$(\omega_a - \Omega/2)(\omega_b - \Omega) = 2\Gamma^2 I \tag{26}$$

for calculation of the frequency Ω. This equation has two roots

$$\Omega_{1,2} = \omega_a + \omega_b/2 \pm \sqrt{(\omega_a - \omega_b/2)^2 + 8\Gamma^2 I}. \tag{27}$$

We see that our classical system has two "eigenfrequencies" depending on intensity I of A-mode. In quantum terms, the intensity I is proportional to the number of a quanta, $n_a = \hat{a}^\dagger\hat{a} = I/\hbar$, $n_b = \hat{b}^\dagger\hat{b} = I/(2\hbar)$, which in the classical limit are much greater than unity, $n_a = 2n_b \gg 1$. However in usual infrared experiments the lines observed correspond to the excitation of the state with $n_a = 2$, $n_b = 1$. Let us try to apply our classical formula (27) to this quantum region, i.e., we substitute in it $I = A^*A = 2\hbar$ (two a-quanta). As a result we obtain that the energies of these two states are equal to

$$E_{1,2} = \hbar\Omega_{1,2} = \hbar\omega_a + \hbar\omega_b/2 \pm \sqrt{(\hbar\omega_a - \hbar\omega_b/2)^2 + 16\hbar^3\Gamma^2}. \tag{28}$$

It is interesting that this semiclassical formula coincides with the exact quantum result. Indeed, the interaction term in quantum Hamiltonian (22) couples the states

$$|\psi_1\rangle = |2_a 0_b\rangle \quad \text{and} \quad |\psi_2\rangle = |0_a 1_b\rangle, \tag{29}$$

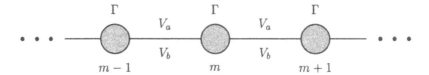

Figure 1: Schematical picture of one–dimensional crystal with intra–molecular Fermi resonance (interaction constant Γ) and inter–molecular coupling of vibrations (interaction constants V_a and V_b).

where $|n_a m_b\rangle = |n\rangle_a |m\rangle_b$ is the oscillators' state with n a–quanta and m b–quanta. Using the well-known relations

$$\hat{a}|n\rangle_a = \sqrt{n}|n-1\rangle_a, \quad \hat{a}^\dagger|n\rangle_a = \sqrt{n+1}|n+1\rangle_a,$$
$$\hat{b}|m\rangle_b = \sqrt{m}|m-1\rangle_b, \quad \hat{b}^\dagger|m\rangle_b = \sqrt{m+1}|m+1\rangle_b, \tag{30}$$

we obtain the matrix elements of the Hamiltonian in the basis (29):

$$H = \begin{pmatrix} 2\hbar\omega_a & 4\hbar^{3/2}\Gamma \\ 4\hbar^{3/2}\Gamma & \hbar\omega_a \end{pmatrix}. \tag{31}$$

The eigenstates' energies are determined by the secular equation

$$(2\hbar\omega_a - E)(\hbar\omega_b - E) = 16\hbar^3\Gamma^2$$

and coincide exactly with the semiclassical result (28). Thus, we see that such a molecule has two energy levels $E_{1,2}$ each connected with the ground state by one–quantum transition due to large b–component in both eigenfunctions. This leads to the observed splitting of the infrared and Raman spectra lines. Just such splitting of the Raman spectra was explained by E. Fermi in his pioneering articles [1]–[2].

1.2. FROM MOLECULES TO SUPERLATTICES

The next natural step is discussion of Fermi resonance effects in molecular crystals (see, e.g., review articles [3]–[5]). Let molecules having Fermi resonance between intramolecular vibrations form a molecular crystal due to weak (van der Waals) forces. Then the individual molecular vibrational excitations discussed above become coupled to each other and form collective Fermi resonance bands. We shall consider here a simple 1D model with intermolecular interaction between only nearest neighbours (see figure 1). The quantum mechanical Hamiltonian can be written in the form

$$\hat{H} = \sum_m \left[\hbar\omega_a \hat{a}_m^\dagger \hat{a}_m + \hbar\omega_b \hat{b}_m^\dagger \hat{b}_m + \hbar^{3/2}\Gamma \left(\hat{a}_m^2 \hat{b}_m^\dagger + \hat{a}_m^{\dagger 2} \hat{b}_m \right) \right]$$
$$+ \sum_m \left[\hbar V_a \left(\hat{a}_{m+1}^\dagger \hat{a}_m + \hat{a}_m^\dagger \hat{a}_{m+1} \right) + \hbar V_b \left(\hat{b}_{m+1}^\dagger \hat{b}_m + \hat{b}_m^\dagger \hat{b}_{m+1} \right) \right], \tag{32}$$

where m is the index of the lattice site, the first sum represents molecular Hamiltonians in each site (see eq. (22)), and the second sum corresponds to the intermolecular interaction of vibrations—the term $\hbar V_a \hat{a}^\dagger_{m+1} \hat{a}_m$ describes a transition of one a–quantum from the site m to the site $m+1$, and analogous interpretation have the other terms. In classical approximation we obtain the Hamiltonian

$$H = \sum_m \left[\omega_a A^*_m A_m + \omega_b B^*_m B_m + \Gamma \left(A^2_m B^*_m + A^{*2}_m B_m \right) \right]$$
$$+ \sum_m \left[V_a \left(A^*_{m+1} A_m + A^*_m A_{m+1} \right) + V_b \left(B^*_{m+1} B_m + B^*_m B_{m+1} \right) \right], \tag{33}$$

and the following equations of motion for complex amplitudes A_m and B_m :

$$i\partial A_m / \partial t = \partial H / \partial A^*_m = \omega_a A_m + V_a \left(A_{m-1} + A_{m+1} \right) + 2\Gamma A^*_m B_m,$$
$$i\partial B_m / \partial t = \partial H / \partial B^*_m = \omega_b B_m + V_b \left(B_{m-1} + B_{m+1} \right) + \Gamma A^2_m. \tag{34}$$

Let us look for the solution in the form of plane wave

$$A_m = A \exp\left[-i\left(\Omega t - Km\right)/2\right], \quad B_m = B \exp\left[-i\left(\Omega t - Km\right)\right] \tag{35}$$

(we assume that the lattice constant is equal to unity). Then the infinite system reduces to a simple system of two algebraic equations

$$\left(\omega_a - \Omega/2 + 2V_a \cos(K/2)\right) A + 2\Gamma A^* B = 0,$$
$$\left(\omega_b - \Omega + 2V_b \cos K\right) B + \Gamma A^2 = 0, \tag{36}$$

which actually coincides with the system (24). If there were no Fermi resonance interaction ($\Gamma = 0$), then we would have two linear modes with well–known dispersion laws

$$\Omega_1 \left(K\right) = 2\omega_a + 4V_a \cos\left(K/2\right), \quad \Omega_2 \left(K\right) = \omega_b + 2V_b \cos K. \tag{37}$$

Fermi resonance coupling between molecular vibrations leads to interaction of these linear modes with each other which gives rise to mixed waves. To obtain their dispersion law, we again introduce the intensity

$$I = |A|^2$$

and reduce the system to the equation

$$\left(\omega_a - \Omega/2 + 2V_a \cos\left(K/2\right)\right)\left(\omega_b - \Omega + 2V_b \cos K\right) = 2\Gamma^2 I \tag{38}$$

with the solutions

$$\Omega_{1,2} \left(K\right) = \omega_a + \omega_b/2 + 2V_a \cos(K/2) + V_b \cos K$$
$$\pm \left[\left(\omega_a - \omega_b/2 + 2V_a \cos(K/2) - V_b \cos K\right)^2 + 8\Gamma^2 I\right]^{1/2}. \tag{39}$$

These expressions define the dispersion laws of normal modes arising from linear plane waves due to nonlinear Fermi resonance interaction. It is important that the nonlinearity leads to the dependence of the dispersion laws on the intensity I of

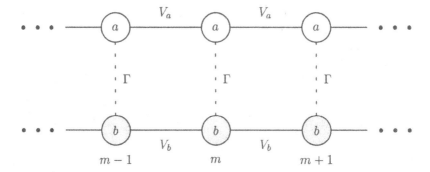

Figure 2: One–dimensional model of a crystal with two molecules in elementary cell. Vibrational modes in a and b molecules satisfy the Fermi resonance condition $2\omega_a \simeq \omega_b$ and interact with vibrations in neighbouring cells with constants V_a and V_b, respectively; constant Γ describes the nonlinear inter–molecular interaction between molecules a and b in the same elementary cell.

vibrations. Such a dependence gives rise to the soliton solutions discussed in the following sections.

The relations (39) can be considered as a good enough approximation only in the limit of large intensity

$$I \gg \hbar. \tag{40}$$

Nevertheless, as we saw in the preceding section, semiclassical formulas give exact enough results even in the quantum region $I \geq \hbar$. More exact relations can be easily derived by means of quantum mechanical treatment [3]–[5].

Thus, we make a conclusion that at certain conditions the bulk molecular crystals have nonlinear waves arising due to Fermi resonance interaction between molecular vibrations. It is clear that this picture is not limited to the case with intramolecular Fermi resonance only. Suppose that each elementary cell of the crystal contains two molecules of different types a and b, and vibrational modes of these molecules satisfy the condition (5) of Fermi resonance. Such a 1D model can be schematically depicted as in figure 2. The Hamiltonian of such a crystal coincides with eq. (32) or (33), and equations of motion coincide with (34), where index m means here the number of the cell, and Γ describes the Fermi resonance interaction between neighbouring molecules in one elementary cell.

Now we make the last step from crystals to superlattices. Modern technique [6] permits one to fabricate structures with any prescribed number of layers built from different types of molecules. For example, figure 2 can be interpreted as a thin film consisting of two layers—one made of a molecules and another of b molecules. We can add infinitely many layers of a molecules from above and of b molecules from below. Then we obtain the interface between two 2D crystals with Fermi resonance interaction through the interface. The question arises which modes can propagate along such an interface. We can fabricate the superlattices with alternating layers of a and b molecules and ask ourselves about types of waves propagating across such a superlattice. These are questions addressed in the following section.

2. Interface Fermi resonance waves

2.1. 2D INTERFACE FERMI RESONANCE WAVES IN TWO LAYER STRUCTURE

For the first time the quantum theory of Fermi resonance interface states for 1D model (with "point" interface) was developed in [21] and classical approach to such states was developed in [22] (1D case) and [23] (2D case). Plane nonlinear waves with constant amplitudes were considered in [24]. In this chapter we consider simple examples of such waves in layered structures of different types.

We shall begin with the lattice structure mentioned in the preceding section: one interface separates two simple cubic 2D crystals composed of different molecules of a and b types interacting across the interface. In two dimensions with line interface (3D generalization is straightforward) the equations of motion for the amplitudes A_{mn} and B_{mn} read (compare with eqs. (34))

$$
\begin{aligned}
&\vdots \qquad\qquad\qquad \vdots \\
n = 2: \quad & i\dot{A}_{2,m} = \omega_a A_{2,m} + V_a \left(A_{2,m-1} + A_{2,m+1} + A_{1,m} + A_{3,m} \right), \\
n = 1: \quad & i\dot{A}_{1,m} = \omega_a A_{1,m} + V_a \left(A_{1,m-1} + A_{1,m+1} + A_{2,m} \right) + 2\Gamma A^*_{1,m} B_{0,m}, \\
n = 0: \quad & i\dot{B}_{0,m} = \omega_b B_{0,m} + V_b \left(B_{0,m-1} + B_{0,m+1} + B_{-1,m} \right) + \Gamma A^2_{1,m}, \\
n = -1: \quad & i\dot{B}_{-1,m} = \omega_b B_{-1,m} + V_b \left(B_{-1,m-1} + B_{-1,m+1} + B_{-2,m} + B_{0,m} \right), \\
&\vdots \qquad\qquad\qquad \vdots
\end{aligned}
$$

$$(41)$$

where the equations with $n > 2$ and $n < -1$ have the same structure as at $n = 2$ and $n = -1$, respectively. They describe the propagation of excitations in bulk a and b crystals above and below the interface. New interface ("surface") modes arise due to Fermi resonance interaction across the interface [21]–[23]. We look for the solution in the form

$$
\begin{aligned}
A_{n,m} &= A e^{-\kappa_a (n-1)} \exp\left[-i \left(\Omega t - Km \right)/2 \right], \quad \text{for} \quad n \geq 1, \\
B_{n,m} &= B e^{\kappa_b n} \exp\left[-i \left(\Omega t - Km \right) \right], \quad \text{for} \quad n \leq 0,
\end{aligned}
$$

$$(42)$$

where the amplitudes of vibrations exponentially decay as we go away from the interface. Then the equations (41) with $n \geq 2$ and $n \leq -1$ are solved by these functions provided

$$
\begin{aligned}
\Omega/2 &= \omega_a + 2V_a \cos(K/2) + 2V_a \cosh \kappa_a, \\
\Omega &= \omega_b + 2V_b \cos K + 2V_b \cosh \kappa_b.
\end{aligned}
$$

$$(43)$$

These equations give us the values of κ_a and κ_b as functions of K and Ω. Then the equations (41) reduce to

$$V_a \exp(\kappa_a) A = 2\Gamma A^* B, \tag{44}$$

$$V_b \exp(\kappa_b) B = \Gamma A^2. \tag{45}$$

We multiply (44) by A and introduce the intensity of vibrations

$$I = |A|^2, \tag{46}$$

so that eqs. (44, 45) give

$$V_a V_b \exp(\kappa_a + \kappa_b) = 2\Gamma^2 I. \tag{47}$$

Elimination of $V_a e^{\kappa_a}$ and $V_b e^{\kappa_b}$ from this equation can be done with the use of eqs. (43), and yields the relation defining implicitly the dependence of Ω on K, i.e., the dispersion laws of the Fermi resonance modes

$$\begin{aligned}
&\left\{ \Omega/2 - \omega_a - 2V_a \cos(K/2) + \operatorname{sgn}\left(\Omega/2 - \omega_a - 2V_a \cos(K/2)\right) \right.\\
&\left. \cdot\sqrt{\left(\Omega/2 - \omega_a - 2V_a \cos(K/2)\right)^2 - 4V_a^2} \right\}\\
&\cdot\left\{ \Omega - \omega_b - 2V_b \cos K + \operatorname{sgn}\left(\Omega - \omega_b - 2V_b \cos K\right) \right.\\
&\left. \cdot\sqrt{\left(\Omega - \omega_b - 2V_b \cos K\right)^2 - 4V_b^2} \right\} = 8\Gamma^2 I.
\end{aligned} \tag{48}$$

We see that they are modified compared to the case of Fermi resonance waves (38, 39) in infinite 1D cubic crystal. At $I = 0$ (or $\Gamma = 0$), when interaction across the interface disappears, the equation (48) reproduces two dispersion relations for surface waves in half-infinite crystals.

Note that addition of a new dimension to 1D cubic crystal with "point" interface leads to the replacements

$$\omega_a \to \omega_a + 2V_a \cos\frac{K}{2}, \quad \omega_b \to \omega_b + 2V_b \cos K, \tag{49}$$

which are a result of translational invariance along the interface following from the interaction terms in the Hamiltonian (we omit index n numbering sites along the axis perpendicular to the interface)

$$\sum_m \left[V_a \left(a_m^\dagger a_{m-1} + a_{m-1}^\dagger a_m \right) + V_b \left(b_m^\dagger b_{m-1} + b_{m-1}^\dagger b_m \right) \right].$$

If we add the third dimension (labelled by index l) along the interface, we can take it into account by means of the following replacements in the formulas for 1D case

$$\omega_a \to \omega_a + 2V_a \left(\cos\frac{K_m}{2} + \cos\frac{K_l}{2} \right), \quad \omega_b \to \omega_b + 2V_b \left(\cos K_m + \cos K_l \right), \tag{50}$$

K_m and K_l being the wave vectors along m and l axes, respectively. Thus, we conclude that in calculations of dispersion laws of waves propagating in superlattices

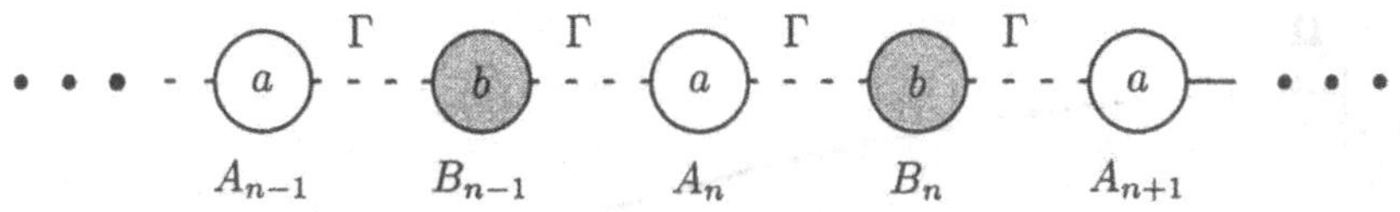

Figure 3: An example of 1D superlattice.

with plane interfaces, it is sufficient to consider first the one–dimensional models with the axis directed perpendicular to the interfaces. Then the general formulas for dispersion laws can be obtained by means of replacements (49) or (50). So we reduce 2D or 3D problems to 1D problem with only one coordinate directed perpendicular to interfaces.

2.2. 3D INTERFACE FERMI RESONANCE WAVES IN SUPERLATTICES

Here we shall consider an example of nonlinear waves propagating in 3D organic superlattices with Fermi resonance interaction across the interfaces between each two neighbouring layers. We assume again that molecules occupy the sites of a simple cubic lattice and each cell of the superlattice consisits of two films made of molecules of two different types, a–molecules and b–molecules, so that there is Fermi resonance interaction across the interfaces between the excitations $2\hbar\omega_a$ and $\hbar\omega_b$ of these molecules.

Actually, it suffices to consider 1D case only since 3D generalization can be achieved by means of simple replacement (50). Therefore we shall investigate only dispersion laws for waves propagating perpendicular to interfaces.

We shall consider here the simplest case of 1D superlattice—an alternating chain of a and b molecules $(\dots ababab \dots$; see figure 3). Let A_n and B_n denote the classical amplitudes of vibrations in the n-th elementary cell:

$$\dots A_{n-1} B_{n-1} A_n B_n A_{n+1} B_{n+1} \dots \tag{51}$$

We assume that each a molecule has a Fermi resonance interaction with two neighbouring b molecules, and, hence, each b molecule has a Fermi resonance interaction with two neighbouring a molecules. Thus, the equations of motion for A_n and B_n amplitudes have the form

$$i\partial A_n/\partial t = \omega_a A_n + 2\Gamma A_n^*(B_{n-1} + B_n), \quad i\partial B_n/\partial t = \omega_b B_n + \Gamma(A_n^2 + A_{n+1}^2). \tag{52}$$

We shall look for the plane wave solution of these equations

$$A = \exp\left[-i(\Omega t - Kn)/2\right] A, \qquad B_n = \exp\left[-i(\Omega t - Kn)\right] B, \tag{53}$$

where Ω and K are the frequency and the wave vector of the wave propagating along the z axis with superlattice's cells labelled by index n. Substitution of (53)

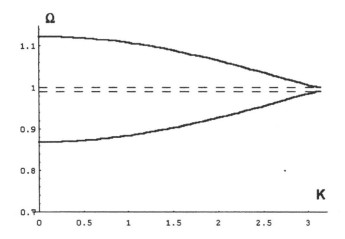

Figure 4: Dispersion curves of nonlinear waves in 1D alternating chain of a and b molecules.

into (52) leads to algebraic linear equations for the variables A^2 and B:

$$(\omega_a - \Omega/2)A^2 + 2\Gamma I(1 + e^{-iK})B = 0, \quad \Gamma(1 + e^{ik})A^2 + (\omega_b - \Omega)B = 0, \quad (54)$$

where $I = |A|^2$ is the intensity of a–vibrations. This system has a nontrivial solution only if its determinant vanishes what gives us the desired dispersion relation

$$(\omega_a - \Omega/2)(\omega_b - \Omega) = 4\Gamma^2(1 + \cos K). \quad (55)$$

In this simplest case we are able to express the frequency Ω as a function of the wave number K in the explicit form:

$$\Omega = \omega_a + \omega_b/2 \pm \sqrt{(\omega_a - \omega_b/2)^2 + 8\Gamma^2 I(1 + \cos K)}. \quad (56)$$

In figure 4 these dispersion curves are shown for the following values of parameters: $\omega_b = 1.0$, $2\omega_a = 0.99$, $\Gamma = 0.01$, $I = 10$. As we see there is a forbidden band between $2\omega_a$ and ω_b. The widths of two allowed bands depend on the intensity I and they grow with increase of I. Nonlinear waves in more complex superlattices are discussed in [24].

2.3.   BISTABLE ENERGY TRANSMISSION THROUGH THE INTERFACE WITH FERMI RESONANCE INTERACTION

Interesting phenomena can take place in systems with Fermi resonance under influence of external electromagnetic field. Here we consider one example of such

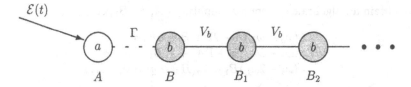

Figure 5: The sketch of 1D interface structure under influence of electromagnetic field.

behaviour—bistable energy transmission through the interface with Fermi resonance interaction [22]. To make calculations easier, we shall consider the following simplified model. Let monomolecular layer of a molecules be deposited on the plane surface of a crystal made of b molecules. For 1D case such a system is shown in figure 5. Let a molecules interact with electromagnetic field

$$\mathcal{E}(t) = E + E^*, \qquad E = E_0 \exp(-i\omega_L t). \tag{57}$$

in resonance with ω_a vibrations, $\omega_L \approx \omega_a \simeq \omega_b/2$, so that we can neglect the direct pumping of ω_b excitations. We take the dipole moment of a molecules to be linear in their coordinates q_a so that the electromagnetic interaction term in Lagrangian is proportional to (see eqs. (10))

$$q_a \mathcal{E}(t) = \frac{1}{\sqrt{2M_a\omega_a}}(A + A^*)(E + E^*) \simeq \frac{1}{\sqrt{2M_a\omega_a}}(AE^* + A^*E),$$

where we have omitted the fast oscillating terms $AE \propto \exp(-2i\omega_L t)$ and $A^*E^* \propto \exp(2i\omega_L t)$ according to our rotating wave approximation. Thus we have

$$L_{int} = \mu\left(AE^* + A^*E\right), \tag{58}$$

where μ up to a constant factor is the dipole moment of a molecules. Correspondingly, the equation of motion of a molecule in our 1D model reads

$$i\partial A/\partial t = \omega_a A + 2\Gamma A^*B + \mu E(t). \tag{59}$$

The b molecules do not interact with electromagnetic field and their equations of motion have usual form

$$i\partial B/\partial t = \omega_b B + \Gamma A^2 + V_b B_1, \quad i\partial B_1/\partial t = \omega_b B_1 + V_b(B + B_2), \quad \dots \tag{60}$$

Now we have driving force $E(t) = E_0 \exp(-i\omega_L t)$ so that the a molecule oscillates with this laser field frequency ω_L and due to Fermi resonance interaction across the interface this leads to oscillations of b molecules with frequency $2\omega_L$. As a result, we obtain an algebraic system for amplitudes A, B, B_1, ...:

$$
\begin{aligned}
(\omega_a - \omega_L)A + 2\Gamma A^* B + \mu E_0 &= 0, \\
(\omega_b - 2\omega_L)B + \Gamma A^2 + V_b B_1 &= 0, \\
(\omega_b - 2\omega_L)B_1 + V_b(B + B_2) &= 0, \quad \dots.
\end{aligned}
\tag{61}
$$

Equations for amplitudes B_1, B_2, ... are linear and solved by

$$
B_n = \exp(ipn)B_{n-1}, \qquad B_1 = \exp(ip)B,
\tag{62}
$$

where the wave vector p is determined by the equation

$$
2\omega_L = \omega_b + 2V_b \cos p.
\tag{63}
$$

Then we get from the first two equations of the system (61)

$$
B = \frac{\exp(ip)\Gamma A^2}{V_b} = \frac{2\Gamma A^2}{2\omega_L - \omega_b + \mathrm{sgn}\,(2\omega_L - \omega_b)\sqrt{(2\omega_L - \omega_b)^2 - 4V_b^2}},
\tag{64}
$$

and

$$
A\left[\omega_L - \omega_a - \frac{4\Gamma^2|A|^2}{2\omega_L - \omega_b + \mathrm{sgn}\,(2\omega_L - \omega_b)\sqrt{(2\omega_L - \omega_b)^2 - 4V_b^2}}\right] = \mu E.
\tag{65}
$$

Now we introduce the "pumping" intensity

$$
I_{pump} = |\mu E|^2 / (\omega_a - \omega_L)^2
\tag{66}
$$

and arrive at the following equation

$$
I(1 - DI)^2 = I_{pump}
\tag{67}
$$

which determins implicitly the intensity of vibrations

$$
I = |A|^2
\tag{68}
$$

as a function of the pumping intensity I_{pump}, where

$$
D = \frac{4\Gamma^2}{(\omega_L - \omega_a)\left[2\omega_L - \omega_b + \mathrm{sgn}\,(2\omega_L - \omega_b)\sqrt{(2\omega_L - \omega_b)^2 - 4V_b^2}\right]}.
\tag{69}
$$

The cubic equation (67) can have three roots which indicates on bistability—two values of I correspond to one pumping intensity (third root corresponds to an unstable state). The plot of the function $I(I_{pump})$ is shown in figure 6, where one can see the bistability region $0 \leq I_{pump} \leq 0.15 I_c$ (with $I_c = 1/D = 1$). More details can be found in [22]. Here we note only that there exists nonzero solution of (67) even at vanishing pumping:

$$
I = 1/D,
\tag{70}
$$

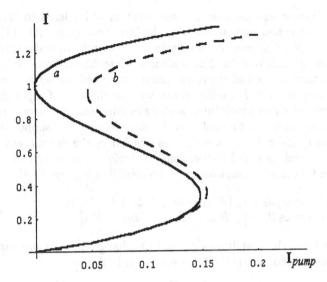

Figure 6: Dependence of intensity I of a vibrations on pumping intensity: (a) without damping; (b) with damping.

i.e., when the oscillation frequency of b molecules Ω satisfies the following equation

$$(\Omega/2 - \omega_a)\left[\Omega - \omega_b + \text{sgn}\,(\Omega - \omega_b)\sqrt{(\Omega - \omega_b)^2 - 4V_b^2}\right] = 4\Gamma^2 I. \qquad (71)$$

This vibrational state existing without pumping is just the Fermi resonance interface state studied in section 2.1.

3. Fermi resonance solitons

3.1. EXACT SOLITON SOLUTIONS

The dependence of dispersion laws on the intensity I of vibrations suggests that a dispersion of wave packet can be compensated by this nonlinearity as it takes place, for example, in the case of pulses in fiber guides described by the nonlinear Schrödinger equation (see, e.g., [20]). Here we shall investigate such a possibility and find that indeed this is the case [23], [25].

We shall consider a long wavelength limit when the width of a pulse is much greater than the lattice constant. It is easy to see that not any dispersion can be compensated by Fermi resonance nonlinearity. For example, constant amplitude waves propagating along one interface (section 2.1) is localized in the z direction perpendicular to the interface (amplitudes decay exponentially as we go away from the interface; see eqs. (42)). But a sum of several such waves (wave packet) is not a solution of the nonlinear equations under consideration which leads to "radiation" of waves in the direction perpendicular to the interface. To get rid of these "radiation" effects, we shall discuss an effectively one–dimensional system where all amplitudes depend on only one coordinate along the interfaces. Evidently, such one–dimensional solitons do not radiate vibrations in the directions perpendicular to the direction of propagation. For definiteness, we take the case of superlattice with alternating layers of a and b molecules, keeping in mind that analogous treatment can be applied to other cases actually on the same footing. We suppose that the pulse propagates in the x direction with sites labelled by the index l, and the amplitudes do not depend on y and z directions. Therefore it is enough to write equations of motion for complex amplitudes in two neighbouring layers only:

$$i\partial A_l/\partial t = \omega_a A_l + V_a \left(A_{l+1} + A_{l-1} + 2A_l\right) + 4\Gamma A_l^* B_l,$$
$$i\partial B_l/\partial t = \omega_b B_l + V_b \left(B_{l+1} + B_{l-1} + 2B_l\right) + 2\Gamma A_l^2. \tag{72}$$

In a long wave limit the amplitudes A_l and B_l change little in one lattice constant and the corresponding terms can be presented in the form

$$A_{l+1} + A_{l-1} \simeq 2A(x) + \frac{\partial^2 A(x)}{\partial x^2}, \qquad B_{l+1} + B_{l-1} \simeq 2B(x) + \frac{\partial^2 B(x)}{\partial x^2},$$

where the lattice constant is taken as a unit of length: $A_l \to A(x)$, $B_l \to B(x)$. Hence, equations (72) reduce to

$$i\partial A/\partial t = \tilde{\omega}_a A + V_a \partial^2 A/\partial x^2 + 4\Gamma A^* B,$$
$$i\partial B/\partial t = \tilde{\omega}_b B + V_b \partial^2 B/\partial x^2 + 2\Gamma A^2, \tag{73}$$

where $\tilde{\omega}_a = \omega_a + 4V_a$, $\tilde{\omega}_b = \omega_b + 4V_b$.

Mathematically equivalent equations have recently attracted much attention in another brunch of physics—in nonlinear optics [8]–[19] where the phenomenon they describe is called "cascaded $\chi^{(2)}$ nonlinearity". This new view at the old problem of second harmonic generation was stimulated by the idea to create large effectively third–order nonlinearity by means of cascading two second–order nonlinearities. In optical terms A means fundamental harmonic of electromagnetic field and B the corresponding second order harmonic. Coincidence of (73) with optical equations has clear physical sense: in both cases we have Fermi resonance of two linear waves with frequencies ω and 2ω, which transform one into another due to the second order nonlinearity. Let us transform equations (73) into the form used in nonlinear optics, where the nonlinear terms are considered as small corrections leading to slow change of the wave packet envelopes. If we neglect the nonlinear interaction terms in (73), then the linear equations have the solutions with constant amplitudes

$$A(x,t) = a \exp\left[-i\left(\Omega_a t - K_a x\right)\right], \qquad B(x,t) = b \exp\left[-i\left(\Omega_b t - K_b x\right)\right], \tag{74}$$

with dispersion relations

$$\Omega_a = \tilde{\omega}_a - V_a K_a^2, \qquad \Omega_b = \tilde{\omega}_b - V_b K_b^2. \tag{75}$$

We suppose now that the amplitudes are slowly varying functions $a = a(x,t)$, $b = b(x,t)$. Substitution of (74) into (73) with taking into account (75) and under the second–order resonance condition

$$2\Omega_a(K_a) = \Omega_b(K_b) \tag{76}$$

yields the evolution equations for amplitudes $a(x,t)$, $b(x,t)$:

$$
\begin{aligned}
i\partial a/\partial t - 2iV_a K_a \partial a/\partial x &= V_a \partial^2 a/\partial x^2 + 4\Gamma a^* b \exp\left(-i\Delta K x\right), \\
i\partial b/\partial t - 2iV_b K_b \partial b/\partial x &= V_b \partial^2 b/\partial x^2 + 2\Gamma a^2 \exp\left(i\Delta K x\right),
\end{aligned}
\tag{77}
$$

where the wave–vector mismatch ΔK at frequency $\Omega_a = \Omega_b/2$ is equal in our case to

$$\Delta K = 2K_a - K_b = 2\sqrt{(\tilde{\omega}_a - \Omega_a)/V_a} - \sqrt{(\tilde{\omega}_b - \Omega_b)/V_b}. \tag{78}$$

The system (77) has the same form as optical equations studied in [8]-[19], which describe the interaction of fundamental (a) and second (b) harmonics.

Now we want to demonstrate that the system (73) has exact soliton solutions. To this end, we shall seek the solution in the form

$$A = F \exp\left[-i(\Omega t - Kx)/2\right], \quad B = \beta F \exp\left[-i(\Omega t - Kx)\right], \quad F = F(x - vt), \tag{79}$$

where β is a constant and v is a velocity of a pulse. The equations (79) imply that the exact second order resonance condition $\Omega_a = \Omega_b/2$ and the phase matching condition $\Delta K = 2K_a - K_b = 0$ are fulfilled. Substitution of these expressions into eqs. (73) gives

$$
\begin{aligned}
i\left(v + V_a K\right) F' - \left(\Omega/2 - \tilde{\omega}_a + V_a K^2/4\right) F + V_a F'' + 4\Gamma \beta F^2 &= 0, \\
i\left(v + 2V_b K\right) F' - \left(\Omega - \tilde{\omega}_b + V_b K^2\right) F + V_b F'' + 2\Gamma F^2/\beta &= 0,
\end{aligned}
\tag{80}
$$

where prime denotes a derivative with respect to x.

The imaginary parts of these equations vanish when

$$
\begin{aligned}
&\text{(i)} \quad K = 0, \qquad v = 0; \\
&\text{(ii)} \quad V_a = 2V_b, \qquad v = -V_a K = -2V_b K.
\end{aligned}
\tag{81}
$$

These equations have simple physical sense that both A and B wave packets propagate with the same group velocity v determined by the linear dispersion relations (75). These group velocities are equal to

$$v_a = \frac{d\Omega_a}{dK_a} = -2V_a K_a, \qquad v_b = \frac{d\Omega_b}{dK_b} = -2V_b K_b,$$

and since in our case $K_a = K_b/2 = K/2$ (see eq. (79)), they are equal to each other if any of the conditions (81) is fulfilled.

Let us first consider the case (i). Equations (80) for function F are compatible if

$$\frac{\Omega/2 - \tilde{\omega}_a}{\Omega - \tilde{\omega}_b} = \frac{V_a}{V_b} = 2\beta^2, \tag{82}$$

which determines β and Ω

$$\beta = \pm\sqrt{\frac{V_a}{2V_b}}, \qquad \Omega = \frac{2\left(\tilde{\omega}_a V_b - \tilde{\omega}_b V_a\right)}{V_b - 2V_a}. \tag{83}$$

Under these conditions F satisfies the equation

$$F'' - \frac{2\tilde{\omega}_a - \tilde{\omega}_b}{V_b - 2V_a}F \pm \frac{2\sqrt{2}\Gamma}{\sqrt{V_a V_b}}F^2 = 0. \tag{84}$$

Its first integral has the form

$$\left(\frac{dF}{d\xi}\right)^2 = (\alpha \mp F)F^2 - (\alpha \mp \gamma)\gamma^2, \tag{85}$$

where

$$\xi = \left(\frac{4\sqrt{2}\Gamma}{3\sqrt{V_a V_b}}\right)^{1/2} x, \qquad \alpha = \frac{3\sqrt{V_a V_b}}{4\sqrt{2}\Gamma}\frac{2\tilde{\omega}_a - \tilde{\omega}_b}{V_b - 2V_a}, \tag{86}$$

and γ is an integration constant. Zeros of the polynomial on the right hand side of eq. (85) are

$$F_1 = \gamma, \qquad F_2 = \frac{1}{2}\left[\alpha - \gamma \pm \sqrt{(\alpha + \gamma)^2 - 4\gamma^2}\right], \tag{87}$$

for the upper sign choice, and

$$F_1 = \gamma, \qquad F_2 = \frac{1}{2}\left[-\alpha - \gamma \pm \sqrt{(\alpha - \gamma)^2 - 4\gamma^2}\right], \tag{88}$$

for the lower sign choice. Their locations on the F axis depend on the values of α and γ. There are different cases in which F oscillates between the zeros where this polynomial is positive. In all these cases the solution of eq. (85) can be expressed in terms of elliptic functions. However, when the two zeros of the polynomial, between which it is negative, coincide, we obtain the soliton solutions. Simple analysis leads to the "bright" soliton solution (i.e., higher intensity pulse on the zero background)

$$A = \frac{\alpha\,\text{sgn}(\beta)\exp(-i\Omega t/2)}{\cosh^2 \kappa x}, \qquad B = \frac{\alpha|\beta|\exp(-i\Omega t)}{\cosh^2 \kappa x} \tag{89}$$

and the "dark" soliton solution (i.e., lower intensity pulse on the constant nonzero background)

$$A = -\alpha\,\text{sgn}(\beta)\left(\frac{1}{\cosh^2 \kappa x} - \frac{2}{3}\right)e^{-i\Omega t/2}, \qquad B = -\alpha|\beta|\left(\frac{1}{\cosh^2 \kappa x} - \frac{2}{3}\right)e^{-i\Omega t}, \tag{90}$$

where the soliton's width is equal to

$$\kappa = \frac{1}{2}\sqrt{\left|\frac{2\tilde{\omega}_a - \tilde{\omega}_b}{V_b - 2V_a}\right|}, \tag{91}$$

whereas β and α are given by eqs. (83) and (86), respectively. These solutions do not depend on any free parameters. The bright soliton solution (90) was found in [8], and the dark soliton solution in [10].

In the same way a particular one–parameter solution can be obtained in the case (ii) (see eqs. (81)). In this case we have

$$\beta = \pm 1, \qquad \Omega = \frac{2}{3}(2\tilde{\omega}_b - \tilde{\omega}_a) - \frac{V_a}{2}K^2 \tag{92}$$

and integration of the eq. (85) for the function F leads to the bright soliton solution

$$A = \frac{\alpha\beta \exp\left[-i(\Omega t - Kx)/2\right]}{\cosh^2\left[\kappa(x - vt)\right]}, \qquad B = \frac{\alpha \exp\left[-i(\Omega t - Kx)\right]}{\cosh^2\left[\kappa(x - vt)\right]}, \tag{93}$$

which was found in [23] and a little later in optical terms in [14]. In similar way the moving "dark" soliton can be derived (see [14], [19])

$$A = -\alpha\beta \left(\frac{1}{\cosh^2\left[\kappa(x - vt)\right]} - \frac{2}{3}\right) \exp\left[-i(\Omega t - Kx)/2\right],$$
$$B = -\alpha \left(\frac{1}{\cosh^2\left[\kappa(x - vt)\right]} - \frac{2}{3}\right) \exp\left[-i(\Omega t - Kx)\right]. \tag{94}$$

In these expressions we have for the parameters' values

$$\alpha = \frac{\tilde{\omega}_b - 2\tilde{\omega}_a}{4\Gamma}, \qquad \kappa = \sqrt{\left|\frac{\tilde{\omega}_b - 2\tilde{\omega}_a}{6V_a}\right|}, \tag{95}$$

$(\tilde{\omega}_b - 2\tilde{\omega}_a)/V_a \geq 0$ for the bright soliton and $(\tilde{\omega}_b - 2\tilde{\omega}_a)/V_a \leq 0$ for the dark soliton. Note that, as K (and, hence, v) approachs zero, we return to solutions (89) with $V_a = 2V_b$. The solutions (93) and (94) depend on one free parameter K.

There is another possibility to find an asymptotically exact solution of eqs. (73). As many authors have shown (see, e.g., [11], [12]) in the limit of large mismatch $|2\tilde{\omega}_a - \tilde{\omega}_b| \gg |V_a|, |V_b|$ the wave propagation is approximately described by the nonlinear Schrödinger (NLS) equation. We shall not discuss here this NLS limit.

Thus, we have found that there exist several families of exact or asymptotically exact solutions of eqs. (73). One may expect that in the system under consideration there are more general solitonic excitations depending on two parameters and not constrained by any conditions. This possibility will be discussed in the next section by means of variational approach.

3.2. VARIATIONAL APPROACH TO FERMI RESONANCE SOLITONS

In quantum physics variational approach is widely used for calculations of approximate values of atomic energy levels. The ground state corresponds to the minimal value of the Hamiltonian averaged over some "trial" form of wave functions of electrons, and an accuracy of such a calculation depends on the choice of these trial wave functions.

On the other hand, everybody knows that classical equations of motion can be formulated as conditions of extremal values of action for real trajectories of particles or for real space–time evolution of classical fields. Therefore one can hope to find some reasonable approximations to real behaviour by means of minimization of action with the use of some trial field functions (see, e.g., [26]). And again an accuracy of such a procedure depends on the choice of the trial functions. In our case we know some exact solutions of evolutional equations and these solutions prompt the form of the trial functions for variational calculation which, as we shall see, leads to a quite good approximation.

The discussed below variational approach [27] to soliton–like excitations in molecular crystals with Fermi resonance interaction between vibrations is based on the generalization of Lagrangian (17) on the "field" case, when the variables $A(x,t)$ and $B(x,t)$ are functions of space x and time t coordinates:

$$L = \int_{-\infty}^{\infty} \mathcal{L}\, dx \tag{96}$$

with the Lagrangian density equal to

$$\mathcal{L} = \frac{i}{2}\left(-A^*A_t + A_t^*A - B^*B_t + B_t^*B\right) + \tilde{\omega}_a A^*A + \tilde{\omega}_b B^*B$$
$$-V_a A_x^* A_x - V_b B_x^* B_x + 2\Gamma\left(A^2 B^* + A^{*2}B\right) \tag{97}$$

where $A_t = \partial A/\partial t$, $A_x = \partial A/\partial x$, etc., denote the partial derivatives of "field" variables with respect to x and t. According to the well–known rules, minimization of action

$$S = \int L dt \tag{98}$$

leads to the Euler–Lagrange equations

$$\frac{d}{dt}\frac{\partial \mathcal{L}}{\partial A_t^*} + \frac{d}{dx}\frac{\partial \mathcal{L}}{\partial A_x^*} - \frac{\partial \mathcal{L}}{\partial A^*} = 0, \qquad \frac{d}{dt}\frac{\partial \mathcal{L}}{\partial B_t^*} + \frac{d}{dx}\frac{\partial \mathcal{L}}{\partial B_x^*} - \frac{\partial \mathcal{L}}{\partial B^*} = 0, \tag{99}$$

(and their complex conjugate), which after substitution of (97) reproduce the "field" equations (73).

Let us suppose now that $A(x,t)$ and $B(x,t)$ have the form of such trial functions

$$A = A\{x, q_i(t)\}, \qquad B = B\{x, q_i(t)\}, \tag{100}$$

depending on parameters $q_i(t)$, $i = 1,\ldots,m$, and space coordinate x, that the integral in (96) can be calculated explicitly in closed form. Then we get the Lagrangian

$$L = L(q_i, \dot{q}_i) \tag{101}$$

analogous to that of mechanical system with m degrees of freedom corresponding to "generalized coordinates" q_i, $i = 1, \ldots, m$. Evolution of these "coordinates" (parameters q_i) is governed by the standard Lagrange equations

$$\frac{d}{dt}\frac{\partial L}{\partial \dot{q}_i} - \frac{\partial L}{\partial q_i} = 0, \qquad i = 1, \ldots, m. \tag{102}$$

If we solve these equations, we will find the field variables A and B in the form (100) with known parameters. As was mentioned above, the accuracy of such a procedure depends on the choice of the trial functions. Fortunately, we know some exact solutions of our problem found in the preceding section. For example, we have the solution (93) correct at $V_a = 2V_b$. Suppose that $|V_a - 2V_b|$ is small enough ($|V_a - 2V_b| \ll |V_a|$). Then we may expect that the form of solution does not differ considerably from (93), but the parameters α, β, Ω, K, κ, v have to satisfy the equations (102) which at $V_a = 2V_b$ must reduce to (81), (92) and (95). Let us put this consideration into practice.

Expressions (93) suggest the following form for the trial functions

$$A = \frac{a\exp(i\varphi/2)}{\cosh^2[\kappa(x - \zeta)]}, \quad B = \frac{b\exp(i\varphi)}{\cosh^2[\kappa(x - \zeta)]}, \tag{103}$$

where

$$\varphi = K(x - \zeta/2) + \delta, \tag{104}$$

so that $a(t)$, $b(t)$, $K(t)$, $\kappa(t)$, $\zeta(t)$, $\delta(t)$ are the above mentioned parameters depending, in principle, on time t. We substitute these trial functions (103) into the Lagrangian density (97) and then integration over x in (96) yields the Lagrangian

$$\begin{aligned}
L &= (4a^2/3\kappa)\left[\tilde{\omega}_a - V_a\left(K^2/4 + 4\kappa^2/5\right) - (K\zeta_t/2 - \delta_t)/2 + \zeta K_t/4\right] \\
&\quad + (4b^2/3\kappa)\left[\tilde{\omega}_b - V_b\left(K^2 + 4\kappa^2/5\right) - (K\zeta_t/2 - \delta_t) + \zeta K_t/2\right] + 64\Gamma a^2 b/15\kappa,
\end{aligned} \tag{105}$$

where subscript t denotes the time derivative.

The Lagrange equations for the variables a, b, δ, ζ, K, κ read

$$\tilde{\omega}_a - V_a\left(K^2/4 + 4\kappa^2/5\right) - (K\zeta_t/2 - \delta_t)/2 + \zeta K_t/4 + 16\Gamma/5b = 0, \tag{106}$$

$$\tilde{\omega}_b - V_b\left(K^2 + 4\kappa^2/5\right) - (K\zeta_t/2 - \delta_t) + \zeta K_t/2 + 8\Gamma a^2/5b = 0, \tag{107}$$

$$\left((a^2 + 2b^2)/\kappa\right)_t = 0, \tag{108}$$

$$\left((a^2 + 2b^2)/\kappa\right)_t (K/2) + \left((a^2 + 2b^2)/\kappa\right)\cdot K_t = 0, \tag{109}$$

$$\left(\zeta(a^2 + 2b^2)/\kappa\right)_t + (a^2/\kappa)(\zeta_t + 2KV_a) + (2b^2/\kappa)(\zeta_t + 4KV_b) = 0, \tag{110}$$

$$\begin{aligned}
a^2\left[\tilde{\omega}_a - V_a\left(K^2/4 + 4\kappa^2/5\right) - (K\zeta_t/2 - \delta_t)/2 + \zeta K_t/4\right] + 8V_a a^2\kappa^2/5 + 8V_b b^2\kappa^2/5 \\
+ b^2\left[\tilde{\omega}_b - V_b\left(K^2 + 4\kappa^2/5\right) - (K\zeta_t/2 - \delta_t) + \zeta K_t/2\right] + 16\Gamma a^2 b/5 = 0.
\end{aligned} \tag{111}$$

From eqs. (108) and (109) we get $dK/dt = 0$, i.e. $K = $ constant. Then differentiation of the phase (103) with respect to time t yields the expression for frequency

$$\Omega = K\zeta_t/2 - \delta_t. \tag{112}$$

From eq. (110) we find velocity

$$v = \zeta_t = -K \cdot \frac{V_a a^2 + 4V_b b^2}{a^2 + 2b^2}, \tag{113}$$

and, hence,

$$\Omega = -\delta_t - \frac{V_a a^2 + 4V_b b^2}{a^2 + 2b^2} \cdot \frac{K^2}{2}. \tag{114}$$

Note that at $V_a = 2V_b$ the dependence on amplitudes a and b dissappears and eqs. (113) and (114) reproduce the formula (ii) of eq. (81) and eq. (92), correspondingly, provided

$$\delta_t|_{V_a=2V_b} = 2(2\tilde{\omega}_b - \tilde{\omega}_a)/3.$$

This simplification explains to some extent the origin of exact solution (93).

In general case there is dependence on amplitudes a and b. Substitution of (106) and (107) into (111) yields the relation

$$(V_a a^2 + V_b b^2) \kappa^2 = \Gamma a^2 b. \tag{115}$$

Equations (106) and (107) can then be written in the form

$$c = \frac{5}{16\Gamma} \left(\frac{\Omega}{2} - \tilde{\omega}_a + V_a \left(\frac{K^2}{4} + \frac{4}{5}\kappa^2 \right) \right), \tag{116}$$

$$a^2 = \frac{25}{64\Gamma^2} \left(\Omega - \tilde{\omega}_b + V_b \left(K^2 + \frac{4}{5}\kappa^2 \right) \right) \left(\frac{\Omega}{2} - \tilde{\omega}_a + V_a \left(\frac{K^2}{4} + \frac{4}{5}\kappa^2 \right) \right). \tag{117}$$

If we substitude these expressions into (115), we obtain the equation for Ω. Its solution gives us the frequency Ω as a function of two parameters K and κ:

$$\Omega(K,\kappa) = \omega_a + \frac{\omega_b}{2} + \frac{2}{5}(6V_a + V_b)\kappa^2 - \frac{1}{4}(V_a + 2V_b)K^2$$

$$\pm \sqrt{\left[\omega_a - \frac{\omega_b}{2} + \frac{2}{5}(6V_a - V_b)\kappa^2 - \frac{1}{4}(V_a - 2V_b)K^2 \right]^2 + \frac{256}{25}V_a V_b \kappa^4}. \tag{118}$$

If we choose the sign before radical so that eq. (118) reproduces (83) at $K = 0$ and κ given by eq. (91), we obtain the interpolation formula for Ω which reduces to (83) and (92) in both limiting cases when exact solutions exist. Let us note a simple and useful formula presenting the two first terms of the series expansion of eq. (118) in powers of K^2 at κ given by eq. (91):

$$\Omega \simeq \frac{2(\tilde{\omega}_a V_b - \tilde{\omega}_b V_a)}{V_b - 2V_a} - \frac{3V_a V_b}{2(V_a + V_b)} K^2. \tag{119}$$

It reproduces the exact solutions (83) and (92) in both limiting cases $K = 0$ and $V_a = 2V_b$. If the parameters V_a and V_b are small compared to $|\omega_b - 2\omega_a|$, eq. (119) is a good approximation for the exact dependence (118) of $\Omega(K)$ on K. Substitution of (119) and (91) into (116) and (117) yields the expressions for the amplitudes.

This variational calculation and analogous consideration applied to the NLS type solution were published in [27] and for the case of cascaded $\chi^{(2)}$ nonlinearity in [19]. Variational calculation with Gaussian trial functions was suggested in [18]. Such a choice of trial functions leads to the approximation which does not reproduce the exact solutions of initial equations in appropriate limits.

3.3. CONCLUSION: INTERACTION OF FERMI RESONANCE SYSTEMS WITH ELECTOMAGNETIC FIELD

We have discussed several problems demonstrating some interesting features of the Fermi resonance phenomena in nonlinear spectroscopy of new organic materials. Here we would like to point out some directions of future possible development.

As was mentioned above, there are close links between Fermi resonance theory and cascaded $\chi^{(2)}$ nonlinearity and these interconnections can be rather fruitful for both fields of research. Maybe, it is even more important that Fermi resonance materials provide new mechanism of $\chi^{(2)}$–type nonlinearity. Hence we come to problems in which an interplay between vibrational A and B fields on one hand and resonant electromagnetic fields $\mathcal{E}_1$ and $\mathcal{E}_2$ on the other hand becomes most important. Let us write here the corresponding equations which describe these new phenomena.

We take into account only resonant modes of the electromagnetic field, so that the interaction term in the Lagrangian for a and b molecules have the form

$$L_{int} = d_a q_a \mathcal{E}_1 + d_b q_b \mathcal{E}_2,$$

where $d_a q_a$ and $d_b q_b$ are dipole moments of the molecules arising due to their vibrations, and we suppose for simplicity that both electric fields are directed along corresponding dipole moments. Simple calculation leads to the following equations describing interaction of A and B fields with resonant electromagnetic fields $\mathcal{E}_1(x,t) = E_1(x,t) + E_1^*(x,t)$ and $\mathcal{E}_2(x,t) = E_2(x,t) + E_2^*(x,t)$:

$$i\partial A/\partial t = \omega_a A + V_a \partial^2 A/\partial x^2 + 2\Gamma A^* B + \mu_a E_1,$$
$$i\partial B/\partial t = \omega_b B + V_b \partial^2 B/\partial x^2 + \Gamma A^2 + \mu_b E_2,$$
$$(120)$$

where $\mu_a = d_a/\sqrt{2M_a \omega_a}$ and $\mu_b = d_b/\sqrt{2M_b \omega_b}$.

Evolution of electromagnetic field in our two mode approximation is described by the equations

$$\partial^2 E_{1,2}/\partial x^2 - (1/c^2)\,\partial^2 E_{1,2}/\partial t^2 = (4\pi/c^2)\,\partial^2 P_{1,2}/\partial t^2,$$

where polarizations are equal to $P_1 = \sum d_a q_a$, $P_b = \sum d_b q_b$ (summation over molecules in physically small volume element is implied). Their second derivatives with respect to time can be expressed in terms of A and B fields with the help of (9–11), and we arrive at the system

$$\partial^2 E_1/\partial x^2 - (1/c^2)\,\partial^2 E_1/\partial t^2 = -i(4\pi\mu_a\omega_a/c^2)\,\partial A/\partial t,$$
$$\partial^2 E_2/\partial x^2 - (1/c^2)\,\partial^2 E_2/\partial t^2 = -i(4\pi\mu_b\omega_b/c^2)\,\partial B/\partial t. \tag{121}$$

The equations (120) and (121) give the desired system describing our four interacting fields. Their consequences are the subject of further publications.

At last, we would like to note that some qualitative effects discussed above for vibrational excitations can be observed in the case of electronic or vibronic excitations.

Acknowledgment

This work was supported by INTAS grant 93-0461.

References

[1]   Fermi, E. (1931) *Zs. f. Phys.* **71**, 250 ; (1932) *Memoire Acad. d'Italia* **3** (1), Fis., 239.

[2]   Fermi, E. and Rasetti, F. *Zs. f. Phys.* **71**, 682.

[3]   Agranovich, V.M. (1983) Biphonons and Fermi Resonance in Vibrational Spectra Crystals, in V.M. Agranovich and R.M. Hochstrasser (eds.) *Spectroscopy and Excitation Dynamics of Condensed Molecular Systems*, North-Holland, Amsterdam, p.83.

[4]   Agranovich, V.M. and Lalov, I.I. (1985) *Uspekhi Fiz. Nauk.* **146**, 267 [(1985) *Sov. Phys. Uspekhi* **28**, 484].

[5]   Agranovich, V.M. and Dubovsky, O.A. (1988) Phonon Multimode Spectra: Biphonons and Triphonons in Crystals with Defects, in R.J. Elliott and I.P. Ipatova (eds), *Optical Properties of Mixed Crystals*, North-Holland, Amsterdam, 297.

[6]   See for instance: So, F.F., Forest, S.R., Shi, Y.Q. and Steier, W.H. (1990) *Appl. Phys. Lett.* **65**, 674; Karl, N. Sato, N. (1992) *Mol. Cryst. Liq. Cryst.* **218**, 79; Akimichi, H., Inoshita, T., Hotta, S., Noge, H. and Sakaki, H. (1993) *Appl. Phys. Lett.* **71**, 2098; Nonaka, T., Mori, Y., Nogai, N., Nakagawa, Y., Saeda, M., Takahagi, T. and Ishitani, A. (1994) *Thin Solid Films* **239**, 214; Haskal, E.I., Shen, Z., Burrows, P.E. and Forrest, S.R. (1995) *Phys. Rev.* **B51**, 4449; Umbach, E., Seidel, C., Taborski, J., Li, R. and Soukopp, A. (1995) *phys. stat. sol.* (b) **192**, 389.

[7]   Agranovich, V.M. (1993) *Physica Scripta* **T49**, 699; (1995) *Nonlinear Optics* **9**, 87.

[8]   Karamzin, Y.N. and Sukhorukov, A.P. (1975) *Zh. Exp. Teor. Fiz.* **68**, 834 [(1976) *Sov. Phys. JETP* **41**, 414].

[9]   DeSalvo, R., Hagan, D.J., Sheik-Bahae, M., Stegeman, G.I. and Vanherzeele, H. (1992) *Opt. Lett.* **17**, 28.

[10] Hayata, K. and Koshiba, M. (1993) *Phys. Rev. Lett.* **71**, 3275.

[11] Schiek, R. (1993) *J. Opt. Soc. Am* **B10**, 1848.

[12] Guo, Q. (1993) *Quantum Optics* **5**, 133.

[13] Werner, M.J. and Drummond, P.D. (1993) *J. Opt. Soc. Am.* **B10**, 2390.

[14] Menyuk, C.R., Schiek, R. and Torner, L. (1994) *J. Opt. Soc. Am.* **B11**, 2434.

[15] Buryak, A.V. and Kivshar, Y.S. (1995) *Phys. Lett.* **A197**, 407.

[16] Buryak, A.V. and Kivshar, Y.S. (1995) *Phys. Rev.* **A51**, R41.

[17] Buryak, A.V., Kivshar, Y.S. and Steblina, V.V. (1995) *Phys. Rev.* **A52**, 1670.

[18] Steblina, V.V., Kivshar, Y.S., Lisak, M. and Malomed, B.A. (1995) *Opt. Commun.* **118**, 345.

[19] Agranovich, V.M., Darmanyan, S.A., Kamchatnov, A.M., Leskova, T.A. and Boardman, A.D. (1997) *Phys. Rev.* **E55**, 1894.

[20] Newell, A.C. and Moloney, J.V. (1992) *Nonlinear Optics*, Addison-Wesley, Redwood City.

[21] Agranovich, V.M. and Dubovsky, O.A. (1993) *Chem. Phys. Lett.* **210**, 458.

[22] Agranovich, V.M. and Page, J.B. (1993) *Phys. Lett.* **A183**, 395.
[23] Agranovich, V.M. and Kamchatnov, A.M. (1994) *Pis'ma Zh. Exp. Teor. Fiz.* **59**, 397 [(1997) *JETP Lett.* **59**, 425].
[24] Agranovich, V.M., Dubovsky, O.A. and Kamchatnov, A.M. (1995) *Chem. Phys.* **198**, 245.
[25] Agranovich, V.M., Dubovsky, O.A. and Kamchatnov, A.M. (1994) *J. Phys. Chem.* **98**, 13607.
[26] Anderson, D. (1987) *Phys. Rev.* **A27**, 3135.
[27] Agranovich, V.M., Darmanyan, S.A., Dubovsky, O.A., Kamchatnov, A.M., Ogievetsky, E.I. Reineker, P. and Neidlinger. Th. (1996) *Phys. Rev.* **B53**, 15451.

CASCADED PROCESSES IN GYROTROPY MEDIA AND NOVEL ELECTRO-OPTICAL EFFECT ON $\chi^{(2)}$ NONLINEARITY

L. I. PAVLOV, L. M. KOVACHEV
Institute of Electronics, Bulgarian Academy of Sciences
Tzarigradsko chaussee 72, 1784 Sofia, Bulgaria

1. Introduction

Nonlinear interaction of the electromagnetic field with dielectrics and semi-conductors is possible when an optical field of high intensity impacts on the medium. While most of the processes and principles of Nonlinear Optics now appear to have been identified (self-induced transparency, self-focusing and self-defocusing, wave mixing etc.) but up to date the Nonlinear Optics remains still a very active field because of its relevance to optical information processing and its use to all-optical devices and switches. We should note that all opto- electronic devices are based on some nonlinear optical effect.

Although most of these effects are realizable by optical cubic nonlinearity $\chi^{(3)}$, the main problem now is connected with the efficiency of the nonlinear processes. That is why the "cascaded" nonlinearity in quadratic media pays great attention. Especially, the "cascaded" nonlinearity is of several order greater than the direct electron nonlinearity. That gives a great advantage concerning the efficiency of the nonlinear process and it offers additional applications of the phenomena which are uncovered by all- optical devices for the control of light by light. It is well known that it is preferable to use two cascaded processes in a medium of quadratic nonlinearity

$$\omega + \omega = 2\omega, \omega + 2\omega = 3\omega$$

instead of one direct process for THG

$$\omega + \omega + \omega = 3\omega$$

in view of their efficiency. The importance of the cascaded nonlinear effects has been reported much earlier by E.Yablonovich, C.Flytzanis [1] and also

A. D. Boardman et al. (eds.),
Advanced Photonics with Second-order Optically Nonlinear Processes, 277–292.
© 1999 *Kluwer Academic Publishers. Printed in the Netherlands.*

in [2]. Contribution of the two-step processes [1] was remarkably stressed as well as the resonant interaction of focusing beams in $\chi^{(2)}$ media [2]. We should pay attention also to the work of J.Reintjes [3], who tested experimentally the spectra of the interacted optical waves. According to their results [3] a considerable structure appear on the pump and on second harmonic generation pulses, which is due to the self-phase modulation and cross-phase modulation for longer nonlinear crystals. But only in last several years this topics is investigated in detail. We should note that recently the cascaded nonlinear processes in centrosymmetric media are also analyzed.This corresponds to our results [4, 5], when including the experimentally obtained much greater efficiency for the cubic cascaded processes in comparison to the direct higher harmonic generation and the multiphoton processes (see also [6]). In our lecture, we present results on cascaded nonlinear processes in gyrotropic $\chi^{(2)}$ media and we analyze the circumstances when exact soliton solutions of the amplitude nonlinear equations are possible. The second part of the lecture is devoted to novel electro-optical effect in $\chi^{(2)}$ media, when a significant polarization charge arises. Because of the unvanishing divergency of the electric $\vec{E}$ field, new amplitude equations for nonlinear second-order processes in optical crystals are derived, and analytical soliton solutions for fundamental as well as for second harmonics wave component are obtained too.

2. Propagation of laser light in quadratic gyrotropic media.

If we consider the cascaded nonlinear polarization effects, we can obtain an information about combined effects of spatial dispersion and optical nonlinearity. We would also pay attention on some its advantages. Recently, polarization self-action is of great interest. Such phenomenon allows a development of light by light control elements, codding and information transferring elements through the polarization state of the radiation. When a light beam of high intensity propagates in media, nonlinear optical activity takes place and such a phenomena is nonlinear analogue of some known effects as Kerr or Faradey linear optical activity. If we look at the expansion of electrical induction in series of $\vec{E}$ field:

$$D_i = \chi_{ij}^{(1)} E_j + \chi_{ijk}^{(2)} E_j E_k + \chi_{ijkl}^{(3)} E_j E_k E_l + .. \tag{1}$$

$$+\gamma_{ijk}^{(1)} \nabla_k E_j + \gamma_{ijkl}^{(2)} E_l \nabla_l E_k + \gamma_{ijklm}^{(3)} E_j E_k \nabla_m E_l + ... \tag{2}$$

the gyrotropy tensor γ is induced in the terms with spatial derivatives of the electric $\vec{E}$ field. The γ tensor is connected with magneto-dipole and with electrical quadrupole transitions. The linear natural optical activity

$\gamma_{ijk}^{(1)}$ is typical for all non-centrosymmetric media and also for some centrosymmetric crystals. But we should note, that even when we have no linear gyrotropy for some medium (cubic susceptibility, for example), it will arise nonlinear optical activity (NOA) in strong field which we should rather refer to the self- induced gyrotropy. We will give the expression for the nonlinear rotation angle in the form:

$$\beta^{nl} = \rho_0 z + \rho_2 I z$$

where ρ_0 is a constant of usual natural activity, and ρ_2 is a constant of nonlinear optical activity in isotropic media. The β^{nl} angle is proportional to the incident I intensity and this dependence has resonant and nonresonant parts. We should distinguish two mechanisms of nonlinear polarization rotation in the medium. First, NOA 1 is due to the orientation and it is caused by spatial dispersion of the nonlinearity which is of nonresonant character. Second NOA 2, is due to anisotropy of the nonlinear absorption and it is most like to a dissipative mechanism. We can describe the nonlinear anisotropy in centrosymmetric media by the tensor

$$\chi_{xxxx}^{i,j} - 3\chi_{xxyy}^{i,j}$$

In [7], an intensity-dependent polarization rotation has been observed in KTP crystal, but it is only due to the effect of nonlinear anisotropy. Thus, polarization rotation in $\chi^{(2)}$ media is studied both experimentally and theoretically by L.Lefort et all [7] when the gyrotropy tensor is vanished $\gamma = 0$.

2.1. BASIC EQUATIONS.

From the nonlinear wave equation:

$$\nabla \times \nabla \times \vec{E} + \frac{1}{c^2}\frac{\partial^2}{\partial t^2}(\vec{E} + 4\pi\vec{P_l} + 4\pi\vec{P_{nl}}) = 0 \tag{3}$$

we have an expansion:

$$\vec{P} = \alpha\vec{E} - \beta\frac{\partial\vec{B}}{\partial t}; \tag{4}$$

$$\nabla \times \vec{E} = -\frac{\partial\vec{B}}{\partial t} \tag{5}$$

where the components of dielectric susceptibility tensor are defined by:

$$\alpha_i = \delta_{ij}\alpha_j + \delta_{ij}\chi_{ijk}^{(2)}E_k + \delta_{ij}\chi_{ijkl}^{(3)}E_kE_l + \dots$$

The gyrotropy of linear, quadratic and cubic media are described by:

$$\beta_i = \delta_{ij}\gamma_j^{(1)} + \delta_{ij}\gamma_{ijk}^{(2)}E_k + \delta_{ij}\gamma_{ijkl}^{(3)}E_kE_l + \dots. \tag{6}$$

The expansion of gyrotropy γ tensor in series of $\vec{E}$ field gives possibility to separate the symmetrical part from the asymmetrical part. In a coordinate system connected with main optical axes of the crystal the symmetrical part add a nonlinear walk-of part to the amplitude equations and it has essential influence on a cubic nonlinear media. As it will be mentioned later for some quadratic nonlinear materials this part is vanishing because of the phase-matching conditions. The asymmetrical part $(rot\vec{E})$ in (4) gives detail description of the linear and nonlinear optical activity by rotation of the polarization plane. But we should note, that the analysis in (4) and (6) takes into account only the antisymmetrical part of the gyrotropy tensor. As it is well known the ratio for γ and χ tensor components, is defined by:

$$\gamma \approx a\chi$$

and a is a dimension of the local structure in the medium. The meaning of this ratio is, that the transverse section of the light beam should be of the order of the characteristic "a" length of the medium, i.e. "a" may correspond to the crystal lattice size.

2.2. SINGLE FREQUENCY LINEAR PLANAR WAVES.

We have linear optical activity for the amplitudes:

$$\vec{A}_x = \vec{x}\frac{1}{2}[A_{ox}expi(\omega t - \vec{k}\cdot\vec{z}) + c.c.]$$

$$\vec{A}_y = \vec{y}\frac{1}{2}[A_{oy}expi(\omega t - \vec{k}\cdot\vec{z}) + c.c.]$$

where

$$A_{ox} = A_{oy} = const.$$

Then for the solutions when the propagation is along the main optical axis we have:

$$-k^2 E_{ox}\vec{x} + \omega^2(1 + 4\pi\alpha E_{ox} + ik\beta E_{oy})\vec{x} = 0$$

$$-k^2 E_{oy}\vec{y} + \omega^2(1 + 4\pi\alpha E_{oy} + ik\beta E_{ox})\vec{y} = 0$$

and for the refractive index we obtain the well known relation:

$$n^2 = \bar{n}^2 \pm \beta$$

The optical activity exists for such configuration and it corresponds to a circular birefringence. Both waves $n_1 \neq n_2$ propagate along z and they are circular polarized.

2.3. LINEAR PROPAGATION OF TWO PLANAR WAVES WITH DIFFERENT FREQUENCIES.

Now we will investigate the linear gyrotropy regime of both waves of fundamental frequency and of second harmonic. When the intensity of the light beams is very small the waves propagate along the crystal in linear regime. The beams propagate along the main optical z-axis and there are type 1 *ooe* $90°$ phase-matching when the gyrotropy is absent. The amplitude of main frequency A_x is collinear with x axis, and A_y will be polarized collinear with y-axis. We search the solutions of type:

$$\vec{A}_x = \vec{x}\frac{1}{2}[A_{ox} exp\, i(\omega_1 t - \vec{k}_1 \cdot \vec{z}) + c.c.]$$

$$\vec{A}_y = \vec{y}\frac{1}{2}[A_{oy} exp\, i(\omega_1 t - \vec{k}_2 \cdot \vec{z}) + c.c.]$$

These relations are the following:

$$E_{ox}[-k_1^2 + \omega_1^2(1 + 4\pi\alpha)]exp(i\omega_1 t) + ik_2\beta\omega_2^2 E_{oy} exp(i\omega_2 t)\vec{x} = 0 \quad (7)$$

$$E_{oy}[-k_2^2 + \omega_2^2(1 + 4\pi\alpha)]exp(i\omega_2 t) + ik_1\beta\omega_1^2 E_{ox} exp(i\omega_1 t)\vec{y} = 0 \quad (8)$$

Because of the gyrotropy, it will appear another wave of different frequency ω_2 along the direction of first polarized wave ω_1. Thus on the x axis we have both frequencies. Since

$$n(\omega_1) \neq n(\omega_2)$$

the change of the refraction index will break the phase matching and it will be not fulfilled any more.

2.4. NONLINEAR OPTICAL ACTIVITY IN SLAB WAVEGUIDES. GYROTROPIC SOLITONS.

Now we should investigate the nonlinear wave equations in slab waveguide for quadratic media including also the nonlinear gyrotropy. There are two geometries for investigating these processes in quadratic media. The first one is for nonlinear crystals with cubic symmetries of kind $\bar{4}3m$ and 23

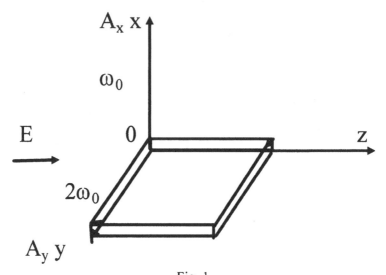

Fig. 1.
Geometry configuration for a nonlinear gyrotropy
observation in quadratic media.
The ordinary wave of ω_0 fundamental frequency is
orthogonal to the plane of the slab $\chi^{(2)}$ waveguide.

as GaSb,InAs,CuBr where: $\gamma^{(1)} = 0$ but $\gamma^{(2)} \neq 0$ In this case there are a constant vector mismatch

$$\Delta k = k^{(2\omega)} - 2k^{(\omega)} = 2\omega(n^{2\omega} - n^{omega}).$$

The second one is for the crystals with ordinary and extraordinary axis and we choose such a geometry mentioned above and in addition the ordinary wave is orthogonal to the plane of slab waveguides (Fig.1.). Otherwise, if we have no exact 90^0 phasematching then the gyrotropy will turn the electric vectors without any coincidence with the main axis and the phase matching:

$$2\vec{k}_2 = \vec{k}_1 - \Delta\vec{k}$$

will be totally broken.

From the nonlinear wave equation we search the solutions of type:

$$\vec{E}_x = \vec{x}\frac{1}{2}[A_x(y,z,t)expi(\vec{k}_1 \cdot \vec{z} + \vec{\beta}_1 \cdot \vec{y} - \omega_1 t) + c.c.]$$

$$\vec{E}_y = \vec{y}\frac{1}{2}[A_y(y,z,t)expi(\vec{k}_2 \cdot \vec{z} + \vec{\beta}_2 \cdot \vec{y} - \omega_2 t) + c.c.]$$

For such a geometries we obtain from the wave equations (3) and polarization (4) the following system of amplitude equations for the main frequency and second harmonic:

$$i\frac{\partial A_x}{\partial z} + i\frac{\beta_1}{k_1}\frac{\partial A_x}{\partial y} + \frac{1}{2k_1}\frac{\partial^2 A_x}{\partial y^2} + (\sigma_x A_x^* A_y + \gamma_x A_x^*\frac{\partial A_y}{\partial z})exp(i\Delta kz) = 0 \quad (9)$$

$$i\frac{\partial A_y}{\partial z} + i\frac{\beta_2}{k_2}\frac{\partial A_y}{\partial y} + \frac{1}{2k_2}\frac{\partial^2 A_y}{\partial y^2} + (\sigma_y A_x^2 - \gamma_y A_x\frac{\partial A_x}{\partial z})exp(-i\Delta kz) = 0 \quad (10)$$

where

$$\sigma_x = k_1\chi_x^{(2)} - ik_2\gamma_x^{(2)}$$

$$\sigma_y = k_2\chi_y^{(2)} + ik_1\gamma_y^{(2)}$$

We should simplify the equations. It is useful to redefine A_x and A_y setting:

$$\hat{A}_x(y,z,t) = A_x(y,z,t)expi(\frac{\Delta k.z}{2})$$

$$\hat{A}_y(y,z,t) = A_y(y,z,t)$$

After such a substitution, we obtain:

$$i\frac{\partial A_x}{\partial z} + i\frac{\beta_1}{k_1}\frac{\partial A_x}{\partial y} - \frac{\Delta k}{2}A_x + \frac{1}{2k_1}\frac{\partial^2 A_x}{\partial y^2} + \hat{\sigma}_x A_x^* A_y + \gamma_x A_x^*\frac{\partial A_y}{\partial z} = 0 \quad (11)$$

$$i\frac{\partial A_y}{\partial z} + i\frac{\beta_2}{k_2}\frac{\partial A_y}{\partial y} + \frac{1}{2k_2}\frac{\partial^2 A_y}{\partial y^2} + \hat{\sigma}_y A_x^2 - \gamma_y A_x\frac{\partial A_x}{\partial z} = 0 \quad (12)$$

For the nonlinear optical activity, we get again phasematching, when gyrotropy is taken into account. The nonlinear coefficients are:

$$\hat{\sigma}_x = k_1\chi_x^{(2)} - ik_2\gamma_x^{(2)}$$

$$\hat{\sigma}_y = k_2\chi_y^{(2)} + ik_1\gamma_y^{(2)} + i\frac{\Delta k\gamma_y^{(2)}}{2}$$

First, we will note, that in case of gyrotropy absence $\gamma = 0$ this is the well known system of amplitude equations which has been investigated by [2, 8] and were found soliton solutions:

We point, that from the gyrotropy will be appear a z-component of the electrical field but as:

$$\beta_i \ll k_i$$

and

$$\frac{\partial A_x}{\partial y} \ll \frac{\partial A_x}{\partial z}$$

this z component will be neglected.

We will solve now the amplitude equations at $\gamma \neq 0$, when including the nonlinear gyrotropy which was not done before. There are soliton solutions of the type of (Fig.2 and 3.) when $\beta_i = 0$:

$$A_x(y, z) = A_{ox}sech(y)^2 \cdot exp(i\phi z)$$

$$A_y(y, z) = A_{oy}sech(y)^2 \cdot exp(i\psi z)$$

and

$$A_x(y, z) = A_{ox}sech(y)th(y) \cdot exp(i\phi z)$$

$$A_y(y, z) = A_{oy}sech(y)^2 \cdot exp(i\psi z)$$

It is evident that new dispersion relation appears. We obtaine additional dispersion relations for the complex part of the nonlinear coefficients which connect the wavevectors $\vec{k}$ and phase mismatch Δk : Due to such compensation of the complex nonlinear coefficients we keep the same form of the soliton solutions even at $\gamma = 0$. But this compensation will not be possible for arbitrary wave vectors, except of those which satisfy the relations $\phi = k_1$

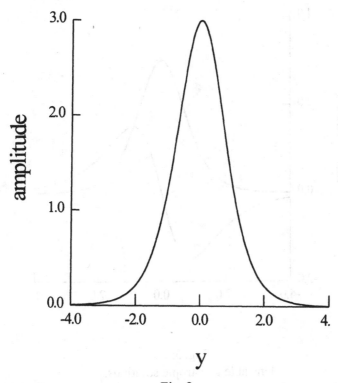

Fig. 2.

Soliton solution of the amplitude equations, when the nonlinear

gyrotropy is taken into account $\gamma \neq 0$.

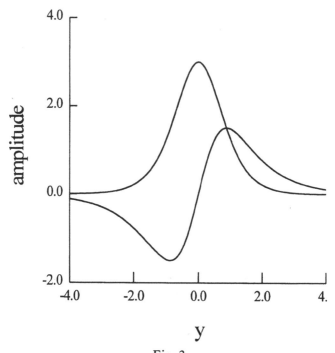

Fig. 3.
Unstable gyrotropic solutions.

and $\psi = k_2 + \delta k$. As a matter of fact the system means, that if we have two wave vectors k_1 and k_2, there are only a fixed Δk which satisfy the system. This is a main difference between our solutions restricted within the relations and the family of soliton solutions of the system at $\gamma = 0$ when following the classification of Y.Kivshar [9] for the parameter $\alpha = 2 + \frac{\Delta k}{\beta}$.

3. Electro-optical effect in $\chi^{(2)}$ media

3.1. LINEAR ELECTRO-OPTICAL EFFECT.

When introducing the linear electro-optical (Pockels) effect we investigate the propagation of a planar wave in an electrical field of $\vec{E} = const$. If $\vec{E} = 0$, in main coordinate system we have:

$$\frac{x^2}{n_x^2} + \frac{y^2}{n_y^2} + \frac{z^2}{n_z^2} = 1$$

The electro-optical constants at $\vec{E} = const$ are:

$$\eta_{ij}(\vec{E}) - \eta_{ij}(0) = \Delta\eta_{ij} = r_{ijk}E_k$$

where r_{ijk} are well-known Pockels constants. We will consider as an example the electrooptical effect in potassium dihydrophosphate KDP [7]. This crystal is of orthorhombical symmetry $\bar{4}2m$ and the electro-optical tensor is:

$$r_{ij} = \begin{pmatrix} 0 & 0 & 0 \\ 0 & 0 & 0 \\ 0 & 0 & 0 \\ r_{42} & 0 & 0 \\ 0 & r_{41} & 0 \\ 0 & 0 & r_{63} \end{pmatrix}$$

When we apply direct current electric field collinear with the x axis $\vec{E}(\vec{x}, 0, 0)$, we have:

$$\frac{x^2}{n_x^2} + \frac{y^2}{n_y^2} + \frac{z^2}{n_z^2} + 2yzr_{41}E_x = 1$$

In order to obtain new main axis, we will turn to the yz plane by angle of θ :

$$y = \hat{y}cos\theta - \hat{z}sin\theta$$

$$z = \hat{z}sin\theta + \hat{z}cos\theta$$

and

$$\frac{x^2}{n_0^2} + (\frac{1}{n_0^2} + r_{41}E_x tg\theta)y'^2 + (\frac{1}{n_e^2} - r_{41}E_x tg\theta)z'^2 = 1$$

where

$$tg2\theta = 2r_{41}E_x(\frac{1}{n_0^2} - \frac{1}{n_e^2})$$

When $E = 10^6 V/m$ is applied along x axis, the main axis of the optical indicatrix is turned in respect to the previous axis in KTP by:

$$\theta = 0.04^0$$

Practically, the turn to the main axis due to the linear electrooptical effect for KDP is very small. The θ is of essential value only when $n_0 \approx n_e$.

3.2. CHARGE DENSITY FOR DIFFRACTED BEAMS AND WAVEGUIDES.

Now we will consider the impact of external DC electrical field onto the polarization charge. We should note, that the approximation $div\vec{E} = 0$ is not valid for anysotropic media. This leads to an additional term in the amplitude equations. In the case of type I second harmonic generation (or scalar ooe phasematching), this term is:

$$\beta\frac{\partial A_2}{\partial y}$$

where β is the walk-off angle of the extraordinary wave. Then, the amplitude equations are:

$$-i\frac{\partial A_x}{\partial z} + \frac{1}{2k_1}(\frac{\partial^2 A_x}{\partial x^2} + \frac{\partial^2 A_x}{\partial y^2}) + \sigma_x A_x^* A_y exp(i\Delta kz) = 0 \qquad (13)$$

$$-i\frac{\partial A_y}{\partial z} + \frac{1}{2k_2}(\frac{\partial^2 A_y}{\partial x^2} + \frac{\partial^2 A_y}{\partial y^2}) + \sigma_y A_x^2 exp(-i\Delta kz) = 0 \qquad (14)$$

These amplitude equations are obtained from the first two Maxwell equations in $\chi^{(2)}$ media. From the third one we have:

$$\nabla \cdot (\vec{x}A_x) = \frac{1}{1+\epsilon}\sigma_x(A_x^*\frac{\partial A_y}{\partial x} + A_y\frac{\partial A_x^*}{\partial x})exp(i\Delta kz) + c.c. = \rho_{nl}^x$$

$$\nabla \cdot (\vec{y}A_y) = \frac{1}{1+\epsilon}\sigma_y(\frac{\partial A_x^2}{\partial y})exp(-i\Delta kz) + c.c. = \rho_{nl}^y$$

Our remark is, that this effect is obtained in the case when $\vec{E}$ is not a planar wave and it depends on the gradients of the amplitude envelopes on x and y only.

3.3. BRIGHT SOLITONS

Recently two dimmensional spatial solitary waves in quadratic media were observed experimentally by G.I.Stegeman and all [10] and theoretically by Y. Kivshar [9]. If bright solitons take place, the charges periodically change their sign with a period of Δk , and

$$\rho^x \cong -\rho^y$$

3.4. BRIGHT AND DARK SOLITONS.

In the case of bright and dark solitons there are the solutions with a charge:

$$\rho = \rho^x - \rho^y \neq 0 \qquad .$$

These soliton solutions propagate together along z axis periodically and the sing of the charge is periodically changed because of the Δk.

3.5. SLAB WAVEGUIDES

In the case of slab waveguides the amplitude $A_i(y, z)$ does not depend on x axis and then:

$$\nabla \cdot (\vec{x} A_x) = 0$$

Hence, the derivative $\frac{\partial A}{\partial x} = 0$ and the charge of the fundamental waves $\rho_{nl}^x = 0$. However, there is a polarization charge for second harmonic:

$$\nabla \cdot (\vec{y} A_y) = \frac{1}{1 + \epsilon} \chi^{(2)} \left(\frac{\partial A_x^2}{\partial y}\right) exp(-i\Delta kz) + c.c. = \rho_{nl}^y \tag{15}$$

Various experimental circumstances are possible, and they mainly depend on the medium we use for second harmonic generation. We can estimate the effect of charge density getting in mind that:

$$\frac{div\vec{E}}{\rho_{nl}} \approx \chi^{(2)} \frac{E_0}{2} \tag{16}$$

E_0 is the laser field at half maximum. For most of the crystals we have:

$$\epsilon \approx 1; \chi^{(2)} =\approx 10^{-12} - 10^{-11} m/V$$

If the electric field is of the order of:

$$E_0 \sim 10^7 V/m$$

Then, for the crystals we can estimate:

$$\frac{div\vec{E}}{\rho_{nl}} \approx 10^{-4}$$

For usual lasers and crystals this effect is not possible to obtain, except of the situation when we use the focused beams and we increase in this way the electric field applied. Another possibility is the use of quasi-phasematching techiques which leads to a substantial growth of second harmonic field up to $10^{11} V/m$ [11]. We have a new possibility for semiconductors, where the quadratic $\chi^{(2)}$ nonlinearity is much greater.

$$\chi^{(2)} \approx 10^{-5} - 10^{-7} m/V$$

Then

$$\frac{div\vec{E}}{\rho_{nl}} \approx \chi^{(2)}\frac{E_0}{2} \approx 1$$

Since $\rho_{nl} \neq 0$ we can change the direction of the optical beams in the space by an external dc electrical field. Thus, we can control the 2ω beam in the space.

3.6. EXACT SOLITON SOLUTIONS FOR SLAB WAVEGUIDE WITH A POLARIZATION CHARGE

Since $div\vec{E} = 0$ is not valid for $\chi^{(2)}$ media, it is correct to start with:

$$\nabla \times \nabla \times \vec{E}$$

The electric field is polarized along x axis for fundamental wave and along z axis for the 2ω wave. The slab waveguide plane is (y, z), so, the derivatives $\frac{\partial E_i(y,z)}{\partial x} = 0$. We choose for example E_x to be ordinary n_0 wave, and E_z to be extraordinary n_e wave. It is possible to polarize along z axis, no matter that there is an evolution of the intensity along the axis of propagation. The last evolution prevents only the diffraction. In such a geometry we will search solution for a field of the type of :

$$\begin{pmatrix} x & y & z \\ 0 & \frac{\partial}{\partial y} & \frac{\partial}{\partial z} \\ E_x(y,z) & 0 & E_z(y,z) \end{pmatrix} \tag{17}$$

We will ask for solutions of kind:

$$\vec{E}_x = \vec{x}\frac{1}{2}[A_x(y,z,t)expi(\vec{k}_1 \cdot \vec{z} - \omega_1 t) + c.c.]$$

$$\vec{E}_z = \vec{z}\frac{1}{2}[A_z(y,z,t)expi(\vec{k}_2 \cdot \vec{z} - \omega_2 t) + c.c.]$$

For the amplitude equations we obtain:

$$i\frac{\partial A_x}{\partial z} + \frac{1}{2k_1}\frac{\partial^2 A_x}{\partial y^2} + \sigma_x A_x^* A_z exp(i\Delta kz) = 0 \qquad (18)$$

$$\frac{\partial^2 A_z}{\partial y^2} + k_2^2 A_z + \sigma_z A_x^2 exp(-i\Delta kz) = 0 \qquad (19)$$

It is ineresting to mention, that at the chosen geometry configuration, there are exact solutions of the (18) equations only when $\sigma_x < 0$ and $\sigma_z > 0$. Because of the using of double rotation of $\vec{E}$ vector , we obtain an additional term for the $\vec{E}$ field in y polarization , which should be equal to zero. Then, the following conditions should be fulfilled:

$$\frac{\partial^2 A_z}{\partial z \partial y} + ik_2 \frac{\partial A_z}{\partial y} = 0$$

As a distinction from the xy geometry, here in (18) we have $\vec{A}\|\vec{k}$. Also, we obtain a nonlinear term with k_2^2 in the second equation. Concerning the nonlinear coefficients, we determine

$$\sigma_z = \frac{k_2^2 \chi(2)}{n^2(2\omega)}$$

The (18) equations give soliton solutions of the kind of:

$$A_x(y,z) = -iA_{ox}sech(y)th(y) \cdot exp(i\phi z)$$
$$A_z(y,z) = -iA_{oy}sech(y)^2 \cdot exp(-ik_2 z)$$

The difference from the standart soliton solutions in $\chi^{(2)}$ media is that the phase of the second harmonic soliton is connected with the wave vector k_2 and also there is a polarization charge in 2ω wave. Having soliton solutions in this case, it is quite easy to solve the problem with an efficient electrooptical modulation of the intensive coherent light.

4. Conclusion

As a result of recent studies, the nonlinear gyrotropy in $\chi^{(2)}$ media leads to a new phase-matching conditions and to a new type of amplitude equations.

Such equations give exact soliton solutions ("gyrotropic solitons") which are restricted within specific dispersion relations.

We should note, that the localized light waves in $\chi^{(2)}$ media carry substantial nonlinear polarization charge. Using suitable configuration of the fundamental and second harmonic fields, we obtain new class of amplitude equations. These equations allow exact soliton solutions with a nonlinear polarization charge only for the second harmonic wave. When we apply an electric field, it is possible to control the second harmonic beam in the space. At the same time, we have no any charge for the fundamental wave in such geometry. Since the quadratic $\chi(2)$ nonlinearity is much greater in value than the cubic $\chi(3)$ susceptibility, we find that it is quite important to observe such novel efficient electro-optical effect namely in $\chi(2)$ media because of possible application for codding of the optical information and for the light control.

5. Acknowledgements.

We acknowledge Prof. M.M.Fejer for useful discussion of these results.

References

1. E.Yablonovich, C.Flytzanis, N.Bloembergen,(1972), Anisotropic Interference of Three-Wave and Double Two-Wave Frequency Mixing in GaAs, *Phys. Rev. Lett.*, **vol.29**, pp. 865-867.
2. Y.Karamzin,A.Sukhorukov,(1974), Nonlinear interaction of diffracted light beams; mutual focusing of beams and efficiency restriction of optical frequency converters *JETP Letters*, **vol.20**, pp. 734-739.
3. J.Reintjes, R.C.Eckardt,(1977), Efficient harmonic generation from 532 to 266 nm in ADP and KDP,*Apll.Phys.Lett.*,vol. **30**, p. 91.
4. D.I.Metchkov, V.M.Mitev,L.I.Pavlov,K.V.Stamenov(1977), Fifth harmonic generation in sodium vapor,*Opt.Communications*,**vol. 21**, pp. 391-394.
5. M.G.Grozeva, V. M. Mitev,L.I.Pavlov,K.V.Stamenov (1977), Seventh harmonic generation in metal vapor, *Phys.Letters*,**vol.61A**, pp. 41-42.
6. J.Reintjes,(1984),*Nonlinear optical parametric processes in liquids and gases*, Academic Press Inc.(Harcourt B.J.Publishers), S.F.-New York.
7. L.Lefort,A.Barthelemy,(1995),Intensity-dependent polarization rotation associated with type II phase matched second-harmonic generation:application to self-induced transparency, *Optics Letters*,vol.20, pp. 1749-1751.
8. M.J.Werner,P.D.Drummond,(1993),Simulton solutions for the parametric amplifier, *J.Opt.Soc.Amer. B*,**vol.10**, pp. 2390-2393.
9. Y.Kivshar,(1997), Invited lecture in the same Proceedings of NATO ASI, Sozopol,1997.
10. W.E.Torruellas, Z.Wang, D.J.Hagan, E.W.VanStryland, G.I.Stegeman, L.Torner, C.R.Menyuk, Observation of Two-Dimensional Spatial Solitary Waves in Quadratic Medium: experimental results,*Phys.Rev. Lett.*,vol.74, pp.5036-5039.
11. J.A.Armstrong, N.Bloembergen, J.Ducuing, P.S.Pershan, (1962),Interactions between Light Waves in a Nonlinear Dielectric, *Phys.Rev.*,**vol.127** ,p. 1918.

CLASSICAL AND QUANTUM ASPECTS
OF C.W. PARAMETRIC INTERACTION IN A CAVITY

C. FABRE
Laboratoire Kastler Brossel[†]
Université Pierre et Marie Curie
Tour 12 - Case 74
75252 PARIS Cedex 05
France

1. Introduction

Optical Parametric Oscillators (OPOs) were invented in the early days of nonlinear optics, shortly after the advent of lasers, first in the pulsed regime [1],[2] then in the c.w. regime[3]. They raised a considerable interest in the years 65-75 because of their broad tunability[4], but were superseded by the dye lasers as sources of tunable coherent light. After a decade of almost complete oblivion, they recently enjoyed an impressive revival, because of technological advances : availability of better crystals, better laser sources, progress of electronic control techniques. OPOs are now encountered in many optics laboratories, mainly in their pulsed version. In the case of c.w. OPOs, another boost came from the quantum optics community, where it was realized that these devices provided very efficient sources of light having pure quantum features (squeezed light, twin beams, quantum non demolition measurement devices...). Because of their narrow bandwidth and broad range of operating frequencies, c.w. OPOs are also of great interest in metrology and spectroscopy, and are sometimes used to inject high power pulsed OPOs[5].

The purpose of this paper is to give a general introduction to this rapidly evolving domain. It will first derive the main properties of c.w. OPOs in the simplest case of a parametric coupling between plane waves in a ring cavity made of plane mirrors. It will then give a detailed account of the properties of more realistic OPOs, using a linear cavity with curved mirrors, finite

---

[†]Laboratoire de l'UPMC et de l'ENS, associé au CNRS

A. D. Boardman et al. (eds.),
Advanced Photonics with Second-order Optically Nonlinear Processes, 293–318.
© 1999 *Kluwer Academic Publishers. Printed in the Netherlands.*

pump size, etc... It will finally briefly account for the specific quantum features of these devices, stressing on the most recent developments.

2. Main properties of OPOs in the simplest case

2.1. PARAMETRIC INTERACTION BETWEEN COLINEAR PLANE WAVES

Let us first recall the basic equations describing three-wave mixing in a $\chi^{(2)}$ crystal. The total complex electric field in the crystal, propagating along the z direction, is written as :

$$E\left(\mathbf{r}, \mathbf{t}\right) = \sum_i \mathcal{E}_i\left(z\right) e^{i(k_i z - \omega_i t)} \tag{1}$$

where the sum extends over the indices $i = 0$ (pump mode), $i = 1$ (signal mode) and $i = 2$ (idler mode). One has $\omega_0 = \omega_1 + \omega_2$, and $k_i = n_i \omega_i/c$, n_i being the index of the crystal at frequency ω_i. In the slowly varying envelope approximation, the propagation equations for the three interacting modes are :

$$\begin{array}{l} \frac{d\mathcal{E}_0}{dz} = \frac{i\omega_0}{n_0 c} \chi^{(2)} \mathcal{E}_1 \mathcal{E}_2 e^{-i\Delta kz} \\ \frac{d\mathcal{E}_1}{dz} = \frac{i\omega_1}{n_1 c} \chi^{(2)} \mathcal{E}_0 \mathcal{E}_2^* e^{i\Delta kz} \\ \frac{d\mathcal{E}_2}{dz} = \frac{i\omega_2}{n_2 c} \chi^{(2)} \mathcal{E}_0 \mathcal{E}_1^* e^{i\Delta kz} \end{array} \tag{2}$$

with $\Delta k = k_0 - k_1 - k_2$. We make now the following change in variables :

$$\alpha_i\left(z\right) = \sqrt{\frac{n_i c \varepsilon_0}{2\hbar\omega_i}} \mathcal{E}_i\left(z\right) \tag{3}$$

so that $N_i = |\alpha_i|^2$ gives the photon flux through the transverse plane at position z for the mode i (in photons $\text{m}^{-2}\text{s}^{-1}$). Then one gets the simplified following set of equations :

$$\begin{array}{l} \frac{d\alpha_0}{dz} = i\xi\alpha_1\alpha_2 e^{-i\Delta kz} \\ \frac{d\alpha_1}{dz} = i\xi\alpha_0\alpha_2^* e^{i\Delta kz} \\ \frac{d\alpha_2}{dz} = i\xi\alpha_0\alpha_1^* e^{i\Delta kz} \end{array} \tag{4}$$

where $\xi = \chi^{(2)} \frac{2\hbar\omega_0\omega_1\omega_2}{\epsilon_0 c^3 n_0 n_1 n_2}$. These equations have two propagation invariants: the total energy flux, $\hbar\left(\omega_0 N_0 + \omega_1 N_1 + \omega_2 N_2\right)$, as expected from a pure parametric interaction, with no energy transfer to the medium itself, and the photon flux difference, $N_1 - N_2$. This last property, which is also known as the Manley Rowe relation, is derived from the classical equations of nonlinear mixing and has no real quantum origin. It enforces nevertheless the common picture of parametric interaction as a parametric splitting of a pump photon into a signal photon and an idler photon, which accounts for the energy conservation ($\omega_0 = \omega_1 + \omega_2$) and the equality in the number of signal and idler photons created or destroyed in the process (Manley-Rowe relation). This will have important consequences when we go to the quantum description of the same process.

2.2. OPO CAVITY EQUATIONS

To begin our exploration of the main properties of the OPO we will assume perfect phase matching ($\Delta k = 0$). Let us assume that the pump beam is a c.w. laser delivering a power in the 100 mW range. With available nonlinear crystals having a length of the order of one cm, the parametric gain is very small (typically a few %). As a result, to a good approximation, the amplitudes of the different fields do not vary much in the nonlinear crystal, and one can use a linear approximation of their variation and write for example

$$\alpha_0(z) = \alpha_0(0) + i\xi z \, \alpha_1(0)\alpha_2(0) \tag{5}$$

Thus, in a crystal of geometrical length L extending between $z = 0$ and $z = L$, the mean value of the different mode amplitudes will be obtained in the middle of the crystal, i.e. for $z = L/2$. It turns out to be very convenient to take this mean value as the variable describing the corresponding mode. We will then write :

$$\alpha_i = \alpha_i \left(L/2 \right) \tag{6}$$

The amplitudes of the fields at the exit of the crystal $\alpha_i \left(L \right)$ $(i = 0, 1, 2)$ can be expressed in terms of the corresponding input field $\alpha_i \left(0 \right)$, and of the variables α_i, by the following set of equations, obtained by neglecting terms proportional to χ^n, with $n \geq 2$:

$$\left\{ \begin{array}{l} \alpha_1 \left(L \right) = \alpha_1 \left(0 \right) + 2i\chi\alpha_0\alpha_2^* \\ \alpha_2 \left(L \right) = \alpha_2 \left(0 \right) + 2i\chi\alpha_0\alpha_1^* \\ \alpha_0 \left(L \right) = \alpha_0 \left(0 \right) + 2i\chi\alpha_1\alpha_2 \end{array} \right. \left\{ \begin{array}{l} \alpha_1 = \alpha_1 \left(0 \right) + i\chi\alpha_0\alpha_2^* \\ \alpha_2 = \alpha_2 \left(0 \right) + i\chi\alpha_0\alpha_1^* \\ \alpha_0 = \alpha_0 \left(0 \right) + i\chi\alpha_1\alpha_2 \end{array} \right. \tag{7}$$

with $\chi = \frac{\xi L}{2}$.

The nonlinear crystal is now inserted in a ring optical cavity of geometrical length L' and made of plane mirrors (see Figure (1)). The coupling mirror has amplitude reflexion and transmission coefficients at frequency ω_i noted r_i and t_i $(i = 0, 1, 2)$. We will restrict ourselves to the case of high finesse cavities for which, $r_i = 1 - \rho_i$ (with $\rho_i \ll 1$), and $t_i \approx \sqrt{2\rho_i}$

2.3. SINGLY RESONANT OPO (SROPO)

Let us first consider the case where the cavity finesse is high for the signal mode, and the idler and pump fields are not reflected at all by the cavity

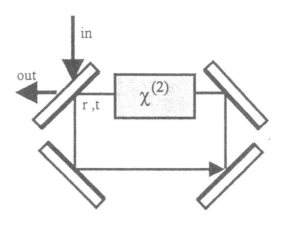

Figure 1. scheme of the OPO optical cavity

mirrors. It is straightforward to determine the value of the signal mode amplitude at the center of the crystal, $(\alpha_1)_{RT}$, after one round trip in the cavity, as a function of the initial signal field amplitude :

$$(\alpha_1)_{RT} = [(\alpha_1 + i\chi\alpha_0\alpha_2^*) r_1 + i\chi\alpha_0\alpha_2^*] e^{i\varphi_1} \qquad (8)$$

where φ_1 is the round trip propagation phase shift :

$$\varphi_1 = \frac{\omega_1}{c} \left((n_1 - 1) L + L'\right) \qquad (9)$$

In steady state operation, one must have $(\alpha_1)_{RT} = \alpha_1$, which leads to the following cavity equation for the signal field α_1:

$$(-\rho_1 + i\delta_1) \alpha_1 + 2i\chi\alpha_0\alpha_2^* = 0 \qquad (10)$$

This equation has been obtained by assuming the cavity is nearly resonant with the cavity :

$$\varphi_1 = 2\pi p_1 + \delta_1 \qquad (11)$$
$$\delta_1 \ll 1 \qquad (12)$$

(p_1 being an integer) and neglecting all terms which are not linear with respect to the small quantities $\chi\alpha_0$, ρ_1, δ_1. The equation for the pump and idler fields are just the linear propagation equations in the case of no input on the idler mode α_2 :

$$\alpha_0 = \alpha_0^{in} + i\chi\alpha_1\alpha_2 \qquad (13)$$
$$\alpha_2 = i\chi\alpha_0\alpha_1^*$$

where α_0^{in} is the input pump beam amplitude. From equations (10) and (13), one finds the following equation for α_1 :

$$\left(-\rho_1 + i\delta_1 + 2\chi^2 |\alpha_0|^2\right) \alpha_1 = 0 \tag{14}$$

Therefore $\alpha_1 = 0$ is a possible solution (OPO below threshold). There also exists a nonzero steady state solution for α_1 when the following conditions are fulfilled:

$$2\chi^2 |\alpha_0|^2 = \rho_1 \quad ; \quad \delta_1 = 0 \tag{15}$$

Let us notice that the first of conditions (15) sets a *fixed value* to the intracavity pump field, independent of the input pump field. The output photon fluxes on the signal and idler modes N_1^{out} and N_2^{out} are respectively equal to $t_1^2 |\alpha_1|^2$ and $|\alpha_2(L)|^2$. these two quantities are equal, due to the twin character of the photons created by the parametric splitting of the pump light, and their common value is :

$$N_1^{out} = N_2^{out} = 4N_t^{(SROPO)} \left[\sqrt{N_0^{in}/N_t^{(SROPO)}} - 1\right] \tag{16}$$

$N_t^{(SROPO)} = t_1^2/4\chi^2$ is the SROPO threshold for the pump photon flux. From (16), one can calculate the conversion power efficiency, equal to :

$$\left(\omega_1 N_1^{out} + \omega_2 N_2^{out}\right)/\omega_0 N_0^{in} \tag{17}$$

One easily sees that this quantity peaks to 100 % when $N_0^{in} = 4N_t^{(SROPO)}$: one of the main interests of such a cw (idealized) OPO is that *one can have perfect conversion from the pump to the signal and idler modes*, even with a very low single pass parametric gain.

The frequency tuning of the device is governed by the condition $\delta_1 = 0$, which implies :

$$\omega_1 = \frac{2\pi c}{(n_1 - 1)L + L'} p_1 \tag{18}$$

This equation shows that, when the cavity length L' changes, ω_1 varies in a way which is exactly like in a laser. In this model, the output power does not depend on the frequency, and the overall frequency bandwidth on which the SROPO oscillates is infinite. This is because we have assumed perfect phase matching, which is not true at any frequency. The actual bandwidth is limited by the phase matching requirement, and is usually very broad. Unfortunately, the threshold of such SROPO lies in the several Watt range[6], which make them difficult to operate. The new periodically poled nonlinear materials which have higher nonlinearities, have been recently used to make SROPO [12].

2.4. DOUBLY RESONANT OPO (DROPO)

In this case, one must write two cavity equations similar to (10), and only one propagation equation like (13). Two different situations can be found : signal-pump double resonance, and signal-idler double resonance, that we will consider successively.

2.4.1. *Pump signal DROPO*

In this case, the equation for the nearly resonant pump field α_0 is similar to (10), with a pump term $t_0 \alpha_0^{in}$ added, so that the three coupled mode equations are in this case :

$$
\begin{aligned}
(-\rho_0 + i\delta_0)\,\alpha_0 + 2i\chi\alpha_1\alpha_2 &= -t_0\alpha_0^{in} \\
(-\rho_1 + i\delta_1)\,\alpha_1 + 2i\chi\alpha_0\alpha_2^* &= 0 \\
-\alpha_2 + i\chi\alpha_0\alpha_1^* &= 0
\end{aligned}
\tag{19}
$$

where the signal fiel detuning δ_1 is defined by an equation similar to (12). From these equations, it is easy to show that a nonzero solution is obtained for a null detuning of the signal mode ($\delta_1 = 0$), and for any detuning δ_0 of the pump mode with a cavity resonance. The output photon flux is given by :

$$
N_1^{out} = N_2^{out} = 4N_t^{(DROPO)} \left[\sqrt{N_0^{in}/N_t^{(DROPO)} - (\delta_0^2/\rho_0^2)} - 1 \right]
\tag{20}
$$

where $N_t^{(DROPO)} = t_1^2 t_0^2 / 16\chi^2$. The pump threshold is now given by :

$$
N_t^{(DROPO)} \left(1 + \delta_0^2/\rho_0^2\right)
\tag{21}
$$

It is minimum at exact pump resonance, and lowered by a factor $t_0^2/4$ with respect to the previous case : this is simply due to the fact that the cavity resonance enhances the intracavity pump power. The tuning properties of this DROPO come from the condition $\delta_1 = 0$, and therefore are "laser-like", as for the SROPO, except that the output power now depends on the pump detuning, and therefore on the cavity length, so that the cavity length range on which the oscillation occurs is now limited to a pump cavity resonance. Minimum thresholds of these devices are in the range of 100 mW[13],[14].

2.4.2. *Signal idler DROPO*

The three coupled equations are in this case :

$$
\begin{aligned}
-\alpha_0 + i\chi\alpha_1\alpha_2 &= -\alpha_0^{in} \\
(-\rho + i\delta_1)\,\alpha_1 + 2i\chi\alpha_0\alpha_2^* &= 0 \\
(-\rho + i\delta_2)\,\alpha_2 + 2i\chi\alpha_0\alpha_1^* &= 0
\end{aligned}
\tag{22}
$$

For simplicity, we have assumed equal reflectivities for the signal and idler modes ($r_1 = r_2 = 1 - \rho$ with $\rho \ll 1$). The last two equations are linear homogeneous equations with respect to α_1 and α_2^*. A nonzero solution for the signal and idler modes is obtained when the determinant of this system is zero, which leads to the relations :

$$\begin{aligned} \delta_1 = \delta_2 &= \delta \\ 4\chi^2 \left| \alpha_0 \right|^2 &= \delta^2 \end{aligned} \tag{23}$$

For the OPO oscillation to take place, *there must therefore exist a definite relation between the cavity detunings of the signal and idler modes.* Note that the two detunings must be equal, but not necessarily simultaneously zero. The output photon flows are given by

$$N_1^{out} = N_2^{out} = 4N_t^{(DROPO')} \left[\sqrt{N_0^{in}/N_t^{(DROPO')} - (\delta^2/\rho^2)} - 1 \right] \tag{24}$$

where $N_t^{(DROPO')} = t_1^4/16\chi^2$. The threshold is now equal to :

$$N_t^{(DROPO')} \left(1 + \delta^2/\rho^2 \right) \tag{25}$$

It is minimum at exact double resonance and has the same order of magnitude as in the other DROPO. Because the threshold value depends on the cavity detuning δ, and therefore on the cavity length, the signal-idler DROPO oscillates only within a finite range of length values, which have width that increase when the pump power increases. This kind of DROPO is the most commonly used of c.w. OPOs [7],[8],[9],[10],[11].

Let us now consider the tuning characteristics of this kind of OPO : The condition $\delta_1 = \delta_2$ implies that :

$$\frac{\omega_1}{c} \left((n_1 - 1) L + L' \right) - 2\pi p_1 = \frac{\omega_2}{c} \left((n_2 - 1) L + L' \right) - 2\pi p_2 \tag{26}$$

where p_1 and p_2 are integers. This relation, together with the constraint $\omega_0 = \omega_1 + \omega_2$, *completeley determines the oscillation frequency of the OPO.* For example, the signal frequency is given by[15],[16] :

$$\omega_1 = \frac{\omega_0}{2} - \frac{\omega_0 (n_1 - n_2) L + 2\pi c (p_2 - p_1)}{(n_1 + n_2 - 2) L + 2L'} \tag{27}$$

To derive this formula, we have assumed that the indices where independent of frequency. As a result the possible oscillation frequencies depend only on one integer $p_2 - p_1$. This is no longer true when index dispersion is taken into account. When the cavity length L' changes, the variation rate of the signal frequency is equal to :

$$\frac{d\omega_1}{dL'} = \frac{1}{L_{cav}} \left(\omega_1 - \frac{\omega_0}{2}\right) \qquad (28)$$

where $L_{cav} = \left(\frac{n_1 + n_2}{2} - 1\right)L - L'$ is the "mean" cavity length for the signal and idler wavelengths. $\frac{d\omega_1}{dL'}$ is proportional to the distance to degeneracy : Far from degeneracy, the frequency tuning rate tends to the one encountered in lasers, where

$$\frac{d\omega_1}{dL'} = -\frac{1}{L_{cav}} \omega_1 \qquad (29)$$

Close to degeneracy, the OPO frequency does not vary much with the cavity length. This behaviour is very different from what is found in lasers : a length change induces in a nearly degenerate DROPO mostly a power variation, not a frequency variation. In all cases, the frequency tuning range is limited by the range of cavity length where the system is above threshold. The tunability of DROPOs can always be obtained by mode hopping, corresponding to changes of the integers p_1 and p_2 in eq (26).

2.5. TRIPLY RESONANT OPO (TROPO)

One writes now three cavity equations, in the simple case of equal mirror transmission for the signal and idler fields :

$$\begin{aligned}
(-\rho_0 + i\delta_0)\,\alpha_0 + 2i\chi\alpha_1\alpha_2 &= -t_0\alpha_0^{in} \\
(-\rho + i\delta_1)\,\alpha_1 + 2i\chi\alpha_0\alpha_2^* &= 0 \\
(-\rho + i\delta_2)\,\alpha_2 + 2i\chi\alpha_0\alpha_1^* &= 0
\end{aligned} \qquad (30)$$

The signal and idler equations are the same as in the previous case, and therefore a non zero solution for α_1 and α_2 only exists when :

$$\delta_1 = \delta_2 = \delta \quad ; \quad 4\chi^2 |\alpha_0|^2 = \rho^2 + \delta^2 \qquad (31)$$

The output photon flux N_1^{out} is now given by an implicit equation [17] :

$$\left(1 - \delta\delta_0/\rho\rho_0 + N_1^{out}/4N_t^{(TROPO)}\right)^2 + (\delta/\rho + \delta_0/\rho_0)^2 = N_0^{in}/N_t^{(TROPO)} \qquad (32)$$

with $N_t^{(TROPO)} = t_0^2 t^4/64\chi^2$, where t is the common value of the transmission factors t_1 and t_2. Three cases are encountered, depending on the detunings δ_0 and δ :

(1) $\delta_0\delta < \rho\rho_0$:

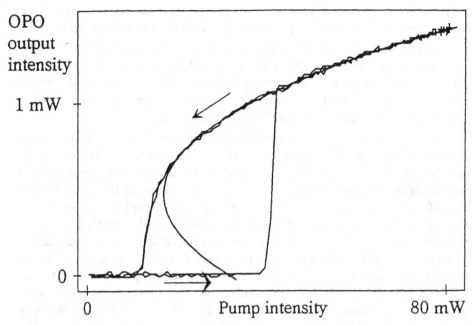

Figure 2. TROPO output power as a function of pump power in the region of bistability. The parabolic curve is a theoretical fit.

The TROPO behaves like a DROPO, with a threshold equal to :

$$N_t^{(TROPO)} \left(1 + \delta^2/\rho^2\right) \left(1 + \delta_0^2/\rho_0^2\right) \tag{33}$$

which is minimum at exact triple resonance ($\delta = \delta_0 = 0$). The threshold is lowered by a factor $t_2^2/4$ with respect with the DROPO case : it lies now in the mW range or less.

(2) $\delta_0\delta > \rho\rho_0$:

The TRO is bistable between a zero solution and a non zero solution. This situation has been recently observed experimentally[18]. Figure (2) shows the hysteresis cycle experimentally observed in the output signal intensity when the pump intensity is scanned back and forth while the cavity length is kept constant.

(3) $\delta_0\left(\delta_0 + 2\delta\right) \leq -\rho_0\left(\rho_0 + 2\rho\right)$:

One is in the monostable case (1), but there is a second threshold corresponding to the onset of a self-pulsing behaviour. Numerical simulations performed in this case [17] show that a further increase in power leads to a chaotic behaviour through a sequence of period doubling bifurcations.

The frequency tuning of a TROPO, governed by the condition $\delta_1 = \delta_2$, is similar to the one of a DROPO : length changes mainly induce power changes, and for a given input pump power, the OPO oscillates only around specific length values, and on a length range which is roughly equal to the narrowest of the pump and signal cavity resonances.

Figure (3) gives an experimental signal observed in a TROPO comprising a KTP crystal, pumped at 0.53 nm by a frequency doubled YAG laser[18]. The cavity finesses are around 30 for the pump mode and 1000 for the signal and idler modes. As a result, the minimum threshold is very low (400 μW). The figure gives the power of the reflected pump beam, and the power of the signal and idler modes when the cavity length is scanned. One observes that the OPO operates at nonzero values only in small ranges of length values, for which pump depletion is also observed in the reflected pump beam. The wide and shallow dip in the pump power is due to the cavity resonance with the pump mode. One also observes that the outer peaks, corresponding to a pump detuning δ_0 of the order of γ_0 are strongly asymmetrical, which is another evidence for the bistable operation of the TROPO.

Practically, to get a stable c.w. output intensity from signal-idler DROPOs and TROPOs, one needs a very good control of the OPO cavity length. This is obtained by using the same techniques as for frequency stabilizing single mode lasers : design of compact, vibration free, small size structures, and use of servo-loops techniques reacting on the OPO cavity length (or sometimes on the pump frequency). Semi-monolithic or monolithic structures[18],[19],[20],[21] are often preferred, because of their small losses and good mechanical stability.

It is easy to show that in the steady state regime, the sum of the phases of the signal and idler fields is locked to the phase of the pump field, and that the difference of these phases is a parameter which is not fixed by the OPO equations. This situation is somehow similar to the case of the laser, where the phase of the output field is not determined, and undergoes a phase diffusion process under the influence of the input noise, mainly the spontaneous emission phase noise. The same diffusion process occurs in the OPO on the difference between the signal and idler phases, and as a result, the ultimate linewidth measurable on the difference between the signal and idler frequency (beatnote $\omega_1 - \omega_2$) is the Schawlow-Townes linewidth :

$$\Delta(\omega_1 - \omega_2) = \frac{\Delta\omega_{cav}}{N_1^{cav}} \qquad (34)$$

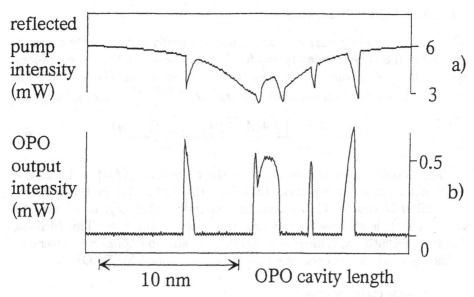

Figure 3. Variation of the reflected pump intensity (a), and OPO output intensity (b) as a function of the OPO cavity length in the vicinity of a pump-cavity resonance

where $\Delta\omega_{cav} = \frac{c\tau_0}{L_{cav}}$ is the cavity bandwidth and N_1^{cav} the number of intracavity signal (or idler) photons. The signal mode linewidth has therefore two origins[22] : the pump laser frequency jitter affecting the phase sum between the signal and idler modes, and the Schawlow-Townes diffusion process affecting the phase difference. One finds [23] :

$$\Delta\omega_1 = \frac{1}{2}\sqrt{\left(\frac{\Delta\omega_{cav}}{N_1^{cav}}\right)^2 + \Delta\omega_{pump}^2} \qquad (35)$$

3. Towards a more realistic OPO model

So far, we have described a "perfect" OPO with exact phase matching, no losses, plane waves for the three modes, and using a ring cavity. We will now consider briefly what modifications in the OPO properties appear when one tries to describe the OPO by a more realistic model.

3.1. EXTRA CAVITY LOSSES

The output coupling mirror transmission is usually not the only cause of losses for the different cavity modes. Let A_i be the intensity extra-loss coefficient for each one. One can easily show that, in the TROPO case for example, the oscillation threshold parameter $N_t^{(TROPO)}$ is now given by :

$$N_t^{(TROPO)} = \frac{\left(t_0^2 + A_0\right)^2}{t_0^2} \frac{\left(t_1^2 + A_1\right)\left(t_2^2 + A_2\right)}{64\chi^2} \tag{36}$$

The output power is now equal to the proportion $t_1^2 / \left(t_1^2 + A_1\right)$ of the value obtained using equation (32) (or (16), (20), (24) in the SROPO and DROPO cases). The maximum conversion efficiency, equal now to $t_1^2 / \left(t_1^2 + A_1\right)$, is still reached at four times above threshold. This formula has been recently experimentally checked[24], and an impressive conversion efficiency of 81 % has been observed in a semi monolithic DROPO.

3.2. LINEAR CAVITY CASE

In a linear cavity the crystal is used twice per round trip, but the fact that the parametric gain depends on the relative phase between the three fields must be taken into account, especially in the TROPO case. By using the same analysis as in the previous section, one shows that in a linear cavity, the ring cavity equations can be used, provided that the nonlinear coefficient χ appearing in the different equations is replaced by an effective coefficient χ_L equal to :

$$\chi_L = \chi \left(1 + e^{i\Delta\theta}\right) \tag{37}$$

where $\Delta\theta = \theta_0 - \theta_1 - \theta_2$. θ_i is the propagation phase accumulated by the field α_i between two successive paths in the crystal. It includes the effect of index dispersion of the medium (air usually), the propagation phase shift of the gaussian beam due to its curvature, and the mirror phase shift, related to its exact multilayer structure. χ_L ranges therefore between 0 and 2π and in principle the threshold for a linear cavity ranges between one fourth of the ring cavity threshold ($\Delta\theta = 0$) and infinity ($\Delta\theta = \pi$) .

3.3. NON PERFECT PHASE MATCHING : RING CAVITY CASE

When Δk is not equal to zero, it is well known that one must replace, in equation (5) and in the following ones, χ by $\chi \left(e^{i\Delta kL} - 1\right)/i\Delta kL$. This change has two effects:

(1) an amplitude effect : One must replace χ^2 by $\chi^2 sinc^2 \left(\Delta kL/2\right)$ in the threshold expressions of the ring cavity case, and the threshold value

increases. For a given pump power above the minimum threshold, this leads to a "gain bandwidth" of the OPO inside which one can find the different oscillation modes, obtained for different values of the integers p_1 and p_2, and therefore for different values of the cavity length.

(2) a phase effect : One must introduce a term proportional to the mode power $|\alpha_i|^2$ in the detunings. This effect is in many respects similar to a cross-Kerr effect[25] and is a manifestation of a cascading nonlinear effect in the parametric interaction. It leads also to a bistable behaviour of the OPO.

When one varies for example the crystal temperature, the different indices vary, and the couple of frequencies ω_1, ω_2 corresponding to perfect phase matching is changed. There is therefore a shift of the OPO gain bandwidth, as defined above. In the type I case, the variation of the signal and idler frequencies close to degeneracy has a parabolic variation with temperature, and there exists a temperature range where the OPO cannot oscillate[4]. In the type II case, the same variation is linear with temperature, and the OPO oscillates at any temperature value.

3.4. NON PERFECT PHASE MATCHING : LINEAR CAVITY CASE

In the case of a linear cavity, the dephasing effects of imperfect phase matching and double passage in the crystal add coherently. One can show that χ must now be replaced by χ'_L equal to

$$\chi'_L = \chi \left(1 + e^{i(\Delta\theta + \Delta kL)}\right) \left(e^{i\Delta kL} - 1\right) / i\Delta kL \qquad (38)$$

The threshold expression is thus modified by a factor equal to[15],[26] :

$$4 \left(\sin c\frac{\Delta kL}{2} \cos \frac{\Delta\theta + \Delta kL}{2}\right)^2 \qquad (39)$$

The two dephasing effects may compensate each other in some respect: when $\Delta\theta$ is non-zero, the minimum threshold is now reached for a nonzero value of Δk. It has a value larger than the threshold obtained when $\Delta\theta = 0$, but with a smaller increase than the one calculated at $\Delta k = 0$. As a result, when $\Delta\theta$ increases up to π, the minimum threshold increases only by 90% [15], but to obtain this threshold, one must change one parameter in the OPO (for example the crystal temperature), so that ΔkL reaches the value ± 2.32. This feature is one of the consequences of the fact that, in an OPO, the system is "free to choose" one frequency (ω_1 or ω_2) to reduce as much as possible the oscillation threshold .

3.5. MULTIMODE OPERATION

When the pump power is increased, the oscillating ranges (in terms of cavity length) of two different couples of signal and idler modes (α_1, α_2) and (α_3, α_4) broaden, and may even overlap (See figure (4), where two oscillation peaks are mixed around some length value). As a result, there exists a cavity length domain where these two couples of longitudinal modes may simultaneously oscillate.

In most cases, the frequencies of these different signal and idler modes are not equal ($\omega_1 \neq \omega_3$, $\omega_2 \neq \omega_4$), and therefore the signal and idler modes of different pairs, (α_1, α_4) and (α_3, α_2), cannot be coupled by the parametric interaction. The cavity equations are in this case :

$$
\begin{aligned}
(-\rho_0 + i\delta_0)\,\alpha_0 + 2i\chi\alpha_1\alpha_2 + 2i\chi'\alpha_3\alpha_4 &= -t_0\alpha_0^{in} \\
(-\rho_1 + i\delta_1)\,\alpha_1 + 2i\chi\alpha_0\alpha_2^* &= 0 \\
(-\rho_2 + i\delta_2)\,\alpha_2 + 2i\chi\alpha_0\alpha_1^* &= 0 \\
(-\rho_3 + i\delta_3)\,\alpha_3 + 2i\chi'\alpha_0\alpha_4^* &= 0 \\
(-\rho_4 + i\delta_4)\,\alpha_4 + 2i\chi'\alpha_0\alpha_3^* &= 0
\end{aligned}
\tag{40}
$$

with notations analogous to the TROPO case studied previously. The mode coupling mechanism is thus only due to the pump depletion (first equation in (40)). One can show[27] that when χ is different from χ', which is the most common situation, there is no possibility of simultaneous oscillation of the OPO on the two pairs and that the only stable stationary mode is the one corresponding to the largest $|\chi|$, i.e. to the lowest threshold. When one varies the cavity length, and therefore the detunings, the OPO hops from one pair to the other at the point when the two threshold values cross. This situation is exactly the one encountered in a homogeneously broadened laser.

The situation is a bit more complex in the transient regime, because the switching time between the two modes is inversely proportional to $|\chi| - |\chi'|$, which can be very small if the two couples of modes have frequencies close to each other. A transient multimode operation may occur in these conditions. This is the case for instance in pulsed OPOs.

4. Transverse characteristics of the OPO modes

In c.w. OPOs, the cavity, and the transverse characteristics of the input pump field, impose the transverse structure of the oscillating modes. In most cases, one uses a cavity with curved mirrors, or even lenses, and a Gaussian TEM_{00} input pump mode. As a result the natural basis for the study of the transverse characteristics of an OPO are the set of $\text{TEM}_{\ell m}$

Laguerre-Gauss modes, and we will write the complex fields as :

$$E_i\left(\overrightarrow{r},z\right)=\sum_{p,\ell}u_{p\ell}^i\left(\overrightarrow{r},z\right)e^{ik_iz}A_{p\ell}^i\left(z\right) \tag{41}$$

where the Laguerre-Gauss modal functions $u_{p\ell}^i\left(\overrightarrow{r},z\right)$ are given by :

$$u_{p\ell}^i\left(\overrightarrow{r},z\right)=\sqrt{\frac{2p!}{\pi w^2(z)(p+|\ell|)!}}\left(\sqrt{2}\frac{r}{w_i(z)}\right)^{|\ell|}L_p^{|\ell|}\left(\frac{2r^2}{w_i^2(z)}\right) \\ \exp i\left(\frac{k_ir^2}{2\overline{z}}-(2p+|\ell|+1)\tan^{-1}\frac{z}{z_R}+\ell\theta\right) \tag{42}$$

where r denotes the modulus of the transvers position $\overrightarrow{r}$, θ is the angular variable and :

$$w_i\left(z\right)=w_0^i\left(1+\frac{z^2}{z_R^2}\right)^{1/2}\qquad w_0^i=\sqrt{\frac{2z_R}{k_i}}\qquad \overline{z}=z-iz_R \tag{43}$$

In eqs (41) and (42), p is a non negative integer, and ℓ can take any integer value. One can show that the propagation equations for the Gaussian field envelope $A_{p\ell}^i\left(z\right)$ are :

$$\begin{cases}\frac{dA_{p\ell}^0(z)}{dz}=i\chi^{(2)}\frac{\omega_0}{n_0c}\sum_{qm,rn}\left(\Lambda_{pqr}^{\ell mn}\left(z\right)\right)^*A_{qm}^1\left(z\right)A_{rn}^2\left(z\right)e^{i\Delta kz}\\ \frac{dA_{q\ell}^1(z)}{dz}=i\chi^{(2)}\frac{\omega_1}{n_1c}\sum_{p\ell,rn}\left(\Lambda_{pqr}^{\ell mn}\left(z\right)\right)A_{p\ell}^0\left(z\right)A_{rn}^2\left(z\right)^*e^{-i\Delta kz}\\ \frac{dA_{rn}^2(z)}{dz}=i\chi^{(2)}\frac{\omega_2}{n_2c}\sum_{p\ell,rn}\left(\Lambda_{pqr}^{\ell mn}\left(z\right)\right)A_{p\ell}^0\left(z\right)A_{qrn}^2\left(z\right)^*e^{-i\Delta kz}\end{cases} \tag{44}$$

where the coupling coefficient $\Lambda_{pqr}^{\ell mn}$ is the overlap integral between the Gaussian modes associated with the three interacting fields.

4.1. COUPLING COEFFICIENTS

If the input pump field is in a TEM$_{00}$ mode having the same Rayleigh length z_R as for the signal and idler modes, if the system is operating near degeneracy ($\omega_1\approx\omega_2\approx\frac{\omega_0}{2}$) and in the perfectly phase matched configuration ($\Delta k=0$), one finds[27] :

$$\Lambda_{0qr}^{0mn}\left(z=0\right)=\frac{2}{\sqrt{\pi w_0^2}}\frac{(|m|+q+r)!}{[q!r!\left(q+|m|\right)!\left(r+|m|\right)!]^{1/2}2^{|m|+q+r}}\delta_{m-n} \tag{45}$$

(where w_0 is the common value of the signal and idler beam waists $w_1\left(0\right)$ and $w_2\left(0\right)$). In particular, the coupling constant between the fundamental TEM$_{00}$ modes is equal to :

$$\Lambda_{000}^{000}\left(z=0\right)=\frac{2}{\sqrt{\pi}w_0} \tag{46}$$

It is inversely proportional to the beam waist, which means that tightly focused light modes ensure a higher local nonlinear coupling coefficient.

Let us now consider the coupling between higher order transverse modes, i.e. the coefficient :

$$C_{qr}^m = \frac{\Lambda_{0qr}^{0m-m}(z=0)}{\Lambda_{000}^{000}(z=0)} \tag{47}$$

C_{qr}^m is maximum when $m = 0$, and also when $q = r$, i.e. for identical transverse variations of the signal and idler modes. For large values of q, it is roughly equal to $(\pi q)^{-1/2}$: it therefore decreases very slowly with the order of the transverse mode. This means that when one uses a Gaussian TEM$_{00}$ pump mode matched to the OPO cavity, a significant number of signal and idler transverse modes are coupled together, and are likely to oscillate above some pump threshold.

4.2. CAVITY EQUATIONS

The same kind of analysis as the one outlined at the beginning of the first section leads to the following set of coupled equtions, for example in the TROPO case :

$$\begin{cases} \left(-\rho_1 + i\delta_{qm}^1\right)\alpha_{qm}^1 + i\sum_r \chi_{qrm}\alpha_{00}^0\alpha_{r-m}^{2*} = 0 \\ \left(-\rho_2 + i\delta_{r-m}^2\right)\alpha_{r-m}^2 + i\sum_q \chi_{qrm}\alpha_{00}^0\alpha_{qm}^{1*} = 0 \\ \left(-\rho_0 + i\delta_{00}^0\right)\alpha_{00}^0 + i\sum_{qmr} \chi_{qrm}\alpha_{qm}^1\alpha_{r-m}^2 = -t_0\alpha_0^{in} \end{cases} \tag{48}$$

α_{pl}^i, proportional to the field amplitude of the transverse modes at the center of the crystal, $A_{pl}^i(L/2)$, is such that $\left|\alpha_{pl}^i\right|^2$ gives the *numeric photon current* (photons$\times s^{-1}$) in the corresponding transverse mode (and no longer the numeric flow of photon flux (photons m$^{-2}\times s^{-1}$) like in the previous plane wave model of eq(4)). The coupling coefficients χ_{qrm} in (48) are given by

$$\chi_{qrm} = \pi\chi^{(2)}\frac{z_R}{c}\sqrt{\frac{\hbar\omega_0\omega_1\omega_2}{2\varepsilon_0cn_0n_1n_2\tau_0\tau_1\tau_2}}I_{q+r}\left(\frac{L}{2z_R}\right)\Lambda_{0qr}^{0m-m}(0) \tag{49}$$

where $I_{q+r}\left(\frac{L}{2z_R}\right)$ is a coefficient taking into account the longitudinal variation of the transverse properties of the interacting modes integrated over the length of the crystal. It is equal to :

$$I_p(\xi) = (-1)^p\frac{2}{\pi}\left[\tan^{-1}\xi + \sum_{k=1}^p(-1)^k\frac{\sin\left(2k\tan^{-1}\xi\right)}{k}\right] \tag{50}$$

In the limit of a thin medium ($L \ll z_R$), $I_p(\xi) \simeq L\pi z_R$, and one retrieves the usual equations. For a very long crystal ($L \gg z_R$), I_p tends to the constant $(-1)^p$.

4.3. MULTIMODE OPERATION FOR DIFFERENT TRANSVERSE MODES

The different sums appearing in (48) extend over the set of transverse modes which are nearly resonant with the cavity. More precisely, the round-trip phase shift $\varphi^i_{p\ell}$ for the mode $A^i_{p\ell}$ is such that :

$$\varphi^i_{p\ell} = \frac{\omega_i}{c} \left[(n_i - 1) L' + L \right] + (2p + |\ell| + 1) \varphi_{cav} = 2\pi s^i_{p\ell} + \delta^i_{p\ell} \qquad (51)$$

with $s^i_{p\ell}$ integers, $\delta^i_{p\ell} \ll 1$ $\quad i = 0, 1, 2$, and φ_{cav} being the TEM$_{00}$ mode propagation phase shift, which depends on the exact cavity geometry. When $\varphi_{cav} = 0$ (plane cavity), 2π (spherical cavity), or π (confocal cavity), a very large number of low order transverse modes are coupled together. But in most cases ("nondegenerate cavities") the different transverse modes are resonant with the cavity for different cavity lengths. As a result, the oscillation ranges of the different transverse modes do not overlap at low pumping regimes, and a single transverse mode operation is possible in such nondegenerate cavities around definite length values. At higher pump levels, these ranges overlap and a simultaneous oscillation of two pairs of signal and idler transverse modes may occur.

4.3.1. *Single mode operation with three TEM$_{00}$ modes*

Let us first consider the case where only one transverse mode for each field is nearly resonant for a given cavity length. For the sake of simplicity, we assume that these modes are the lowest order TEM$_{00}$ transverse modes for the signal idler and pump modes. Then the plane wave analysis outlined previously holds, if χ is replaced by χ_{00}, equal to :

$$\chi_{00} = \chi \frac{2 \tan^{-1} \left(\frac{L}{2z_R} \right) w_0 w_1 w_2}{\pi \left(w_0^2 w_1^2 + w_1^2 w_2^2 + w_2^2 w_0^2 \right)} \qquad (52)$$

Let us consider as an example the case of a signal-idler resonant DRO, for which the cavity geometry imposes the signal and idler mode structure. It imposes more precisely the value of the Rayleigh length z_R, in a way that is independent of the mode frequency ω_i. As a result, the equation :

$$z_R = \frac{1}{2c} \omega_1 w_1^2 = \frac{1}{2c} \omega_2 w_2^2 \qquad (53)$$

relates the mode sizes to their frequencies. It is then easy to show, using (52), that the threshold will be minimized by using a pump mode with a waist w_0 corresponding to the same Rayleigh length. In other words, the best coupling is achieved when the three interacting modes can be sustained by the same cavity geometry, like in a TROPO.

4.3.2. *Multimode operation on two pairs of signal and idler transverse modes*

Let us now consider the case where two transverses mode for each field are nearly resonant for a given cavity length. Eq (48) shows that the two transverse signal modes α^1_{qm} and $\alpha^1_{q'm'}$, having the same frequency ω_1, are simultaneously coupled to the two transverse idler modes $\alpha^2_{r,-m}$ and $\alpha^2_{r',-m'}$ of frequency ω_2, with four different coupling constants given by relation (45). To simplify the notations, we will write $\alpha_1 = \alpha^1_{qm}$, $\alpha'_1 = \alpha^1_{q'm'}$, $\alpha_2 = \alpha^2_{r,-m}$ and $\alpha'_2 = \alpha^2_{r',-m'}$. Equations (48) now reduce to the following set of equations

$$
\begin{aligned}
\left(-\rho + i\delta^1_{qm}\right)\alpha_1 + i\chi\alpha_0\alpha^*_2 + i\chi_c\alpha_0\alpha'^*_2 &= 0 \\
\left(-\rho + i\delta^1_{q'm'}\right)\alpha'_1 + i\chi'_c\alpha_0\alpha^*_2 + i\chi'\alpha_0\alpha'^*_2 &= 0 \\
\left(-\rho + i\delta^2_{r,-m}\right)\alpha_2 + i\chi\alpha_0\alpha^*_1 + i\chi'_c\alpha_0\alpha'^*_1 &= 0 \\
\left(-\rho + i\delta^2_{r',-m'}\right)\alpha'_2 + \chi_c\alpha_0\alpha^*_1 + \chi'\alpha_0\alpha'^*_1 &= 0 \\
\left(-\rho_0 + i\delta^0_{00}\right)\alpha_0 + i\chi\alpha_1\alpha_2 + i\chi'\alpha'_1\alpha'_2 + i\chi_c\alpha_1\alpha'_2 + i\chi'_c\alpha'_1\alpha_2 &= -t_0\alpha^{in}_0
\end{aligned}
$$

$$(54)$$

(We assumed $\gamma_1 = \gamma_2 = 1 - \rho$). The coupling constants are given by

$$
\chi = \chi_{qrm} \qquad \chi' = \chi_{q'r'm'} \qquad \chi_c = \chi_{qr'm}\delta_{mm'} \qquad \chi'_c = \chi_{q'rm}\delta_{mm'} \quad (55)
$$

There is a simple case when the two pairs of modes are decoupled, i.e. when χ_c and χ'_c vanish : this occurs when $m \neq m'$. We are led back to the set of eqs (41), and to the same conclusions as in subsection (3.5). When $m = m'$, The situation is more complex and has been studied in ref[27]. Let us first notice that if a given set of solutions $(\alpha_1\alpha_2\alpha'_1\alpha'_2)$ has been obtained, then $(\alpha_1 e^{i\varphi}, \alpha_2 e^{-i\varphi}, \alpha'_1 e^{i\varphi}, \alpha'_2 e^{-i\varphi})$ is also a solution of (54) : there is a global phase indeterminacy in the system, like in a laser, to which will be associated a phase diffusion process. However, the relative phase of α_i and α'_i is fixed due to the mode cross coupling. This implies a dynamically stable intensity-independent spatial pattern of the signal (idler) field, if a stationary solution of this type exists (in the more general case of more than one pump mode contributing, this feature is lost and the field pattern can change with pump strength). Eq.(54) can be simplified for certain special values of the nonlinear couplings, namely when :

$$
\chi\chi' = \chi_c\chi'_c . \tag{56}
$$

This is only possible when $k = k'$ (or $l = l'$) i.e. when the same signal (or idler) transverse mode is coupled to two different transverse idler (or signal) modes. In this case, one shows[27] that there exist some combinations of modes $(\alpha_1 \alpha_1')$ and $(\alpha_2 \alpha_2')$ leading to decoupled equations. The oscillation then takes place on a new fixed mode having a transverse pattern given by the linear combination of the two coupled transverse modes which has the lowest threshold. The general case is more difficult to study, and is detailed in ref[27] : in the perfectly resonant case for all the modes, one finds a modification of the thresholds, an oscillation on a combined mode, and a possibility of multistability between the different steady state solutions. The situation is quite changed if one allows some of the modes to be detuned. In this case the combined oscillation leads to an self-pulsing behaviour involving all the modes.

4.3.3. *Multimode operation in a degenerate cavity*

The situation is more tricky in a degenerate cavity (for example a confocal cavity) where many transverse modes are likely to oscillate simultaneously. This situation is studied in [28]. In this case it is the pump beam structure, and no longer the cavity geometry, which imposes the signal and idler mode transverse patterns. In particular, these patterns now depend on the pump intensity.

5. Quantum properties of Optical Parametric Oscillators

5.1. NONCLASSICAL PROPERTIES OF LIGHT

The theoretical description for most experimental situations in optics is the "semiclassical" approach of light-matter interaction : classical electromagnetic fields, obeying Maxwell equations, interact with atoms or solids obeying quantum mechanics. This is what is done generally in nonlinear optics, for example in the description of $\chi^{(2)}$ interaction. But the semiclassical approach is of course an approximation : the electromagnetic field is actually quantized, and special quantum features, impossible to describe within the semi-classical theory, must show up in some experimental situations.

When one deals with light beams of the order of a mW or more, the flow of photons is very large (typically $10^{16}/$s), and it is impossible to count separately each photon impinging on the detector. The quantized nature of light does not appear directly in the photocurrent, which is not formed of distinct "clicks". One can show that in such a regime, the mean values of the different quantities measured in optics are well described by the semi-classical theory. But this is no longer true for the *fluctuations* of these quantities around their mean values and for the *correlations* existing between the different observables. In a rather paradoxical way, the quantized nature of light indeed appears in such "macroscopic" situations, in which a very large number of photons is involved. The reader will find in refs[29] a detailed account of these nonclassical properties of light and of the corresponding experiments.

The semiclassical approach predicts in particular that there is always a noise in the photocurrent I resulting from the detection of a light beam of perfectly constant intensity : this noise is the well-known "shot noise" ($\Delta I_{shot} = \sqrt{2qIB}$, where q is the electron charge and B the detection bandwidth). It arises in this theory from the random character of the photodetection process. The full quantum description of the same situation gives a different answer : it shows that there exists special light beams, called nonclassical states of light (for example *squeezed light*, or *sub-Poissonian light*) leading to a photocurrent noise which is below the shot noise limit ΔI_{shot}, and can be even exactly null, at least in principle. This feature can be important in some applications, where the shot noise turns out to be a limiting factor for the accuracy or sensitivity of measurements. The quantum theory of light also shows that in an optical measurement, even though there is no *a priori* lower limit for the variance of a single measurement, there is a fundamental lower limit for the *product* of the variances of two conjugate observables, like in the well-known Heisenberg inequality $\Delta x.\Delta p \geq \hbar/2$. One can show for example that, for any intense light beam[30] :

$$\Delta I.\Delta\varphi \geq 2qB \qquad (57)$$

where $\Delta\varphi$ is the uncertainty in the measurement of the phase of the light beam. As a result a nonclassical beam giving a photocurrent noise well below the shot noise limit (*sub-Poissonian beam*) will have a large phase uncertainty and will yield fringes with poor contrast in an interferometric measurement.

Quantum theory shows that *nonlinear processes* are necessary to produce such nonclassical beams. In particular, the $\chi^{(2)}$ interaction is very efficient to produce these states[29] : for example, it has been shown, first theoretically then experimentally, that a degenerate OPO operating below threshold generates a nonclassical state of light of zero mean value called "squeezed vacuum"[31].

In a similar way the quantum theory of light contradicts the semiclassical theory for the predictions about the *correlations* between the fluctuations of different light beams. As a simple example, let us consider the problem of producing *two light beams with exactly equal intensities* : in classical optics, this is easily done by separating a light beam using a perfect semi-reflecting plate. The semiclassical theory of light predicts that in such a configuration, the fluctuations of the two resulting photocurrents will be uncorrelated, because they arise from uncorrelated photodetection processes : as a result, there will be residual fluctuations in the difference between the two phocurrents, equal to ΔI_{shot}, in which I is now the photocurrent corresponding to the beam before the beamsplitter. This residual noise is a limitation for any differential measurement in optics. The quantum theory of light confirms this prediction, and shows that this ΔI_{shot} is actually the noise measured in such an experimental situation, whatever the quantum state of the beam impinging on the beamsplitter. It also shows that there exist pairs of nonclassical beams (*twin beams*), produced by other ways than by using a beamsplitter, which give fluctuations on the photocurrent difference below the "standard quantum limit" ΔI_{shot}. These fluctuations can even be in principle totally cancelled. The counterpart is that for such a pair of beams, the fluctuations on the conjugate quantity, namely the difference between the phases of the two beams, is increased to infinity in this limiting case.

Like in the single beam case, nonlinear optics is the way to generate such twin beams. In particular a nondegenerate OPO operating above threshold produces signal and idler beams having perfectly correlated intensity fluctuations [7],[32],[33]. This is related to the fact already mentionned that in the parametric interaction signal and idler photons are produced in a perfectly correlated way.

5.2. OBSERVATION OF SQUEEZING IN THE PUMP BEAM REFLECTED BY A TRIPLY RESONANT OPO

As an example of generation of nonclassical light by an OPO, I will describe a recent experiment[34] using the triply resonant, ultralow threshold OPO already described in section (2.5). The full-quantum analysis of this device[35] predicts that the pump light which is going out of the cavity after having interacted inside the optical cavity with the signal and idler beams can be significantly squeezed.

The TROPO is pumped at a wavelength $\lambda_0 = 0.53$ μm and generates signal and idler beams at non degenerate wavelengths λ_1 and λ_2 which are close to 1.064 μm and can be adjusted by temperature tuning. As already mentionned, the cavity finesse is roughly 1000 around 1.06 μm and 45 around 0.53 μm, ensuring a threshold equal to 400 μW at exact triple resonance[18]. The reflected pump beam is mixed with a part of the pump laser beam used as a local oscillator, then sent onto a balanced homodyne detection device. This setup enables us to measure the quadrature noise of the reflected pump beam. The local oscillator phase is scanned by moving the mirror M_1. A servomechanism reacting on the OPO cavity length locks it at a value corresponding to a nonzero, but small, cavity detuning with the pump, signal and idler modes while the local oscillator phase is scanned. Figure (4) displays the resulting noise level, recorded at a noise frequency of 30 MHz, and for an incident power of 8 mW.

314

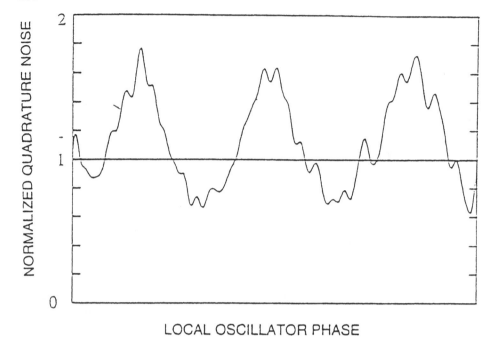

NORMALIZED QUADRATURE NOISE

LOCAL OSCILLATOR PHASE

Figure 4. Signal of the homodyne detection as a function of the local oscillator phase, normalized to shot noise

The noise level oscillates with the local oscillator phase, showing that the noise area is now distorted with respect to the coherent state case. It goes below the shot noise level at a given phase value. This shows that the measured light is squeezed : the quadrature noise is reduced by 30 % below the shot noise level. It corresponds to a level of 48 % squeezed light at the output of the OPO, when one corrects for the effect of losses in the propagation from the OPO output to the detection and for the effect of detection imperfect quantum efficiency.

The OPO can thus be seen as a simple and compact "quantum noise eater", which transforms in a passive way any incoming light beam in the several mW range into a squeezed beam. The squeezing performances of this device are presently limited by the non negligible optical losses of our KTP crystals for the green light beam, and by the low quantum efficiency of detectors at this wavelength. Better results will be obtained in devices working with pump beams in the near infrared, thus generating signal and idler waves around 2 μm.

5.3. SUB SHOT NOISE SPECTROSCOPY USING OPO SIGNAL AND IDLER BEAMS

Non classical states of light can be used to improve the signal to noise ratio in various optical experiments. This has been shown in a recent experiment[36] in which the sensitivity of a spectroscopic measurement of a very weak transition is increased beyond the shot noise limit using signal and idler "twin beams" generated by an OPO. Other similar experiments have been performed, using either squeezed vacuum[37] or sub-Poissonian diode lasers[38, 39].

Among different spectroscopic techniques, absorption spectroscopy is often used, because of its simplicity. If the absorption dip is directly recorded on the intensity of the transmitted beam, the sensitivity of the experiment is limited by the intensity noise of the light source, which is usually above the shot noise level. The background noise can be further reduced by performing a differential measurement : a beamsplitter placed before the absorbing region creates a reference beam which is directly measured, and a transmitted beam which interacts with the absorbing medium. The absorption signal is then measured on the difference between the intensities of the transmitted beam and the reference beam. In this case, the sensitivity of the measurement is limited by the shot noise ΔI_{shot} of the incoming beam (before the beamsplitter). It is possible to go below this shot noise limit by replacing the beamsplitter by an OPO : the idler beam, for example is used as a reference beam, and the signal beam crosses the absorbing medium. The noise in the difference between the intensity of these two beams, which limits the sensitivity of the experiment, can be zero in principle, and is well below the shot noise in actual experiments.

As stated previously, DROPOs or TROPOs are difficult to frequency tune continuously without mode hoppings. We have circumvented this difficulty by using a two-photon transition : the two-photon resonance is induced by the signal beam from the OPO and by an auxiliary laser beam, which can be easily frequency scanned and modulated ; at resonance, the upper state population, and therefore the absorption of the signal beam, is also modulated. It is then possible to record the modulated absorption of the signal beam transmitted through the medium on the difference between the signal and idler beam intensities, which has a squeezed background noise.

The experiment is performed on the $4S_{1/2} \to 5S_{1/2}$ two-photon transition in atomic potassium, which is induced by the OPO signal beam at $\lambda_{opo} \simeq 1.064\mu m$ and by an auxiliary diode laser at $\lambda_{aux} \simeq 0.859\mu m$. The twin beams are produced by a triply resonant semi-monolithic OPO pumped at $532nm$. The nonlinear crystal used in the OPO (KTP) is of type II, so that the signal and idler beams have orthogonal polarizations

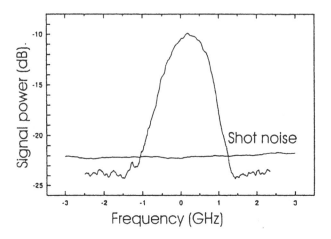

Figure 5. Two-photon absorption signal as a function of the diode laser frequency in logarithmic power units. The lower trace shows the shot noise level recorded with a YAG laser of identical power

and are separated at the output by a polarizing beamsplitter. The idler beam is directly detected by a high quantum efficiency photodetector, while the signal beam is sent through the cell containing the atomic potassium and then detected by a photodetector identical to the previous one. The auxiliary beam used to induce the two-photon transition is provided by a semiconductor laser which is amplitude modulated at ν_{HF} =3MHz by an electro-optic modulator, and then chopped by a mechanical chopper at a frequency ν_{LF} of 650Hz. It is then superimposed with the signal beam before entering the cell using a dichroic mirror. The two beams are focused at the center of the potassium cell. After the cell, the auxiliary beam is rejected by two dichroic mirrors. This ensures that no residual background modulation is detected at the 10^{-8} level. In these conditions, the optical noise of the auxiliary laser beam plays no role in the background noise of our signal, detected on the twin beam intensity difference. The resolution bandwidth for the total detection system of about 3Hz.

The quantum noise reduction in the intensity difference measured directly at the output of the OPO is 3dB around 3MHz. This value, smaller than in previous experiments[32], is limited by the non-negligible losses in the KTP crystal that we used, and by the stability of the set-up. When measured in the actual spectroscopic set-up with all its components, the quantum noise reduction is further reduced by the losses in the different optical elements to a value of 1.9dB at a noise frequency of 3MHz.

Fig. (5) shows the amplitude signal output of the lock-in amplifier in logarithmic scale and in power units, recorded using a 20mW diode laser beam and a 5.8mW signal OPO signal beam. The resonant two-photon modulation transfert signal corresponds to a relative absorption of 2.5×10^{-7} of the signal beam. It is observed on a noise background which allows a minimum measured signal of 5×10^{-8} (signal to noise ratio equal to 1). The lower trace corresponds to the shot noise level recorded by using in the differential measurement a YAG laser beam of same intensity instead of the signal and idler beams generated by the OPO. The signal to noise ratio changes from 12 to 13.9dB, when one uses the OPO instead of the YAG laser. This corresponds to an improvement equal to 1.9dB (35%) beyond the shot noise limit.

These results illustrate the potentiality of c.w. OPOs as light sources for spectroscopy in the visible and near infrared region and show that the quantum correlations between twin beams can be used to improve the sensitivity beyond the shot noise limit in a *real* spectroscopy experiment. The present performances represent only a moderate improvement of the signal to noise ratio. They are limited by the losses in the twin beam production and in their propagation in the spectroscopy experimental set-up, and also by the poor stability of the total set-up. Much better results are expected if one uses stabler pump sources for the OPO and better optical components.

Acknowledgements

I would like to express my special thanks to Drs Gao JiangRui, K. Kasai, A. Maitre, C. Schwob and P. Souto-Ribeiro for their essential contributions to the work presented in this paper, and also A. Gatti, L.Lugiato, M. Marte and H. Ritsch for a fruitful collaboration. This work has been partly funded by the E.C. TMR contract QSTRUCT (Project ERB FMRX-CT96-0077).

References

1. Giordmaine, J., Miller, R. (1965) *Phys Rev Letters* **14**, 973.
2. Akhmanov, A., Kovrigin, A., Kolosov, A., Piskarskas, A., Fadeev, V., Khoklov, R. (1966) *Sov. Phys JETP* **3**, 241.
3. Smith, R.G., Geusic, J., Levinstein, H., Rubin, J., Singh, S., Van Uitert, L. (1968) *Appl. Phys. Lett.* **123**, 308.
4. See for example: Byer, R.L. (1975) Treatise in Quantum Electronics, H. Rabin, C.L. Tang, (eds), p.587, Academic Press.
5. See for example the special issue of JEOSB, Quant. Semiclass. (1997) Opt., **9**, 131-295.
6. Yang, S., Eckardt, R., Byer, R.L. (1993) *J. Opt. Soc. Am*, **B10**, 1684.
7. Heidmann, A., Horowicz, R.J., Reynaud, S., Giacobino, E., Fabre, C. and Camy, G. (1987) *Phys. Rev. Letters* **59**, 2555.
8. Colville, F., Padgett, M., Henderson, A., Zhang, J., Sibbett, W. and Dunn, M. (1993) *Optics Letters* **18**, 1065.

9. Serkland, D., Eckardt, R. and Byer, R. (1993) *Optics Letters* **19**, 1046.
10. Lee, D. and Wong, N. (1993) *J. Opt. Soc. Am.* **B10**, 1659.
11. Gerstenberger, D. and Wallace, R. (1993) *J. Opt. Soc. Am.* **B10**, 1681.
12. Arbore, M. and Fejer, M. (1997) *Optics Letters* **22**, 151.
13. Scheidt, M., Bejer, B., Boller, K. and Wallenstein, R. (1997) *Optics Letters* **22**, 1287.
14. Schneider, K., Kramper, P., Schiller, S. and Mlynek, J. (1997) *Optics Letters* **22**, 1293.
15. Debuisschert, T., Sizmann, A., Giacobino, E. and Fabre, C. (1993) *J. Opt. Soc. Am.* **B10**, 1668.
16. Eckardt, R., Nabors, C., Kozlovsky, W. and Byer, R.L. (1991) *J. Opt. Soc. Am.* **B8**, 646.
17. Lugiato, L., Oldano, C., Fabre, C., Giacobino, E. and Horowicz, R. (1988) *Nuovo Cimento* **D10**, 959.
18. Richy, C., Petsas, K.I., Giacobino, E., Fabre, C. and Lugiato, L. (1995) *J. Opt. Soc. Am.* **B12**, 456.
19. Nabors, C., Eckardt, R., Kozlovsky, W. and Byer, R.L. (1989) *Opt. Lett.* **14**, 1134.
20. Gerstenberger, D. and Wallace, R. (1993) *J. Opt. Soc. Am.* **B10**, 1681.
21. Schiller, S. and Byer, R.L. (1993) *J. Opt. Soc. Am.* **B10**, 1696.
22. Graham, R. and Haken, H. (1968) *Z. Phys.* **210**, 276.
23. Courtois, J.Y., Smith, A., Fabre, C. and Reynaud, S. (1991) *J. Mod. Opt.* **38**, 177.
24. Breitenbach, G., Schiller, S. and Mlynek, J. (1995) *J. Opt. Soc. Am.* **B12**, 2095.
25. See for example G. Stegeman's contribution in this issue.
26. Bjorkholm, J., Ashkin, A., Smith, R. (1970) *IEEE J. Quantum. Electr.* **QE6**, 797.
27. Schwob, C., Cohadon, P.F., Fabre, C., Marte, M., Ritsch, H., Gatti, A. and Lugiato, L. *"Transverse Effects and Mode Couplings in OPOs"*, Preprint.
28. Leduc, D., Maître, A. and Fabre, C. *"Transverse Patterns in a Confocal OPO above Threshold"*, Preprint.
29. See for example: Kimble, H.J. (1992) Fundamental Systems in Quantum Optics, Les Houches session LIII, J. Dalibard, J.M. Raymond, J. Zinn-Justin (eds), Elsevier Science Publisher pp.545; Kimble, H.J. (1992) *Physics Reports* **219**, 227; Fabre, C. (1992) *Physics Reports* **219**, 215; Reynaud, S., Heidmann, A., Giacobino, E. and Fabre, C. (1992) *Progress in Optics* **30**, 1; Walls, D., Milburn, G. (1995) *Quantum Optics*, Springer.
30. Fabre, C. (1997) Quantum Fluctuations, Les Houches Session LXIII, S. Reynaud, Giacobino, E, J Zinn-Justin (eds), Elsevier Science, pp.181.
31. Wu, LingAn, Kimble, J., Hall, J. and Wu. H. (1986) *Phys. Rev. Letters* **57**, 2520.
32. Mertz, J., Debuisschert, T., Heidmann, A., Fabre, C. and Giacobino, E. (1991) *Optics Letters* **16**, 1234.
33. Reynaud, S., Heidmann, A., Giacobino, E. and Fabre, C. (1993) *Progress in Optics* **30**, 1.
34. Kasai, K., JiangRui Gao, and Fabre, C. (1997), *Europhys. Lett.* **40**, 25.
35. Fabre, C., Giacobino, E., Heidmann, A., Lugiato, L., Reynaud, S., Vadacchino, M. and Kaige Wang, (1990) *Quantum Optics* **2**, 159.
36. Souto Ribeiro, P., Schwob, C., Maître, A. and Fabre, C. *Optics Letters* (in press).
37. Polzik, E., Carri, J. and Kimble, J. (1992) *Phys. Rev. Letters* **68**, 3020.
38. Kasapi, S., Lathi, S. and Yamamoto, Y. (1997) *Optics Letters* **22**, 478.
39. Marin, F., Bramati, A., Jost, V. and Giacobino, E. (1997) *Optics Commun.* **140**, 146.

ALTERNATIVE MEDIA FOR CASCADING

Second Order Nonlinearities and Cascading in Plasma Waveguides

M. GEORGIEVA-GROSSE and A. SHIVAROVA
Faculty of Physics, Sofia University,
BG-1164 Sofia, Bulgaria

Abstract. Cascading to third order nonlinear effects of self-phase modulation and waveguided channel formation as well as wave propagation at second harmonics are considered in waveguides containing weakly ionised gaseous plasmas. The nonlinear plasma response is in the inhomogeneous field of surface waves. An unified procedure for consideration of cascading through second order nonlinear high-frequency and low-frequency plasma responses is demonstrated by taking as an example simplified cases of wave propagation along a single interface and of a local and instantaneous nonlinear plasma response. The study aims to show that: (i) cascading through low-frequency nonlinear response may be more important than cascading through high-frequency response at second harmonics, (ii) the nonequal to zero $divE$ – term (where E is the wave electric field) in the nonlinear wave equation which is associated with appearance of inhomogeneous space charge density self-consistently induced by the nonlinearity may play role in both second and third order effects and, (iii) the effects of the nonlinear changes of the modal field are part of the third order effect of self-modulation and self-focusing.

1. Introduction

The common basis of the nonlinear (NL) wave phenomena and the tremendous wealth of information accumulated at their studies in different media and different fields of physics are sources of knowledge at the treatment of each particular problem. Formation and propagation of NL waves in optical waveguide systems and plasma waveguide configurations is a particular example which offers possibilities for transfer of experience from one field to another. The presentation of the study here is subordinated to such an aim: subjects involved in treatments of third order NL efects which go through cascading in plasma waveguides are discussed in connection with their possible relevance to analogous problems in nonlinear optics. The study is on effects of self-action of surface waves (SWs) – envelope solitons and formation of waveguided channels – in plasma waveguides consisting of weakly-ionised low-temperature gas-discharge plasmas bounded by a dielectric. With the low density

319

A. D. Boardman et al. (eds.),
Advanced Photonics with Second-order Optically Nonlinear Processes, 319–343.

$n_0 = (10^{15} - 10^{18})$ m^{-3} of the gas-discharge plasmas, the frequency range of the wave propagation is up to hundreds MHz – few GHz.

Investigations on NL SWs have been carried out for many years starting with the pioneering works by Alanakyan [1] and Boev and Prokopov [2] where dispersion relations of strongly NL SWs in semi-bounded plasmas are derived at a nonlinearity due to the ponderomotive force (p. f.) [3,4] and at ionisation nonlinearity. Cases of waveguides of different geometries [5-7] have been also covered. Experimental evidence of effects of second harmonic (SH) generation [8-10] has involved interest in studying effects of weak nonlinearities. The coupled mode theory [11,12] has been applied. However the main extension of the studies has been directed to SW envelope solitons. Nonlinearity due to the p. f. effects has been usually in the focus of the studies [13-16]. However it appears that cascading through high frequency (HF) plasma response at SH frequency [15,17-19] could be more efficient. The ionisation nonlinearity involved comparatively later [20-23] in the studies on self-phase modulation (SPM) of SWs and waveguided channel formation gains in the competition. In fact, this is the mechanism involved in explanations of experimental results [24-26] on modulation instability and tendency to bright envelope soliton formation. The theoretical results on NL SWs are reviewed in [20, 27, 28].

Since the second order nonlinearity is the lowest order one in plasmas, the third order effects of self-action of waves (of frequency ω and wavenumber k_ω) goes always through cascading. Both HF NL plasma response at SH frequency 2ω and low-frequency (LF) NL plasma response at zero (0) frequency which could originate either from ionisation nonlinearity or p. f. effect are involved in the process (Figure 1). This

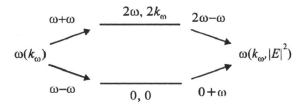

Figure 1. Channels - HF and LF - of cascading to third order effects of self-action.

means that the realization of the third order effects passes through two consequent stages of effects of second order. At the first stage the HF ($\omega + \omega \to 2\omega$) and LF ($\omega - \omega \to 0$) NL plasma responses to the action of the wave field are formed. At the second stage, the self-action of the wave appears: $2\omega - \omega \to 0$, $0 + \omega \to \omega$.

The presentation here on self-action of SWs in a plasma waveguide – a single plasma-dielectric interface – aims at stressing on: (*i*) the importance of the LF plasma response in the scheme of the cascading, (*ii*) the contribution of the nonlinearly induced – by the wave field – inhomogeneity of the medium and, (*iii*) the NL changes of the modal wave field. SW propagation at SH frequency is treated also separately. Experimental evidence of some of the effects is shown.

2. Initial Set of Equations

An isotropic plasma is in the HF field of an electromagnetic wave:

$$\vec{E}(\vec{r},t) = (1/2)\left\{\vec{\mathcal{E}}_\omega(\vec{r})\exp(-i\omega t) + c.c\right\}, \quad \vec{B}(\vec{r},t) = (1/2)\left\{\vec{\mathcal{B}}_\omega(\vec{r})\exp(-i\omega t) + c.c\right\}. \quad (1)$$

The Maxwell equations written in their standard form

$$rot\vec{E} = -\frac{\partial\vec{B}}{\partial t}, \quad rot\vec{H} = \frac{\partial\vec{D}}{\partial t}, \quad div\vec{D} = 0 \quad \text{or} \quad div\vec{E} = \frac{\rho}{\varepsilon_0} \quad (2)$$

with the material relations for the displacement

$$\vec{D}(\vec{r},t) = \varepsilon_0\vec{E}(\vec{r},t) + \int_{-\infty}^{t} dt'\vec{j}(t',\vec{r}) \quad \text{or} \quad \vec{D}(\omega,k_\omega) = \varepsilon_0\varepsilon(\omega)\vec{E}(\omega,k_\omega), \quad (3a,b)$$

which involve the plasma response to the field, and the plasma model equations close the initial set of equations. In (2), (3), $\vec{B} = \mu_0\mu\vec{H}$ ($\mu = 1$), $\rho = e(n_i - n_e)$ is the charge density with n_α ($\alpha = e,i$) being electron and ion concentrations, ε_0 and μ_0 are the vacuum permittivity and susceptibility, $\vec{j}(\vec{r},t) = -en\vec{v}$ (with $n \equiv n_e$) is the (electron) current density, e is the electron charge, $\vec{v}$ is the electron velocity and

$$\varepsilon_\omega = 1 - \left(\omega_p^2/\omega^2\right) \quad (4)$$

with ω_p being the plasma frequency, is the plasma permittivity at frequency ω. The treatment is based on the wave equation

$$\vec{\nabla}^2\vec{\mathcal{E}}_\omega - \vec{\nabla}\left(\vec{\nabla}.\vec{\mathcal{E}}_\omega\right) = -\frac{\omega^2}{\varepsilon_0 c^2}\vec{\mathcal{D}}_\omega \quad (5)$$

in which a presentation in the form (1) of the displacement $\vec{D}$ is used.

When HF NL plasma response at SH frequency and LF NL plasma response related to p. f. effects are described, the continuity equation and the electron motion equation

$$\frac{\partial n}{\partial t} + div(n\vec{v}) = 0, \quad mn\left[\frac{\partial\vec{v}}{\partial t} + \left(\vec{v}.\vec{\nabla}\right)\vec{v}\right] = -en\vec{E} - en\left(\vec{v} \times \vec{B}\right) - nmv\vec{v} - T_e\vec{\nabla}n \quad (6a,b)$$

complete the initial set of equations. Here T_e is the electron temperature (in energy unis), m is the electron mass and v is the electron-neutral elastic collision frequency.

When LF plasma response related to ionisation nonlinearity is described, the particle balance equation replaces (6a,b) and the electron energy balance equation has to be added:

$$-\vec{\nabla}^2(D_A n) + \rho_r n^2 = \nu_i(T_e)n + \rho_{si}(T_e)n^2, \qquad -\frac{5}{2}\vec{\nabla}(D_e n\vec{\nabla}T_e) + \frac{3}{2}n\nu_*(T_e)U_* = Q. \text{(7a,b)}$$

The left hand side of (7a) describes nonlocal losses of charged particles through ambipolar diffusion (D_A is the diffusion coefficient) and local losses through recombination (with ρ_r being the recombination coefficient). The right hand side of (7a) gives particle gain through direct and step ionisation (ν_i is the frequency of direct ionisation and ρ_{si} is the rate coefficient of step ionisation). The left hand side of (7b) describes nonlocal losses of electron energy through thermal conductivity (D_e is the electron diffusion coefficient) and local losses through collisions. The inelastic collisions for excitation with ν_* and U_* being respectively the collision frequency and the excitation energy of the first atom excited state, are usually taken as predominant ones. The Joule heating term is $Q = \vec{j}.\vec{E} \equiv -en\vec{v}.\vec{E}$.

The nonlinearity in (6), (7) is of the second order. The NL force in (6b):

$$\vec{F}_{NL} = -m(\vec{v}.\vec{\nabla})\vec{v} - e(\vec{v} \times \vec{B}) \tag{8}$$

responsible both for LF response at 0-frequency ($\omega - \omega \to 0$) and HF response at 2ω-frequency ($\omega + \omega \to 2\omega$) is one of the sources of nonlinearity. Taken at zero frequency it gives the p. f. [3,4] which results in perturbations of the plasma density $\delta n_0^{(\omega-\omega)}$. In the case of a HF plasma response, the NL force (8) leads to perturbations of both the plasma density $\delta n_{2\omega}^{(\omega+\omega)}$ and the electron velocity $v_{2\omega}^{(\omega+\omega)}$. Moreover, in the latter case, the NL force (8) is involved also in the second stage of second order NL interactions which goes through cascading to a NL response at the fundamental frequncy: $2\omega - \omega \Rightarrow \omega$. In this case it gives third order perturbations of the electron velocity $v_{(2\omega-\omega)}$ which are at the fundamental frequency. The meaning of the ionisation nonlinearity as a second order effect stems directly from the form of the Joule heating term in (7b). The latter determines an electron temperature disturbance $\delta T_{e0}^{(\omega-\omega)}(|E|^2)$ at zero frequency and this is an effect of a thermal nonlinearity which is the starting point of the ionisation nonlinearity. The final expression of the latter is a density perturbation $\delta n_0^{(\omega-\omega)}(|E|^2)$ at zero frequency. Therefore, (i) the NL force (8) and, (ii) the Joule heating term in (7b) considered together with the NL terms in (7a), are the sources of nonlinearity treated further on.

At weak nonlinearity, $n = n_0 + \delta n$ is composed by the unperturbed – by the field – density n_0 and the perturbation $\delta n = \delta n_L^{(\omega)} + \delta n_0^{(\omega-\omega)} + \delta n_{2\omega}^{(\omega+\omega)} + \delta n_\omega^{(2\omega-\omega)} + \delta n_\omega^{(0+\omega)}$ caused by the field. The electron velocity $\delta v \equiv v = v_L^{(\omega)} + v_0^{(\omega-\omega)} + v_{2\omega}^{(\omega+\omega)} + v_\omega^{(2\omega-\omega)}$ $+ v_\omega^{(0+\omega)}$ contains only perturbations. The disturbances δn, v include: (i) linear plasma responses at frequency ω, (ii) second order NL responses at zero frequency, caused by ionisation nonlinearity and p. f. effects, and at 2ω – frequency and, (iii) third order NL perturbations at the fundamental frequency ω formed through cascading both via SH and zero frequencies (third harmonic generation is not included in the considerations). After having the density and velocity perturbations of the given order determined, the plasma nonlinear response at frequency ω, i.e. the effect of the self-action of the wave, can be obtained by specifying the nonlinear current $\vec{j}_{NL}(\vec{r},t) = -en\vec{v}$ in the material relation (3a):

$$\vec{D}_\omega(\vec{r}) = \varepsilon_0 \vec{E}_\omega + (i/\omega)\vec{j}(\vec{r})\big|_{(\omega)}, \tag{9}$$

in which

$$\vec{j}(\vec{r},t)\big|_\omega = -e\Big[n_0 \vec{v}_{L\omega} + n_0 \vec{v}_{2\omega-\omega} + n_0 \vec{v}_{0+\omega}$$
$$+ \delta n_{L\omega} \vec{v}_{\omega-\omega} + (1/2)\delta n_{L\omega}^* \vec{v}_{\omega+\omega} + \delta n_{\omega-\omega} \vec{v}_{L\omega} + (1/2)\delta n_{\omega+\omega} \vec{v}_{L\omega}^*\Big] \tag{10}$$

The first term in (10) is the current density in linear approximation and the other terms are NL contributions. Then the NL part $\delta\varepsilon_{NL}(|E|^2)$ of the plasma permittivity involved in (3b) can be also found.

3. Surface Mode

The consideration involves surface waves (SWs) [29] which are proper modes of plasma waveguides. The simplest case of a single interface between two semi-spaces occupied by a plasma ($x > 0$) and a dielectric/vacuum ($x < 0$; permittivity ε_d or $\varepsilon_v \equiv 1$, in the case of vacuum) is taken. The wave propagation is in z-direction. The wave is a TM mode: E_x, H_y, $E_z \neq 0$. The field variation is of the type: $\propto \exp(-i\omega t + ik_\omega z)$. The

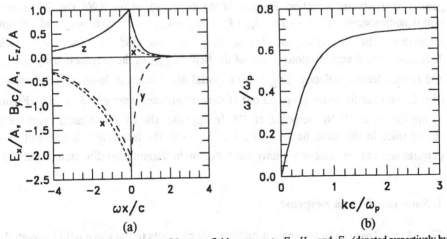

Figure 2.(a): Transverse variation of the wave field components E_x, H_y and E_z (denoted respectively by x, y and z) in linear approximation (Figure 2(1) from Ref. [22]). (b): Dispersion diagram of SWs.

field amplitude is maximum at the interface $(x = 0)$ and decays away from the boundary in the two semi-spaces (Figure 2a). The wave electric field components in linear approximation are:

$$E_z^{(\omega)}(x > 0) = A_\omega \exp(-\kappa_{p\omega} x), \qquad E_z^{(\omega)}(x < 0) = A_\omega \exp(\kappa_{d\omega} x)$$

$$E_x^{(\omega)}(x > 0) = i\left(k_\omega / \kappa_{p\omega}\right) E_z^{(\omega)}(x > 0), \qquad E_x^{(\omega)}(x < 0) = -i\left(k_\omega / \kappa_{d\omega}\right) E_z^{(\omega)}(x < 0)$$

(11)

where $A_\omega \equiv E_z^{(\omega)}(x = 0)$. The magnetic field is related to them trough $\vec{B}_\omega = -(i / \omega) rot \vec{E}_\omega$.

The quantities $\kappa_{p\omega} = \left[k_\omega^2 - \left(\omega^2 / c^2\right)\varepsilon_\omega\right]^{1/2}$, $\kappa_{d\omega} = \left[k_\omega^2 - \left(\omega^2 / c^2\right)\varepsilon_d\right]^{1/2}$ describe the field decay in transverse $(x$-$)$ direction. The velocity of the electron oscillations in the wave field is $\vec{u}_{L\omega} = -ie\vec{E}_\omega / m\omega$. In the linear approximation, there is no density perturbations caused by the wave field in the plasma volume $(\delta n_{L\omega} = 0)$. Therefore the two terms in (10) which contain $\delta n_{L\omega}$ drop. Moreover, according to (2), $\delta n_{L\omega} = 0$ gives $divE_\omega^L = 0$ and, thus, in linear approximation, the second term in (5) is equal to zero. The solution $k_\omega = (\omega / c)\left(\varepsilon_\omega \varepsilon_d / (\varepsilon_d + \varepsilon_\omega)\right)^{1/2}$ of the linear dispersion relation of the waves $\left(\Delta_L(\omega, k_\omega) = (\varepsilon_\omega / \kappa_{p\omega}) + (\varepsilon_d / \kappa_d) = 0\right)$ is given in Figure 2b. The wave propagation is at frequencies $\omega / \omega_p \leq 1/\sqrt{1 + \varepsilon_d}$, i.e. in a range in which the plasma permittivity (4) is negative $(\varepsilon_\omega < -\varepsilon_d)$.

The SW field is inhomogeneous in transverse (x-)direction. Therefore NL density perturbations at zero frequency due to p. f. effects are present and they, as well as, the density perturbations caused by the ionisation nonlinearity induce plasma inhomogeneity in transverse direction. Because of the dispersion of the SWs, the SH generation is nonresonant: $\omega + \omega \rightarrow 2\omega$, $k_\omega + k_\omega = 2k_\omega \equiv k_{2\omega} - \Delta k$ where $k_{2\omega}$ and Δk are respectively the linear wavenumber at frequency 2ω and the linear mismatch. Therefore, forced perturbations at 2ω of the field $(E_{2\omega})$, of the plasma density $(\delta n_{2\omega})$ and of the electron velocity $(\delta v_{2\omega})$ with spatial distribution in longitudinal direction (the direction of the wave propagation) of their amplitudes determined by $2k_\omega$ (Figure 1) are the local HF NL response at SH. In opposite, the LF NL plasma response is strong since in this case $(\omega - \omega \rightarrow 0$, $k_\omega - k_\omega \rightarrow 0)$ the interaction is resonant. The perturbantion at zero frequency have not variation in longituginal direction (Figure 1).

4. Nonlinear Plasma Response

The NL response of the plasma – a medium which reacts to the wave field through the motion of its free charged particles – is formed via the current density (10).

In the case of a second order interaction (Figure 1) of the type $(\omega - \omega)$, no matter that only the electrons react directly to the HF field, both the electrons and the ions are included in the formation of the LF NL plasma response. In the case of p. f. effects, the ions are involved through their reaction to the steady-state electric field arising at the electron separation from the ion background. In the case of ionisation nonlinearity, the ions are involved because they appear together with the electrons at the atom ioni-sation. In both cases, a perturbation $\delta n_{\omega-\omega}$ of the stationary density is the plasma response to the HF field; there is no disturbance of the velocity $(v_0^{(\omega-\omega)} = 0)$. Therefore the third term in the right hand side of (10) is zero and there is only one term (the sixth one $\delta n_{\omega-\omega} v_{L\omega}$) which is responsible for the wave self-action.

When the second order interaction (Figure 1) is of the type of $(\omega + \omega)$, the plasma reaction to the HF field is completely in the HF range and only the electrons are involved in the formation of the HF NL plasma response. Both the electron density $(\delta n_{\omega+\omega})$ and velocity $(v_{\omega+\omega})$ have perturbations at frequency 2ω. Moreover, the NL force (8) acts at the two stages $(\omega + \omega, \; 2\omega - \omega, \; \text{Figure 1})$ of the cascading to a NL response at the frequency ω of the fundamental wave, and two terms (the second and the last one) in (10) are responsible for the wave self-action.

4.1. IONISATION NONLINEARITY: LF SECOND ORDER RESPONSE

In the case of weak nonlinearity, the effect of the ionisation nonlinearity originates from additional heating of the electrons in the HF field $|E|^2$. The increase δT_e $\left(T = T_{e0} + \delta T_e\right)$ of the electron temperature causes additional ionization, i.e. an increase $\delta v_i \; \left(v_i = v_{i0} + \delta v_i\right)$ of the ionisation frequency, and results into a contribution δn_0 $(n = n_0 + \delta n_0^{(\omega-\omega)})$ to the plasma density (Figure 3a). When the field is inhomoge-neous like in the case of SWs, the density becomes inhomogeneous. The complete

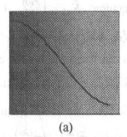 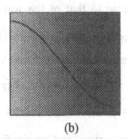

(a)   (b)

Figure 3. Illustration for plasma density changes resulting from effects in inhomogeneous fields of ionisation nonlinearity (a) and a nonlinearity due to the ponderomotive force action (b).

picture of the ionisation nonlinearity is complicate because the losses of electron energy and charged particles can be local or nonlocal as described by (7). In the simplest case of local losses of electron energy through collisions and local losses of charged particles through recombination, the plasma response is completely local. With an excitation frequency $\nu_* = \nu_{*0} + \delta\nu_*$ approximated by $\nu_* = \dot{\nu}_* \exp(-U_* / T_e)$ where $\dot{\nu}_* \approx const.$, the temperature increase $\delta T_{e0} = T_{e0}\left(|\vec{\mathcal{E}}(\vec{r})|^2 / E_{wth}^2\right)$ in the field can be easily obtained from (7b). (Here $E_{wth} = \left[3m\left(\omega^2 + \nu^2\right)\nu_{*0}\, U_*^2 / \left(e^2 \nu T_{e0}\right)\right]^{1/2}$ is the normalizing field which describes the effeciency of the electron heating in the field.) The particle balance (7a), simplified to a form which accounts only for particle production by direct ionisation (with ionization frequency $\nu_i = \dot{\nu}_i \exp\left(- U_i / T_e\right)$ where U_i is the ionisation energy, $\dot{\nu}_i \approx const.$ and δT_{e0} is taken into account in T_e) and particle losses through recombination (with $\rho_r = const.$), gives the result for the plasma density response:

$$\delta n_0^{(\omega-\omega)} = n_0\left(|\vec{\mathcal{E}}(\vec{r})|^2 / E_{n(i)}^2\right) \tag{12}$$

Here $E_{n(i)} = \sqrt{T_{e0} / U_i}\, E_{wth}$ characterieses the efficiency of the ionisation nonlinearity. Expression (12) is the final result at the first stage of the cascading (i.e. the $\omega - \omega \to 0$ interaction, Figure 1). It gives local increase of the density with the increase of the field intensity $|\vec{\mathcal{E}}(\vec{r})|^2$ (Figure 3(a)).

4.2. NONLINEAR FORCE: LF AND HF SECOND ORDER RESPONSES

The NL force (8) in the equation of motion (6b) is involved in the two channels of the cascading (Figure 1). Its realization $\left(\vec{F}_{NL(\omega-\omega)}^{(0)} = -(1/4)\left\{\left[m\left(\vec{u}.\vec{\nabla}\right)\vec{u}^* + e\left(\vec{u} \times \vec{\mathcal{B}}^*\right)\right] + c.c.\right\}\right)$ at zero frequency through ($\omega - \omega$)-second order NL interaction results in the ponderomotive force (p. f.) $\vec{F}_{NL}^{(0)} = -\left(e^2 / 4m\omega^2\right)\vec{\nabla}\left|\vec{\mathcal{E}}_\omega(\vec{r})\right|^2$ which acts on the electrons by pushing them out of regions of higher field intensity. In the case of SW fields (Figure 2a) the p. f. acts in a direction transverse to that of the wave propagation: $\vec{F}_{NL} = \left(\vec{F}_{NLx}^{(0)}, 0, 0\right)$. The same force (8), with its effect $\left(\vec{F}_{NL(\omega+\omega)}^{(2\omega)} = -(1/4)\left\{\left[m\left(\vec{u}_\omega.\vec{\nabla}\right)\vec{u}_\omega + e\left(\vec{u}_\omega \times \vec{\mathcal{B}}_\omega\right)\right]e^{-2i\omega t} + c.c.\right\}\right)$ at SH frequency ensures cascading through HF plasma response. In this case, the space variations of the NL force $\vec{F}_{NL}^{(2\omega)} = (1/2)\left\{\vec{\mathcal{F}}_{NL}^{(2\omega)} e^{-2i\omega t} + c.c.\right\}$ are given by $\vec{\mathcal{F}}_{NL}^{(2\omega)} = \left(e^2 / 4m\omega^2\right)\vec{\nabla}\vec{\mathcal{E}}_\omega^2$. In the case of SWs, the NL force at SH frequency has components both in transverse and longitudinal directions: $\vec{\mathcal{F}}_{NL}^{(2\omega)} = \left(\mathcal{F}_{NLx}^{(2\omega)}, 0, \mathcal{F}_{NLz}^{(2\omega)}\right)$.

4.2.1. Ponderomotive Force Effects: LF Second Order Response

The final result for the plasma response to the action of $\vec{F}_{NL}^{(0)}$ is illustrated in Figure 3(b). The force moves the electrons in the direction in which the amplitude of the HF field decreases. The deviation from the quasineutrality and the density inhomogeneity caused by $\vec{F}_{NL}^{(0)}$ create a steady state field of uncompensated charge $\vec{E}_{ch}$ and a pressure force which act on the electrons in an opposite direction. However the effect of these forces, even in their combination, is weaker than that of $\vec{F}_{NL}^{(0)}$ because the ions react to the field $\vec{E}_{ch}$ and move in the direction of the p. f. The ion density becomes also inhomogeneous, and thus, a pressure force is the other force acting on them. Therefore the plasma as a whole become inhomogeneous when it is imposed in an inhomogeneous HF field. The result for the plasma density response is:

$$\delta n_0^{(\omega-\omega)} = -n_0\left(\left|\vec{\mathcal{E}}_\omega(\vec{r})\right|^2 / E_{n(s)}^2\right). \tag{13}$$

Here $E_{n(s)} = (\omega/e)\sqrt{4mT_e}$ is the effective field in the case of a nonlinearity caused by the p. f. It has the meaning of a field intensity in which the velocity of the electron oscillations is equal to their thermal velocity. The plasma response (13) is local causing density redistribution in the wave field: the density decrease with the increase of $|\vec{\mathcal{E}}(\vec{r})|^2$.

In cold plasmas ($T_e = 0$), the p. f. action results in:

$$\delta n_0^{(\omega-\omega)} = -\left(\varepsilon_0/4m\omega^2\right)\vec{\nabla}^2\left|\vec{\mathcal{E}}_\omega(\vec{r})\right|^2. \tag{14}$$

4.2.2. Plasma Response at Second Harmonic Frequency: HF Second Order Response
Although going in parallel with the treatment of the p. f. effects in cold plasmas, the HF plasma response at SH frequency shows essential differences. Firstly, the NL force has two components in this case: in transverse and longitudinal directions. Also, besides of density perturbations at frequency 2ω, it causes perturbations of the electron velocity and creates a magnetic field at 2ω :

$$\delta n_{2\omega}^{(\omega+\omega)} = \frac{\varepsilon_0}{4m\omega^2}\vec{\nabla}^2\vec{\mathcal{E}}_\omega^2 , \quad \vec{u}_{2\omega} = \frac{ie^2}{8m^2\omega^3}\left(\vec{\nabla}\mathcal{E}_\omega^2\right), \quad \vec{\mathcal{B}}_{NL}^{(2\omega)} = \frac{ie}{8m\omega^3}\vec{\nabla}\times\left(\vec{\nabla}\mathcal{E}_\omega^2\right). \tag{15}$$

This means that the cascading to third order effects requires application once more of the NL force (8).

4.3. NONLINEAR RESPONSE AT THE FUNDAMENTAL FREQUENCY

The description of effects of self-action of SWs – SPM and waveguided channel formation – aimed further on, is based on (5) in which the displacement (3b), (9) at the frequency ω of the fundamental wave and the NL current density (10) are involved. With $\delta n_{L\omega} = 0$ in the case of SWs and $\vec{u}_0^{(\omega-\omega)} = 0$ in the case of LF plasma responses, (10) reduces to $\vec{j}(\vec{r})|_\omega = \vec{j}_L^{(\omega)} + \vec{j}_{NL}^{(\omega)}$ where $\vec{j}_L^{(\omega)} = -en_0\vec{u}_{L\omega}$ is the linear response and

$\vec{j}(\vec{r})\big|_{NL}^{(\omega)} = \vec{j}_{NL}\big|_{\omega}^{0+\omega} + \vec{j}_{NL}\big|_{\omega}^{2\omega-\omega}$, the NL response at the fundamental frequency, is composed by:

$$\vec{j}_{NL}\big|_{\omega}^{0+\omega} = -e\delta n_0^{(\omega-\omega)}\vec{u}_{L\omega} , \qquad \vec{j}_{NL}\big|_{\omega}^{2\omega-\omega} = -e\left[n_0\vec{u}_{(2\omega-\omega)} + (1/2)\delta n_{2\omega}\vec{u}_{L\omega}^*\right]. \quad (16a,b)$$

The latter are the responses realized through LF $(0+\omega)$ and HF $(2\omega-\omega)$ NL second order interactions, respectively.

In the cases of cascading through LF plasma response (ionisation nonlinearity (12) and p. f. effect in warm plasmas (13)), $\delta n_0^{(\omega-\omega)}$ is directly related to $|\mathcal{E}_\omega(\vec{r})|^2$:

$$\delta n\big|_0^{(\omega-\omega)}\big/n_0 = \pm\left|\mathcal{E}_\omega(\vec{r})\right|^2\big/E_{n(i,s)}^2 . \quad (17)$$

The NL plasma permittivity can be easily obtained from (17) and (9), and the result for the displacement comes out in its form (3b):

$$\vec{D}_\omega(\vec{r}) = \varepsilon_0\varepsilon_{NL}\vec{\mathcal{E}}_\omega. \quad (18)$$

The NL plasma permittivity $\varepsilon_{NL} = \varepsilon_L + \delta\varepsilon_{NL}$ is composed by its linear part (4) and a NL contribution:

$$\delta\varepsilon\big|_{NL(i,s)}^{(0:\omega-\omega)} = -\frac{\omega_{p0}^2}{\omega^2}\frac{\delta n_0^{(\omega-\omega)}}{n_0} = \mp\frac{\omega_{p0}^2}{\omega^2}\frac{|\mathcal{E}_\omega(\vec{r})|^2}{E_{n(i,s)}^2} . \quad (19)$$

Cases of ionisation nonlinearity and p. f. effects are related respectively to upper and lower signs in (17), (19). Expression (19) directly shows cubic type of nonlinearity formed in cascading through $(\omega-\omega)$–interaction. In the field of SWs [20-22] the NL contribution to the plasma permittivity is:

$$\delta\varepsilon_{NL}^{(0)} = \pm\frac{(\varepsilon_L-1)^2}{\varepsilon_L}e^{-2\kappa_p x}\frac{|A_\omega|^2}{E_{n(i,s)}^2} . \quad (20)$$

In the case of cascading through LF (14) and HF (15) second order nonlinearity in cold plasmas a result of the type of (17), i.e. an expression for a cubic type of nonlinearity, comes out after specification of the wave field configuration. Since in the general case the relations of $\delta n_0^{(\omega-\omega)}$ (14) and $\delta n_{2\omega}^{(\omega+\omega)}$ (15) to the field intensity are in an operator form, it is more convenient to use expressions (3a), (9) for the displacement. Therefore,

$$\vec{D}(\vec{r}) = \vec{D}_{L\omega} + \delta\vec{D}_{NL} \quad (21)$$

with a first term $\vec{D}_{L\omega} = \varepsilon_0\varepsilon_{L\omega}\vec{\mathcal{E}}_\omega$ being the displacement in linear approximation and a secont one $\delta\vec{D}_{NL} = (i/\omega)\vec{j}_{NL}^{(\omega)}$ giving the NL contribution. In the case of cascading through $(\omega-\omega)$ -NL interaction associated with p. f. effects, expessions (14), (16a) result in:

$$\delta\vec{\mathcal{D}}_{NL}\Big|_\omega^{0+\omega} = \left(e^2\varepsilon_0/4m^2\omega^4\right)\left(\vec{\nabla}^2\big|\vec{\mathcal{E}}_\omega(\vec{r})\big|^2\right)\vec{\mathcal{E}}_\omega. \tag{22}$$

In order to have the corresponding result for the NL contribution to the displacement in the case of cascading through HF plasma response at SH frequency completed from expression (16b), an expression for $\vec{\upsilon}_{(2\omega-\omega)}$ should be obtained. Equation (6b) written at frequency ω involves a NL force $\vec{F}_{NL}^{(\omega=2\omega-\omega)} = (1/2)\left\{\vec{\mathcal{F}}_{NL}^{(\omega=2\omega-\omega)}e^{-i\omega t} + c.c.\right\}$ with $\vec{\mathcal{F}}_{NL}^{(\omega)} = -(1/2)\left\{m\left[\left(\vec{\upsilon}_L^*\cdot\vec{\nabla}\right)\vec{\upsilon}_{2\omega} + \left(\vec{\upsilon}_{2\omega}\cdot\vec{\nabla}\right)\vec{\upsilon}_L^*\right] + e\left(\vec{\upsilon}_L^*\times\vec{\mathcal{B}}_{2\omega} + \vec{\upsilon}_{2\omega}\times\vec{\mathcal{B}}_L^*\right)\right\}$. With the use of (15), the latter transforms into $\vec{\mathcal{F}}_{NL}^{(\omega)} = \left(e^3/16m^2\omega^4\right)\vec{\nabla}\left[\vec{\mathcal{E}}_\omega^*\cdot\left(\vec{\nabla}\vec{\mathcal{E}}_\omega^2\right)\right]$ and gives the NL contribution $\vec{\upsilon}_{NL}^{(\omega)} = i\left(e^3/16m^3\omega^5\right)\vec{\nabla}\left[\vec{\mathcal{E}}_\omega^*\cdot\left(\vec{\nabla}\vec{\mathcal{E}}_\omega^2\right)\right]$ to the electron velocity at frequency ω obtained through $(2\omega-\omega)$–interaction. Finally, the result for the NL contribution to the displacement (21) in the case of cascading through HF plasma response at SH frequency obtained from (15) and (16b) is:

$$\delta\vec{\mathcal{D}}_{NL}\Big|_\omega^{2\omega-\omega} = \left(e^2\varepsilon_0/16m^2\omega^4\right)\left\{\left(\omega_{p0}^2/\omega^2\right)\vec{\nabla}\left[\vec{\mathcal{E}}_\omega^*\cdot\left(\vec{\nabla}\vec{\mathcal{E}}_\omega^2\right)\right] + 2\left(\vec{\nabla}^2\vec{\mathcal{E}}_\omega^2\right)\vec{\mathcal{E}}_\omega^*\right\}. \tag{23}$$

The first term in (23) stems from the first term in (16b) which includes $\vec{\upsilon}_{2\omega-\omega}$ and the second one correspond to the second term which contains $\delta n_{2\omega}$.

5. Self-Action of Surface Waves

The effects of modulation instability/envelope solitons and of waveguided channel formation at SW propagation are solutions of equation (5) in which the displacement can be presented either by (18), i.e. through the NL plasma permittivity (e.g., (19), (20)), or by (21), i.e. through the NL current density (e.g. (22), (23)) involved in $\delta\vec{\mathcal{D}}_{NL}$. The presentation here starts with NL permittivity involved in treatment of self-action which passes through LF plasma response (Figure 1). Then on the basis of NL current density, third order effects resulting from the two channels of interactions (LF and HF second order NL responses) are given.

5.1. CASCADING THROUGH LOW-FREQUENCY PLASMA RESPONSE

The cases of ionisation nonlinearity and p. f. effects in warm plasmas which results in presentation of the displacement in the form (18) with a NL plasma permittivity involved in it, lead to the following form of the NL wave equation (5):

$$\vec{\nabla}^2\vec{\mathcal{E}}_\omega - \vec{\nabla}\left(\vec{\nabla}\cdot\vec{\mathcal{E}}_\omega\right) = -\left(\omega^2/c^2\right)\varepsilon_{NL}\vec{\mathcal{E}}_\omega. \tag{24}$$

Although having $\vec{\nabla}.\vec{\mathcal{E}}_\omega = 0$ for SWs in linear approximation, in the NL wave equation the term $\left(\vec{\nabla}.\vec{\mathcal{E}}_\omega\right)$ is not equal to zero and it appears as a second NL term in addition to the NL part of the second time derivative of the displacement (the term in the right hand side of (24)). The $\left(\vec{\nabla}.\vec{\mathcal{E}}_\omega\right)$–term in the NL equation (24) is associated with inhomogeneity of the stationary density which is nonlinearly induced – by the inhomo-geneity of the wave field – through the mechanism of the ionisation nonlinearity and of the p. f. effects (subsections 4.1. and 4.2.1).

Introduction of (18) in equation $\vec{\nabla}.\vec{\mathcal{D}}_\omega = 0$ (see the set (2)), leads directly to $\vec{\nabla}.\vec{\mathcal{E}}_\omega = -\left(1/\varepsilon_L\right)\vec{\nabla}.\left(\delta\varepsilon_{NL}\vec{\mathcal{E}}_\omega^L\right)$, and even in the case of waves for which $\vec{\nabla}.\vec{\mathcal{E}}_\omega^L = 0$, one has:

$$\vec{\nabla}.\vec{\mathcal{E}}_\omega = -\left(1/\varepsilon_L\right)\vec{\mathcal{E}}_\omega^L.\vec{\nabla}\delta\varepsilon_{NL}. \tag{25}$$

In a way, the spatial inhomogeneity of the density, self-consistently introduced by the field through the nonlinearity, gives a NL term in (5) which is related to the spatial rate of changes of $\delta\varepsilon_{NL}$.

Therefore, in the case of weakly NL processes, the wave equation reduces to:

$$\vec{\nabla}^2\vec{\mathcal{E}}_\omega +\left(\omega^2/c^2\right)\varepsilon_L\vec{\mathcal{E}}_\omega = -\left(\omega^2/c^2\right)\delta\varepsilon_{NL}\vec{\mathcal{E}}_\omega - \varepsilon_L^{-1}\vec{\nabla}\left[\vec{\mathcal{E}}_\omega.\vec{\nabla}\delta\varepsilon_{NL}\right]. \tag{26}$$

The left hand side of (26) has the form of the equation which describes wave propagation in linear approximation. The two NL terms in the right hand side of (26) come out respectively from the time derivative of $\vec{\mathcal{D}}$ and from the $\left(\vec{\nabla}.\vec{\mathcal{E}}_\omega\right)$-term.

Applied to SWs, i.e. at space variation $\vec{\mathcal{E}}_\omega(\vec{r}) = \vec{E}_\omega(x)\exp\left(ik_\omega z\right)$ of the wave field, (26), reduces to the following equation:

$$\frac{d^2 E_{\omega z}}{dx^2} - \kappa_p^2 E_{\omega z} = -\frac{\omega^2}{c^2}\left(1+\frac{2}{1+\varepsilon_L}\right)E_{\omega z}^L\delta\varepsilon_{NL} \tag{27}$$

for the E_z-field component at $x > 0$. The quantity κ_p^2 (introduced in section 3) includes the NL wavenumber k_ω^{NL}. The right hand side of (27) shows that with $\varepsilon_L < -1$ in the frequency range of the SWs, the two NL terms in it give contribu-tions which are opposite in signs. In the long wavelength range the first term coming from the time derivative of the displacement gives larger contribution. At $\varepsilon_L = -3$ the sum of the two NL terms is equal to zero and the nonlinearity vanishes. In the short wavelength range the second NL term in (27), i.e. that coming from the $\left(\vec{\nabla}.\vec{\mathcal{E}}_\omega\right)$-term, predominates. The solution of (27) is:

$$E_{\omega z}(x) = C_\omega e^{-\kappa_p x} \pm \frac{1}{8}(3+\varepsilon_L)\frac{(\varepsilon_L-1)^2}{\varepsilon_L^3}\frac{|A_\omega|^2}{E_{n(i,s)}^2}A_\omega e^{-3\kappa_{pL}x}. \tag{28}$$

Here $C_\omega \equiv E_z(x=0)$ is the constant in the solution for the NL waves. The partial solution of the inhomogeneous equation for the E_z-component in the plasma semi-space is equal to zero at $\omega/\omega_{p0} = 0.5$ (i.e. $\varepsilon_L = -3$) and changes its sign there. The solution for the H_y-component of the SWs has no peculiarities.

The evolution equation for the amplitude C_ω of the NL SW and the NL dispersion relation of the waves

$$\Delta_{NL}\left(\omega, k_\omega^{NL}, |A_\omega|^2\right)C_\omega = 0, \quad \Delta_{NL}\left(\omega, k_\omega^{NL}, |A_\omega|^2\right) = 0 \tag{29a,b}$$

results from the boundary conditons $\{E_z\}_{|x=0} = 0$, $\{H_y\}_{|x=0} = 0$ at the plasma-vacuum interface. In $\Delta_{NL}\left(\omega, k_\omega^{NL}, |A_\omega|^2\right) \equiv \Delta_L\left(\omega, k_\omega^{NL}\right) + \delta\Delta_{NL}\left(\omega, k_\omega^L, |A_\omega|^2\right)$, $\Delta_L\left(\omega, k_\omega^{NL}\right) = \left(\varepsilon_L/\kappa_{pL}\right) + \left(1/\kappa_{vL}\right)$ has the form of the linear dispersion relation with k_ω^{NL} included in it, and $\delta\Delta_{NL}\left(\omega, k_\omega^L, |A_\omega|^2\right) = \pm\left[(\varepsilon_L-1)^3/4\kappa_{pL}\varepsilon_L^2\right]\left(|A_\omega|^2/E_n^2\right)$ involves directly the field amplitude $|A_\omega|^2$. The zero order approach of (29b) is the dispersion relation of the linear SWs and its first order approach gives the NL contribution to the wavenumber

$$\delta k_\omega^{NL} = \mp\frac{\kappa_{pL}^2(\varepsilon_L-1)^2}{4k_L\varepsilon_L^3(\varepsilon_L+1)}\frac{|A_\omega|^2}{E_{n(i,s)}^2} \tag{30}$$

shown in Figure 4(a). Therefore cascading through LF NL plasma response could leads either to positive or negative NL wavenumber shift. The ionisation nonlinearity causes a decrease of the wavenumber whereas the p. f. effects leads to its increase.

5.1.1. Envelope Solitons of Surface Waves

The nonlinear Schrödinger equation which describes the envelope solitons results directly [30] from a two term Taylor expansion of the evolution equation (29a) with replacements made in it by using the linear dispersion relation and the wave energy concervation law. However conclusions about the type of the soliton solutions can be made also directly from the results (Figure 4(b)) for the dispersion behaviour of the SWs. Having in mind the negative group-velocity dispersion ($\partial v_{gr}/\partial k_L < 0$) of the SWs obtained from the linear dispersion law (Figure 4(a)), the Ligthhill's criterion $\delta\omega_{NL}(\partial v_{gr}/\partial k_L) < 0$ for modulation instability [31] shows that a NL response through ionisation nonlinearity leads to bright solitons whereas a nonlinearity associated with the p. f. effect takes care for formation of dark solitons ($\delta\omega_{NL}$ here is the NL frequency shift).

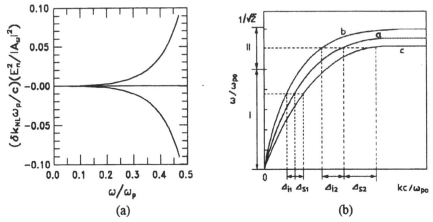

Figure 4.(a) Frequency dependence of the NL contribution to the wavenumber; $\delta k_{NL} < 0$: ionisation nonlinearity, $\delta k_{NL} > 0$: ponderomotive force effects. *(b)* Dispersion behaviour of SWs; a: linear regime, b: NL regime at ionisation nonlinearity and c: nonlinearity due to ponderomotive force effects (Figure 4(1) from [22]).

5.1.2 Waveguided Channel Formation at Surface Wave Propagation

Now attention is paid to the influence of the nonlinearity on the transverse distribution of the SW field [22]. Separation of the NL contributions from the linear terms ($k_\omega^{NL} = k_\omega^L + \delta k_{NL}$, $\vec{E}_\omega = \vec{E}_\omega^L + \delta \vec{E}_\omega^{NL}$, $\vec{H}_\omega = \vec{H}_\omega^L + \delta \vec{H}_\omega^{NL}$) transforms, e.g., (27) in the following equation for $\delta E_{\omega z}$:

$$\frac{d^2 \delta E_{\omega z}}{dx^2} - \kappa_{pL}^2 \delta E_{\omega z} = -\frac{\omega^2}{c^2}\left(1 + \frac{2}{1 + \varepsilon_L}\right) E_{\omega z}^L \delta \varepsilon_{NL} + 2 k_L \delta k_{NL} E_{\omega z}^L . \tag{31}$$

The last term in (31) is also present in the corresponding equation for the wave field in the vacuum region ($x < 0$). Therefore a formation of a waveguided channel should be expected also there. The solutions for the longitudinal field component of the NL wave are:

$$E_{\omega z}(x) = E_{\omega z}^L(x) + \frac{1}{8} A_\omega \begin{Bmatrix} \mathbf{Z} \\ \mathbf{Z}_v \end{Bmatrix} \frac{|A_\omega|^2}{E_{n(i,s)}^2} \tag{32}$$

where $\mathbf{Z} = \pm Z$ and $\mathbf{Z}_v = \pm Z_v$ describe the field distribution in transverse (x-) direction respectively in the plasma ($x > 0$) and vacuum ($x < 0$) semi-spaces (Figure 5).

As it is known, ionisation nolinearity in the field of a TM mode is associated with pushing of the wave energy outside the region of the higher density plasma whereas the p. f. effects lead to self-focussing. Therefore in the case of SWs (Figure 6) it should be expected that conditions of ionisation nonlinearity should move the waveguide channel to the vacuum region and conditions of p. f. effects should move it to the plasma region.

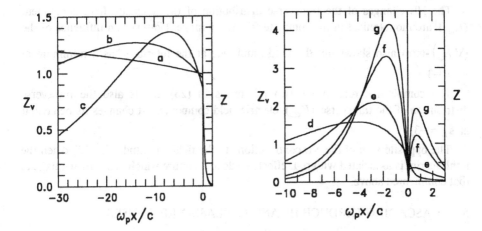

Figure 5. Transverse structure of the NL contribution to the $E_{z\omega}$ –field component in the case of ionisation nonlinearity: ω/ω_p = 0.1(a), 0.2(b), 0.3(c), 0.4(d), 0.5(e), 0.577(f) and 0.6(g).

The results in Figure 5 where the case of ionisation nonlinearity is presented, are in agreement with such conclusions: the NL contribution to the E_z -component of the SW field in the case of ionisation nonlinearity is larger in the vacuum semi-space and the maximum which shows up there forms a NL channel in this space region. However the influence of the $(\nabla.E)$ -term in the wave equation makes the picture more complicate. The NL contribution $\delta E_z^{NL}(x=0)$ at the interface has a minimum at $\varepsilon = -2$ and this is associated with the competitive role of the two NL terms in the wave equation (26). In the frequency range $\omega/\omega_p < 0.577$ in which the NL term with the time derivative of the displacement in (26) is larger, the field in the plasma region decays monotonically. The maximum of δE_z^{NL} in the vacuum region shift with the frequency increase towards the interface. In the frequency range $\omega/\omega_p > 0.577$ where the $(\nabla.E)$ -NL term in (26) predominates, the variation of δE_z^{NL} in the plasma region forms a hump also there. The maximum in the plasma region moves – with increasing ω – at first away and than towards the interface.

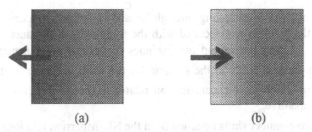

Figure 6. Illustration for the waveguided channeling of the energy of the SWs (a TM mode) at ionisation nonlinearity (a) and ponderomotive force effects (b).

The NL changes of the transverse distribution of the magnetic field component $(H_{\omega y})$ are monotonical in the complete frequency range. However an influence of the $\left(\nabla.\vec{E}_{\omega}\right)$-term also shows up: the field amplitude at the interface has a minimum at $\varepsilon_L = -3$.

The competitive role of the two NL terms in (26) affects also the transverse distribution of the transverse $(E_{x\omega})$ electric field component. It changes its behaviour at $\varepsilon_L = -3$.

The opposite sign of the NL contributions to the field amplitudes in (32) when the nonlinearity is associated with p. f. effects leads to a picture which is a mirror-image to that commented above.

5.2. CASCADING THROUGH HF AND LF PLASMA RESPONSES

Presentation of the displacement in the form (21) in which the NL contribution to it is given by expressions (22) and (23) respectively in the cases of cascading in cold plasmas through LF NL response due to p. f. effects and through HF response at SH frequency reduces (5) to:

$$\vec{\nabla}^2\vec{\mathcal{E}}_{\omega} + \frac{\omega^2}{c^2}\varepsilon_{L\omega}\vec{\mathcal{E}}_{\omega} = -\frac{\omega^2}{\varepsilon_0 c^2}\delta\vec{\mathcal{D}}_{NL} - \frac{1}{\varepsilon_0\varepsilon_L}\vec{\nabla}\left(\vec{\nabla}.\delta\vec{\mathcal{D}}_{NL}\right). \qquad (33)$$

The last term in (33) stems directly from $\vec{\nabla}.\vec{\mathcal{D}} = 0$. Applied to SW field, (33) gives an equation in the form of (27). The left hand side of the equation is the same. Its right hand side is $Q|A_{\omega}^2|A_{\omega}\exp\left(-3\kappa_{pL}x\right)$ with:

$$Q\Big|_{\omega}^{|0+\omega} = \frac{e^2\kappa_{pL}^4}{m^2\omega^4\varepsilon_L}\left(1+\frac{k_L^2}{\kappa_{pL}^2}\right)\left(1-\frac{3k_L^2}{\kappa_{pL}^2}\right), \qquad (34)$$

$$Q\Big|_{\omega}^{|2\omega-\omega} = \frac{e^2\kappa_{pL}^4}{2m^2\omega^4\varepsilon_L}\left(1-\frac{k_L^2}{\kappa_{pL}^2}\right)\left[4\frac{\omega_{p0}^2}{\omega^2}\frac{k_L^2}{\kappa_{pL}^2}+\left(1-\frac{k_L^2}{\kappa_{pL}^2}\right)\left(1+\frac{3k_L^2}{\kappa_{pL}^2}\right)\right] \qquad (35)$$

respectively, in cases of cascading through LF and HF plasma responses. In (35), the first term in the brackets is associated with the third order disturbance of the electron velocity $(v_{2\omega-\omega})$ and the second one includes the second order density disturbance $\delta n_{2\omega}$ (see, expression (23)). In the solution for the wave magnetic field component a term appers in which the NL contribution related to these two terms are opposite in signs and competitive.

The NL wavenumber shifts obtained from the NL dispersion relations, are:

$$\delta k_{NL}^{(0+\omega)} = \frac{\kappa_{pL}^2}{3k_L} \frac{1-\varepsilon_L}{(1+\varepsilon_L)^2} \left[1 - \frac{1}{4}\frac{\varepsilon_L+3}{\varepsilon_L}\right] \frac{V^2}{c^2} \tag{36}$$

$$\delta k_{NL}^{(2\omega-\omega)} = \frac{\kappa_{pL}^2}{4k_L(1+\varepsilon_L)} \left[\frac{\varepsilon_L-1}{\varepsilon_L} - \frac{(\varepsilon_L+1)^2}{2\varepsilon_L^2}\right] \frac{V^2}{c^2} \tag{37}$$

respectively in the cases of cascading through LF and HF plasma responses: $(\omega - \omega) + \omega \Rightarrow (0+\omega)$ and $(\omega + \omega) - \omega \Rightarrow (2\omega - \omega)$ –interactions. The positive wave-number shift (Figure 7(a)) obtained from NL effects associated with the p. f. action shows that this NL mechanism is responsible for dark soliton formaion. Figure 7(b) shows that a nonlinearity due to a HF plasma response leads to formation of bright solitons. The term with the third order velocity disturbance $(v_{2\omega-\omega})$ earns in the competition with the term which contains the second order density disturbance. In the two cases of cascading in cold plasmas - through HF and LF plasma response - the ration v_E / v_{ph} (where $v_E = e|A_\omega|/m\omega$ is the velocity of the electron oscillations in the field and $v_{ph} = \omega / k$ is the phase velocity) is the small parameter of the interaction. Involvment of V/c (with $V = e|A_\omega|/ m\omega_p$) as a small parameter $(V/c \sim 10^{-4})$ is convenient for presentation of the frequency dependence of the NL wavenumber shift.

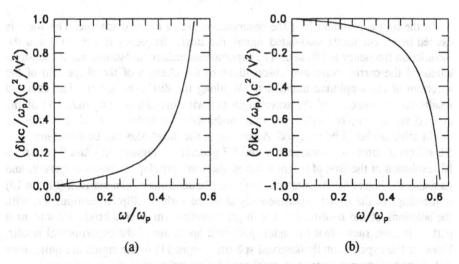

(a)   (b)

Figure 7. Frequency dependence of the NL wavenumber shift obtained in cascading through LF (a) and HF (b) plasma responses.

5.3. EXPERIMENTAL EVIDENCE

Experimental provement of effects resulting from self-action of SWs in gas-discharge plasmas has been demonstrated in two experiments. In general, the obtained results show development of modulation instability of SWs and tendency to bright soliton formation at their propagation.

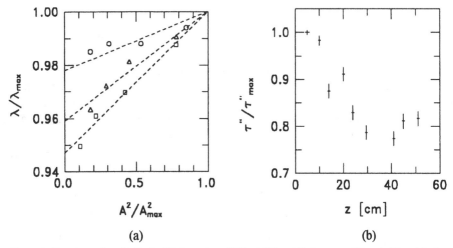

(a)   (b)

Figure 8. Experimental results for the NL dispersion of SWs at different discharge conditions in (a) and for the changes of the width of the envelope with increasing distance from the wave launcher in (b). (Figures 6 and 7 from Ref. [24]).

In the first experiment [24] the observations are in a d.c. discharge. The wave is excited by an amplitude modulated signal: the carrier frequency is 126 MHz and the modulation frequency is 100 kHz. The experiments include measurements of the wavelength of the carrier wave and observations of the changes of the shape and of the spectrum of the amplitude modulated SW along the discharge length. The obtained results for an increase of the wavelength (λ) with increasing amplitude (A) of the applied signal (Figure 8(a)) show that modulation instability and NL evolution to bright solitons should be expected. According to the theory this can be associated with NL effects of ionisation nonlinearity and SH generation. Figures 8(b) and 9 confirm a NL evolution of the type of bright solitons: the wave envelope narrows (Figure 9) and its width τ'' (Figure 8(b)) decreases along the plasma column. Comparison (Figure 10) of the shape of the signal experimentally observed with dn-elliptic funstions, i.e. with the solutions of the nonlinear Schrödinger equation which give bright solitons as a particular case, shows that the latter quite well approximate the experimental results. However the experimentally observed spectra (Figure 11) of the signal are quite more rich than those corresponding to the shape of the amplitude modulated signals (Figure 10). This is explained (Figure 12) by involving a model of phase modulation in addition to the effects of the amplitude modulation.

A second experiment [25,26], which also shows observations of a development of modulation instability of SWs, has been recently performed in pulsed discharges

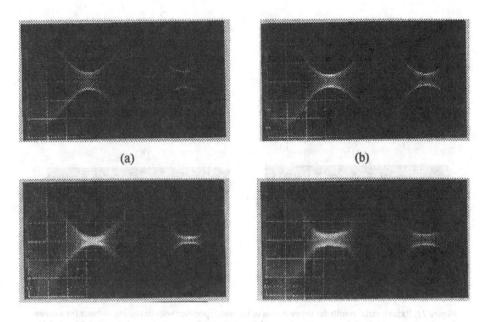

Figure 9. Evolution of the shape of the wave envelope with increasing distance (from (a) to (d)) from the launcher (Figure 7 from Ref. [24]).

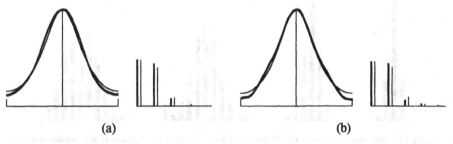

(a)   (b)

Figure 10. Experimentally observed (thick curves) shapes of the wave envelope and corresponsing spectra with theoretical dn-function (thin curves) and their spectra. The parameter of the dn-function is a fitting parameter (Figure 10 from Ref. [24]).

produced by high-amplitude ionising SWs. The frequency of the carrier wave is 2.45 GHz. The development of the shapes along the discharge length of the microwave pulses which produce the discharge and of the total light emission of the discharge which is an indication about the plasma density has been observed. The modulation instability is observed on the stationary level of the pulses. Its development splits the total pulses of the wave energy and of the plasma density to consequences of pulses of simultaneous propagation of charged particle bunches and wave energy packets. The experimental results are explained in terms of self-action of SWs under the conditions of weak ionisation nonlinearity ($(\omega - \omega) + \omega$ -interaction).

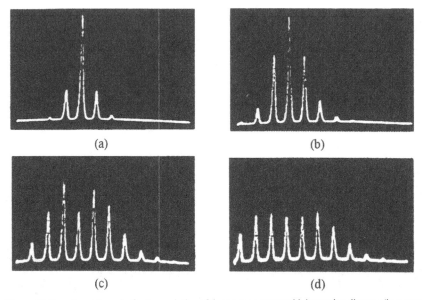

Figure 11. Experimental results for the evolution of the wave spectrum with increasing distance (in a consequence from (a) to (d)) from the wave launcher at high amplitude of the applied signal (Figure 8 from Ref. [24]).

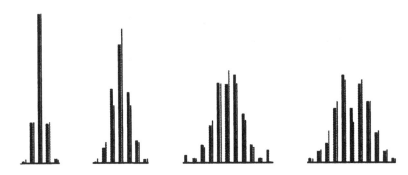

Figure 12. Experimentally observed (thick lines) and theoretical (thin lines) spectra at different distance (in a consequence from left to right) away from the wave launcher (Figure 11 from Ref. [24]).

6. Second Order Nonlinear Response at Second Harmonic Frequency

In this section, theoretical and experimental results on the propagation of SWs at SH frequency are briefly discussed. The theoretical results are obtained in terms of two models: the coupled mode model and the interference model.

In the frameworks of the coupled mode model, the consideration starts with an assumption for propagation of two SWs (at frequencies ω and 2ω). With the NL force (8) in the equation of motion (6b) as a source of the nonlinearity, the plasma response at SH frequency, presented in the form (9) which involved the NL curernt density at frequency 2ω $(\vec{j}_{2\omega} = -en_0\vec{u}_{2\omega})$, is:

$$\vec{\mathcal{D}}_{2\omega} = \varepsilon_0 \left[\varepsilon_{2\omega}^L \vec{\mathcal{E}}_{2\omega} + \frac{\omega_p^2}{2\omega(2\omega + i\nu)} \frac{\vec{\mathcal{F}}_{NL}^{(2\omega)}}{e} \right] \tag{38}$$

where $\vec{\mathcal{F}}_{NL}^{(2\omega)} = \dfrac{e^2(\omega - i\nu)}{4m\omega^3} \vec{\nabla}\vec{\mathcal{E}}_\omega^2 - \dfrac{ie^2\nu}{m\omega^3}(\vec{\mathcal{E}}_\omega \cdot \vec{\nabla})\vec{\mathcal{E}}_\omega$ and $\varepsilon_{2\omega} = 1 - \omega_p^2/2\omega(2\omega + i\nu)$. The wave equation (5) written for the wave at 2ω reduces to the following equation for the $E_{2\omega z}$ -field component in the plasma semi-space ($x > 0$):

$$\frac{\partial^2 \mathcal{E}_{2\omega z}}{\partial x^2} + \frac{\partial^2 \mathcal{E}_{2\omega z}}{\partial z^2} + \frac{(2\omega)^2}{c^2} \varepsilon_{2\omega} \mathcal{E}_{2\omega z} = G_{NL} A_\omega^2 e^{-2\kappa_{p0}x} e^{2ik_\omega z} \tag{39}$$

where $G_{NL} = -\dfrac{\omega_p^2}{2\omega(2\omega + i\nu)e} \left\{ \dfrac{1}{\varepsilon_{2\omega}} \dfrac{\partial}{\partial z}(\vec{\nabla} \cdot \vec{\mathcal{F}}_{NL}^{2\omega}) + \dfrac{(2\omega)^2}{c^2} \mathcal{F}_{NLz}^{2\omega} \right\}$. In the case of SWs at SH

frequency, the contribution of the NL $\left(\nabla.\vec{\mathcal{E}}_\omega\right)$ -term to the wave equation (5) being opposite in sign to the NL part of the time derivative of the displacement earns in the competition with it. The solution of (39) and the corresponding solution for the wave magnetic field introduced in the boundary conditions leads to an equation of the type of $\Delta_{2\omega} A_{2\omega} = -i\Lambda A_\omega^2 \exp(-i\Delta kz)$. After a Taylor expansion applied, the evolution equation for the amplitude of the wave at SH frequency can be easily obtained:

$$\frac{dA_{2\omega}}{dz} - \alpha_{2\omega} A_{2\omega} = \lambda A_\omega^2 \exp(-i\Delta kz). \tag{40}$$

Here $\Delta_{2\omega} = (\varepsilon_{2\omega}/\kappa_{p2\omega}) + (\varepsilon_d/\kappa_{d2\omega})$, $\lambda = \Lambda/(\partial\Delta_{2\omega}/\partial\beta_{2\omega})$, $\Delta k = k_{2\omega} - 2k_\omega$ is the linear wavenumber mismatch, $k_{2\omega} = \beta_{2\omega} + i\alpha_{2\omega}$ and

$$\lambda = -\frac{ek_\omega(\omega - i\nu)}{2m\omega^3}\left(1 - \frac{k_\omega^2}{\kappa_{p\omega}^2}\right) \frac{(1 - \varepsilon_{2\omega})}{2\kappa_{p\omega}\beta_{2\omega}\left(\dfrac{\varepsilon_{2\omega}}{\kappa_{p2\omega}^3} + \dfrac{\varepsilon_d}{\kappa_{d2\omega}^3}\right)} \left\{ 3\frac{\omega_p^2}{c^2\kappa_{p2\omega}^2}\frac{1}{\left(\dfrac{2\kappa_{p\omega}}{\kappa_{p2\omega}} + 1\right)} + 1 \right\} \tag{41}$$

is the coupling coefficient. At $\nu = 0$, $\alpha_{2\omega} = 0$, (40) has the well-know [32] solution:

$$A_{2\omega} = \frac{2\lambda}{\Delta\beta} A_\omega^2 \sin\left(\frac{\Delta\beta z}{2}\right) \exp\left(-i\frac{\Delta\beta z}{2}\right) \tag{42}$$

which describes the general behaviour of energy transfer from the fundamental wave to the SH one in the case when the SH generation is a nonresonant process. The case of SWs in semi-bounded plasmas is specified by the coupling coefficient (41). Its variation with ω determines the changes of the coherence length z_{coh} and the maximum amplitude $A_{2\omega}$ of the wave at 2ω with the frequency of the fundamental wave (Figure 13).

The other model with which SW propagation at SH frequency has been treated in [9,10], is known as an interference model. Although being simple, it gives more transparent picture of the wave generation and propagation. A wave at frequency ω

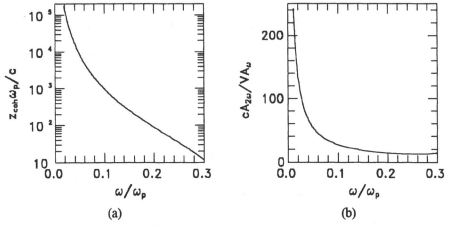

Figure 13. Variation with the frequency of the fundamental wave of the coherence length (a) and the amplitude (b) of SWs generated at 2ω.

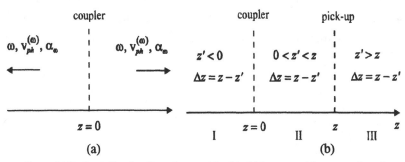

Figure 14. Regions of local exciters of waves at 2ω (b) which appears at the propagation of a wave externally (a) excited at ω (Figures 5 and 6 from Ref. [10]).

(with propagation characteristics: phase velocity $v_{ph}^{(\omega)}$ and space damping rate α_ω) is excited by a signal externally applied at $z = 0$. The signal at given z' position is

$$a_\omega(z',t) = A_\omega(z=0)\exp\left(\mp\alpha_\omega z'\right)\cos\left\{\omega\left[t \mp \left(z'/v_{ph}^{(\omega)}\right)\right]\right\}.$$ Because of the nonlinearity of the medium, each point z' is an exciter of a signal of frequency 2ω:

$$a_{2\omega}(z',t) \propto a_\omega^2(z',t) \equiv \left(A_\omega^2/2\right)\exp\left(\mp 2\alpha_\omega z'\right)\cos\left[2\beta_\omega z' \mp 2\omega t\right].$$ Since in the case of SWs the SH generation is a nonresonant process, the waves at frequency 2ω propagate according to the linear dispersion law with a phase velocity $v_{ph}^{(2\omega)} = 2\omega / k_{2\omega}$. The signal at frequency 2ω detected by a pick-up at given z-position is a result of the interference of the waves which are generated at each z'-position and reach the pick-up at a given moment t. The final result of the interfarence

341

$$E_{2\omega}(z,t) \propto M \exp(-\alpha_{2\omega}z)\sin(\beta_{2\omega}z - 2\omega t + \psi_1) - N\exp(-2\alpha_\omega z)\sin(2\beta_\omega z - 2\omega t + \psi_2) \quad (43)$$

where M, N and $\psi_{1,2}$ are amplitudes and phases, gives a propagation of a wave at frequency 2ω with characteristics - wavenumber $\beta_{2\omega}$ and $\alpha_{2\omega}$ - given by the linear dispersion law and a second signal at frequency 2ω which can be considered effectively as a wave with characteristics $2\beta_\omega$, $2\alpha_\omega$. The latter satisfy the synchronism conditions. Since for SWs $\alpha_{2\omega} > 2\alpha_\omega$, the two signals at frequency 2ω can be observed separately. The experimental results presented in terms of time-space diagrams in Figure 15 show such a behaviour. (The horizontal axis is the time, the vertical one is the distance away from the launcher whose position is at the upper side of each diagram and the vertical distance between two trace is equal to the wavelength.) The complete picture of the wave propagation at frequency 2ω as given by expression (43) is evident in Figure 15(2c): the first wave ($\beta_{2\omega}$, $\alpha_{2\omega}$) predominates in the region close to the launcher and after its damping the second one ($2\beta_\omega$, $2\alpha_\omega$) becomes pronounced. When the fundamental frequency is lower and the damping of the first wave at 2ω is weaker its prodominant propagation is over the whole column. When the fundamental frequency is higher and the first wave at 2ω is strongly damped, the second one covers the complete length of the plasma column.

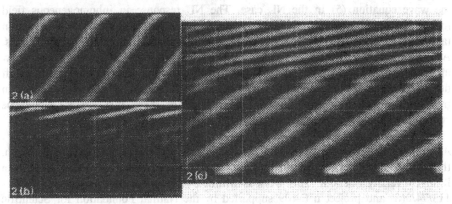

Figure 15. Time-space diagrams of: (2a) a wave excited directly at the fundamental frequency 80 MHz. (2b) a wave excited directly at double frequency (160 MHz), and (2c) SH waves generated by a fundamental wave. (Figure 1 from Ref. [10]).

The comparison of the final results (42) and (43) of the coupled mode model and the interference model shows that they are identical. With the contribution of region II in Figure 14(b) as the only one taken into account, one has $M \equiv N = 1/\Delta\beta$ and $\psi_1 \equiv \psi_2 = \pi/4$, and (43) reduces to:

$$A_{2\omega} \propto \left(A_\omega^2/\Delta\beta\right)\sin\left(\frac{\Delta\beta z}{2}\right)\cos\left(\frac{\beta_{2\omega} + 2\beta_\omega}{2}z - 2\omega t + \frac{\pi}{4}\right). \quad (44)$$

Expression (42) can be transformed to the same expression (44). However it gives in addition to the result of the interference model also the value 2λ of the coefficient of proportionality in (43).

7. Conclusions

The results presented here for effect of self-action of SWs are in gas-discharge plasmas. However the same procedure can be applied also to solid state waveguide structures of semiconductor materials whose treatment is analogous to that of weakly collisional $(\omega \gg \nu)$ gaseous plasmas. Since in both cases a second order nonlinearity is the lowest order one, the processes of wave self-action go through cascading. The same force couses cascading via HF (at SH frequency) and LF (at zero frequency) responses. Therefore the two channels of cascading (Figure 1) should be considered simultaneously and the choice of one of them should be made after estimations for their efficiency. E.g., in plasma waveguides with gas-discharge plasmas, the ionisation nonlinearity is usually the most efficient NL mechanism, and therefore, the cascading to third order effects is through $(\omega - \omega) + \omega$ –interaction. In the discussions of the results on wave channel formation through cascading via LF response and on wave propagation at SH frequency it has been stressed on the influence of the $divE$ – term in the wave equation (5) in the NL case. The NL response in inhomogeneous field induces an inhomogeneity in the medium and the $divE$ – term in the NL wave equation takes care for accounting for the rate of the spatial changes of medium characterisrtics. As it has been shown recently [30,33-35], effects associated with the same term show up also in optical waveguides. It should be also stressed that the NL wavenumber shift determines both the NL evolution in longitudinal direction and the NL changes in transverse direction of the modal field. Therefore these two effects should be considered together in treatments of time and space solitons.

Acknowledgments. Discussions with Mr. D. Grozev are highly acknowledged. A. Sh. thanks the Alexander von Humboldt Foundation for the possibilities for a research in Ruhr-University Bochum. This work is within DFG co-operation project 436 BUL - 113/74/0(S) and project n°F-409 supported by the Natinal Foundation for Scientific Research in Bulgaria.

References

1. Alanakyan, Yu.P. (1967) To the theory of surface waves in plasma, *ZhTF* **37**, 817–821 [in Russian].
2. Boev, A. G. and Prokopov, A.V. (1975) On the nonlinear theory of electromagnetic surface waves in ionising plasmas, *ZhETF* **69**, 1208–1217 [in Russian].
3. Gaponov, A.V. and Miller, M.A. (1958) Potential wells for changed particles in a high-frequency electromagnetic field, *ZhETF* **34**, 242–243 [in Russian].
4. Al'pert, Ya.L., Gurevich, A.V., and Pitaevski, L.P. (1965) *Space Physics with Artificial Satelites*, Consultauts Bureau, New York.
5. Shivarova, A. and Dimitrov, N. (1984) Nonlinear surface waves in a plasma slab, in: *Proc. Int. Conf. on Plasma Phys.*, Lausanne, v.1, p.43.

6. Shivarova, A. and Dimitrov, N. (1984) Nonlinear anti-symmetric mode in a plasma slab, *Plasma Phys. Cont. Fusion* **27**, 219–224.
7. Shivarova, A. and Yu, M. (1986) Nonlinear TM azimuthally symmetric mode in layered structures, *Opt. Commun.* **57**, 171–174.
8. Ivanov, B.I. (1967) Nonlinear effects in plasma waveguides (dependence of the phase velocity on the amplitude) *ZhTF* **37**, 1233–1238.
9. Shivarova, A. and Stoychev, T. (1980) Harmonic surface wave propagation in plasma. I. Second order harmonic waves generated by one fundamental wave, *Plasma Phys.* **22**, 517–528.
10. Grozev, D., Shivarova, A., and Stoychev, T. (1981) Harmonic surface wave propagation in plasma. II. Second order harmonic waves generated by two fundamental waves, *Plasma Phys.* **23**, 1093–1105.
11. Lindgren, T., Larsson, J., and Stenflo, L. (1982) Three-wave interaction in plasmas with sharp boundaries, *Plasma Phys.* **24**, 1177–1182.
12. Kostov, N. and Shivarova, A. (1983) Coupled-mode equations for nonresonant interaction of high-frequency surface waves, *Plasma Phys.* **25**, 891–900.
13. Yu, M.Y. and Zhelyaskov, I. (1978) Solitary surface waves, *J. Plasma Phys.* **20**, 183–188.
14. Gradov, O.M. and Stenflo, L. (1982) Solitary surface waves, *Phys. Fluids* **25**, 983–984.
15. Grozev, D. and Shivarova, A. (1984) Nonlinear dispersion of surface waves in a plasma column, *J.Plasma Phys.* **31**, 177–191.
16. Vladimirov, S.V. and Tsitovich, V.N. (1985) Nonlinear interaction of surface waves in plasmas with electromagnetic effects taken into account, *Fizika Plasmy* **11**, 1458–1468 [in Russian].
17. Stenflo, L. and Gradov, O.M. (1986) Equations for solitary surface waves on a plasma cylinder, *IEEE Trans. Plasma Sci.* **14**, 554–556.
18. Grozev, D., Shivarova, A., and Boardman, A.D. (1987) Envelope solitons of surface waves in a plasma column, *J. Plasma Phys.* **38**, 427–437.
19. Vladimirov, S.V. and Tsitovich, V.N. (1988) Electronic nonlinearities and their influence on the interaction of surface waves in plasmas, *Fizika Plasmy* **14**, 593–599. [in Russian].
20. Shivarova, A. (1992) Nonlinear surface modes, in P.Halevi (ed.), *Spatial Dispersion in Solids and Plasmas*, Elsevier, Amsterdam, pp. 557–616.
21. Georgieva, M. and Shivarova, A. (1993) Non-linear behaviour of surface wave propagation in plasma waveguides, in C.M. Ferreira and M. Moisan (eds.), *Microwave Discharges: Fundamentals and Applications*, Plenum, New York, pp. 65–74.
22. Georgieva, M. and Shivarova, A. (1994) Self-action of surface waves in a planar plasma waveguide, *Phys. Scripta* **50**, 523–531.
23. Georgieva, M., Shivarova, A., and Urdev, I. (1994) Self-focusing of surface waves in a cylindrical plasma waveguide, *J. Plasma Phys.* **52**, 391–407.
24. Grozev, D., Shivarova, A., and Tanev, S. (1991) Experiments on the nonlinear evolution of surafec waves in a plasma waveguide, *J. Plasma Phys.* **45**, 297–322.
25. Grozev, D., Kirov, K., Makasheva, K., and Shivarova, A. (1997) Modulation instability in pulsed surface wave sustained discharges, *IEEE Trans. Plasma Sci.* **25**, 415–422.
26. Grozev, D., Kirov, K., and Shivarova, A. (1997) Pulsed waveguided discharges, in J. Marec (ed.), *Microwave Discharges: Fundamentals and Applications*, in press.
27. Vladimirov, S.V., Yu, M.Y., and Tsytovich, V.N. (1994) Recent advances in the theory of nonlinear surface waves, *Phys. Rep.* **241**, 1–63.
28. Stenflo, L. (1996) Theory of nonlinear plasma surface waves, *Phys. Scripta* **T63**, 59–62.
29. Landau, L.D. and Lifshits, E.M. (1982) *Electrodynamics of Continuous media*, Nauka, Moscow [in Russian].
30. Shivarova, A. and Tanev, S. (1994) Effective area of the LP_{01} mode in non-linear optical fibres, *Pure Appl. Opt.* **3**, 725–730.
31. Kadomstev, B.B. (1979) *Collective phenomena in plasmas*, Nauka, Moscow [in Russian].
32. Vinogradova, M.B., Rudenko, O.V., and Sukhorukov, A. P. (1979) *Theory of Waves*, Nauka, Moscow [in Russian].
33. Boardman, A.D., Popov, T., Shivarova, A., Tanev, S., and Zyapkov, D. (1992) Nonlinear dispersion coefficient of a cylindrical optical waveguide, *J. Modern Opt.* **39**, 1083–96.
34. Boardman, A.D., Shivarova, A., Tanev, S., and Zyapkov, D. (1995) Nonlinear coefficients and the effective area of cross-phase modulation coupling of LP_{01} optical fibre modes, *J. Modern Opt.* **42**, 2361–2371.
35. Boardman, A.D., Marinov, K., Pushkarov, D.I., and Shivarova, A. (1997) New type of optical spatial solitary waves due to nonlinearity induced diffraction, in preparation for submission.

FREQUENCY CONVERSION WITH SEMICONDUCTOR HETEROSTRUCTURES

V. BERGER

THOMSON CSF Laboratoire Central de Recherches
Domaine de Corbeville, 91400 ORSAY, FRANCE.

Abstract.

In this lecture different aspects of frequency conversion in semiconductor heterostructures are reviewed. Thanks to the very high degree of control of growth and technology of thin layers of semiconductors, both electronic wavefunction and optical mode properties can be tailored, through band gap engineering and refractive index engineering. These two aspects lead to the possibility of optimization of nonlinear susceptibilities on the one hand, and nonlinear phase matching on the other hand, which are the two most important parameters for nonlinear frequency conversion.

1. Introduction

Optical frequency conversion[1, 2] by second order nonlinear interaction is a way to obtain coherent light in various spectral regions. The frequency doubling (or second harmonic generation, SHG) process is used for instance to obtain green light from the very efficient near infrared YAG laser, or blue light from semiconductor laser diodes, whereas difference frequency generation (DFG) is the basic process for high power mid infrared sources such as optical parametric oscillators. The most famous nonlinear materials used for frequency conversion are KDP, KTP or LiNbO$_3$[2], but several other crystals can be used: AgGaSe$_2$, GaAs, synthetic materials as organic molecules or semiconductor quantum wells (QWs)... The simplest nonlinear frequency conversion process is SHG. In this case, under simple approximations as the non depletion of the pump or the assumption of plane waves, the second harmonic power scales typically as[3]:

A. D. Boardman et al. (eds.),
Advanced Photonics with Second-order Optically Nonlinear Processes, 345–374.
© *1999 Kluwer Academic Publishers. Printed in the Netherlands.*

$$P^{2\omega} \propto \frac{\omega^2 \left(\chi^{(2)}\right)^2 L^2}{n^3} (P^\omega)^2 \frac{\sin^2\left(\frac{\Delta k L}{2}\right)}{\left(\frac{\Delta k L}{2}\right)^2} \tag{1}$$

L is the interaction length of the non linear process (usually the length of the non linear crystal), P^ω the pump power at the frequency ω, and n the refractive index of the nonlinear medium. In equation (1), two other parameters are of paramount importance, among the different characteristics of nonlinear crystals: The first one is the nonlinear coefficient $\chi^{(2)}$, which reflects the strength of the nonlinear interaction, and is related to the degree of asymmetry of the electronic potential at the microscopic level, as we will see later. The second one is the possibility to match the phase velocities between the different frequencies[1]. Indeed, due to the optical dispersion in the nonlinear materials, different waves do not travel at the same velocity in the material. This results in a momentum mismatch $\Delta k = k^{2\omega} - 2k^\omega$ between the propagating harmonic wave and the non linear polarization, the latter being source of the harmonic wave. After a distance $L_{coh} = \frac{\pi}{\Delta k}$ called 'coherence length', the nonlinear polarization and the generated wave acquire a phase lag of π. This results in a destructive interference, and in small SHG efficiencies for $L \gg L_{coh}$, as it appears in Eq. (1) with the $\frac{\sin x}{x}$ function. High SHG effeciencies require small Δk processes, that is to compensate for the dispersion. Reaching or approaching $\Delta k = 0$ is called "phase matching". This is possible for example by means of waves having different polarizations in birefringent crystals. It appears that the phase matching condition $\Delta k < \frac{\pi}{L}$ is more severe when the crystal length increases. This is because the phase matching condition expresses the photon momentum conservation, in the reciprocal space, and this conservation is required with an accuracy inversely proportional to the interaction length, in accordance with the Heisenberg principle. Other parameters can have a strong importance in the choice of a nonlinear crystal, as the refractive index (with a cubic dependence) or the damage threshold for high power applications. However, in this lecture, we will focus only on $\chi^{(2)}$ and Δk because they can be changed by orders of magnitude between two different nonlinear crystals ($\chi^{(2)}$) or two different temperatures or angles (Δk). They are then the two most crucial basic requirements for efficient frequency conversion. Furthermore, they correspond to the two basic requirements for high power coherent electromagnetic emission in general (for instance by an array of microwave antennas): High power emission of individual emitters, and phase coherence between all the emitters, according to the Huyghens-Fresnel principle.

Semiconductors (and especially GaAs) are very interesting materials for nonlinear optics, because the high degree of control of the technology of this material widely used for optoelectronics gives the opportunity to create artificial structures in which these two key features (the nonlinear susceptibility and the phase matching) can be controlled. Thanks to the progress of molecular beam epitaxy or metal organic chemical vapor deposition, different materials such as GaAs, AlGaAs or InGaAs, GaSb or AlGaSb, and a great number of related alloys, can be grown in very thin adjacent layers, with a high control of the interface between the different materials up to the atomic layer. Due to the different band gaps of these different semiconductors, electrons and holes in these heterostructures experience artificial potentials. At the interface between the two materials indeed, the difference between the band gaps of the two materials is shared out into two discontinuities, one for the valence band and one for the conduction band. For instance, a thin GaAs layer (let us say 100 Å typically) between AlGaAs layers results in a QW potential, with associated energy quantization (see figure 1). The study of this kind of semiconductor heterostructures has been in the 30 last years one of the main fields in applied and fundamental solid state physics. This has led to major breakthroughs in applied physics as various types of QW lasers, or high electron mobility transistors. Basics of this field can be found in [4, 5, 6]. Through an optimization of the widths and compositions of the different layers, it is possible to engineer the energy levels of electrons in these artificial potentials in order to get the desired electronic properties. This has been called band gap engineering or quantum design[7, 8]. The energy between different quantized levels can be tuned and also the position of the levels with respect to the barrier potential. This has been used for example in QW infrared photodetectors[9]. The dipole matrix elements describing the strength of the interaction of the material with an electromagnetic field can also be engineered, and this has been used for engineering the $\chi^{(2)}$ in semiconductor heterostructures. This will be addressed in part 2. Moreover, even the relaxation rates between energy levels can be engineered, and this has given birth to another class of devices: the quantum cascade lasers[10], which are beyond the scope of the present review.

In addition to band gap engineering for the control of the behavior of electrons, semiconductor heterostructure growth and technology have given birth to a great number of structures controlling the motion of photons[11]: Integrated waveguides [12], Fabry Perot cavities [13, 14], or more exotic objects as whispering gallery structures [15], pillar microcavities [16], photonic wires [17], air bridges [18] or photonic band gap materials [19, 20, 21]. By analogy to band gap engineering, this field can be called "refractive index engineering" [21], and its success from applied optoelectronics to fundamental studies about field-matter interaction is again based on the high level of control of thin multilayer structure growth and technology. Among

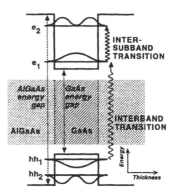

Figure 1. Conduction band and valence band quantum well potentiel resulting from a double heterostructure GaAs/AlGaAs. The intersubband transitions fall into the mid IR (2 to 20 μm typically), whereas the interband transitions are in the near infrared (around 1 μm), for the most widely used GaAs based materials.

all photonic applications, the possibility of phase matching nonlinear interactions will be the subject of section 3. If band gap engineering enables one to engineer the microscopic $\chi^{(2)}$ of the material, it will be shown how refractive index engineering enables one to engineer the macroscopic $\chi^{(1)}$ of the composite material, to get the phase matching condition.

The course is split in two: In section 2, the design of semiconductor QWs for building artificial $\chi^{(2)}$ will be reviewed. Resonant $\chi^{(2)}$ enhancements based on intersubband and interband transitions will be studied separately, in part 2.1 and 2.2, respectively. In section 3, two possibilities of phase matching in semiconductor multilayer heterostructures will be presented. The first one is microcavity phase matching (part 3.1), and the second one is phase matching using form birefringence in a composite multilayer material (part 3.2). These two possibilities are not exhaustive, and in particular the reader is invited to refer to the lecture of Martin Fejer for the description of quasi phase matching with semiconductors.

2. Engineering the nonlinear coefficients with asymmetric quantum wells

The theoretical calculation of the second order susceptibility can be performed using standard density matrix calculations [22, 23]. In the most general form the second order susceptibility $\chi^{(2)}$ appears as a sum of different terms which can be written typically as:

$$\frac{<\psi_l|r_i|\psi_m><\psi_m|r_j|\psi_n><\psi_n|r_k|\psi_l>}{(\omega_a - \omega_{ml} + i\Gamma_{ml})(\omega_b - \omega_{ln} + i\Gamma_{ln})} \tag{2}$$

ω_a and ω_b are two of the frequencies involved in the nonlinear process, and the space variable r_i comes from the Hamiltonian describing the interaction of light with matter in the electric dipole approximation. This sum has to be performed on all electronic states in the system. Each term in the sum describes the contribution of a set of three electronic states $|\psi_l>, |\psi_m>, |\psi_n>$ together with their probability of occupation; these states are involved in the dipole matrix elements present in the numerator. ω_{ml} is the frequency of the transition $|\psi_l> \to |\psi_m>$ and Γ_{ml} is a phenomenological damping factor[22]. For an even potential, the electronic wavefunctions are odd or even, and as a consequence at least one of these three dipole matrix elements involves electronic states with the same parity, and therefore vanishes. We find here the classical result in nonlinear optics which stipulates that the second order susceptibility of a centrosymmetric system is equal to zero[23]: an asymmetry is required to get some $\chi^{(2)}$, and this will be fully illustrated in the following. The denominators of the second order susceptibility contain terms $(\omega - \omega_{ml} + i\Gamma_{ml})$, these terms lead to a large increase of the nonlinear susceptibility if a frequency ω involved in the nonlinear process is resonant with the frequency of the transition between states m and l. This is very important since these resonances simplify the calculations. Most of the time, the overall $\chi^{(2)}$ is dominated by a few resonant terms and the very complex sum on all the possible triplets of electronic states is highly simplified. For a given system, depending on the frequency, different transitions will be dominant in the calculation of the $\chi^{(2)}$. For second harmonic generation in doped semiconductor QWs for instance, the nonlinear susceptibility in the middle infrared will be dominated by the resonance of intersubband transitions. On the other hand, in the near infrared interband transitions are the main processes. This distinction is important since calculations are highly different in these two cases, as it will be shown. $\chi^{(2)}$ resonant with intersubband transitions will be addressed in part 2.1 and interband based $\chi^{(2)}$ in part 2.2. However, one has to keep in mind that in each case, all the transitions involving all electronic states are involved, and the distinction between interband and intersubband based $\chi^{(2)}$ is nothing but the result of a simplification of the calculations, with the assumption of keeping the resonant terms only. Furthermore, in both cases the asymmetric QW potential have roughly the same shape, and the same structures can even be considered in both spectral domains for nonlinear experiments.

2.1. INTERSUBBAND TRANSITIONS (ISBTS) IN QUANTUM WELLS

Considerable work has been devoted to intersubband transitions in general, for a review the reader is invited to references [24, 25, 8]. As shown in figure 1, an intersubband transition occurs between two confined electronic states in the same band (that is the conduction band or the valence band). For this, it is assumed of course that the ground level in the conduction band (valence band), is populated by electrons (holes), by an appropriate doping of the structure. Let us recall first that the QW is a one dimensional square potential, and that the motion of electrons is free in the plane of the layers[4]. Each energy "level" represented in figure 1 is in fact a complete subband with all the possible wavevectors in the plane of the layers, associated with a two-dimensional in plane dispersion and density of states. In the conduction band, the dispersion of the different subbands are parallel (see figure 2), and the transitions are vertical in the k space (if we neglect the photon wave vector), as a consequence of the wave vector conservation. Thus, all electrons in the subband 1 give rise to the same intersubband transition (as far as frequency or dipole matrix elements are concerned), and the overall intersubband transition corresponds simply to a one electron transition multiplied by the number of electrons populating the ground subband. In particular, ISB absorption appears experimentally as a lorentzian line, exactly as an atomic absorption line. This model has the advantage of being very simple and describes the QW as an assembly of independent artificial giant molecules[26, 27], the dipole matrix elements being calculated between the simple one dimensional wavefunctions resulting from the solution of the Schrödinger equation in the one dimensional potential. This very simple presentation neglects non parabolicity effects or collective effects[28], which can be included at a more sophisticated level of description, but are beyond the scope of this lecture.

A consequence of the pure one dimensionality of this system is that only one element of the $\chi^{(2)}$ tensor is not equal to zero. From equation 2, it is obvious that taking only z-dependent wavefunctions in the dipole matrix elements leads to a $\chi^{(2)}_{zzz}$ only. This z-oriented dipole has also important consequences for other experiments or devices: All optical phenomena relying on ISBTs (absorption, SHG, photoconductivity...) are vanishing at normal incidence with respect to the plane of the layers. This is referred to as the ISBT selection rule. The particular $\chi^{(2)}$ tensor kills the efficiency of surface emitting SHG from two counter propagating waves[30]. Note also that this selection rule can be partially removed by working with ISBTs in the valence band[31, 32].

To get some $\chi^{(2)}$, an asymmetry is required, and several schemes have been proposed and realized: square QW under an electric field, asymmetric step QW, asymmetric double QW. These different shapes are summarized in figure 3. After the early proposal of Gurnick and DeTemple of building asymmetric QWs to obtain "synthetic nonlinear semiconductors" [33](preceding the first observation of an intersubband absorption in a QW[34]),

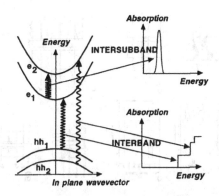

Figure 2. Intersubband and interband transitions in QWs. Since the two subbands are parallel in the in-plane wavevector space, all the populated electronic states in the first subband give the same contribution in the intersubband absorption (same matrix elements, same energy...). The overall intersubband absorption appears then as a simple lorentzian absorption. On the opposite, for interband transitions, the negative mass in the valence band leads to bands with dispersion of opposite signs. The overall absorption spectrum is thus very wide, with an absorption threshold corresponding to the gap of the semiconductor. From reference [29]

calculations performed by Tsang and co-workers[35], Khurgin[36, 37] or Ikonic and co-workers[38], expected very large $\chi^{(2)}$ resonant with ISBTs (several thousands pm/V) in square QWs under an electric field. The main reasons for these huge values are the resonance of the transitions (this is not the case in usual nonlinear optical materials), and the triple product of dipole matrix elements in equation 2, each of these elements scaling as the size of the QW, that is one order of magnitude greater than the atomic size. The first observation of SHG was realized by Fejer and co-workers, in a square GaAs/AlGaAs QW under an electric field[39]. $\chi^{(2)} \approx 20000$pm/V was measured. Higher values were obtained afterwards using asymmetric step QWs[40, 41, 42, 26], using two different compositions of Aluminium in a GaAs/$Al_xGa_{1-x}As$/$Al_yGa_{1-y}As$ structure(see figure 3). In particular, it was shown that a doubly resonant structure, with 1-2 and 2-3 transitions resonant at the frequency ω, (and as a consequence 1-3 is resonant at 2ω) exhibited a record $\chi^{(2)}$, due to the presence of doubly vanishing denominators in equation 2. Among the several possible designs of doubly resonant structures, an optimization of the dipole product was performed to get the maximum $\chi^{(2)}$[26]. $\chi^{(2)}$ of several hundreds of thousands pm/V were measured. Coupled QWs were also designed for the same kind of structures[43, 44, 45], and this scheme of multiple resonance was extended to the case of third harmonic generation by Sirtori and co-workers[46], demonstrating the generality of the concept of band gap engineering. Triply resonant coupled QWs were realized in AlInAs/GaInAs heterostructures, lead-

352

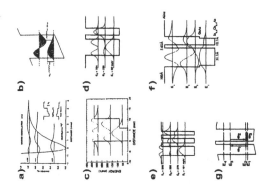

Figure 3. Various asymmetric structures that have been used for frequency conversion in semiconductor quantum wells. All these structures exhibit very large nonlinear susceptibilities due to the resonant character of the nonlinear process. a) Initial scheme from Gurnik and De Temple (From ref. [33]). b) Square quantum well under an electric field, used by Fejer and co-workers (From ref. [35]) c) Doubly resonant asymmetric quantum well used by Rosencher and co-workers (From refs. [40, 26]) d) Double quantum well structure in the AlInAs/GaInAs on InP system, by Sirtori and co-workers (From refs. [45, 43]) e) Triply resonant coupled well structure used by Sirtori and co-workers for third harmonic generation (From ref. [46]) f) Coupled quantum wells used for high photon energy SHG by Chui and co-workers (From ref. [47]) g) Coupled quantum wells used by Sirtori and co-workers for difference frequency generation in the far infrared (From ref. [50]). The energy scales are not identical for all these figures.

ing to a measured $\chi^{(3)}$ of $4.10^{-14}(m/V)^2$. Finally, high energy transitions were explored by Chui, Martinet and co-workers [47, 48], working on the GaAs/InAlAs system presenting a very high conduction band offset. SHG at $2\mu m$ was achieved in this system. On the other side of the spectrum, harmonic generation experiments were pushed in the far IR with a free electron laser ($\lambda \approx 200\mu m$ or $340\mu m$), using large wells or heterojunctions [28, 49]. All these experiments show the great spectral range covered by band gap engineering of ISBTs.

The great number of possibilities offered in the quantum design of QW heterostructures was also illustrated by sum[51] or difference[50, 52, 53] frequency generation experiments. In ref. [50], doubly resonant structures were designed to present two transition energies resonant with two different lines of the CO_2 laser (see fig. 3), showing again the possibility of building structures with the desired electronic level spectrum, and with optimized dipoles. The DFG experiments showed far infrared generation (in the $60\mu m$ range), with $\chi^{(2)}$ of the order of the $\mu m/V$[50].

Going from these beautiful physics experiments with record $\chi^{(2)}$ to real devices is however a difficult challenge. First of all, the layers available by molecular beam epitaxy have a limited thickness (of the order of 10 μm). As the efficiency of the nonlinear frequency conversion scales as L^2, this means that a waveguided geometry in the plane of the layers is required, in order to increase the interaction length. This is of course a source of difficulty, especially in the middle infrared. For example in the case of CO_2 laser doubling, most of the time high power is required, which is incompatible with waveguides.

A second problem comes from the absorption resonant with the $\chi^{(2)}$, as shown in figure 4. Impressive values of $\chi^{(2)}$ have been obtained thanks to this resonance. However several simulations have shown that the related absorption kills the overall efficiency of frequency conversion[54, 55, 56, 57]. To overcome this problem, structures detuned from a perfect resonance, but still presenting a large $\chi^{(2)}$ have to be designed[55, 54, 58, 59]. Different publications have studied the trade off between high $\chi^{(2)}$ and low absorption, and it appears that in this optimization, orders of magnitude are lost in the $\chi^{(2)}$.

Last but not least, the problem of phase matching in this non birefringent system has to be solved. Several schemes have been proposed: quasi phase matching [60, 53, 61], zig zag multipass schemes[62], or Cerenkov configurations[63]. The first one suffers from rather high losses and the efficiency of the two latter is limited by the short effective interaction length. These proposals are presented in figure 5. Let us cite an original way to get phase matching, again based on band gap engineering[64, 57, 59]: It consists of introducing an ISBT between the pump and the harmonic frequency in a SHG experiment. The idea is to benefit from the negative dispersion introduced by this ISBT[65] to compensate for the linear dispersion in the bulk material. This elegant method seems however difficult to realize, because an accurate control of the population in the QWs is required, to get exactly the good refractive index correction. For this, the electron population in the QW can be controlled with an electric field[66, 67, 57, 68] or with an optical pump beam[69] can be envisaged.

Another promising method using form birefringence in waveguides will be presented later in the lecture. Altough this method has been demonstrated only with simple bulk GaAs as nonlinear material, it seems to be very well adapted to frequency conversion with QWs. Let us recall that the lack of efficient devices based on ISBTs may also be due to the lack of interest of frequency conversion in the mid IR, except for very high power applications, where thin QWs are not very well adapted, as explained before.

For this reason, a strong interest has appeared for near infrared nonlinear susceptibilities of QWs, that is resonant on interband transitions. In this spectral range, important low power applications in wavelength division multiplexing systems for example, motivated this research. This is the subject of the next part.

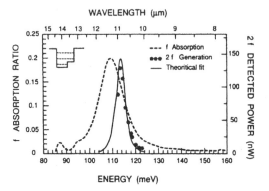

Figure 4. The very large $\chi^{(2)}$ is obtained with resonant structures (here a doubly resonant asymmetric quantum well). In these systems, absorption (dotted line) is at the same energy as the SHG efficiency (full line). This limits the efficiency of a possible device (From ref. [26]).

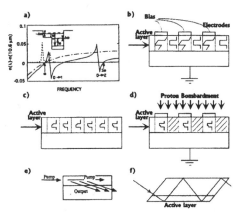

Figure 5. Different schemes that have been proposed for phase matching with intersubband transitions. a) The dispersion associated with an ISBT can be used, compensating for the positive dispersion of the bulk (From ref. [64]). b) Quasi phase matching with an electric field dependent $\chi^{(2)}$[54] c) Quasi phase matching obtained with two growths. One half of the structure is removed by reactive ion etching, and a consecutive growth of QWs with the reverse orientation gives the final structure, with a periodic nonlinear susceptibility d) Quasi phase matching with a periodic supression of the nonlinear susceptibility by proton bombardment[60] e) Cerenkov configuration (From ref. [53]) f) Multipass geometry[62].

2.2. INTERBAND TRANSITIONS

The interest for frequency conversion using resonant interband (and not intersubband) transitions comes from the fact that the wavelengths are in the near infrared, where a lot of applications (and laser sources) exist, whereas ISBTs fall into the mid IR, because of the limited band offsets in the conduction band. As in the case of ISBTs, with interband transitions in QWs new elements of the $\chi^{(2)}$ tensor appear with respect to bulk GaAs; in addition the possibility of controlling $\chi^{(2)}$ (by changing the shape of the QWs[70, 71, 72] or by applying an electric field[73, 74]) can offer the opportunity of quasi-phase matching schemes. Applications include mid IR generation by DFG from two near infrared pump beams, or frequency shifting near 1.55μm by DFG with a 0.75μm pump, for wavelength division multiplexing systems.

The calculation of $\chi^{(2)}$ based on interband transitions is a difficult problem (see also the lecture of Arnold in these proceedings), and the solutions are not analytical. With respect to ISBTs, the difficulties come from the following problems:

- As explained before, in the case of ISBTs, the structure can be described as a set of identical dipoles. This comes from the parallelism of the subbands in the conduction band, which is a very good approximation in GaAs QWs, for example. The overall $\chi^{(2)}$ appears then finally as the electron density times a single electron $\chi^{(2)}$. In the case of interband transitions, the subbands in the valence and in the conduction band have opposite curvatures (see figure 2), and each electron state in the valence band gives a different contribution in the total $\chi^{(2)}$. An infinite summation has to be performed, which has unfortunately a very slow convergence, with consequent numerical problems.

- In the case of interband transitions, the dispersion of the band structures are far more complicated. In the valence band indeed, the simple models of parabolic subbands are not valid anymore, and mixing between light holes and heavy holes have to be taken into account[75]. These details are beyond the scope of this paper.

- Exciton effects may be taken into account[76, 77, 78, 79], as confirmed experimentally[80, 81, 82].

Due to these theoretical difficulties, large discrepancies between different calculations have been published [83, 73, 84, 79]. It seems finally that calculated $\chi^{(2)}$ fall in a few pm/V range[85], which has been confirmed by experimental measurements[86]. These calculations are too complicated to be detailed here, and for a deeper understanding of these problems, the reader is encouraged to consult the references. I will just give here qualitative arguments explaining why the second order nonlinear susceptibilities are so disappointing in the interband regime, compared to the intersubband regime. First of all, I have already stressed that the dipole matrix elements related to ISBTs are greater than the interband ones, by one order of magnitude. As the second order nonlinear susceptibility depends on the product of three dipole matrix elements, this leads to an increase of several orders of magnitude [1]. A second important effect which explains the low interband based $\chi^{(2)}$ is the opposite curvature of the two subbands. This has been illustrated in figure 2. In the ISBT case, all the contributions from different states at different wavevectors are identical and thus give a constructive interference in the overall $\chi^{(2)}$. This is not the case in the interband case, where the frequency depends on the wavevector (see figure 2). There is a destructive interference[85, 87, 75] between the different $\chi^{(2)}$ arising from different states in the subbands, and also from different subbands. A sum rule has even been demonstrated in [85] concerning this destructive interference. This point is crucial, since it was shown that great care has to be taken on the summation over the different electronic states: In the sum for the calculation of the $\chi^{(2)}$, very high energy states have to be taken into account, in order to describe correctly this destructive interference. In particular, states in the continuum above the QW barrier have to be considered[88, 85, 87], which lead to numerical difficulties. The last difference between ISBTs and interband transitions is the difficulty of designing doubly resonant structures in the latter case. Except smart structures realized in the AlSb/GaSb/InAsSb system[89], doubly resonant structures are impossible because the band gap is much greater than the band offset between the two materials (that is the depth of the QWs). In the case of ISBTs, double resonance was one of the key features of the record $\chi^{(2)}$s.

All these reasons explain why so disappointing $\chi^{(2)}$s have been predicted for interband resonances. Since these $\chi^{(2)}$s are finally one order of magnitude smaller than bulk GaAs itself, this is somewhat discouraging as far as a device is concerned. These values (10 to 20 pm/V) were confirmed experimentally[86, 87], in the frequency region near half the band gap. This caracterization was made difficult by the presence of bulk, giving a far much higher signal. That is why a lock in technique was used, with an electric field applied on the QWs, in a transmission geometry[86]. The interest of a transmission experiment is the direct measure of the nonlinear susceptibility, whereas the reflection measurements are complicated by the mixing between interface, bulk and QWs contributions[90]. The agreement between theory and experiments on these low $\chi^{(2)}$s seems to put an end to the realization of efficient frequency converters. However, beautiful experiments have been performed, making use of QWs specificities. Two of them will be reported here.

---

[1]One has to be careful about this point: The increase of the dipole corresponds simultaneously to a decrease of the frequency of the transition. (The oscillator strength of an ISBT $f = \frac{2m^*}{\hbar}\omega_{ij}| < \psi_i|z|\psi_j > |^2$ is always roughly equal to 1 in a square QW, whatever the frequency of the transition). As the efficiency of the SHG depends on the square of the frequency (see equation 1), the large $\chi^{(2)}$ is in fact partially compensated by the decrease of the frequency, also by one order of magnitude.

The first one is a SHG experiment in a quasi-phase matched structure[82]. Quasi phase matching was obtained by reversing the asymmetric QW orientation, as proposed by Khurgin[73] or Harshman and Wang[74]. This is very easy to do by molecular beam epitaxy, since it consists only of reversing the growth order of the different layers in the asymmetric QW, every coherence length. The experiment was done in a counterpropagating surface emitting geometry, so that the coherence length was simply equal to half the SHG wavelength[82]. The efficiency of the quasi phase matching process was clearly demonstrated and the SHG spectra showed a resonantly enhanced $\chi^{(2)}$ at the excitonic energy, that is for a pump corresponding to half the first electron-heavy hole transition energy. From this kind of structures, efficient and smart autocorrelators[91, 92] have been designed[93].

A second interesting experiment was a middle infrared DFG performed by the team of Joffre[94]. DFG was obtained between the different frequencies of a single near infrared beam arising from a 10 fs oscillator. The short pulse presented a large enough spectrum to generate mid IR through this optical rectification process up to 5 μm wavelengths. These experiments extend in the mid IR previous experiments performed in the THz range[95, 96]. Using a nonlinear medium having a non resonant $\chi^{(2)}$ (as bulk GaAs), this experiment was able to produce ultra short pulses (of the order of 10 fs) in the mid IR (broad band from 5 to 15 μm). Using a resonant $\chi^{(2)}$ as a semiconductor QW with an ISBT resonant with a frequency difference[97, 98], this experiment was able to measure simultaneously the non radiative relaxation times between subbands and the coherence time of the ISBT. It has shown a great sensitivity for QWs characterization, measuring the nonlinear susceptibility of structures presenting very low asymmetries[94, 99]. On the other hand, this kind of transient optical rectification provides a unique, very short source in the mid IR(in the 100 fs range), much more monochromatic.

3. Phase matching with semiconductors

As already said in the introduction, another key feature of a nonlinear material is the possibility of phase matching. Let us forget in this part the artificial $\chi^{(2)}$s obtainable in multilayers, which are the subject of the preceding section. We will here assume that the nonlinear material is simply bulk GaAs, which is indeed a very attractive nonlinear material ($\chi^{(2)} \simeq 100$ to 250 pm/V). This large $\chi^{(2)}$ gives about one order of magnitude greater intrinsic efficiency that the commonly used nonlinear materials. To get phase matching, or to increase the coherence length, birefringent crystals as KDP or $LiNbO_3$ can be used; in that case the polarizations of the different waves are carefully chosen in order to adjust their different phase velocities. However, since GaAs is an optically isotropic (cubic) semiconductor, it is non birefringent and phase matching is impossible. In order to use this highly nonlinear material in spite of the problem of phase matching, quasi phase matching[1, 100] has been used [101, 61]. In this case, the sign of the nonlinear interaction is changed periodically by reversing the orientation of the material, in order to compensate for the periodic destructive interference due to phase mismatch. Although quasi phase matching has been demonstrated in GaAs for both second harmonic [101] and difference frequency [61] generations, the technological steps necessary to obtain the material are tricky and the final materials suffer from detrimental losses. In this section, the purpose is the study of phase matching in GaAs. Quasi phase matching will not be treated here and the reader is invited to refer to the latter references or to the lecture of M. Fejer. Phase matching means playing with the velocity of waves in the nonlinear material, therefore it means engineering the refractive indexes in the material. This is again possible using heterostructures. Two ways will be presented here: The first (part 3.1) is frequency conversion in a doubly resonant microcavity. In that case, the beams are propagating perpendicularly to the layers, and phase shifts in Bragg mirrors are used to compensate for the phase lag between the nonlinear polarization and the harmonic wave. The second way (part 3.2) consists of frequency conversion in a waveguide with a well designed birefringence. In that case light propagates in the plane of the layers. This second method seems to be very promising, and will be illustrated by the most recent results.

3.1. FREQUENCY CONVERSION IN SEMICONDUCTOR MICROCAVITIES

During the last years, many optoelectronic semiconductor devices have appeared or have been improved thanks to their implementation into microcavities: vertical cavity surface emitting lasers[13], resonant intracavity photodetectors[102], light emitting diodes[103]... All these devices are based on the high quality factors Q that can be obtained in microcavities, thanks to the progresses of thin film deposition, and especially molecular beam epitaxy or metal organic chemical vapor deposition. The intracavity power enhancement resulting from the high Q can also be used in non linear optics.

It is the purpose of this part to prospect the possibility of using microcavities for frequency conversion in a semiconductor as GaAs. No phase matching is used in the intracavity non linear material; as a consequence the cavity length has to be less or equal than the coherence length of the non linear process L_{coh}. The non linear material length is thus several orders of magnitude smaller than in usual frequency conversion experiments. As frequency conversion efficiency depends on the square of the interaction length, this important loss has to be compensated by a competitive increase. This is why a high Q doubly resonant cavity is used. We will consider here the case of SHG, but a discussion on DFG in a semiconductor microcavity can be found in [104].

After the proposals of Bloembergen and co-workers [1] and the pioneering work of Ashkin or Smith [105, 106] A considerable work has been devoted to intracavity frequency conversion in general, and especially to doubly resonant schemes. A large set of references can be found in ref. [107], and a more general approach of $\chi^{(2)}$ processes in cavities is presented in the contribution of C. Fabre, in this book. Let us consider here the simple case of a completely monolithic planar cavity. Indeed, as far as a device is concerned, a monolithic cavity is highly desirable since its internal losses are considerably lower than those of conventional cavities and also because of the practical aspects of compact size. In addition, the cavity stability is greater in a monolithic geometry, because the cavity length fluctuations lead to excessively noisy doubly resonant resonators. Since the maximum useful cavity length L_{cav} that can be used is L_{coh}, L_{cav} can thus be very short (a few microns), depending on the pump wavelength and the material dispersion. That is why the word "microcavity" is used. However, there are no fundamental changes in the theoretical calculations due to the small cavity length, and no characteristic or special effects are expected from the use of a microcavity instead of a usual cavity, in the principles of the efficiency calculation. Calculations of the SHG efficiency in such a system have been detailed in ref.[107]. A particularly interesting situation is obtained when the cavity length is exactly equal to one coherence length of the nonlinear process: $L_{cav} = L_{coh} = \frac{\lambda^\omega}{4(n^{2\omega} - n^\omega)}$.

The choice of a one-coherence-length cavity has two consequences:

First, after half a round trip the pump and SH waves have opposite phases. However, it was shown [107] that the positive interference between the two waves can be preserved by choosing multilayer mirrors with well designed phases at the reflection. In this case of a one-coherence-length cavity, the mirror reflection phases have to fulfill the condition:

$$\phi^{2\omega} - 2\phi^\omega = \pi \tag{3}$$

This relation ensures that the reflected pump electric field generates a harmonic field in phase with the reflected harmonic field. In other words, the interaction between the nonlinear polarization on the one hand, and of the harmonic field on the other hand, is constructive during all the cavity round trip. This scheme of cavity phase matching is in fact very analogous to quasi phase matching, as it is shown in figure 6. In both cases, the harmonic field and the non linear polarization have to be set back in phase every coherence length. In the former case, the sign of the harmonic field is

360

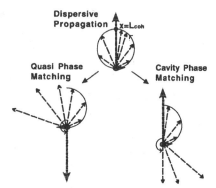

Figure 6. The thick striped arrows represent the nonlinear polarization, fixed in the scheme rotating at $2k^\omega$. It creates a SH field in quadrature (small dashed triangles at the origin). The thin dashed black arrows represent the SH field, which is dephased with respect to the polarization during the dispersive propagation. After one coherence length, the SH field decreases back to zero. Quasi phase matching consists of changing the sign of the polarization (by reversing $\chi^{(2)}$) after L_{coh}. Cavity phase matching studied here consists of changing the sign of the SH field only thanks to metallic (or multilayer acting as metallic) mirrors. From reference [107].

changed with respect to the non linear polarization at the mirror, thanks to relation 3, and in the latter the sign of the non linear polarization is changed through the inversion of the non linear susceptibility.

Relation 3 can be fulfilled by a perfect metallic mirror. However, due to the important losses of semiconductor/metallic in the infrared, high finesse microcavities should better use multilayer dielectric mirrors. A conventional Bragg mirror cannot be used for this application, because the reflection of a perfectly balanced $(\frac{\lambda}{4}, \frac{\lambda}{4})$ Bragg mirror vanishes at the SH frequency for a reason of symmetry. A classical way to obtain bicolor multilayer mirrors is simply to stack two mirrors, one for each frequency. However, the accurate control of both phases in such a structure is impossible since the phase at the reflection of the frequency ω_1 on the bottom mirror depends on the thickness of the entire multilayer stack on top reflecting at ω_2, which cannot be controlled accurately. It is therefore better to design a single multilayer stack reflecting at ω and 2ω. This can be done by breaking the symmetry of the $(\frac{\lambda}{4}, \frac{\lambda}{4})$ Bragg mirror : it was shown[107] by an optimisation of the Fourier transform of the optical path in the multilayer stack that the most efficient fundamental and SH mirrors are obtained with a $(\alpha \frac{\lambda}{2}, (1-\alpha)\frac{\lambda}{2})$ multilayer stack, where $\alpha = \alpha_{max} = 0.304$. The phase at ω and 2ω can finally be fixed by adjusting the thicknesses of the first two layers, in such a way that relation 3 can be fullfilled easily. It is clear that the reflection coefficient at the fundamental and at the SH can be tuned independently, by choosing α between α_{max} and $\frac{1}{2}$. It is then possible to design a doubly resonant

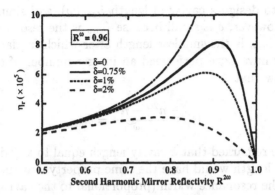

Figure 7. Double resonance with only one tuning parameter: The SHG is collected in reflection (the bottom mirror is assumed to present a perfect reflectivity). The SHG efficiency enhancement is plotted as a function of the top mirror SH reflectivity, for several SH detunings. For small $R^{2\omega}$, the efficiency decreases and tends to the single resonance enhancement. For high $R^{2\omega}$, the SH is out of resonance when the fundamental is resonant (The calculations have been performed in the plane wave approximation, non depleted wave approximation, and absorption losses are neglected). From reference [107].

microcavity with two independent quality factors at the fundamental and at the harmonic frequency, Q_ω and $Q_{2\omega}$, respectively. This point is crucial, as it will be shown below. An interesting alternative realization of dual wavelengths Bragg mirrors can also be found in [108].

The choice $L_{cav}=L_{coh}$ has a second consequence which is of paramount importance as far as a device is concerned : This is the cavity length for which the system can be set as near as possible to double resonance with only one tuning parameter. For an application point of view, it is absolutely necessary to have only one tuning parameter in the system (this can be the pump wavelength, the temperature or the cavity length through the electro-optic effect, the angle of incidence[109] ...). However, it is strictly speaking impossible to get double resonance with one parameter only : When the cavity is in resonance for the fundamental frequency, there is no reason for the SH to be resonant at the same time. In that case, the SH intracavity power build-up is decreased, especially for a high SH quality factor $Q_{2\omega}$. However, we will show that this power decrease is minimized when $L_{cav} = L_{coh}$. For this cavity length, let us put the fundamental frequency at resonance by changing the temperature, for instance. At perfect resonance, L_{cav} is not exactly equal to L_{coh} since the refractive index has been changed to reach the resonance, but we have still $L_{cav} \simeq L_{coh}$. Let us call p the order of the longitudinal mode, resonant at the fundamental : $k^\omega L_{cav} = p\pi$. After a $L_{cav} \simeq L_{coh}$ trip, by definition of the coherence length, we have $k^{2\omega} L_{cav} = (2p + 1 + \delta)\pi$, the origin of the undesirable non vanishing δ is the difference between L_{cav} and L_{coh}, which comes from the impossibility to design a cavity of length $L_{coh}(\omega)$, and simultaneously resonant at ω. However, δ is small because L_{cav} is the resonant cavity length nearest L_{coh}, L_{coh} is the smallest length after which fundamental and second harmonic have both performed an integer number of oscillations. We can easily show that:

$$|\delta| \leq \frac{n^{2\omega} - n^{\omega}}{n^{\omega}} = \delta_{max} \qquad (4)$$

It could be remarked that a cavity length equal to an odd integer times the coherence length would have the same property, but the corresponding reduction of the resonance width (proportional to the cavity length) would decrease the SH intracavity power, for a given δ. In conclusion, the coherence length L_{coh} is the optimal cavity length for doubly resonant SHG with only one tuning parameter; a satisfactory approximation of double resonance can be obtained, if $Q_{2\omega}$ is not to high of course. Figure 7 shows the maximum doubly resonant SHG enhancement as a function of the SH reflectivity, with $L_{cav} = L_{coh}$ and with only one tuning parameter. It is clear that for high coherence length processes, the maximal detuning δ_{max} is small and the SH reflectivity can be increased (up to 0.9 in the case of $10.6\mu m$ doubling in GaAs, where $\delta_{max} = 0.75$), leading to very high SHG efficiencies. This appears in Figure 8, where the efficiency using double resonance with one tuning parameter is plotted with respect to the case where the pump only is resonant (and mirrors transparent at 2ω). It is clear that double resonance is interesting only for high coherence length processes, for which double resonance with only one tuning parameter is obtainable at a rather high $Q_{2\omega}$. Typically, if all the phase factors and resonances are optimized, the SHG enhancement due to the double resonant cavity is equal to $\frac{4}{(1-R_\omega)^2(1-R_{2\omega})}$, whereas it is only $\frac{1}{(1-R_\omega)^2}$ for a single resonant cavity. The latter expresses only the square of the pump power intracavity enhancement, and in the former expression, the factor 4 accounts for the interaction length doubling (SHG occurs in both counterpropagating directions), and the factor $\frac{1}{(1-R_{2\omega})}$ accounts for the second harmonic field enhancement. Let us finally note that the efficiency of this geometry could be increased using an intracavity quasi phase matched stack of GaAs plates, that has already been demonstrated at $10.6\mu m$[110].

A crucial point has to be emphasized : GaAs/AlAs microcavities with high Q values have been obtained relatively easily on (100) GaAs by molecular beam epitaxy[14]. However, the non-linear coefficient of (100) GaAs vanishes at normal incidence. A first solution to overcome this problem is to work with an angle of incidence, as shown in the first experimental demonstration of SHG in doubly resonant cavities[109]. In that case, however, due to the high refractive index of GaAs, the angle of propagation

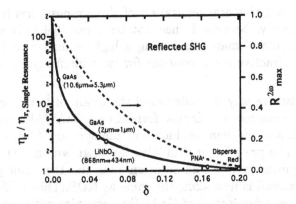

Figure 8. Gain due to double resonance, with respect to single resonance of the pump only. Only one tuning parameter is required. High coherence length processes can benefit from double resonance with one tuning parameter, and with high $R^{2\omega}$.

inside the material is always small, as the resulting $\chi^{(2)}$. Another way is to grow (111) or (110) GaAs microcavities. Thick layer growth on (111) oriented GaAs substrates is a challenge and no high-quality GaAs/AlAs Bragg mirrors have been reported yet on these orientations. Growth on (111) orientation is more difficult than growth on the conventional (100) orientation, since standard growth conditions used for the (100) orientation usually results in layers with a high density of twins and stacking faults associated with a faceted surface. Furthermore, the morphology is known to depend critically on the type of substrate orientation miscut[111]. It was shown that the microscopic stability of the steps on the (111)B surface is the decisive factor determining the adequate growth parameters[112]. In particular, high-quality atomically smooth GaAs layers can be grown starting with (111)B GaAs misoriented 2° off towards (2 -1 -1) (i.e. with stable steps). These conditions are not appropriate for the growth of AlAs : it is necessary to increase the growth temperature for each layer of AlAs in order to maintain a smooth morphology. Preliminary results of (111) GaAs/AlAs Bragg structures have been reported in [104].

3.2. FORM BIREFRINGENCE IN ARTIFICIAL HETEROSTRUCTURES

Another way of *perfect phase matching* (as opposite to quasi phase matching) will be demonstrated here using an isotropic material as microscopic nonlinear source. Phase matching can be obtained by using a built-in *artificial birefringence* in a new composite multilayer material: the isotropy of bulk GaAs is broken by inserting thin oxidized AlAs (Alox) layers in GaAs [113]. The concept of this so-called form birefringence[114, 115] was proposed in 1975 by Van der Ziel[116] for frequency conversion phase matching.

However, the experimental realization of this proposal has been achieved only very recently, because it has not been possible before to find the well suited couple of materials having a high nonlinear coefficient and a high enough refractive index contrast for form birefringence phase matching[113].

A first intuitive way to understand the origin of form birefringence is to consider the macroscopic crystal formed by a GaAs/Alox multilayer system. GaAs is a cubic semiconductor of point group $\bar{4}3m$, therefore non birefringent. The presence of thin Alox layers grown on a (100) substrate breaks the symmetry of 3-fold rotation axes and the point group of the composite material is now $\bar{4}2m$, the same as KDP. This artificial material has the nonlinear properties of GaAs (in particular with the same tensorial character and roughly the same nonlinear coefficient, if we neglect the small zero contribution of thin Alox layers), but the linear optical symmetry of KDP. This is possible by taking advantage of the microscopic nature of the nonlinear polarization given by the ionicity of GaAs and the macroscopic engineering of the refractive index on the scale of the extended electromagnetic wavelengths. Let us note that with (111) oriented GaAs, the introduction of Alox layers switches the point group from $\bar{4}3m$ to $3m$, which is that of another nonlinear material: $LiNbO_3$; the same birefringence properties are of course expected.

In the pioneering paper of Van der Ziel[116], form birefringence was calculated in the multilayer system directly from Maxwell's equations. Another physical explanation can be given using modal wavefunction considerations. Let us consider an infinite periodic multilayer material. Following Joannopoulos[117], for a given wavevector the frequency of an allowed electromagnetic mode in a composite medium increases with the fraction of electric field in the low index material. The difference between the Transverse Electric(TE) and Transverse Magnetic(TM) polarizations arises then from the continuity equations at the boundaries between the two materials, as illustrated in figure 9. In the TM polarization, the continuity of the electric displacement forces the electric field to have an important value in the low index material. This mode thus has a higher frequency for a given wavevector. For a large wavelength compared to the unit cell, the light experiences an effective medium. This is why the dispersion relation $\omega(k)$ in figure 9 is linear at the origin. Form birefringence appears then as the difference between the slopes of the dispersion relations for TE and TM waves. In this long wavelength approximation, the two dielectric constants of the uniaxial composite material are given by:

$$\epsilon_{TE} = \alpha_1 \epsilon_1 + \alpha_2 \epsilon_2 \qquad (5)$$

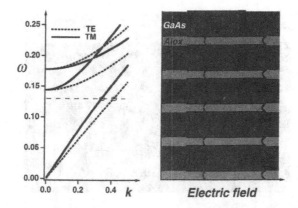

Figure 9. Dispersion relation for an in-plane propagation in a periodic composite material which consists of 25% of Alox (n≈1.6) and 75% of GaAs (n≈3.5), for TM modes (full line) and TE modes (dotted lines). The physical origin of form birefringence appears in the mode wavefunction, pictured to the left for a frequency $\omega = 0.13 \times \frac{2\pi}{d}$ (they correspond to the open circles in the dispersion relation). These Bloch waves have been calculated using standard periodic multilayer theory[115]. The direction of propagation is perpendicular to the plane of the figure. Due to the continuity of the electric displacement ϵE normal to the layers, the TM mode (full line) has an important overlap with the low ϵ layer (Alox, in light gray), and a lower average dielectric constant. The continuous TE electric field (dotted line) has a higher value in GaAs (dark gray), and a higher average dielectric constant. From reference [113].

$$\frac{1}{\epsilon_{TM}} = \frac{\alpha_1}{\epsilon_1} + \frac{\alpha_2}{\epsilon_2} \qquad (6)$$

where α_i and ϵ_i are the filling factors $(\alpha_1 + \alpha_2 = 1)$ and the dielectric constant of the two constitutive materials. These equations are analogous to electrical series and parallel capacitors. This is obvious, since the charge equality $C_1 V_1 = C_2 V_2$ between series capacitors is nothing but the static limit of the electric displacement continuity relation for TM waves $\epsilon_1 E_1 = \epsilon_2 E_2$, and the bias equality $V_1 = V_2$ for parallel capacitors is equivalent to the electric field continuity for TE waves $E_1 = E_2$.

It appears from (1) and (2) that form birefringence $(\sqrt{\epsilon_{TE}} - \sqrt{\epsilon_{TM}})$ increases with the refractive index contrast between the two materials in the multilayer, as for photonic band gap effects in photonic crystals[117, 21]. Although form birefringence in a GaAs/AlGaAs multilayer structure has been proposed for phase matching[116], the refractive index contrast between GaAs (n≈3.5) and AlAs (n≈2.9) is too low to provide the birefringence required to compensate for the dispersion. This is the reason why thin film layers of Alox (n≈1.6) in GaAs have been used to get sufficient form birefringence. Alox results from selective oxidation at 400-500 °C in a water vapour atmosphere of AlAs layers embedded in GaAs. This technology of AlAs oxidation has emerged in the early 90s[118], and since then

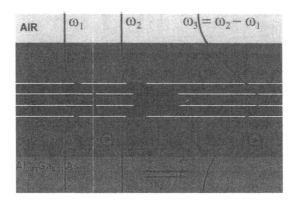

Figure 10. DFG process in the sample. Three periods of the composite material GaAs(325 nm)/Alox(40 nm) constitute the core of the waveguide. The birefringence of the composite material was engineered to compensate for the dispersion arising from both the natural dispersion in bulk GaAs and the optical confinement dispersion in the waveguide. The sample was grown by molecular beam epitaxy on a GaAs (100) substrate and consists of: 2800nm $Al_{0.97}Ga_{0.03}As$; 1500nm $Al_{0.70}Ga_{0.30}As$ (waveguide cladding layers), three periods of birefringent composite material (40 nm Alox; 325nm GaAs)×3 and 40 nm Alox; 1500nm$Al_{0.70}Ga_{0.30}As$ and a final 30 nm GaAs cap layer. The oxidation process is described in detail in reference [123]. The three modes involved in the DFG process are pictured together with their polarization (↑ for TM, ⊙ for TE). The higher overlap of the TM mode with the low refractive index Alox layers is apparent, which is the origin of form birefringence. The arrows recall the "phase matching" momentum conservation. From reference [113].

Alox has led to breakthroughs in the field of semiconductor lasers[119] or Bragg mirrors[120] -thanks to its refractive index contrast with GaAs-.

An example of a structure used for DFG is presented in figure 10, together with the three modes involved in the nonlinear interaction. The details of the experiments can be found in [121, 113, 122]. A form birefringence n(TE)-n(TM)=0.154 has been measured on the structure presented in figure 10[123]. Even higher birefringences of the order of 0.2 have been obtained with different samples. This birefringence is sufficient to phase match mid infrared generation between 3μm and 10μm by DFG from two near infrared beams. Note that by increasing the width of Alox layers (as in the example of figure 9), much higher birefringences up to 0.65 could be achieved, in principle.

Given the $\chi^{(2)}_{xyz}$ tensor of GaAs, the DFG process (1.035μm, TM) - (1.32μm, TE) $\mapsto$ (4.8μm, TE) is represented schematically in figure 10. A typical infrared signal for this process is shown in figure 11 as a function of the Ti:Sa wavelength. This function has the well known $\left(\frac{\sin x}{x}\right)^2$ shape, which is a clear evidence of phase matching. This kind of experiment has been the first achievement of perfect phase matching with a cubic nonlinear

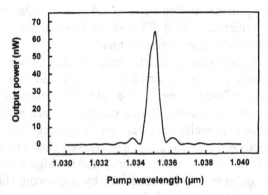

Figure 11. Mid-Infrared DFG signal measured by an InSb detector, as a function of the Ti:Sa laser wavelength. The pump powers were 0.2mW and 1.6mW for the YAG and the Ti:Sa lasers, respectively. This function has the well known shape of a phase matching resonance: It is a $\left(\frac{\sin x}{x}\right)^2$ function, characteristic of a momentum conservation in a fixed interaction length.

material. This was also the first realization of Van der Ziel's 22 year old proposal.

Typical mid IR output powers of 120 nW were obtained for 0.4 mW and 17 mW of Nd:YAG and Ti:Sa pump powers, respectively[122]. This result can easily be pushed into the μW range by increasing pump powers and reducing scattering losses originating from processing[122]. Those are power levels interesting for mid infrared spectroscopic applications. Note also that generated wavelengths up to 5.3μm at room temperature and 5.6μm by increasing the temperature of the sample have been demonstrated. These wavelengths are in the absorption range of LiNbO$_3$, where few nonlinear optical materials exist. Waveguided Fourier transform infrared spectroscopic measurements show that absorption losses in Alox start for wavelengths greater than 7.5μm, opening a very large spectral range for tunable non linear frequency conversion. The tunability of the mid infrared wavelength was also demonstrated by varying the temperature. A linear dependence of the signal wavelength from 5.2 to 5.6μm was obtained with a temperature scan from 0°C to 150°C[122]. No degradation of the sample was observed during temperature cycles.

The birefringence of the composite structure is sufficient not only to phase match DFG in the mid infrared, but also for SHG around 1.55μm. I give these two examples because of their interest for applications: Continuously tunable mid infrared compact sources are desirable for pollutant detection in the molecular fingerprint region or for process monitoring, and a 1.55μm signal can be shifted by mixing it with a 0.75μm pump. This function is required in wavelength division multiplexing systems. The experiment showing phase matched SHG at 1.55μm has also been performed and is reported in reference [124]. This shows the generality of the concept of phase matching in the composite material.

Some more technological characteristics of this system are of critical importance, in particular characterization of near and mid infrared losses, due to absorption and scattering on the ridge inhomogeneities introduced during the technological process. These parameters are very important for the success of this composite material for frequency conversion devices, and have been analyzed in ref.[122]. They showed that rather high mid IR losses (of the order of 50 cm^{-1}) were mainly due to ridge sidewall scattering. Experiments are underway to reduce losses by improving the technological process and by an optimization of the geometrical parameters of the ridge.

Interesting perspectives are opened by this system, which go beyond the simple passive nonlinear material: GaAs is also the top material for QW lasers, and the possibility to integrate QWs in the core of the nonlinear waveguide is under study. In a first step, this would enable DFG with only one external source, the other frequency being given by the "internal" QW laser. In a second step, parametric fluorescence from this laser would make a completely monolithic micro optical parametric oscillator, on a GaAs chip, tunable with temperature, a realistic possibility.

4. Conclusion

Semiconductor heterostructures form a system where both linear and non-linear optical properties can be engineered at will. Among the different possibilities explored in this review, two of them are particularly attractive: First, intersubband transitions in QWs have shown high nonlinear coefficients, and give a very attractive nonlinear material. A great care has to be taken to detune the structure from a possible resonant absorption. Second, phase matching has been obtained for a propagation in the plane of the layers, using artificial birefringence in GaAs/Alox multilayers. The latter experiments have been demonstrated very recently, using bulk GaAs as nonlinear material. Using intersubband based $\chi^{(2)}$ with such phase matched structures is easily feasible in principle by growing asymmetric QWs in the heart of a GaAs/Alox structure; this may lead to highly efficient devices. For such a purpose, deep QWs for high energy intersubband transitions have to be choosen, to avoid the absorption range of Alox.

5. Acknowledgements

A great number of discussions have to be acknowledged during the last seven years of nonlinear optics, in particular with D. Delacourt, A. Fiore and E. Rosencher. A great number of results presented in this review have been obtained in (or in collaboration with) the Laboratoire Central de Recherches of Thomson CSF. People involved in this work are P. Bravetti, P. Bois, E. Costard, S. Crouzy, N. Laurent, E. Martinet, B. Vinter and D. Weill (LCR Thomson CSF); A. Alexandrou, A. Bonvalet, M. Joffre, J. L. Martin and A. Migus (Laboratoire d'Optique Appliquée, ENSTA Palaiseau); P. Boucaud, O. Gauthier-Lafaye, F. Julien, J. M. Lourtioz and D. D. Yang (Institut d'Electronique Fondamentale, Orsay); H. C. Liu (in the LCR for a short time), Y. Beaulieu, L. Delobel, S. Janz, J. P. McCaffrey, P. Van der Meer, Z. R. Wasilewski and D. X. Xu (National Research Council of Canada, Ottawa). The samples used in the experiments have been grown by S. Delaître, X. Marcadet and J. Nagle. Finally, the author is deeply indebted to A. Fiore, C. Sirtori and B. Vinter, for a critical reading of the manuscript.

References

1. J. A. Armstrong, N. Bloembergen, J. Ducuing, and P. S. Pershan, "Interactions between light waves in a non linear dielectric," Phys. Rev. **127**, 1918–1939 (1962).
2. M. M. Fejer, "Nonlinear optical frequency conversion," Phys. Today May **1994**, 25–32 (1994).
3. A. Yariv, *Quantum electronics* (John Wiley & sons, 1989).
4. G. Bastard, *Wave mechanics applied to semiconductor heterostructures* (Les Editions de Physique CNRS, 1988).
5. C. Weisbuch and B. Vinter, *Quantum Semiconductor Structures: Fundamentals and Applications* (Academic Press, 1991).
6. C. Weisbuch, "The Future of Physics of Heterostructures: A Glance Into the Crystal (Quantum) Ball," Physica Scripta **T68**, 102–112 (1996).
7. F. Capasso, Science **235**, 172 (1987).
8. F. Capasso, J. Faist, and C. Sirtori, "Mesoscopic phenomena in semiconductor nanostructures by quantum design," J. of Math. Phys. **37**, 4775–4792 (1996).
9. B. F. Levine, "Quantum Well Infrared Photodetectors," J. of Appl. Phys. **74**, R1–R81 (1993).
10. J. Faist, F. Capasso, D. L. Sivco, C. Sirtori, A. L. Hutchinson, and A. Y. Cho, "Quantum Cascade Laser," Science **264**, 553–556 (1994).
11. *Microcavities and Photonic Band Gaps: Physics and Applications*, J. Rarity and C. Weisbuch, eds., (Kluwer Academic Publishers, Dordrecht, 1996).
12. T. Tamir, *Guided-Wave Optoelectronics* (Springer-Verlag, 1990).
13. J. L. Jewell, J. P. Harbison, A. Scherer, Y. H. Lee, and L. T. Florez, "Vertical-Cavity Surface-Emitting Lasers: Design, Growth, Fabrication, Characterization," IEEE J. of Quant. Elec. **27**, 1332 (1991).
14. R. P. Stanley, R. Houdré, U. Oesterle, M. Gailhanou, and M. Ilegems, "Ultrahigh finesse microcavity with distributed Bragg reflectors," Appl. Phys. Lett. **65**, 1883 (1994).
15. U. Mohideen, W. S. Hobson, S. J. Pearton, F. Ren, and R. E. Slusher, "GaAs/AlGaAs microdisks lasers," Appl. Phys. Lett. **64**, 1911 (1994).
16. J. M. Gérard, D. Barrier, J. Y. Marzin, R. Kuszelewicz, L. Manin, E. Costard, V. Thierry-Mieg, and T. Rivera, "Quantum boxes as active probes for photonic microstructures: The pillar microcavity case," Appl. Phys. Lett. **69**, 449 (1996).
17. J. Zhang, D. Y. Chu, S. L. Wu, S. T. Ho, W. G. Bi, C. W. Tu, and R. C. Tiberio, "Photonic-Wire Laser," Phys. Rev. Lett. **75**, 2678 (1995).

370

18. J. S. Foresi, P. R. Villeneuve, J. Ferrera, E. R. Thoen, G. Steinmeyer, S. Fan, J. D. Joannopoulos, L. C. Kimerling, H. I. Smith, and E. P. Ippen, "Photonic-Band-Gap Microcavities in Optical Waveguides," Nature, accepted for publication (1997).

19. D. Labilloy, H. Benisty, C. Weisbuch, V. Bardinal, T. Krauss, R. Houdré, U. Oesterle, D. Cassagne, and C. Jouanin, "Quantitative measurement of transmission, reflexion and diffraction of two-dimensional photonic bandgap structures at near-infrared wavelengths," Phys. Rev. Lett., submitted for publication (1997).

20. C. C. Cheng, A. Sherer, V. Arbet-Engels, and E. Yablonovitch, "Lithographic band gap tuning in photonic band gap crystals," J. Vac. Sci. Technol. B **14**, 4110 (1996).

21. V. Berger, "From photonic band gaps to refractive index engineering," Optical Materials, to be published (1997).

22. N. Bloembergen, *Nonlinear Optics* (Benjamin, 1977).

23. Y. R. Shen, *The Principles of Nonlinear Optics* (Wiley, 1984).

24. E. Rosencher, B. Vinter, and B. Levine, *Intersubband Transitions in Quantum Wells* (Plenum, New York, 1992).

25. H. C. Liu, B. F. Levine, and J. Y. Andersson, *Quantum Well Intersubband Transition Physics and Devices* (Plenum, Dordrecht, 1994).

26. E. Rosencher and P. Bois, "Model system for optical nonlinearities: Asymmetric quantum wells," Phys. Rev.B **44**, 11315 (1991).

27. F. Capasso, C. Sirtori, and A. Y. Cho, "Quantum well quasimolecules with nonlinear optical properties and new heterostructures with bound states in the continuum," Optoelectronics. Devices and Technology **8**, 479 (1993).

28. J. N. Heyman, K. Craig, B. Galdrikian, M. S. Sherwin, K. Campman, P. F. Hopkins, S. Fafard, and A. C. Gossard, "Resonant harmonic generation and dynamic screening in a double quantum well," Phys. Rev. Lett. **72**, 2183 (1994).

29. E. Rosencher and B. Vinter, *Optoélectronique* (Masson, 1997).

30. V. Berger, P. Bois, and E. Rosencher, "Comment on surface emitting second harmonic generator by intersubband transition in asymmetric quantum wells with slab waveguide (Appl. Phys. Lett. 62, 1502 (1993))," Appl. Phys. Lett. **64**, 800 (1994).

31. S. Li and J. Khurgin, "Second order nonlinear optical susceptibility in p-doped asymmetric quantum wells quantum wells," Appl. Phys. Lett. **62**, 1727 (1993).

32. X. Qu and H. Ruda, "Structure dependence of second harmonic generation in asymmetric quantum well structures," IEEE J. of Quant. Electron. **31**, 228 (1995).

33. M. K. Gurnick and T. A. DeTemple, "Synthetic nonlinear semiconductors," IEEE J. Quantum Electron. **QE-19**, 791 (1983).

34. L. C. West and S. J. Eglash, "First observation of an extremely large dipole infrared transition within the conduction band of a GaAs quantum well," Appl. Phys. Lett. **46**, 1156–1158 (1985).

35. L. Tsang, D. Ahn, and S. L. Chuang, "Electric field control of optical second harmonic generation in a quantum well," App. Phys. Lett. **52**, 697 (1988).

36. J. Khurgin, "Second-order intersubband nonlinear optical susceptibilities of asymmetric quantum well structures," In *Quantum Wells for Optics and Optoelectronics*, (Washington DC, 1989).

37. J. Khurgin, "Second order intersubband nonlinear optical susceptibilities of asymmetric quantum well structures," J. of Opt. Soc. Am. B **6**, 1673 (1989).

38. Z. Ikonic, V. Milanovic, and D. Tjapkin, "Resonant second harmonic generation by a semiconductor quantum wells in electric field," IEEE J. of Quant. Elec. **25**, 54 (1989).

39. M. M. Fejer, S. J. B. Yoo, R. L. Byer, A. Harwit, and J. S. Harris, "Observation of extremely large quadratic susceptibility at 9.6 - 10.8 μm in electric-field-biased quantum wells," Phys. Rev. Lett. **62**, 1041 (1989).

40. E. Rosencher, P. Bois, J. Nagle, and S. Delaître, "Second harmonic generation by intersubband transitions in compositionally asymmetrical MQWs," Electron. Lett. **25**, 1063 (1989).

41. P. Boucaud, F. H. Julien, D. D. Yang, J. M. Lourtioz, E. Rosencher, P. Bois, and J. Nagle, "Detailed analysis of second harmonic generation near 10.6 μm in GaAs/AlGaAs asymmetric quantum wells," App. Phys. Lett. **57**, 215 (1990).

42. P. Boucaud, F. H. Julien, D. D. Yang, J. M. Lourtioz, E. Rosencher, and P. Bois, "Saturation of the second harmonic generation in GaAs/AlGaAs asymmetric quantum wells," Opt.. Lett. **16**, 199 (1991).

43. C. Sirtori, F. Capasso, D. L. Sivco, S. N. G. Chu, and A. Y. Cho, "Observation of large second order susceptibility via intersubband transitions at $\lambda = 10\mu m$ in asymmetric coupled AlInAs/GaInAs quantum wells," Appl. Phys. Lett. **59**, 2302 (1991).

44. C. Sirtori, F. Capasso, D. L. Sivco, A. L. Hutchinson, and A. Y. Cho, "Resonant stark tuning of second-order susceptibility in coupled quantum wells," Appl. Phys. Lett. **60**, 151 (1992).

45. F. Capasso, C. Sirtori, and A. Y. Cho, "Coupled quantum well semiconductors with giant electric field tunable nonlinear optical properties in the infrared," IEEE Electron Device Lett. **30**, 1313 (1994).

46. C. Sirtori, F. Capasso, D. L. Sivco, and A. Y. Cho, "Giant, Triply Resonant, Third-Order Nonlinear Susceptibility $\chi^{(3)}_{3\omega}$ in Coupled Quantum wells," Phys. Rev. Lett. **68**, 1010 (1992).

47. H. C. Chui, E. L. Martinet, G. L. Woods, M. M. Fejer, J. S. Harris, C. A. Rella, B. I. Richman, and H. A. Schwettman, "Doubly resonant second harmonic generation of 2.0μm light in coupled InGaAs/AlAs quantum wells," Appl. Phys. Lett. **64**, 3365 (1994).

48. E. Martinet, H. C. Chui, G. L. Woods, M. M. Fejer, J. S. Harris, C. A. Rella, B. A. Richman, and H. A. Schwettman, "Short wavelength (5.36-1 .85μm) nonlinear spectroscopy of coupled InGaAs/AlAs intersubband quantum wells," Appl. Phys. Lett. **65**, 2630 (1994).

49. W. W. Bewley, C. L. Felix, J. J. Plombon, M. S. Sherwin, M. Sundaram, P. F. Hopkins, and A. C. Gossard, "Far-infrared second harmonic generation in GaAs/AlGaAs heterostructures: perturbative and nonperturbative response," Phys. Rev. B **48**, 2376 (1993).

50. C. Sirtori, F. Capasso, J. Faist, L. N. Pfeiffer, and K. W. West, "Far-infrared generation by doubly resonant difference frequency mixing in a coupled quantum well two-dimensional electron gas system," Appl. Phys. Lett. **65**, 445 (1994).

51. H. C. Liu, E. Costard, E. Rosencher, and J. Nagle, "Sum frequency generation by intersubband transition in step quantum wells," IEEE J. of Quant. Electr. **31**, 1659 (1995).

52. H. C. Chui, G. L. Woods, M. M. Fejer, E. L. Martinet, and J. S. Harris, "Tunable mid-infrared generation by difference frequency mixing of diode laser wavelengths in intersubband InGaAs/AlAs quantum wells," Appl. Phys. Lett. **66**, 265 (1995).

53. H. C. K. Chui, Ph.D. thesis, Stanford University, 1994.

54. S. J. B. Yoo, Ph.D. thesis, Standford University, 1991.

55. E. Rosencher, "Two photon optical nonlinearities in a resonant quantum well system," J. of Appl. Phys. **73**, 1909 (1992).

56. G. Almogy and A. Yariv, "Second harmonic generation in absorptive media," Optics Lett. **19**, 1828 (1994).

57. J. R. Meyer, C. A. Hoffman, F. J. Bartoli, and L. R. Ram-Mohan, "Intersubband second harmonic generation with voltage controlled phase matching," Appl. Phys. Lett. **67**, 608 (1995).

58. J. R. Meyer, C. A. Hoffman, F. J. Bartoli, E. R. Youngdale, and L. R. Ram-Mohan, "Momentum space reservoir for enhancement of intersubband second harmonic generation," IEEE J. of Quant. Electron. **31**, 706 (1995).

59. I. Vurgaftman, J. Meyer, and L. R. Ram-Mohan, "Optimized second-harmonic generation in asymmetric double quantum wells," IEEE J. of Quant. Electr. **32**, 1334 (1996).

372

60. S. J. B. Yoo, M. M. Fejer, and R. L. B. ans S. J. Harris, "Second order susceptibility in asymmetric quantum wells and its control by proton bombardment," Appl. Phys. Lett. **58**, 1724 (1991).

61. S. J. B. Yoo, C. Caneau, R. Bhat, M. A. Koza, A. Rajhel, and N. Antoniades, "Wavelength conversion by difference frequency generation in AlGaAs waveguides with periodic domain inversion achieved by wafer bonding," Appl. Phys. Lett. **68**, 2609 (1996).

62. K. Vodopyanov, Appl. Phys. Lett., submitted for publication (1998).

63. G. L. Woods, Ph.D. thesis, Stanford University, 1997.

64. G. Almogy, M. Segev, and A. Yariv, "Intersubband transitions induced phase matching," Optics Lett. **19**, 1192 (1994).

65. E. B. Dupont, D. Delacourt, and M. Papuchon, "Mid-infrared Phase Modulation via Stark Effect on Intersubband Transitions in GaAs/GaAlAs Quantum Wells," IEEE J. Quant. Electron. **29**, 2313 (1993).

66. V. Berger, E. Dupont, D. Delacourt, B. Vinter, N. Vodjdani, and M. Papuchon, "Triple Quantum Well Electron Transfer Infrared Modulator," Appl. Phys. Lett. **61**, 2072 (1992).

67. E. B. Dupont, D. Delacourt, V. Berger, N. Vodjdani, and M. Papuchon, "Phase and Amplitude Modulation Based on Intersubband Transitions in Electron Transfer Double Quantum Wells.," Appl. Phys. Lett. **62**, 1907 (1993).

68. V. Berger, N. Vodjdani, D. Delacourt, and J. Schnell, "Room-temperature quantum well infrared modulator using a Schottky diode," Appl. Phys. Lett. **68**, 1904 (1996).

69. V. Berger, N. Vodjdani, B. Vinter, E. Costard, and E. Böckenhoff, "Optically induced intersubband absorption in biased double quantum wells.," Appl. Phys. Lett. **60**, 1869–1871 (1992).

70. J. Khurgin, "Large-scale quantum well domain structures," J. Appl. Phys. **64**, 5026–5029 (1988).

71. S. Janz, F. Chatenoud, and R. Normandin, "Quasi phase matched second harmonic generation from asymmetric coupled quantum wells," Optics Letters **19**, 622 (1994).

72. X. Qu, H. Ruda, S. Janz, and A. J. S. Thorpe, "Enhancement of second harmonic generation at $1.06\mu m$ using a quasi phase matched AlGaAs/GaAs asymmetric quantum well structure," Appl. Phys. Lett. **65**, 3176 (1994).

73. J. Khurgin, "Second-order nonlinear effects in asymmetric quantum-well structures.," Phys. Rev. B **38**, 4056–4066 (1988).

74. P. J. Harshman and S. Wang, "Asymmetric AlGaAs quantum wells for second-harmonic generation and quasiphase matching of visible light in surface emitting waveguides," Appl. Phys. Lett. **60**, 1277–1279 (1992).

75. D. C. Hutchings and J. M. Arnold, "Determination of second order nonlinear coefficients in semiconductors using pseudo spin equations for three level systems," Phys. Rev. B **56**, 4056 (1997).

76. A. Shimizu, "Optical Nonlinearity induced by giant dipole moment of Wannier excitons," Phys. Rev. Lett. **61**, 613 (1988).

77. L. Tsang and S. L. Chuang, "Exciton effects on second-order nonlinear susceptibility in a quantum well with an applied electric field," Phys. Rev. B **42**, 5229 (1990).

78. R. Atanasov, F. Bassani, and V. M. Agranovich, "Second order nonlinear susceptibility of asymmetric quantum wells," Phys. Rev. B **50**, 7809 (1994).

79. H. Kuwatsuka and H. Ishikawa, "Calculation of the second order optical order optical susceptibility in biased $Al_x Ga_{1-x} As$ quantum wells," Phys. Rev. B **50**, 5323 (1994).

80. Y. L. Xie, Z. H. Chen, D. F. Cui, S. H. Pan, D. Q. Deng, and Y. L. Zhou, "Optical second-order susceptibility of GaAs/AlxGa1-xAs asymmetric coupled-quantum-well structures in the exciton region," Phys. Rev. B **43**, 12477–12479 (1991).

373

81. V. Pellegrini, A. Parlangeli, M. Borger, R. D. Atanasov, F. Beltram, L. Vanzetti, and A. Franciosi, "Interband second harmonic generation in ZnCdSe/ZnSe strained quantum wells," Phys. Rev. B **52**, 5527 (1995).

82. A. Fiore, Y. Beaulieu, S. Janz, J. P. McCaffrey, Z. R. Wasilewski, and D. X. Xu, "Quasiphase matched surface emitting second harmonic generation in periodically reversed asymmetric GaAs/AlGaAs quantum well waveguide," Appl. Phys. Lett. **70**, 2655 (1997).

83. J. Khurgin, "Second-order susceptibility of asymmetric coupled quantum-well structures.," Appl. Phys. Lett. **51**, 2100 (1987).

84. L. Tsang, S. L. Chuang, and S. M. Lee, "Second-order nonlinear susceptibility of a quantum well with an applied electric field," Phys. Rev. B **41**, 5942 (1990).

85. A. Fiore, E. Rosencher, B. Vinter, D. Weill, and V. Berger, "Second order optical susceptibilty of biased quantum wells in the interband regime," Phys. Rev. B **51**, 13192 (1995).

86. A. Fiore, E. Rosencher, V. Berger, and J. Nagle, "Electric field induced interband second harmonic generation in GaAs/AlGaAs quantum wells," Appl. Phys. Lett. **67**, 3765 (1995).

87. A. Fiore, Ph.D. thesis, Université d'Orsay, 1997.

88. F. Bogani, S. Cioncolini, E. Lugagne-Delpon, P. Roussignol, P. Voisin, and J. P. André, "Resonant second harmonic generation in type II heterostructures of InP/Al$_{0.48}$In$_{0.52}$As," Phys. Rev. B **50**, 4554 (1994).

89. S. Scandolo, A. Baldereschi, and F. Capasso, "Interband near-infrared second-harmonic generation with very large $|\chi^{(2)}(2\omega)|$ in AlSb/GaSb-InAsSb/AlSb asymmetric quantum wells," Appl. Phys. Lett. **62**, 3138 (1993).

90. X. Qu, D. J. Bottomley, H. Ruda, and A. J. S. Thorpe, "Second harmonic generation from a AlGaAs/GaAs asymmetric quantum well structure," Phys. Rev. B **50**, 5703 (1994).

91. R. Normandin and G. I. Stegeman, **36**, 253 (1980).

92. D. Vakhshoori and S. Wang, Appl. Phys. Lett. **53**, 347 (1988).

93. A. Fiore, private communication.

94. A. Bonvalet, J. Nagle, V. Berger, A. Migus, J. L. Martin, and M. Joffre, "Femtosecond infrared emission resulting from coherent charge oscillations in quantum wells," Phys. Rev. Lett. **76**, 4392 (1996).

95. H. G. Roskos, M. C. Nuss, J. Shah, K. Leo, D. A. B. Miller, A. M. Fox, S. Schmitt-Rink, and K. K. hler, "Coherent Submillimeter-Wave Emission from Charge Oscillations in a Double-Well Potentiel," Phys. Rev. Lett. **68**, 2216–2219 (1992).

96. P. C. M. Planken, M. C. Nuss, I. Brener, K. W. Goossen, M. S. C. Luo, S. L. Chuang, and L. Pfeiffer, "Terahertz emission in single quantum wells after coherent optical excitation of light hole and heavy holes excitons," Phys. Rev. Lett. **69**, 3800 (1992).

97. A. Shimizu, M. Kuwata-Gonokami, and H. Sakaki, "Enhanced second-order optical nonlinearity using inter- and intra-band transitions in low dimensional semiconductors," Appl. Phys. Lett. **61**, 399 (1992).

98. X. Qu and H. Ruda, "Microwave and millimeter wave generation using nonlinear optical mixing in asymmetric quantum wells," J. of Appl. Phys. **75**, 54 (1994).

99. A. Bonvalet, Ph.D. thesis, Ecole polytechnique, 1997.

100. M. M. Fejer, G. A. Magel, D. H. Jundt, and R. L. Byer, "Quasi-Phase-Matched Second Harmonic Generation : Tuning and Tolerances," IEEE J. of Quant. Elec. **28**, 2631 (1992).

101. D. E. Thomson, J. D. McMullen, and D. B. Anderson, "Second-Harmonic Generation in GaAs Stack of Plates using High Power CO_2 Laser Radiation," Appl. Phys. Lett. **29**, 113–115 (1976).

102. M. S. Ünlü, "Resonant cavity enhanced photonic devices," J. Appl. Phys. **78**, 607 (1995).

103. N. Hunt and E. F. Schubert, "High efficiency resonant cavity LED's," In *Microcavities and Photonic bandgaps: Physics and Applications*, J. Rarity and C. Weisbuch, eds., (Kluwer Academic Publishers, Dordrecht, 1996).

104. V. Berger, X. Marcadet, and J. Nagle, "Doubly resonant frequency conversion processes in semiconductor microcavities," Pure and Applied Optics, to be published (1997).

105. A. Ashkin, G. D. Boyd, and J. M. Dziedzic, "Resonant optical second harmonic generation and mixing," IEEE J. of Quant. Electron. **2**, 109–124 (1966).

106. R. G. Smith, "Theory of intracavity optical second-harmonic generation," IEEE J. of Quant. Electron. **6**, 215–223 (1970).

107. V. Berger, "Second-harmonic generation in monolithic cavities," J. of Opt. Soc. Am. B **14**, 1351 (1997).

108. D. J. Lovering, G. Fino, C. Simonneau, R. Kuszelewicz, R. Azoulay, and J. A. Levenson, "Optimisation of dual-wavelength Bragg mirrors," Electron. Lett. **32**, 1782 (1996).

109. C. Simonneau, J. P. Debray, J. C. Harmand, P. Vidakovic, D. J. Lovering, and J. A. Levenson, "Second harmonic generation in a doubly-resonant semiconductor microcavity," submitted for publication (1997).

110. L. A. Gordon, "Diffusion-bonded stacked GaAs for quasi-phase matched second-harmonic generation of a carbon dioxide laser," Electron. Lett. **29**, 1942 (1993).

111. K. Yang and L. J. Schowalter, Appl. Phys. Lett. **60**, 1851 (1991).

112. X. Marcadet, J. Olivier, and J. Nagle, "Stability of the step distribution and MBE growth mechanisms on vicinal GaAs(-1-1-1) substrates," Appl. Surf. Sci., to be published (1997).

113. A. Fiore, V. Berger, E. Rosencher, P. Bravetti, and J. Nagle, "Phase matching using an isotropic nonlinear material," Nature (1997).

114. M. Born and E. Wolf, *Principle of Optics* (Pergamon Press, Oxford, 1980).

115. P. Yeh, *Optical Waves in Layered media* (John Wiley & Sons, New York, 1988).

116. J. V. der Ziel, "Phase-matched harmonic generation in a laminar structure with wave propagation in the plane of the layers," Appl. Phys. Lett. **26**, 60 (1975).

117. J. D. Joannopoulos, P. R. Villeneuve, and S. Fan, "Photonic crystals : putting a new twist on light," Nature **386**, 143 (1997).

118. J. M. Dallesasse, J. N. Holonyak, A. R. Sugg, T. A. Richard, and N. El-Zein, "Hydrolysation Oxidation of AlGaAs-AlAs-GaAs Quantum Well Heterostructures and Superlattices," Appl. Phys. Lett. **57**, 2844 (1990).

119. D. L. Huffaker, D. G. Deppe, and K. Kumar, "Native oxide defined ring contact for low-threshold vertical cavity lasers," Appl. Phys. Lett. **65**, 97 (1994).

120. M. H. MacDougal, H. Zao, P. D. Dapkus, M. Ziari, and W. H. Steïer, "Wide-Bandwidth Distributed Bragg Reflectors Using Oxide/GaAs Multilayers," Electr. Lett. **30**, 1147 (1994).

121. A. Fiore, V. Berger, E. Rosencher, P. Bravetti, N. Laurent, and J. Nagle, "Phase-matched mid-IR difference frequency generation in GaAs-based waveguides," Appl. Phys. Lett., to be published (1997).

122. P. Bravetti, A. Fiore, V. Berger, E. Rosencher, J. Nagle, and O. Gauthier-Lafaye, "5.2 - 5.6 microns tunable source by frequency conversion in a GaAs based waveguide," Optics Lett. (1998).

123. A. Fiore, V. Berger, E. Rosencher, S. Crouzy, N. Laurent, and J. Nagle, "$\Delta n=0.22$ birefringence measurement by surface emitting second harmonic generation in selectively oxidized GaAs/AlAs optical waveguides," Appl. Phys. Lett. **71**, 2587 (1997).

124. A. Fiore, S. Janz, L. Delobel, P. van der Meer, P. Bravetti, V. Berger, E. Rosencher, and J. Nagle, "Second harmonic generation at $\lambda = 1.06\mu m$ in GaAs based waveguides using birefringence phase matching," Appl. Phys. Lett., submitted for publication (1998).

PARAMETRIC INTERACTIONS IN WAVEGUIDES REALIZED ON PERIODICALLY POLED CRYSTALS.

M.P. DE MICHELI, P. BALDI and D.B. OSTROWSKY
Laboratoire de Physique de la Matière Condensée
UMR 6622 CNRS Univervité de Nice Sophia-Antipolis,
Parc Valrose
06108 Nice Cedex 2, FRANCE.
e-mail : demichel@naxos.unice.fr

Abstract

In nonlinear optics, a waveguide configuration presents the advantage of a higher power confinement and a larger number of phase matching schemes. These advantages are counterbalanced by a certain number of extra requirements and a greater technological complexity. In this paper, these issues will be addressed by a quick review of different configurations which have been experimentally tested, and, as several studies are still underway, the state of the art will be presented.

1. Introduction

During the last few years, considerable research has been oriented toward the development of nonlinear solid state coherent sources with tunable output wavelengths. The recent and significant advances in electric-field poling [1] of ferroelectric crystals have made Quasi-Phase Matching (QPM) [2], in which the domains of the material are periodically inverted to compensate for dispersion, of great interest for a number of nonlinear processes [3, 4] and in particular for Optical Parametric Oscillators (OPO) [5, 6]. Realizing waveguides in periodically poled crystals is an important issue to allow benefiting from additional advantages offered by the waveguide configuration: higher power confinement, much longer interaction length, and efficient electrooptic tuning. In Periodically Poled Lithium Niobate (PPLN), Proton Exchange (PE) [7, 8] is a very promising technique for fabricating such waveguides, as long as it can be controlled and used in a way that preserves all the qualities of the material.
In this chapter, we will present $\chi^{(2)}$ interactions in waveguides, illustrated with some realizations, reviewing different configurations such as perfect and quasi phase matching between guided modes and "Cerenkov" configurations using radiation modes. In this field, most of the experimental data have been obtained using $LiNbO_3$ and $LiTaO_3$. Therefore, the second part of the paper will be devoted to the description of the proton exchange process, which is the most frequently used waveguide fabrication technique for nonlinear integrated optics devices. This presentation will allow underlining one

A. D. Boardman et al. (eds.),
Advanced Photonics with Second-order Optically Nonlinear Processes, 375–400.
© 1999 *Kluwer Academic Publishers. Printed in the Netherlands.*

technological difficulty due to the fact that in certain circumstances, the proton exchange process can reduce or even cancel the nonlinearity and/or the domains organization of the crystal in part of the guiding area. We will finish by indicating that we have identified a given set of parameters with which it is possible to create high quality waveguides without perturbing the nonlinear properties of PPLN, a technique which will applied to devices as soon as an appropriate masking process is available.

2. Interest and specificities of waveguides in nonlinear optics.

In nonlinear optics, the waveguide configuration presents several advantages and some extra requirements which we list and illustrate in this paragraph.

The first advantage of Integrated Optics (IO), is obviously power confinement [9] which allows to have high efficiencies with low pump power, or, for a given incident pump power, to dramatically increase the efficiency. For Second Harmonic Generation (SHG), this efficiency improvement can attain two orders of magnitude.

Another advantage of IO is that one has to fulfill the Phase Matching (PM) or Quasi Phase Matching (QPM) [10] conditions with the effective indices of the modes rather than with the material indices. Playing with the waveguide index profile offers the possibility of using different guided modes [11] and even radiation modes [12] to satisfy the conservation conditions. But to fully benefit from these advantages, one has to keep in mind that another key factor of an efficient conversion in a waveguide, is the overlap between the interacting modes in the nonlinear volume. Due to the particular field distribution of the modes, this factor limits the number of efficient combinations.

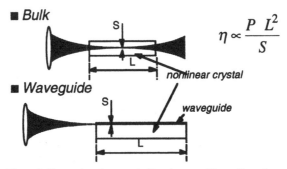

Figure 1. Comparison between bulk and waveguide configurations

2.1. POWER CONFINEMENT AND OVERLAP INTEGRAL

The power confinement advantage in an IO configuration is illustrated in Figure 1. In a bulk configuration, the pump beam is focused at the center of the crystal, and the beam diverges again. The optimization consists in adapting the beam waist S to the crystal length L in the so called confocal configuration $2L=Sn/\lambda$ [13].

In the waveguide case, the pump beam is focused at the input of the waveguide, and remains confined over the entire propagation length. The cross-section of the beam is now independent of the interaction length which can be adapted to the propagation losses and the nonlinear gain.

The requirement on the overlap, is illustrated by Figure 2, in the case of SHG. If the waveguide is monomode at the pump wavelength, it will support several modes at the harmonic frequency.

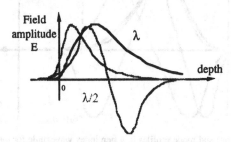

Figure 2. Overlap between different modes.

In Figure 2, we see the field profile of the different modes. The overlap integral I_R between the nonlinear coefficient d and the modes profiles is defined by :

$$I_R \propto \frac{\left(\iint_s d(x,y) \cdot \mathcal{E}_1^i(x,y) \cdot \mathcal{E}_2^i(x,y) \cdot \mathcal{E}_3^t(x,y)\ dxdy \right)^2}{\iint_s \left(\mathcal{E}_1^i(x,y) \right)^2 dxdy\ \iint_s \left(\mathcal{E}_2^i(x,y) \right)^2 dxdy\ \iint_s \left(\mathcal{E}_3^t(x,y) \right)^2 dxdy}$$

Therefore, one sees that when the harmonic is obtained in the first order mode, the overlap term I_R is very small in the case of a constant nonlinearity throughout the waveguide, because of the change of sign of the harmonic field. The corresponding conversion efficiency, which is proportional to I_R, will then be also very small. This is particularly true in a step index waveguide where the shape of the fundamental mode is nearly independent of the frequency and so the overlap between the fundamental modes at both frequencies is by far better than that with the higher order harmonic modes (Figure 3). Howerver, this does not mean that the interaction between fundamental modes is always the most efficient. Most of the useful waveguides have gradient index profiles, and in this case the fundamental mode at the fundamental frequency is quite different from the fundamental mode at the harmonic frequency (Figure 4). In this case one needs to have extra parameters to adjust the profiles. One of these is the thickness of a high index linear layer deposited on top of the waveguide (Figure 5) where the first lobe of the first order harmonic mode is confined allowing a good overlap between its second lobe and the fundamental mode at the pump frequency [14].

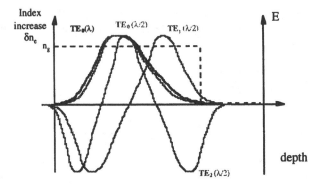

Figure 3. Index (dashed line) and mode profiles in a step index waveguide for the fundamental (in black) and the harmonic (in gray) frequencies.

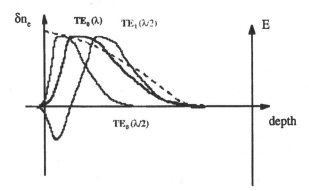

Figure 4. Index (dashed line) and mode profiles in a graded index waveguide for the fundamental (in black) and the harmonic (in gray) frequencies.

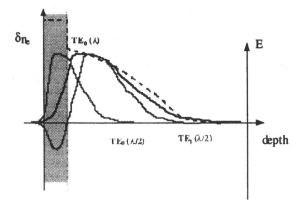

Figure 5. Index (dashed line) and mode profiles in a four layers waveguide for the fundamental (in black) and the harmonic (in gray) frequencies.

2.2. PERFECT PHASE MATCHING IN WAVEGUIDES

In order to determine rapidly the phase matching possibilities of a device, it is useful to plot a phase matching diagram [15]. In the Type I SHG case which we will use to define it, one has to plot $n_i(\lambda)$ and $n_j(\lambda/2)$ versus λ, where i and j represent two directions of polarization.

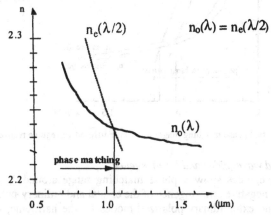

Figure 6. Phase matching diagram in a birefringent material.

The phase matching point is given by the crossing between the two curves (Figure 6). In the case of a waveguide, the situation is more complicated, as several modes can exist at the different frequencies. One should then calculate and plot all the $n^k_{eff}(\lambda)$ and $n^l_{eff}(\lambda/2)$, and look for all the crossing points (Figure 7).

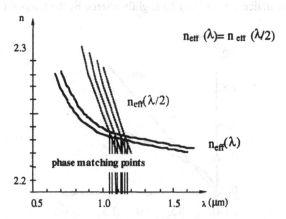

Figure 7. Phase matching diagram in a waveguide.

This can be simplified by noting that the effective indices belong to the interval between the substrate index and the maximum index of the guiding layer, which, in most of the cases, is the surface index, one can plot only $n_{sub}(\lambda)$, $n_{surf}(\lambda)$, $n_{sub}(\lambda/2)$ and $n_{surf}(\lambda/2)$. The overlap between these two sets of curves defines the phase matching range of the waveguide (Figure 8).

380

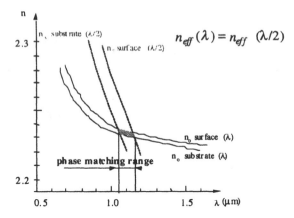

Figure 8. Simplified phase matching diagram for Ti Indiffused waveguide realized on LiNbO₃.

2.2.1 .Ti in-diffused waveguides and birefringence.

Ti in-diffused waveguides show a phase matching range around 1.1μm (Figure 8), obtained using the negative birefringence of the crystal and ordinary polarized modes for the fundamental and extraordinary polarized modes for the harmonic, coupled through the d_{31} (d_{31}=5 pm/V) nonlinear coefficient. This phase matching scheme is compatible with the Ti in-diffusion waveguide fabrication process [16], which increases both indices. The exact position of the phase matching and the efficiency of the interaction then depends on the geometry of the waveguide. In Figure 8, it can be seen that for Ti in-diffused waveguides, in which the index increase is always smaller than 0.02, the phase matching range is very limited around 1.1 μm. Due to the temperature dependence of the indices, it can only be slightly altered by the temperature.

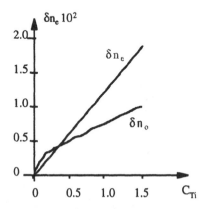

Figure 9. Ordinary and extraordinary index increases in a titanium indiffused waveguide as a function of the titanium concentration.

This configuration has been used by several authors, particularly W. Sohler et al. [17], who carefully studied the dependence of δn_o and δn_e on the Ti concentration (Figure 9), in order to be able to optimize the interaction between fundamental modes at both wavelengths. Using the fact that δn_0 is a nonlinear function of the Ti concentration, and that for small Ti concentration, it is greater than δn_e, they were able to obtain nearly the same confinement for the two interacting modes (Figure 10). Using a resonator [18] to further enhance the power density they were able to get a conversion efficiency as high as 100 %/W, but the harmonic power was limited to only a few microwatt by the photorefractive effect, despite the fact that they were operating around 250 °C.

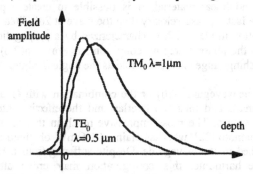

Figure 10. Mode profiled in a Ti-Indiffused waveguide optimized for frequency doubling around 1 µm.

2.2.2. Proton exchanged waveguides and modal dispersion

Since then, a lot of attempts have been made to allow use of the best nonlinear coefficient of LiNbO$_3$, d_{33}=30 pm/V, and to escape the photorefractive damage limitation. Two kinds of attempts were made taking advantage of the large extraordinary index increase (δn_e=0.1) and of the reduced photorefractive sensitivity of Proton Exchanged (PE) waveguides.

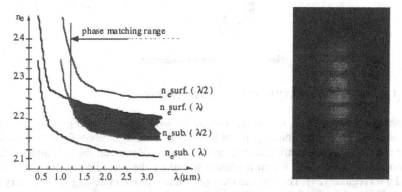

Figure 11. Phase matching diagram of a high index PE waveguide, and near field of the harmonic when phase matching is obtained using the TM$_{50}$ harmonic mode.

Guided modes. In PE waveguides, the extraordinary index increase can be as high as 0.1. This leads to a very important modal dispersion which allows compensation for the chromatic dispersion and gives a broad phase matching range, as it can be seen in Figure 11. But this configuration suffers from a severe limitation : the phase matching condition is satisfied between a low order mode at the fundamental frequency and a high

382

order one at the harmonic frequency. One example of the harmonic field distribution obtained is given in Figure 11, and this clearly demonstrates that, in this case, the value of the overlap integral, and thus of the conversion efficiency, will be very small [19].

"Cerenkov" configuration. One particular case of modal phase matching is the "Cerenkov" configuration, where the harmonic is a radiation mode rather than a guided mode. Indeed, in a nonlinear material, it is possible to create a polarization at 2ω, which travels with a faster phase velocity than the wave at 2ω. Then this polarization is a source for a radiation mode which is characterized by a propagation angle θ (Figure 12), which satisfies the phase matching condition : $n_{eff}(\omega) = n(2\omega) \cos\theta$. In this case also, the phase matching range is considerable as the angle θ adjusts to satisfy the phase matching condition.

A proper design of the waveguide [20] or the combination with Quasi Phase Matching [21], can keep θ small, and thus the overlap and the effective interaction length are maintained at high values. The most impressive result in this configuration has been reported by K. Yamamoto [22], who has claimed 1.6 mW obtained in the blue for 100 mW infrared launched in the waveguide. Despite suffering from a rather unusual field distribution for the harmonic, this configuration may prove attractive for certain applications because of its unique self phase matching possibilities which will be particularly welcome in the case of self doubling of integrated lasers using a rare earth doped nonlinear crystal such as Nd doped LiNbO$_3$ [23].

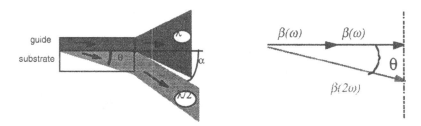

Figure 12. Cerenkov configuration.

2.3. QUASI-PHASE- MATCHING IN WAVEGUIDES.

Quasi phase matching [2] is, nowadays, one of the most popular configurations for $\chi^{(2)}$ interactions in bulk media. Initially, it has been used in waveguide configuration, as the volume necessary in that configuration is much smaller. Moreover, the techniques using a composition gradient and a heat treatment to induce the periodic structure, can also create the waveguide [24]. The main advantage of quasi-phase-matching is that, varying the periodicity of the nonlinear coefficient allows advenement, at room temperature, of phase matched interactions throughout the whole transparency range of the material. This configuration is particularly interesting for ferroelectric crystals where the periodicity of the $\chi^{(2)}$ is obtained by periodically poling the material. Controlling the poling process as been an issue for the last six years, and now several techniques are available on LiNbO$_3$, LiTaO$_3$, KTP, RTA [25, 26, 27, 28, 29, 30, 31]. In this configuration, the waveguides introduces the same advantages and extra requirements as in the perfectly phase matched case, and the quasi phase matching possibilities will be illustrated by the results we obtained in parametric fluorescence.

2.3.1. Influence of the waveguide parameters

In this case a strong pump is launched into the waveguide. Through the nonlinear coupling with the vacuum fluctuations, and if the pump wavelength has been chosen in the proper range, one obtains two waves at lower frequencies, the signal and idler waves, satisfying the energy conservation and the phase matching condition. As for phase matching, quasi phase matching can be obtained between guided modes or can involve radiation modes as well.

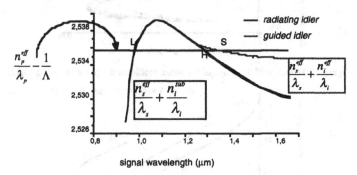

signal wavelength (μm)

Figure 13. Geometrical determination of the signal wavelength in a quasi phase matched parametric interaction. The curve entiled 'radiating idler' coprresponds to the limit where the radiation angle θ=0.

In Figure 13, the wavevector conservation is presented in two configurations where the pump and the signal waves are guided, while the longer wavelength, the idler is either guided or allowed to radiate into the substrate at an angle θ.

In the case where all the waves are guided, it is possible to plot the so call phase matching curve where the idler and signal wavelength are plotted versus the pump wavelength. The numerical simulation shows that despite the fact that the index modifications introduced by the waveguide fabrication processes are generally small, (1 to 2%), their influence on the phase matching curves are important (Figure 14).

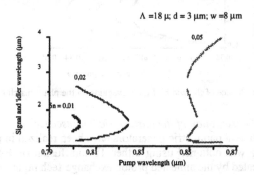

$\Lambda = 18$ μ; $d = 3$ μm; $w = 8$ μm

Figure 14. Influence of the index increase on the phase matching curve.

The detailed analysis, reveals also that the other waveguide parameters play an important role such as the shape of the index profile (Figure 15), its depth (Figure 16) and its width (Figure 17). This sensibility is so important that we have been able to use the experimental phase matching curves to characterize the index profiles of our waveguides with a much better precision than by any other technique.

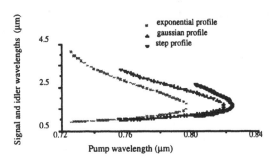

Figure 15. Influence of the shape of the index profile on the phase matching curve.

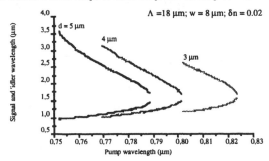

Figure 16. Influence of the depth of the waveguide on the phase matching curve.

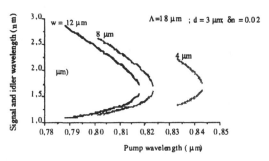

Figure 17. Influence of the width of the waveguide on the phase matching curve.

2.3.2. Experimental observation of the parametric fluorescence

To date, the most efficient parametric generators have been realized in periodically poled LiNbO$_3$. The poling was obtained using either Ti indiffusion or E-field poling. The waveguides were created by the annealed proton exchange technique. When launching a several milliwatt pump beam in the waveguide, one can expect to get signal and idler outputs with a conversion efficiency varying between 10^{-10} and 10^{-8} if the pump is around 0.8 μm, the periodicity of the χ^2 appropriate, and the waveguide properly designed. This very weak fluorescence can nevertheless be analyzed by focusing the near field at the output of the waveguide on the entrance slit of a monochromator after having removed the pump using interference filters and using a nitrogen cooled Ge detector. The spectrum of the collected signal, when launching a 0.8 μm pump in an 8 mm long strip PE waveguide realized in Periodically Poled Lithium Niobate (PPLN) substrate presenting a period of 18 μm (Figure 18), presents a sharp signal peak associated with the interaction between guided modes (the corresponding idler peak cannot be seen because of the limited detection range), and a broad continuum at lower frequencies that corresponds to the signal associated with a radiating idler. On the high frequency side, the signal is limited by the interference filter we used to prevent the pump entering the monochromator [32].

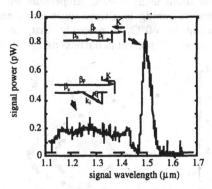

Figure 18. Fluorescence spectrum showing the peak corresponding to a guided idler and the continuum corresponding to a radiated idler.

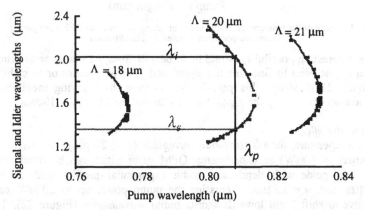

Figure 19. Influence of the domain period on the tuning curves.

By varying the pump wavelength and monitoring the signal peak corresponding to the guided wave interaction, it is possible to plot the phase matching curves which correspond to a given sample. In Figure 19, one can see the influence of the periodicity of the domains on the phase matching curves, and that, by varying the pump by about 20 nm it is possible to cover a range of more than 1 micron for the signal and the idler.

As we have seen numerically and experimentally, the waveguide index profile can dramatically modify the phase matching curve. This fact has two interesting consequences. First, by fitting the experimental phase matching curves it is possible to determine the waveguide parameters with a very high precision ($\delta n_e \pm 0.0001$, depth and width ± 0.1 µm) which is very important to allow predicting the possibilities of a given waveguide before processing it further. Furthermore, this high sensibility to the index variations indicates that, in integrated optics, it is possible to use different processes to fine tune the output wavelength.

Tunability
The first possibility is to use the temperature to tune the output, and Figure 20 indicates the possible variations using a 40°C temperature excursion above the room temperature. But temperature tuning is always quite slow, and the main interest of heating the waveguide up to 60°C is probably that, at this temperature, the photorefractive effect, that we will present in the next paragraph, is suppressed.

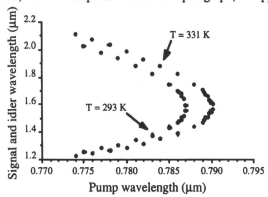

Figure 20. Influence of the temperature on the phase matching curves of an APE waveguide realized on a substrate periodically poled by Ti:Indiffusion .

The most interesting possibility would be to benefit from the planar architecture to use the electro-optic effect to fine-tune the signal and idler frequencies or to stabilize them [33]. Figure 21 illustrates this possibility, showing that varying the indices by $\pm$ 0.0005, allows one to vary the signal by $\pm$ 50nm and the idler by $\pm$ 100nm.

Photorefractive effect
At room temperature, for a 7 µm wide waveguide ($\Lambda = 20$ µm), we observed at low pump power (< 1 mW) an experimental QPM curve which can be fitted numerically using a waveguide with depth and width exponential profiles with $\delta n = 0.0143$, $d = 2.1$ µm and $w = 3.5$ µm. Increasing the pump power up to 20 mW causes the QPM curve to shift 5 nm towards shorter pump wavelengths (Figure 22). The same shift can be obtained numerically by reducing the index increase by 1.8×10^{-3}. This value is compatible with the previously published values of the $LiNbO_3$ photorefractive coefficient [34].
Most of the time, the photorefractive effect is considered to be a perturbation, as it induces instabilities in the operation of a device using this type of waveguide. For

example, at a given pump wavelength, we observed an important change of the signal wavelength (20 nm) by varying the pump power from 1 mW to 30 mW (Figure 23). This appears undesirable for the realization of a source such as an IOPO. However, this effect can be canceled by slightly heating the substrate to 60°C. As shown in Figure 23 we then observed that the signal wavelength is now shifted towards shorter wavelengths but no longer depends on the pump power [33].

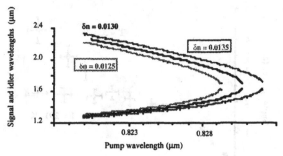

Figure 21. Influence of the electro-optic effect on the phase matching curve.

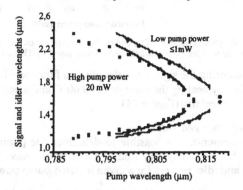

Figure 22. Influence of the photorefractive effect on the phase matching curve of an APE waveguide realized on a substrate periodically poled by Ti:Indiffusion.

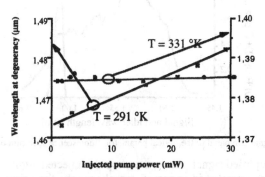

Figure 23. Suppression of the photorefractive effect by heating the substrate of an APE waveguide realized on a substrate periodically poled by Ti:Indiffusion.

The efficiency can also be perturbed by the photorefractive effect. Indeed, as the effective indices are power dependent, the field distribution is also power dependent and at room temperature the overlap integral changes with the injected pump power. This can be observed by plotting the signal power versus the injected pump power and shows that for an injected pump power higher than 30 mW the experimental points deviate from the expected linear dependence (Figure 24).

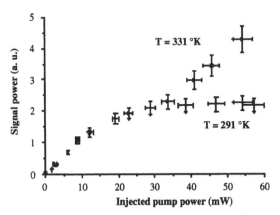

Figure 24. Influence of the photorefractive effect and of the temperature on the output power of an APE waveguide realized on a substrate periodically poled by Ti:Indiffusion.

As we have seen monitoring the signal wavelength as a function of the pump power, we can suppress this effect by heating the waveguide to 60°C and a nearly linear dependence is then observed experimentally (Figure 24).

2.3.3. Difference frequency generators

From the OPF measurements, it is possible to determine the pump wavelength needed to generate a given difference-frequency from a fixed signal wavelength. We can then measure both signal and idler powers for a given injected pump power.

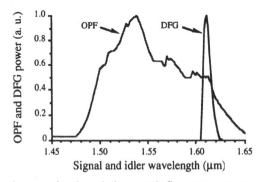

Figure 25. DFG signal compared to the optical parametric fluorescence spectrum of the same sample.

To measure the amplified signal power, we chop the injected signal and pump beams, but to measure the generated idler power, we chop only the injected pump beam to avoid perturbations coming from the signal. Indeed, signal and idler wavelengths are separated by only 60 to 70 nm, and the signal power is 4 orders of magnitude higher than the idler power. Figure 25 shows the difference-frequency generated idler peak compared to the optical parametric fluorescence trace recorded at the output of the same waveguide. Despite these very attractive possibilities and a lot of impressive realizations, some work remains to be done as the waveguide devices in periodically poled substrates, have demonstrated efficiencies which are typically a factor of 2 to 10 bellow theoretical predictions. To understand this problem it is necessary to address the fabrication problems, and as most of the realizations are based on PE waveguides in $LiNbO_3$ or $LiTaO_3$, we will concentrate on these two materials.

3. Waveguides fabrication and characterization in periodically poled $LiNbO_3$ and $LiTaO_3$.

In nonlinear integrated optics , these two materials occupy a very special position and are considered by many authors as a reference. As a matter of fact, most of the efficient components which have been demonstrated so far, are based on them [35]. In order to appreciate the state of the art in nonlinear integrated optics it is then important to know the advantages and the drawbacks of the different techniques that one can use to produce waveguides in these materials.

On $LiNbO_3$, the most commonly used is Ti :indiffusion which increases both ordinary and extraordinary indices. But this technique, which requires heating the crystal near (for $LiNbO_3$) and above (for $LiTaO_3$) the Curie temperature, is hardly compatible with periodic poling.

So, in most of the cases, attempts to get waveguides in PPLN or PPLT, used a proton exchange process to create the waveguide in a previously poled substrate.

Obviously, using E-field poling, one can proceed the other way round, trying to pole the crystal after making the waveguides. The influence of a particular proton exchange process on the poling characteristics of $LiNbO_3$ has recently been pointed out [36, 37], but given the complexity of the proton exchange itself and the number of different crystal modifications which can be introduced by this exchange [38], it is clear that this study is a beginning.

In the following paragraph, we will try to give a flavor of this complexity, summarizing the results we obtained studying the correlation between the optical and the crystallographic properties of the proton exchanged layers.

3.1. PHASE DIAGRAM OF PROTON EXCHANGED LITHIUM NIOBATE AND LITHIUM TANTALATE LAYERS.

Covering a wide range of fabrication parameters and crystal orientation we were able to identify up to 5 crystallographic phases on $H_xLi_{1-x}TaO_3$ and up to 7 in $H_xLi_{1-x}NbO_3$. The exchanges were realized on optical grade X- and Z-cut $LiNbO_3$ and $LiTaO_3$ substrates using, as proton sources, melts and solutions of different acidity such as : $NH_4H_2PO_4$, $KHSO_4$, benzoic acid with and without lithium benzoate, stearic acid, melts of K_2SO_4-Na_2SO_4-$ZnSO_4$-$KHSO_4$ and solutions of LiCl and $KHSO_4$ in glycerin. Some samples were then annealed between 320°C and 400°C, in order to obtain waveguides with an index increase ranging between 0.01 and 0.15 for $LiNbO_3$ and 0.003 and 0.04 for $LiTaO_3$.

390

The obtained planar waveguides were then characterized using a standard prism coupling set-up to measure the modes effective indices at 633 nm and the extraordinary refractive index profiles were reconstructed using the IWKB technique [39]. On the same samples, from different crystallographic planes, we recorded rocking curves that we used to reconstruct the surface layer structure [40]. This work results in the crystallographic phase diagrams of the type shown in Figure 26 and Figure 27.

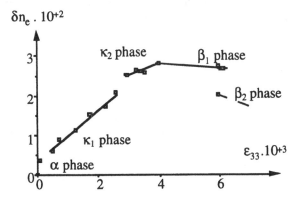

Figure 26. Phase diagram of $H_xLi_{1-x}TaO_3$ layers on top of a Z-cut substrate.

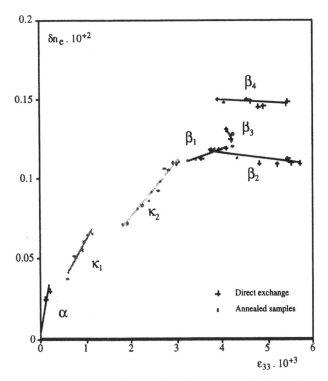

Figure 27. Phase diagram of $H_xLi_{1-x}NbO_3$ layers on top of a Z-cut substrate.

In these diagrams, the phase transitions are characterized by a discontinuity of either the index variation (δn_e), or of the deformations along axis 3 (optic axis : ε_{33}). However, some transitions like the β_1-β_2 transition in Z-cut LiNbO$_3$ or LiTaO$_3$ are not easily observed in these diagrams, but further investigations, such as the measurement of the deformation ε_{23} in order to determine shearing in the planes perpendicular to the surface, reveal the discontinuity.

Moreover, these diagrams show that in each crystalline phase, there is a linear dependence between the index increase (δn_e) and the deformations (ε_{33}) we assume proportional to the proton concentration, but that the slope changes dramatically from one phase to the other.

Based on these facts, we were able to show how these diagrams permit understanding the PE process on LiNbO$_3$ and LiTaO$_3$ and explain the optical properties of the different waveguides obtained with different exchange and annealing conditions [41, 38].

3.1.1. Direct exchange

The α and β phases are obtained by direct exchange using melts of different acidity. When the acidity is varied using a mixture, like lithium benzoate (L.B.) in benzoic acid (B.A.), the β phase is obtained with highly acidic melts, while the α phase is obtained

for higher lithium benzoate concentration $\rho = \dfrac{m_{L.B.}}{m_{B.A.} + m_{L.B.}}$ in the melt.

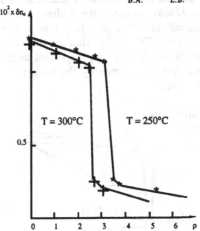

Figure 28. Extraordinary index increase of PE waveguides as a function of the lithium benzoate concentration in the melt.

The concentration limit between the two kinds of exchange is very sharp and also depends on the melt temperature (Figure 28).

3.1.2. Annealing

The effect of annealing on the index profile can be rather complicated, leading sometimes to an increase and sometimes to a decrease of the δn_e. But this complicated behavior can be easily understood by considering the phase diagram and the fact that an annealing always causes a reduction of the proton concentration and thus of the strain ε''_{33}. However, during the annealing, the H$_x$Li$_{1-x}$NbO$_3$ or H$_x$Li$_{1-x}$TaO$_3$ layer may either evolve in a given phase or experience one or several phase transitions. The effect of the

392

annealing on the index profile can then be very complicated but perfectly understood if we know exactly the phase of the original layer. For example, for an initial exchange on LiTaO$_3$ leading to the β_2 phase, index variation (δn_e) and depth will increase during annealing due to an evolution in the β_2 phase followed by a transition to β_1 or κ_2 phase (Figure 26). Continuing the annealing will lead to a decrease of δn_e and an increase of the depth, which corresponds to the evolution through the κ_2, κ_1 and α phases.

Annealing is the only way to obtain layers in the κ phases, but they are not interesting as, especially on LiNbO$_3$, they exhibit very high propagation losses due to a poor crystallographic quality or a perturbed interface between the substrate and the exchanged layer.

3.2. IR ABSORPTION

Another important feature of PE layers can be studied with IR absorption. Indeed, after the exchange the layers produced on X-cut wafers, show a strong absorption band near 3 μm which is due to the vibration of the OH bonds [42]. Depending on the fabrication process, the spectra show a sharp polarized peak alone (σ=3510 cm^{-1}) or followed by a broad absorption band (σ=3280 cm^{-1}) (Figure 29) which tend to disappear when the acidity of the melt is reduced or during the annealing. The precise correlation between the IR absorption spectra and the phase diagram remains to be made. However we have notice [43] that in the case of waveguides in the α phase, the absorption is considerably weaker (this phase corresponds probably to a very low proton concentration) and at a slightly different frequency (σ=3490 cm^{-1}). Though they are still rather qualitative, theses observations of the IR spectra are very important as they reveal the presence of different kinds of phonons which will modify the properties of the crystal or of the dopants introduced in the crystal.

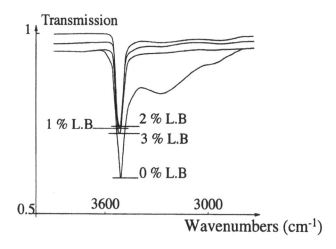

Figure 29. Infrared absorption spectra of the OH bonds created in PE waveguides fabricated at 250°C and with different melts.

3.3. NONLINEAR OPTICAL PROPERTIES OF PROTON EXCHANGED LITHIUM NIOBATE WAVEGUIDES

Studies of the effects of PE on the nonlinear properties of the crystal have lead to some controversy. Some authors claim that it is possible to restore the nonlinearity after certain heat treatments [44, 45] while some others have observed the opposite behavior [46]. In order to clarify this situation, we have investigated the influence of PE on the second order optical nonlinearity [47] using reflected SHG measurements from the polished waveguide end face [48]. This study allows us to show that it is possible to produce, in a one step PE process, waveguides presenting no reduction of the nonlinear coefficient. Moreover, these preliminary results combined with those previously published, tend to indicate that the possibility of restoring the nonlinearity by annealing depends on the fabrication parameters used to produce the initial waveguide and during the annealing.

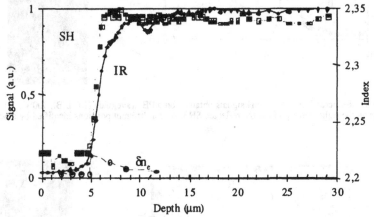

Figure 30. Reflected infrared and SH signals obtained on a PE$_{III}$ waveguide (3% BL, 300°C, 70h).

For the PE$_{III}$ waveguide whose index profile is given in Figure 30, we used a low acidity melt and an exchange duration of 70 h to reach directly the α phase. The registered nonlinear signal is identically continuous from the surface to the bulk. Furthermore, we did not observe any difference between the waveguide and the bulk regions concerning the reflected beam quality. This indicates that this kind of exchange, using a low acidity melt, completely preserves the nonlinear optical properties of the material.

For the PE$_I$ waveguide (Figure 31) prepared in the β_1 phase, using a bath containing 1% of lithium benzoate and an exchange duration of 5h at 300°C, the result is completely different. Indeed, monitoring the SH reflected beam (2ω) while going from the bulk to the waveguide, we observed that the signal practically disappears when the exchanged region is reached. The reflected SH signal measured in the exchanged region is less than 5% of the value registered for the bulk region indicating that in this case the guide nonlinear coefficient is only 20% of the bulk value. At the interface between the exchanged layer and the substrate the nonlinear signal increases, while the quality of the reflected beam severely degrades. This phenomena has also been observed on APE waveguides, with a much higher intensity. Indeed, on our sample which has been annealed in several steps of one or two hours at 330°C and 350°C, the SH signal which was strongly reduced after the initial proton exchange, appears to be two orders of magnitude higher in the exchanged layer than in the substrate!

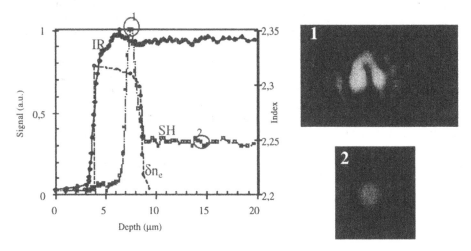

Figure 31. Reflected infrared and SH signals obtained on a PE$_1$ waveguide (1% L.B., 300°C. 5H). The pictures on the right, correspond to the reflected SH spot for different positions identified by a number.

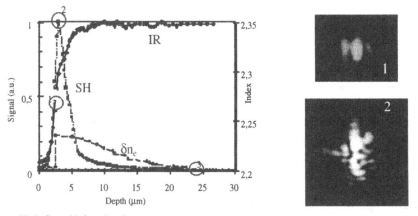

Figure 32. Reflected infrared and SH signals obtained on a APE waveguide. The annealing consists in several steps of one to two hours at 330°C and 350°C. The SH spot corresponding to the position #3 is not represented, but it will be identical to the spot #2 in Figure 31.

In this case also, the increase is accompanied by a strong degradation of the quality of the SH reflected beam. As the examination of the edge of the sample under an optical microscope indicates that the quality of the end face is equivalent to that of the PE$_{III}$ samples, we think that the strong diffusion is due to a poor crystallographic quality of the exchanged layer or of its interface with the substrate. Despite these difficulties this sample suggests that the nonlinearity is at least restored in the initially exchanged layer, and there is no evidence of a linear region at the surface as the fundamental and the second harmonic signals increase simultaneously. We think that this is due to the fact that the initial PE_I waveguide was realized in the β_1 phase and not in the β_2 phase which is used by most other authors [3, 4, 5] using a pure benzoic melt around 200°C. Further studies are underway to clarify these points and to see whether the beam degradation can be avoided using different annealing processes.

In the next section, we will extend our analysis of proton exchange to its effects on the periodic domain structure in periodically poled material.

3.4. DOMAIN STRUCTURE IN PROTON EXCHANGED PPLN WAVEGUIDES.

In this section, we will investigate how the different PE processes affect the domain structure in PPLN crystals using a selective chemical etching technique. We will show that it is possible to produce a waveguide in PPLN without partially erasing the domain structure, by the use a highly diluted melt [49] for the exchange process.

For our study, PE_I and PE_{III} planar waveguides were prepared in the same conditions as those used for the nonlinear measurements. The APE waveguide was exchanged for 25 mn with $\rho=1\%$ and annealed at 350°C for 6h to reach the α phase.

The samples were polished at a wedge angle θ of approximately 0.3 degree to allow, from the top, the observation of both the exchanged and the unexchanged regions (Figure 33). A chemical solution of HF(40%) and HNO3(65%), with the proportions of 1 and 2 respectively, was used to reveal the domains [50] . This mixture etches the Z^- oriented domains much faster than the Z^+ and makes the structure visible through an optical microscope.

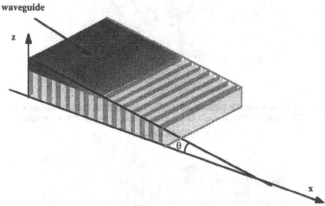

Figure 33. Schematic of the periodically poled samples where the PE surface has been polished at a small angle.

After the polishing, all the samples were etched at room temperature for 10 mn and observed through an optical microscope. Photographs which were taken on each waveguide are shown in the following figures.

PE_I waveguide : After 10 minutes of etching, domains are visible in the region of the sample where the exchanged layer has been completely polished off, while the initial surface layer presents no periodic structure (Figure 34).

APE waveguide : The chemical etching reveals a picture similar to that we observe for the PE_I waveguide (Figure 35). This indicates that the standard annealing process we used on this sample does not allow any regrowth of the domains through the exchanged region. Nevertheless this experiment is not precise enough to say whether the periodic structure disappeared in all the guiding region or only in part of it.

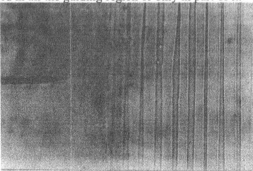

Figure 34. Top surface of periodically poled sample where the PE_I waveguide has been polished away at a small angle. Chemical etching reveals that the exchanged layer shows no periodic structure.

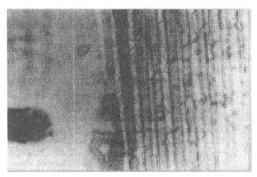

Figure 35. Top surface of periodically poled sample where the APE waveguide has been polished away at a small angle. Chemical etching reveals that the exchanged layer shows no periodic structure.

Figure 36. Top surface of periodically poled sample where the PE_{III} waveguide has been polished away at a small angle. Chemical etching reveals that the periodic structure is an affected by this exchange.

PE$_{III}$ waveguide : Taken after 10 minutes of chemical etching, (Figure 36) shows that the periodic domain structure is present in both the substrate and the exchanged region. This fabrication technique which induces no crystallographic phase transition allows preparing PE waveguides in PPLN without erasing the domain structure.

Table 1. Parametric fluorescence conversion efficiencies of differerent 1 cm long samples prepared using different techniques. In the third column, hc represent the calculated efficiency of a waveguide presenting 0.2 dB/cm propagation losses and no degradation of the domain structure. hm is the measured efficiency. In the fifth column, hc is a calculated efficiency taking into account the defect mentionned in the last column. For sample 4, the high losses are due to the use of an inadequate mask.

Sample	Pump injected in	η^c (10^{-9})	η^m (10^{-9})	η^c (10^{-9})	Defect
Sample 1	TM$_{00}$	0.5			
Poled by Ti diffusion	TM$_{01}$	5	1		
Sample 2 E-field poled	Bulk	1	0.8±0.1	1	Domains erased
Duty cycle ~ 50/50	TM$_{00}$	300	0.1±0.1	0.10	up to 5μm
APE waveguide δn$_e$ = 0.009	TM$_{01}$	100	10±1	10	in the substrate
Sample 3 E-field poled	Bulk	0.1	0.1		Domains erased
Duty cycle ~ 10/90	TM$_{00}$	13	0.05	0.06	up to 7μm
APE waveguide δn$_e$ = 0.009	TM$_{01}$	0.8	0.6	0.5	in the substrate
Sample 4 E-field poled	Bulk				Propagation losses
Duty cycle ~ 50/50	TM$_{00}$	1675	10	10	8.7 dB/cm
PE$_{III}$ waveguide δn$_e$ = 0.016	TM$_{01}$	960	60	62	4 dB/cm

3.5. CONVERSION EFFICIENCY

These results explain why, while bulk experiments have yielded results close to theoretical predictions, the integrated versions of optical parametric generators and optical parametric oscillators [51] have efficiencies below those predicted theorically.

During the last five years, moving from the substrate poled by Ti indiffusion to the e-field poled samples and improving the mode overlap has allowed an improvement of 2 orders of magnitude of the conversion efficiency in parametric fluorescence experiments [52, 53] (Table 1). Nevertheless, the lowest reported threshold of an IOPO is on the order of 1.5W, whereas an optimized IOPO should have a threshold of approximately 10^{-2}W. The best reported difference frequency generator presents a 37%/W conversion efficiency while the theory predicts 800%/W for an optimized device of the same length. In all those devices, the waveguides were realized using the APE technique, which perturbs the domain structure. The degree of perturbation depends strongly on the parameters used to fabricate the waveguides, but with an as yet unpredictable relationship. Our studies concerning the influence of the different kinds of proton exchange on the nonlinear coefficient of the crystal, and on the domain organization [47] indicates that there should be a way to realize waveguide IOPO's presenting the theoretically calculated efficiencies when the technological problem associated with the special photolithography required is solved. Nevertheless to fully understand and control the effect of the PE and APE processes on the shape of the domains further studies are necessary.

4. Conclusion

In this chapter, we have presented the potentialities of Integrated Optics configuration for nonlinear interactions. Today, this field is very active and the state of the art changes very rapidly. To give a deeper understanding of the problems that people realizing demonstrators in PPLN using the proton exchange technique to produce the waveguides, are facing, we have described in detail this process and its interactions with the nonlinear properties of the crystal and the domains organization. This allows us to explain why the IO demonstrators using the Annealed Photon Exchange technique have efficiencies reduced compared to what the theory predicts, and to propose a new fabrication process which should overcome this difficulty.

5. References

1. Webjörn, J., Pruneri, V., Russel, P., Barr, J.R.M. and Hanna, D.C. (1994) Blue light generation in bulk lithium niobate electrically poled via liquid electrod, *Electron. Lett.* **30**, 894.
2. Armstrong, J.A., Bloembergen, N., Ducuing, J. and Pershan, P.S. (1962) Interactions between light waves in nonlinear dielectric, *Phys. Rev.* **127**, 1918-1939.
3. Lim, E.J., Fejer, M.M. and Byer, R.L. (1989) Second harmonic generation of green light in periodically poled lithium niobate waveguide, *Electron. Lett.* **25**, 174-174.
4. Webjörn, J., Laurell, F. and Arvidsson, G. (1989) Blue light generated by frequency doubling of laser diode light in a lithium niobate waveguide, *IEEE Photonics Tech. Lett.* **1**, 316-318.
5. Myers, L.E. Eckardt, R.C., Fejer, M.M., Byer, R.L. and Pierce, J.W. (1995) CW-diode pumped optical parametric oscillator in bulk periodically poled LiNbO₃, *Electronics Letters* **31**, No. 21, 1869-1870.
6. Bosenberg, W.R., Dobshoff, A., Alexander, J.I., Myers, L.E. and Byer, R.L. (1996) Continuous-wave singly resonant optical parametric oscillator based on periodically poled LiNbO₃, *Optics Letters* **21** (10), 713-715.
7. Jackel, J.L., Rice, R.E. and Veslka, J.J. (1982) Proton exchange for high index waveguides in LiNbO₃, *Appl. Phys. Lett.* **41**, 607.
8. De Micheli, M., Botineau, J., Neveu, S., Sibillot, P., Ostrowsky, D.B. and Papuchon, M. (1983) Independent control of index and profiles in proton exchanged lithium niobate guides, *Opt. Lett.* **8**, 114-115.
9. Boyd, J.T. (1972) *J.Q.E.* **8**, 788.

10. Bloembergen, N. (1980) Conservation laws in nonlinear optics, *J. Opt. Soc. Am.* **70** (12) 1429-1436.

11. Stegeman, G.I. and Seaton, C.T. (1985) Nonlinear integrated optics, *J. Appl. Phys.* **58** (12), R-57-R78.

12. Hayata, K., Sugawara, T. and Koshiba, M. (1990) Modal analysis of second harmonic electromagnetic field generated by the Cerenkov effect in optical waveguides, *J. Quant. Elect.* **26** (1), 123-134.

13. Yariv, A. (1975), *Quantum Electronics*, Wiley, New York.

14. Ito, H. and Inaba, H. (1978) Efficient phase-matched second harmonic generation method in four layered optical waveguide structures, *Opt. Lett.* **2** (6), 139-141.

15. De Micheli, M., Botineau, J., Sibillot, P., Neveu, S., Ostrowsky, D.B. and Papuchon, M. (1983) Extension of second harmonic phase-matching range in lithium niobate guides, *Opt. Lett.* **8**, 116-118.

16. Schmidt, R.V. and Kaminov, I.P. (1974) Metal diffused optical waveguides in LiNbO₃, *Appl. Phys. Lett.* **25** (8), 458.

17. Sohler, W. p.449 and Suche, H. p.480 (1984) *New Directions in Guided Waves and Coherent Optics*, D.B. Ostrowsky and E. Spitz (eds), Martinus Nijhoff Ed., The Hague.

18. Hermann, H. and Sohler, W. (1988) Difference-frequency generation of tunable, coherent mid-infrared radiation in Ti:LiNbO₃ channel waveguides, *J. Opt. Soc. Am. B.* **5** (2), 267-277.

19. He, Q., De Micheli, M.P., Ostrowsky, D.B., Lallier, E., Pocholle, J.P., Papuchon, M., Armani, F., Delacourt, D., Grezes-Besset, C. and Pelletier, E. (1992) Self-frequency-doubled high δn proton exchanged Nd:LiNbO₃ waveguides laser, *Opt. Commun.* **89** (1), 54-58.

20. Li, M.J., De Micheli, M.P. and Ostrowsky, D.B. (1990) Cerenkov configuration second harmonic generation in proton exchanged lithium niobate waveguides, *J. Quant. Elect.* **26** (8), 1384-1393.

21. Thyagarajan, K., Mahalakshmi, V. and Shenoy, M.R. (1993) Performance comparison of different configuration for second harmonic generation in planar waveguides, *Int. Journal of Optoelectronics* **8** (4) 319-332.

22. Taniuchi, T. and Yamanoto, K. Mignaturized light sources of coherent blue radiation, *Proc CLEO'87*, 198.

23. He, Q., De Micheli, M.P., Ostrowsky, D.B., Lallier, E., Pocholle, J.P., Papuchon, M., Armani, F., Delacourt, D., Grezes-Besset, C. and Pelletier, E. Self-frequency-doubled high δn proton exchanged Nd:LiNbO₃ waveguide lasers, *Compact Blue-Green Lasers '92*, Santa Fé, New Mexico, USA.

24. Bierlin, J.D., Roelofs, M.G., Brown, J.B., Tohma, T. and Okamoto, S. KTiOPO₄ blue laser using segmented waveguide structures, *Compact Blue-Green Lasers '94*, PDP 7.

25. Yamada, M., Nada, N., Saitoh, M. and Watanabe, K. (1993) First order quasi-phase matched LiNbO₃ waveguides periodically poled applying an external field for efficient blue second harmonic generation, *Appl. Phys. Lett.* **60**, 435-436.

26. Fujimura, M. et al (1992) *Electron Lett.* **28**, 1868-1869.

27. Delacourt, D., Armani, F. and Papuchon, M. (1994) Second harmonic generation efficiency in periodically poled LiNbO₃ waveguides, *J. Quant. Elect.* **30** (4), 1090-1099.

28. Machio, S., Nitanda, F., Ito, K. and Sato, M. (1992) Fabrication of periodically inverted domain structures in LiTaO₃ and LiNbO₃ using proton exchange, *Appl. Phys. Lett.* **61**, (26), 3077-3079.

29. Gupta, M.C., Risk, W.P., Nutt, A.C.G. and Lau, S.D. (1993) Domain inversion in KTiOPO₄ using electron beam scanning, *Appl. Phys. Lett.* **63** (9), 1167-1169.

30. Myers, L.E. (1995) PhD dissertation (G.L. n° 5396, Stanford University).

31. Chen, Q. and Risk, W.P. (1994) Periodic poling of KTiOPO₄ using an applied electric field, *Electron. Lett.* **30** (18), 1516-1517.

32. Baldi, P., Nouh, S., De Micheli, M.P., Ostrowsky, D.B., Delacourt, D., Banti, X. and Papuchon, M. (1993) Efficient quasi phase-matched generation of parametric fluorescence in room temperature lithium niobate waveguides, *Elect. Lett.* **29** (17), 1539.

33. Baldi, P., Aschieri, P., Nouh, S., De Micheli, M. and Ostrowsky, D.B.; Delacourt, D. and Papuchon M. (1995) Modelling and experimental observation of parametric fluorescence in periodically poled lithium niobate waveguides, *J. Quant. Elect.* **31** (6), 997-1008.

34. Mueller, C.T. and Garmire, E. (1984) Photorefractive effect in LiNbO₃ directional couplers, *App. Opt.* **23**, 4348-4351.

35. Fejer, M., this book.

36. Baron, C., Cheng, H. and Gupta, M.C. (1996) Domain inversion in LiTaO₃ and LiNbO₃ by electric field application on chemically patterned crystals, *Appl. Phys. Lett.* **68** (4), 22.

37. Aboud, I., De Micheli, M. and Ostrowsky, D.B.; Smith, P.G.R. and Hanna, D. Etude de l'influence de la fabrication des guides d'ondes sur l'inversion de la polarisation dans le niobate de lithium, *JNOG'97 –* Saine Etienne, France.

38. Korkishko, Yu. N., Fedorov, V.A., De Micheli, M., El Hadi, K., Baldi, P. and Leycuras, A. (1996) Relationships between structural and optical properties of proton-exchanged waveguides on Z-cut lithium niobate, *Applied Optics*, **35** (36), 7056-7060.

39. White, J.M. and Heidrich, P.F. (1976) Optical waveguide refractive index profiles determined from the measurement of mode indices: a simple analysis, *Appl. Opt.* **15**, 151.

40. Fedorov, V.A., Ganshin, V.A. and Korkishko, Yu. N. (1993) New method of double-crystal X-ray diffractometric determination of the strained state in surface-layer structures, *Phys. Status Solidi (a)* **135**, 493.

41. El Hadi, K., Baldi, P., Nouh, S., De Micheli, M.P., Leycuras, A., Fedorov, V.A. and Korkishko, Yu. N. (1995) Control of proton exchange for LiTaO$_3$ waveguides and crystal structure of H$_x$Li$_{1-x}$TaO$_3$, *Opt. Lett.* **20** (16), 1698-1700.

42. Chen, S., De Micheli, M.P., Baldi, P., Ostrowsky, D.B., Leycuras, A., Tartarini, G. and Bassi, P. (1994) Hybrid modes in proton exchanged waveguides realized in LiNbO$_3$, and their dependence on fabrication parameters, *J. Light Tech.* **12** (5), 862-871.

43. Chen, S. (1992) PhD dissertation, Nice.

44. Bortz, M.L., Eyres, L.A. and Fejer, M.M. Depth profiling of the d$_{33}$ nonlinear coefficient in annealed proton exchanged LiNbO$_3$ waveguides, *Appl. Phys. Lett.* **62**, 2012, 2014.

45. Cao, X., Srivastava, R., Ramaswamy, R.V. and Natour, J. (1993) Recovery of second order optical nonlinearity in annealed proton-exchanged LiNbO$_3$, *Photon. Technol. Lett.* **3**, 25-27.

46. Laurell, F., Roelofs, M.G. and Hsiung, H. (1992) Loss of optical nonlinearity in proton-exchanged LiNbO$_3$ waveguides, *Appl. Phys. Lett.* **60**, 301-303.

47. El Hadi, K., Sundheimer, M., Aschieri, P., Baldi, P., De Micheli, M.P. and Ostrowsky, D.B.; Laurell, F. (1997) Quasi-phase-matched parametric interactions in proton exchanged lithium niobate waveguides, *J. Opt. Soc. Am. B* **14** (11), 3197-3203.

48. Ahlfeldt, H. (1994) Nonlinear optical properties of proton-exchanged waveguides inz-cut LiTaO$_3$, *J. Appl. Phys.* **76** (6), 3255-3260.

49. Li, M.J., De Micheli, M., Ostrowsky, D.B. and Papuchon, M. (1987) Fabrication et caractérisation des guides PE présentant une faible variation d'indice et une excellente qualité optique, *J. Optics (Paris)* **18** (3) 139-144.

50. Nassau, K., Levinstein, H.J. and Loïcano (1966) Ferroelectric lithium niobate 1: Growth, domain structure, dislocations and etching, *J. Phys. Chem. Solids* **27**, 983-988.

51. Fejer, M., this book.

52. Baldi, P., Nouh, S., De Micheli, M., Ostrowsky, D.B., Delacourt, D., Banti, X. and Papuchon, M. (1993) Efficient quasi-phase-matched generation of parametric fluorescence in room temperature lithium niobate stripe waveguides, *Electron. Lett.* **29** (17), 1539.

53. Baldi, P., Aschieri, P., Nouh, S., De Micheli, M., Ostrowsky, D.B., Delacourt, D. and Papuchon, M. (1995) Modelling and experimental observation of parametric fluorescence in periodically poled lithium niobate waveguides, *IEEE J. of Quant. Elec.* **31** (6) 997-1008.

Rb:KTP OPTICAL WAVEGUIDES

I. SAVATINOVA [1], I. SAVOVA[1], E. LIAROKAPIS[2],
M.N. ARMENISE [3], V. PASSARO[3]
[1]Institute of Solid State Physics, 72 Tzarigradsko Chausse blvd., 1784
Sofia, Bulgaria
[2]Department of Physics, National Technical University, Athens,
Greece
[3]Dipartimento di Elettrotecnica, Politecnico di Bari, Bari,Italy

1. Introduction

KTiOPO$_4$ (KTP) has unique nonlinear and electrooptic properties, very attractive for a variety of applications in the field of nonlinear and integrated optics. At room temperature, it has a ferroelectric structure and exhibits a second-order displacive phase transition at about 952°C. Waveguide structures in this crystal were produced via exchange of K ions with monovalent ones such as Rb, Tl or Cs [1,2]. It was found that the waveguide parameters are very sensitive to the substrate orientation - it is almost impossible to form a waveguide on X- or Y-cut substrates, while diffusion proceeds intensively on Z-cut plates. The XRD measurements has shown very small change of the lattice constants with the composition [3] and a lower perfection of Rb:KTP crystal lattice, probably due to the difference in the ionic radii (K:1.33, Rb:1.49).

There is an isomorphism of KTP and Rb:KTP, contrary to the case of Cs:KTP. The substitution by Rb causes a small decrease in the axis parameters suggesting that the isomorphic lattice is subjected to a phase transition [2].

2. Optical measurements

Z-cut single domain substrates with refractive indices n_x =1.76404, n_y=1.7724 and n_z=1.86667 (λ=633nm) were immersed in RbNO$_3$ melt at temperatures T=(350 - 400)°C for periods of time t≤2h.

The refractive indices of the substrates and the effective indices of the waveguide modes were measured with the prism coupling technique. The index profiles and the refractive index change on the surface $\Delta n_i(0)$ were calculated by using an improved method for calculating diffused index profiles with parabolic approximation between the successive points ($n_{eff,i}$, z_t). Figure 1 shows the Δn_i (z) profiles (i=x,z) of a typical Rb:KTP waveguide. The profiles are better approximated with exponential function of the form $\Delta n(z)=\Delta n(0)\exp(-z/d_z)$, with the parameters $\Delta n_x(0)$=0.0138, $\Delta n_z(0)$=0.0111 and

401

A. D. Boardman et al. (eds.),
Advanced Photonics with Second-order Optically Nonlinear Processes, 401–404.
© 1999 *Kluwer Academic Publishers. Printed in the Netherlands.*

$d_x \approx d_z = 5\mu m$ (the effective diffusion depth), than with the complementary error function

$n(z) = n(o) \text{erfc} \left(\dfrac{z}{\sqrt{2}d_z} \right)$. As it can be seen from Figure 1 $\Delta n_x(z) > \Delta n_z(z)$ in

disagreement with [1] where $\Delta n_x = \Delta n_z$ is reported.

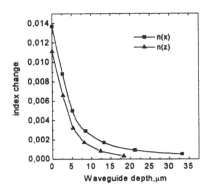

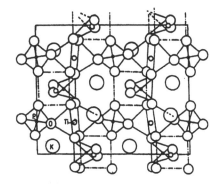

Figure 1. Index profiles Δn_x (z) and Δn_z (z) of a Rb:KTP waveguide.

Figure 2. KTiOPO$_4$ crystal structure.

3. Micro-Raman scattering

KTP crystallizes in the orthorombic system of the nonsymmetric space group C_{2v}^9 (mm2) with eight formula units in the primitive unit cell (Figure 2).

According to the group theoretical analysis there are $47A_1(z) + 48A_2 + 47B_1(x) + 47B_2(y)$ Raman active phonons and LO-TO splitting can also be observed [4]. It is well known that the A_1^{TO} Raman modes are tightly related to the nonlinear susceptibility of the crystal.

The Raman spectra of the KTP single crystals have already been reported [4,5]. The spectral region below 200cm^{-1} is attributed to the external lattice modes in which K ions move with respect to the PO$_4$ and TO$_6$ complexes. The bands in the range (200- 1400)cm^{-1} are attributed to the internal vibrations of the crystal lattice .

Micro-Raman spectra were taken in a back scattering geometry with a triple Jobin-Yvon T64000 spectrometer equipped with a CCD (liquid nitrogen cooled) detector and a microscope with a resolution of 1μm. A 0.5mW 488nm line of Ar$^+$ laser was used.

It can be seen from Figures 3 and 4 that ion diffusion changes the micro-Raman spectra both in the range of the external and TiO$_6$ torsional modes. K$^+$ -Rb$^+$ exchange leads to attenuation of Raman scattering strengths and broadening of some bands. A demonstration of this effect is the Rb:KTP A_1^{TO} (xx) spectrum where the complex spectral structure in the range (100-130) cm^{-1} is almost unresolved (Figure 3). These changes can be attributed to reduction of the translational symmetry of the crystal lattice.

New bands appear in the spectra of the exchanged layers, e.g. in A_1^{LO} (yy) spectrum (Fig. 4) there is a new line at 325 cm^{-1}. After the diffusion some of the bands appear at higher frequencies than in the spectra of the pure crystal. Ion exchange, also, leads to variations in the relative intensity of some bands: in the bulk y(xx)y spectrum (Fig.3), the 122 cm^{-1} band is almost ten orders of magnitude more intensive than the line at 104 cm^{-1}, while in the Rb:KTP spectrum the intensities of the two bands are comparable. The same variations are observed in the range of TiO$_6$ torsional vibrations (Fig.4), which suggest reorientation of the octahedra. Thus, trying to accommodate the larger Rb cations the structure of the exchanged layers substantially changes which leads to changes in the optical and electrooptical properties.

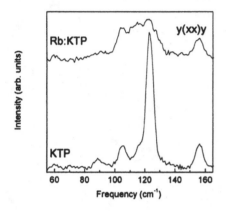

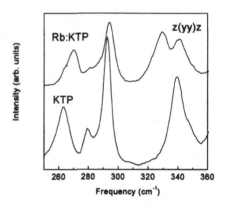

Figure 3. A_1^{TO} Raman spectra in the range of the external lattice modes.

Figure 4. A_1^{LO} Raman spectra in the range of the internal lattice modes.

4. Conclusions

Rb:KTP waveguides have been studied in terms of their mode propagation constants and optical properties. The fabrication of Rb:KTP based devices requires strict control of the ion diffusion parameters. Micro-Raman spectroscopy can be used as an alternative technique for determining and controlling the composition changes.

Acknowledgements

This work was partly supported by the Bulgarian Foundation of Science (Contract no. F-419), the Commission of European Communities (Contract no. COP959) and the NATO grant HTECH.LG 961117.

References

1. Bierlein, J., Ferretti, A., Brixner, L.and Hsu, W. (1987), Appl. Phys. Lett. **50** (18), 1216-1218.
2. Atuchin, V., Bobkov, I., Ziling, C., Plotnikov, A., Semenenko, V.and Terpugov, N. (1993) , Proc. SPIE, Guided-Wave Optics, Proc. SPIE **1932**, 152-172.
3. Lipovskii, A., Lokalov, V. and Nikonorov N. (1993), Proc. SPIE, 6th Europian Conf. on Integrated Opt. ECIO 93, Neuchtel, Switzerland, paper 9/16.
4. Kugel, G., Brehat, F., Wyncke, B., Fontana, M., Marnier, G., Carabatos-Nedelec, C.and Mangin, J. (1988), J. Phys. C **21**, 5565-5583.
5. Tuschel, D., Paz-Pujalt, G., Risk, W. (1995), Appl. Phys. Lett. **66** (9), 1035-1037.

BACKWARD PARAMETRIC INTERACTIONS IN QUASI-PHASE MATCHED CONFIGURATIONS

P. ASCHIERI, P. BALDI, M.P. DE MICHELI, and D.B. OSTROWSKY
Laboratoire de Physique de la Matière Condensée
UMR 6622 CNRS Univervité de Nice Sophia-Antipolis,
Parc Valrose
06108 Nice Cedex 2, FRANCE.
e-mail : aschieri@samoa.unice.fr

G. BELLANCA and P. BASSI
Università di Bologna
Viale Risorgimento 2
I 40 136 Bologna, ITALY

Abstract

We present numerical results on backward interactions in χ^2 media using a Quasi-Phase Matched configuration. Contrary to the classical forward case in which all waves propagate in the same direction, one of the generated waves propagates in the direction opposite to that of the pump. These calculations show that the contrapropagating wave generation configuration is not sensitive to a pump wavelength change over a broad range. This feature may be very useful for realizing a frequency translator. Broadband mirrorless oscillation can also be expected.

Introduction

It is well known that when phase matching is satisfied exciting a χ^2 medium with a pump wave, creates two lower frequency waves called signal and idler[1]. Phase matching can be achieved using crystal birefringence and different polarisations for the interacting waves. However, the largest nonlinear coefficients concern interactions between waves of the same polarisation, QPM technique must be used[2]. The QPM principle is to change periodically the sign of the nonlinear succeptibility. In ferroelectric materials, this can be done using electric field poling[3], concetration gradient due to diffusion[4,5] or e-beam writting[6]. This periodic change of the sign of the nonlinear succeptibility induces a K vector which compensates for the phase mismatch. For a given K, the pump wavelength fixes the signal and idler which satisfy the phase matching conditions. These interactions are governed by the energy and momentum conservation.

405

A. D. Boardman et al. (eds.),
Advanced Photonics with Second-order Optically Nonlinear Processes, 405–408.
© 1999 Kluwer Academic Publishers. Printed in the Netherlands.

It has been shown[7] that the phase matching condition can also be achieved if one of the waves are propagating in the direction opposite to the pump as shown in figures 3 and 4. The K vector is relatively large, so period must be small, typically less than 0.5 µm. This is a major technological problem as it is difficult to realise domain inversion with periods shorter than 1.5 or 2 µm. This may change with the future improvement of this technique, and, higher order interactions can be used.

In the following paragraph we will compare the forward and the backward configuration using the example of *LiNbO3* which is one of the most comonly used material in this field.

Forward configuration.

We show here the classical case of parametric interactions where pump, signal and idler waves are propagating in the same direction.

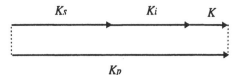

Figure 1: Vector diagram for the forward configuration

where K represents the grating vector and K_p, K_s, K_i are the wave vectors of the pump, signal and idler respectively.

The corresponding QPM curves are presented in figure 2. The plot is done for different poling periods ranging from 17 to 19 µm and one can see that a small change of the pump wavelength induces a large variation of both signal and idler[3].

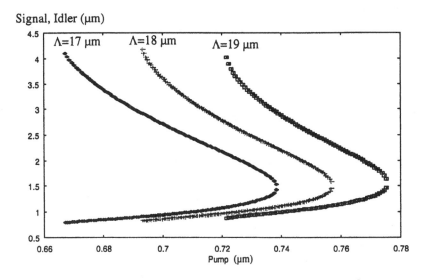

Figure 2: Quasi-Phase-Matching curves in classical configuration

Backward configuration

In the backward configuration, there are two possible wave vector combinations:

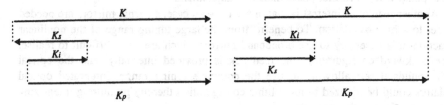

Figure 3: Vector diagram for backward
signal configuration

Figure 4: Vector diagram for backward
idler configuration

The corresponding QPM curves are given for different periods (0.3, 0.4, 0.5μm) as in the forward case. One particularly interesting feature which can be seen on these curves is that, for a wide range of pump wavelength, the backward propagating wave has a nearly constant wavelength while that of the forward wave varies very rapidly. For example with a signal propagating backward, a *100 nm* change in the pump gives a *500nm* change in the idler and only a *1 nm* shift in the signal.

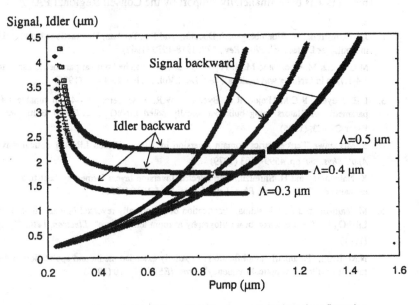

Figure 5: Quasi Phase Matching curves in backward configuration

This feature is particularly intersting to realize a frequency translator. An incoming signal can be translated with the same efficiency in different channels using a tunable pump without further phase matching adjustment.

Another feature of interest is that in the forward case, external mirrors are needed in order to achive oscillation. To benefit from the large tuning range of the nonlinear interaction, it is necessary to use broadband mirrors which are very difficult to realise. In the backward configuration, the feedback is provided internally[7] and the optical cavity builds it self allowing to use the complete tuning range. Integrated optical oscillators could be realized in monolithic configuration thereby permitting integrating other components on the same chip.

Conclusion

We have studied numerically the backward configuration for parametric interactions in QPM scheme. Compared to the forward case, this configuration presents several advantages, such a backward generated wave quasi-insensitive to pump wavelength and the possibility of mirrorless oscillation. A technological effort is still necessary to realized the short period domain inversion required.

Acknowledgement

This work has been financially support by the Conseil Régional PACA.

1. J.A. Armstrong, N. Bloembergen, J. Ducuing, and P.S. Pershan, "Interactions between Light Wave in Nonlinear Dielectric", *Phys. Rev.*, **128**, 1918-1939 (1962)

2. M.L. Bortz, M.Arbore and M.M. Fejer, "Quasi-phasematched optical parametric oscillator between 1.4-1.7μm in LiNbO3 waveguides", *Opt. Lett.*, Vol. **20**, No. **1**, 49-51 (1995)

3. L.E. Meyers, R.C.M. Pejer, R.L. Byer and W.R. Bosenberg, "1.4-4 μm tunable QPM optical parametric oscillators using bulk periodically poled LiNbO3", *CLEO'95* Baltimore, Maryland, Paper CThC3, (1995)

4. S. Miyazawa,"Ferroelectrics domain inversion in Titane diffused LiNbO3 optical waveguide", *J. Appl. Phys.*, **50**, pp. 4599-4603 (1979)

5. K. Nakamura, and H. Shimizu, "Ferroelectric inversion layers formed by heat treatment of proton exchanged LiTaO3", *Appl. Phys. Lett.*, **56**, pp. 1535-1536 (1990)

6. M. Yamada, and K.P. Kishima, "Fabrication of periodically reversed domain structure for SHG in LiNbO3 by direct electron beam litography at room temperature", *Electron. Lett.*, **27**, pp. 828-829 (1991)

7. N.M. Kroll, "Parametric amplification in spatially extended media and application to the design of tunable oscillators at optical frequency", *Proc. IEEE*, **51**, 110 (1963)

MODES OF TM FIELD IN NONLINEAR KERR MEDIA

GRZEGORZ ZEGLINSKI
Laser Technology Department
Institute of Electronics and Computer Science
Technical University of Szczecin
26 Kwietnia 10, 71-062 Szczecin, POLAND

ABSTRACT. The aim of this work is to present waves guidebility at an nonlinear interface. This paper provides solutions of the dispersion equation for two Kerr media (medium 1 - MBBA crystal, medium 2 - YAG crystal) and TM waves at an nonlinear interface. The power for each medium is calculated and showed in this work. This solution can be applied in optical devices.

1. Geometry of the problem

We consider a TM field (E_x, E_z, H_y) and two nonlinear Kerr media (in areas x>0 and x<0 with interface for x=0).

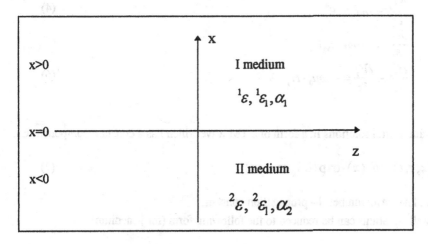

Figure 1. Geometry of the problem

Each of media contain nonlinear uniaxial medium which is describe by the permittivity tensor $^j\hat{\varepsilon}(j=1,2)$

A. D. Boardman et al. (eds.),
Advanced Photonics with Second-order Optically Nonlinear Processes, 409–412.
© 1999 *Kluwer Academic Publishers. Printed in the Netherlands.*

$$
{}^{J}\hat{\varepsilon} = \begin{pmatrix} {}^{J}\varepsilon_{11} & 0 & 0 \\ 0 & {}^{J}\varepsilon_{22} & 0 \\ 0 & 0 & {}^{J}\varepsilon_{33} \end{pmatrix} \tag{1}
$$

where
$$
{}^{J}\varepsilon_{11} = {}^{J}\varepsilon \tag{2}
$$
$$
{}^{J}\varepsilon_{22} = {}^{J}\varepsilon_{33} = {}^{J}\varepsilon_{1} + \alpha_{j} \cdot \left| E_{z} \right|^{2} \tag{3}
$$

We assume a linear part of the refractive index in medium I is larger than in medium II. The media are described by parameters:

- ${}^{1}\varepsilon$, ${}^{1}\varepsilon_{1}$, α_{1} - medium I
- ${}^{2}\varepsilon$, ${}^{2}\varepsilon_{1}$, α_{2} - medium II

${}^{J}\varepsilon$, ${}^{J}\varepsilon_{1}$ -linear permittivities
α_{1}, α_{2}-nonlinear factors
E_{z}- longitunidual component of the electric field

2. Solutions of Maxwell equations

Maxwell equations for the TM field are:

$$
\frac{\partial H_{y}}{\partial z} = i\omega\varepsilon_{0}\varepsilon_{11}E_{x} \tag{4}
$$

$$
\frac{\partial H_{y}}{\partial x} = -i\omega\varepsilon_{0}\varepsilon_{33}E_{z} \tag{5}
$$

$$
\frac{\partial E_{x}}{\partial z} - \frac{\partial E_{z}}{\partial x} = -i\omega\mu_{0} \cdot H_{y} \tag{6}
$$

We want to find solutions in a form of a TM wave which has a constant component Ez.

$$
{}^{J}E_{z}(x,z,t) = {}^{J}E_{z}(x) \cdot \exp\{i(k_{0}hz - \omega t)\} \tag{7}
$$

where: k_{0}- wave number, h- propagation constant
Maxwell equations can be reduced to the following form (for j-medium):

$$
\frac{d^{2}{}^{J}E_{z}}{dx^{2}} - \frac{{}^{J}k^{2}}{{}^{J}\varepsilon}\left({}^{J}\varepsilon_{1} + \alpha_{j}\left|{}^{J}E_{z}\right|^{2}\right){}^{J}E_{z} = 0 \tag{8}
$$

where ${}^{J}k^{2} = k_{0}^{2}(h^{2} - {}^{J}\varepsilon)$ \tag{9}

Solution of the wave equation has a form

$$^jE_z(x) = \delta_j \cdot cn[\gamma_j(x - x_{0j})|m_j)$$ (10)

x_{0j}-second of a constant integration

δ, γ- parameters which determine Jacobi function (cosinus)

$$\delta_j = \frac{^ja + (^ja + 4|^jb|C_1)^{1/2}}{2|^jb|}$$ (11)

$$\gamma_j = (^ja + 4|^jb|C_1)^{1/4}$$ (12)

We assume that this type of wave is propagated in medium 2 (j=2). For medium 1 (j=1) we assume a fading solutions which can be presented in the following form:

$$^jE_z = (\frac{a_j}{b_j})^{1/2}\{ch[a_j(x - x_{0j})]\}^{-1}$$ (13)

where aj and bj :

$$a_j = \frac{^jk^{2\,j}\varepsilon_1}{^j\varepsilon} \quad b_j = \frac{\alpha_j}{2}\frac{^jk^2}{^j\varepsilon}$$ (14)

3. Solutions of the dispersion equation

We apply continuity conditions of tangent components (at interface) and obtain a system of equations which allow detemined of a propagation constant :

$$\delta \cdot cn(\gamma^2 x_0) = (\frac{a_1}{|b_1|})^{1/2}[ch(a_1^{-1}x_0)]^{-1}$$ (15)

$$\frac{2\varepsilon}{k_2^2}\delta\gamma sn(\gamma^2 x_0)dn(\gamma^2 x_0) = \frac{1}{k^2}\frac{\varepsilon}{|b_1|^{1/2}}\frac{a_1^{3/2}}{|b_1|^{1/2}}th(a_1^{-1}x_0)[ch(a_1^{-1}x_0)]^{-1}$$ (16)

The dispersion equation is a second order equation if we omit components which have small values. This equation has 2 solutions:

$$h = \pm\sqrt{\frac{2\varepsilon_2^2 + 2\varepsilon_1\alpha_1}{2\varepsilon_2 + \frac{\varepsilon_1}{\varepsilon_2}\alpha_1 + \alpha_2}} \cong \sqrt{\varepsilon_2}$$ (17)

412

where ε_j - permittivity of medium j ($\varepsilon_j = n_{0j}^2$).

The solutions of the dispersion equation are real and indepent on E0 (field value at interface).

We assume:

- medium 1 - MBBA crystal (n_{01}=1.55, n_{21}=$10^{-9}m^2$/W),
- medium 2 - YAG crystal (n_{02}=1.83, n_{22}=$3*10^{-9}m^2$/W)

We obtain solutions h=±1.8.

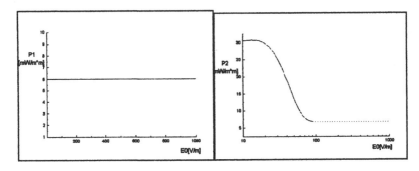

Figure 2. Dependence of the Power (E₀) (for medium 1 x>0 and medium 2 x<0)

4. Conclusions

The solution showed that infuence of nonlinearity on the solutions of the TM waves is very small and influence on the solutions have the linear components. This solution has realizability in: transmision of optic signals, onemodal optical waveguide (h=1.8 - propagation of one mode), polarization devices (possibility of propagation of the TM field, impossibility of propagation of the TE field.)

References

1. W. J. Tomlinson, „Surface wave at a nonlinear interface", *Opt. Lett.*, vol. 5, pp. 323-325,1980.
2. A.A. Maradudin, „Nonlinear surface electromagnetic waves", in *Optical and Aqoustic Waves in Solids-Modern Topics*, M. Borisov, Ed. Singapore: World Scientific Pub., 1983, p.72.
3. N.N. Akhmediev, „Novell class of nonlinear surface waves: Asymmetric modes in symmetric layered structure",*Sov. Phys. JETP.*, vol. 56, pp.299-303, 1982.
4. A.D. Boardman and P. Egan, „Nonlinear electromagnetic surface and guided waves:theory", in. *Proc. 2nd Int. Conf. Surface Waves in Plasmas and Solids*, S. Vukovic, Ed. (Ohrid, Yugoslavia), Singapore: World Pub., pp3-77, 1986.

NUMERICAL SIMULATIONS OF SELF-INDUCED PLASMA SMOOTHING OF SPATIALLY INCOHERENT LASER BEAMS

I. OURDEV[1], D. PESME[1], W. ROZMUS[1], V. TIKHONCHUK[1], C. E. CAPJACK[2] AND R. SENDA[3]

[1] *Department of Physics, University of Alberta, Canada*
[2] *Department of Electrical Engineering*
[3] *Computers and Network Services*

1. Introduction

Stimulated Brillouin Scattering (SBS) is a process in which the incident electromagnetic wave interacts with the ion acoustic fluctuations of the plasma resulting in a scattered electromagnetic wave and a growing resonant ion acoustic wave (IAW). This is a parametric instability which may produce significant level of backscattered light. One way to reduce this level is to introduce a spatial incoherence using laser beam smoothing techniques [1].

The random phase plate (RPP) technique breaks the incident laser light into many beamlets with random phase shifts between 0 and π. In the recent years the interaction of plasma with those large intensity fluctuations, "laser hot spots", were studied both theoretically and numerically (see e.g. [2,3]).

2. Theoretical Model

We employ the general model of ion acoustic wave response to the ponderomotive force of an electromagnetic wave in a two-dimensional (2D) Cartesian geometry. The electric field of the electromagnetic wave is assumed polarized along the direction of propagation (X-axis) and is enveloped in time with respect to the laser frequency ω_0 $\mathcal{E}_x(x,y,t) = (1/2)E(x,y,t)\exp(-i\omega_0 t)+$ c.c.. The electric field amplitude $E(x,y,t)$ satisfies the

$$\left[2i\frac{\omega_0}{c^2}\frac{\partial}{\partial t} + \nabla^2 + \frac{\omega_0^2}{c^2}\left(1 - \frac{n_e}{n_c}\right)\right]E = \frac{\omega_0^2}{c^2}\frac{\delta n_e}{n_c}E \tag{1}$$

A. D. Boardman et al. (eds.),
Advanced Photonics with Second-order Optically Nonlinear Processes, 413–414.

which corresponds to a scalar electric field approximation, $\nabla^2 = \partial^2/\partial x^2 + \partial^2/\partial y^2$. The plasma electron density is described as a constant background n_e and density perturbation, driven by the ponderomotive force which satisfy the following acoustic wave equation:

$$\left[\frac{\partial^2}{\partial t^2} + 2\gamma_s \frac{\partial}{\partial t} - c_s^2 \nabla^2\right] \ln\left(1 + \frac{\delta n_e}{n_c}\right) = \frac{c_s^2}{16\pi n_c T_e}\nabla^2 |E|^2 \tag{2}$$

where c_s is the ion acoustic velocity, n_c is the critical density. The logarithmic term in the l.h.s. of eq.(2) has been introduced to ensure a nonlinear saturation of the density perturbation amplitude.

For the field of the RPP-pump near the focus plane we use:

$$E(x,y) = \sqrt{\frac{P_{inc}}{ia_0}} \frac{1}{2\pi} \sum_{j=1}^{N} e^{i\varphi_j} \exp\left\{-i\theta_j\left(y + \frac{\theta_j}{2}x\right)\right\} \left(2\sin\left(\frac{\pi\xi_j}{N}\right)\bigg/\xi_j\right)$$

with $\xi_j = y + \theta_j x$, $\theta_j = \alpha_j \pi$ being the perpendicular k-vector and φ_j is the random phase of the j-th beamlet.

3. Simulation Results

We studied the propagation of spatially incoherent laser beams in underdense plasma solving numerically eqs.(1,2) with a 2D nonparaxial wave interaction code. The simulations showed an initial co-existence of the density channels created by the ponderomotive force and IAW resulting from SBS. To describe the effects of the self-induced smoothing on SBS we developed a simplified analytical model on the base of the random phase approximation. In the simulations we observed a large angular spread of the transmitted light and an increase of the number of hot spots. The bandwidth and the angular spread increased with the strength of the interaction. The forward spectrum showed evidence of filamentation and forward SBS. The temporal variations of the transmitted light and the changes of the hot spot distribution in time were quantified by laser intensity correlation functions. Comparison of the numerically obtained histograms of hot spot intensities with the theoretical curves [3] justified the applicability of the independent hot spot model.

References

1. Pesme, D. (1987) Effects of temporal and induced spatial incoherence of parametric instabilities in laser plasma interactions, *Annual Report CNRS – LULI* **29**, 1-16.
2. Schmitt,A.J. (1988) The effect of optical smoothing on filamentation in laser plasmas, *Phys.Fluids* **31**, 3079-3101.
3. Rose,H.A., and DuBois, D.F.(1992) Statistical properties of laser hot spots produced by a random phase plate *Phys.Fluids B* **5**, 590-596.

EFFECTS OF IMPURITIES ON EXISTENCE AND PROPAGATION OF INTRINSIC LOCALIZED MODES IN FERROMAGNETIC SPIN CHAINS

S. V. RAKHMANOVA, D. L. MILLS

*Department of Physics and Astronomy and Institute for
Surface and Interface Science, University of California, Irvine,
CA 92697, USA*

The phenomenon of linear localization on the impurities in the periodic structures is very well known. A few decades ago, nonlinear localized waves in ideal periodic lattices have been discovered[1]. Although a great deal of theoretical studies[2, 3, 4] has been devoted to this subject, the existence of intrinsic localized modes (ILM) is still not confirmed experimentally. Thinking about this confirmation we have to take in account the fact that it is impossible to realize an ideal periodic structure in an experiment. There is a certain amount of impurities present in any however clean experimental sample. Therefore, here we investigate the question of coexistence and interaction of localized modes on impurities and intrinsic nonlinear localized modes. We begin by an outline of the properties of nonlinear localized modes in ideal ferromagnet[5]. Linear localization on defects is described in Ref[6].

Let us consider a Hamiltonian of a finite chain of ferromagnetically coupled spins with on-site anisotropy,

$$H = -2J \sum \vec{S}_n \cdot \vec{S}_{n+1} + A \sum (S_n^z)^2 - H_0 \sum S_n^z. \qquad (1)$$

Here J is the strength of the ferromagnetic coupling, A is the hard-axis anisotropy constant, and H_0 is an external magnetic field, applied along the hard axis which is chosen as $\hat{z}$ axis. It is assumed that H_0 is strong so that in the ground state all the spins are aligned along $\hat{z}$. This Hamiltonian is equivalent to the one that describes a magnetic superlattice. In the case of a superlattice spin S_n is the total magnetic moment of an nth film in the lattice. In practice, this magnetic moment is quite large and can be treated as a classical vector. Thus, instead of Heisenberg equations of motion for quantum spins, we use classical Landau-Lifshitz equations:

A. D. Boardman et al. (eds.),
Advanced Photonics with Second-order Optically Nonlinear Processes, 415–418.
© 1999 *Kluwer Academic Publishers. Printed in the Netherlands.*

$$\frac{\partial \vec{S}_n}{\partial t} = \vec{S}_n \times \vec{H}_{eff}, \quad (2)$$

where $\vec{H}_{eff} = H_0 \hat{z} - 2AS_n^z \hat{z} + 2J(\vec{S}_{n-1} + \vec{S}_{n+1})$. We write Eq.(2) separately for $s_n^+ = (S_n^x + iS_n^y)/S$ and $s_n^- = (s_n^+)^*$, where S is the magnitude of the spins. For s_n^+ we assume solution in the form

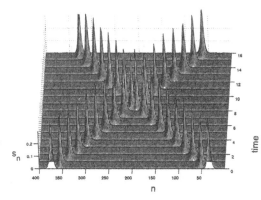

$$s_n^+ = s_n(t)e^{i(kn-\omega t)}, \quad (3)$$

Figure 1. Interaction of two moving ILM with $\Omega = -3.85$.

with $s_n(t)$ complex. Substitution of Eq.(3) into Eq.(2) gives :

$$-i\frac{\partial s_n}{\partial t} = \Omega s_n + 2Bs_n\sqrt{1-|s_n|^2} - s_n(\sqrt{1-|s_{n-1}|^2} + \sqrt{1-|s_{n+1}|^2}) \quad (4)$$

$$+ \cos k(s_{n-1} + s_{n+1})\sqrt{1-|s_n|^2} - i\sin k(s_{n-1} - s_{n+1})\sqrt{1-|s_n|^2},$$

with normalized frequency $\Omega = (\omega - H_0)/2JS$, and normalized anisotropy constant $B = A/2JS$.

Let us assume for the moment that the wave is stationary and circularly polarized. Then, Eq.(4) becomes

$$\Omega s_n = s_n(\sqrt{1-s_{n+1}^2} + \sqrt{1-s_{n-1}^2}) - (s_{n+1} + s_{n-1}) - 2Bs_n\sqrt{1-s_n^2}, \quad (5)$$

where $n = 0, \pm 1, \ldots$ This non-linear equation has solutions in the form of highly localized excitations, if Ω lies above the spin wave band. It is possible to realize one, two or more excitations simultaneously. An example of propagation and interaction of two such localized features is shown in Fig.1. One can see that they have the properties of solitons, i. e. they preserve their shape and speed after the interaction.

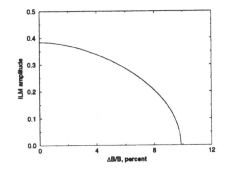

Figure 2. Dependence of an ILM amplitude on the size of the defect. $\Omega = -3.85$.

We now introduce a defect in the system. We write $B = B - \Delta B$ ($\Delta B \geq 0$) in Eq.(4) and Eq(5) for $n = 0$. For small values of ΔB the localized solutions of Eq.(5) still exist, but their amplitude decreases as ΔB increases.

After ΔB reaches some critical value the amplitude of nonlinear localized mode goes to zero and the mode disappears. This process is illustrated in Fig.2 where the amplitude of the ILM with $\Omega = -3.85$ becomes zero when the anisotropy constant of the middle spin is decreased by 10%. One can look at this situation from the following point of view. For any nonzero ΔB there exists a linear mode localized on the defect whose frequency is determined by ΔB, $\Omega_{df} = \Omega_{df}(\Delta B)$,

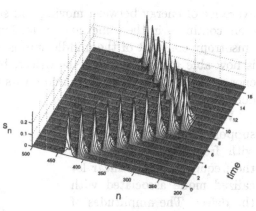

Figure 3. Spin at site number 250 has the value of anisotropy constant about 20% lower than the rest of the spins.

and lies above the extended spin wave band. For an nonlinear ILM to be stable its frequency must be higher than the frequency of any linear mode in the system. Thus, we do not observe ILM with frequencies $\Omega < \Omega_{df}(\Delta B)$.

In the next three Figures we illustrate the propagation of ILM with $\Omega = -3.85$ through the chain with a defect. In Fig.3 the defect in the center of the chain is too strong ($B' = B - \Delta B$ is about 20% less than B) to allow the existence of the nonlinear wave. The wave reflects elastically from it and

travels in the opposite direction with the same amplitude and speed. If the strength of the defect, ΔB, is decreased, then the frequency of the corresponding linear localized mode, $\Omega_{df}(\Delta B)$, becomes comparable to Ω. This case is illustrated in Fig.4, where a traveling ILM hits a defect with B' about 8% less than B, loses part of its energy to initiate a stationary localized wave at the defect site and bounces off it. Due to finite size of the chain the

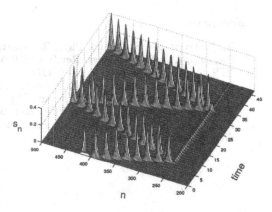

Figure 4. Spin at site number 250 has the value of anisotropy constant about 8% lower than the rest of the spins.

moving mode reflects from the boundary and interacts with the defect again. As one sees from the Figure the traveling ILM actually gains energy after second impact. According to our simulations the

418

exchange of energy between moving and stationary waves at these parameters continues in the same manner for long runtime. Finally, in Fig.5 the anisotropy constant of the middle spin is decreased by 0.75%. The soliton is now able to pass through the defect, but to do so it has to lose some energy in the form of magnons and on its way back it gets trapped on the defect site.

We conclude that defects support stationary excitations with frequencies higher than the frequency of the linear localized mode associated with the defect. The amplitudes of these excitations are smaller than the amplitudes of those on the perfect chain. The presence of defects causes changing of energy and direction of speed of traveling ILM. Sometimes, a moving mode can be pinned to the defect site. Thus, impurities affect strongly the nonlinear localization in ferromagnets and to observe stable low-frequency ILM one need to manufacture relatively clean samples.

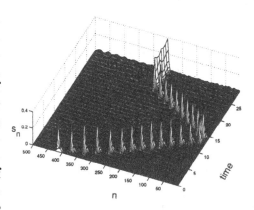

Figure 5. Spin at site number 250 has the value of anisotropy constant about 0.75% lower than the rest of the spins.

References

1. A. J. Sievers and S. Takeno, Phys. Rev. Lett. 61 (1988) 970, S. A. Kiselev, S. R. Bickham, A. J. Sievers, Phys. Rev.B 48 (1993) 13508.
2. Y. S. Kivshar, Phys. Lett. A 161 (1991) 80.
3. R. F. Wallis, D. L. Mills, and A. D. Boardman, Phys. Rev. B 52, R3828 (1995).
4. R. Lai, S. A. Kiselev, and A. J. Sievers, Phys. Rev. B 54, R12 665 (1996).
5. S. Rakhmanova and D. L. Mills, Phys. Rev. B 54, 9225 (1996).
6. A. A. Maradudin, *Theoretical and Experimental Aspects of the Effects of Point Defects and Disorder on the Vibrations of Crystals*, Academic Press, 1966.

OPTICAL PROPERTIES AND HOLOGRAPHIC RECORDING IN Pb₂ScTaO₆ SINGLE CRYSTAL

$$OPTICAL\ PROPERTIES\ AND\ HOLOGRAPHIC\ RECORDING\ IN$$

OPTICAL PROPERTIES AND HOLOGRAPHIC RECORDING IN
Pb$_2$ScTaO$_6$ SINGLE CRYSTAL

V. MARINOVA, S. ZHIVKOVA, D. TONCHEV, N. METCHKAROV
Central Laboratory of Optical Storage and Processing of Information,
Bulgarian Academy of Sciences , Acad.G. Bonchev Str.101,
1113 Sofia, Bulgaria

1. Abstract

Single ferroelecric Pb$_2$ScTaO$_6$ crystals of perovskite stuctural type were obtained by the high temperature solution growth method. The refractive index was measured by the prism method at 10 selected wavelengths. Holographic recording was made in this material for the first time and diffraction efficiency about 20 % was achieved.

Keywords: ferroelecric crystal, refractive index, holographic investigations

2. Introduction

Lead scandium tantalate Pb$_2$ScTaO$_6$ (PST) belongs to the ferroelectic perovskite family with the general formula A(B$'_x$B$''_{1-x}$)O$_3$. What makes it especially interesting is the diffusion phase transition between ferroelectric (FE) and paraelectric (PE) state at room temperature. This transition depends on the degree of structural order of the B-site cations [1]. The temperature dependence of the dielectric permitivity shows that the Curie range of the PST sample is from 10° C to 30° C, the maximum being at 14°C [2]. Photorefractive effect, characteristic for similar materials, makes them attractive as reversible photosensitive medium for holographic recording, holographic interferometry, phase-conjugating mirror, amplification of weak light signals, ect [3-8].

3. Samples and Methods of Investigations

PST crystals were obtained in PbO+PbF$_2$+B$_2$O$_3$ flux in sealed platinum crucibles. Crystals as large as 40 x 40 x 25 mm were produced in the Crystal Growth Laboratory at the Institute of Solid State Physics.
The refractive index was measured using the method of the critical angle determination [9,10], on a crystalline PST prism whose angle at the edge was 16°. The angle of the least deflection was read out at the instant of diffraction pattern disappearing.
Bilaterally polished specimens 1 mm thick and 20 mm in diameter were used to measure transmission spectrum on a Perkin-Elmer 330 spectrophotometer in the visible range and on a Perkin-Elmer 1430 in the infrared.

A. D. Boardman et al. (eds.),
Advanced Photonics with Second-order Optically Nonlinear Processes, 419–422.
© 1999 *Kluwer Academic Publishers. Printed in the Netherlands.*

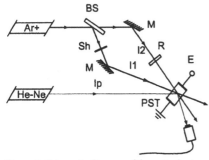

Figure 1. Schematic diagram of the optical set-up.
BS - beam splitter, M - mirrors,
Sh - shutter, R - rotator.

The holographic investigations (Fig.1) were conducted on a PST sample 5x5x5 mm sized and cut in the [100] plane. An Ar^+ laser (λ=488 nm) is used for holographic recording. The interfering waves polarization is perpendicular to the plane of incidence, the angle beteen them is 20°, and their intensities are $I_1 = I_2 = 220$ mW/cm^2. The diffraction efficency is monitored by means of He-Ne beam ($I_p = 2$ mW/cm^2), falling at Bragg's angle. The experiments are conducted at a temperature of 20°C.

4. Results and Discussions

4.1. OPTICAL PROPERTIES

The experimentally determined refractive index values at 10 wavelengths in the visible range are shown in Table 1. Estimated limits of error 0.001.

λ(nm)	457.0	488.0	514.5	530.0	578.2	590.0	610.0	633.0	672.0	700.0
n	2.589	2.547	2.524	2.514	2.493	2.489	2.476	2.471	2.446	2.443

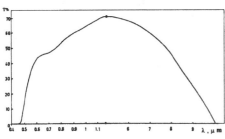

Figure 2. Transmission spectrum of a PST in the visible and inrfared range.

PST crystals are transparent in the 0.38-7.5 µm range (Fig2). Transmission beyond the fundamental absorption edge exhibits a well-pronounced shoulder, which is indicative of high defect density in the crystal structure. From the high-energy end of the transmission region we derived the PST bandgap width $E_g = 3.2$ eV.

4.2. HOLOGRAPHIC RECORDING

The measured diffraction efficiency of the holographic recording in the absence of an external electric field is below 0.008%. The application of a constant field strength of 6 kV/cm results in a drastic increase of the diffraction efficiency. The time dependence of the diffraction efficiency η is shown in Fig.3. After reaching a maximum value η decreases to full erasure of the recording. The curve is asymmetrical - the erasure of the

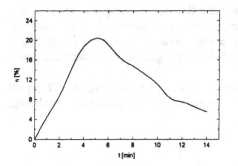

Figure 3. Diffraction efficiency dependence on the exposure time in the presence of an external constant electric field with an intensity E = 6 kV/cm.

grating takes longer than its formation. The maximum diffraction efficiency and the time for reaching it differ considerably in the different cycles recording/erasure - η_{max} varies from 1 to 20%, and $t(\eta_{max})$ - from 1 to 8 minutes. The higher the value of η_{max}, the shorter the time for reaching it. Other peculiarities of the holographic recording in PST become evident in Fig.4. After reaching η_{max}, the behaviour of the curve remains unchanged if one of the recording beams is interrupted (shutter on), i.e. on illuminating the crystal with a spatially uniform light field. Practically, the diffracted beam intensity diminishes at the same rate at the next turn on of the two recording beams (shutter off), the behavior of the curve still remains unchanged (Fig.4a). The situation changes dramatically when the external electric field is switched off (E=0). After the step-wise decrease in the diffraction efficiency by an order of magnitude, the holographic grating continues erasing at the same rate on illuminating the crystal with a spatially uniform light field (Fig.4b), or, the grating erases exponentially on illuminating

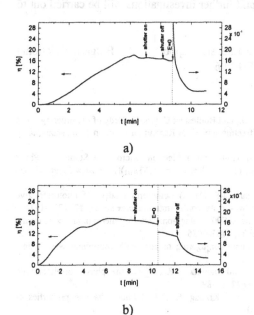

a)

b)

Figure 4. Diffraction efficiency dependence on the exposure time under different conditions during erasure of the holographic grating.

the crystal with an interference pattern (Fig.4a). Such an exponential drop is observed only when the two factors are combined: the two recording beams are turned on (shutter off) and the external electric field is switched off (E=0) (Fig.4a and 4b).

It was found out that the diffraction efficiency depends strongly on the conditions under which the crystal was left just prior to recording. The preliminary spatial uniform illumination with λ=488 nm or 633 nm, simultaneously with the application of a constant external electric field change the recording parameters by orders of magnitude. If the field has one and the same direction before and during the recording, the diffraction efficiency does not exceed 0.1%. However, if the field direction during the recording is opposite to that during the preliminary treatment, the diffraction efficiency is higher than 20% and the time for reaching it is reduced to 1 minute. The application of an external field before the recording in the absence of spatial uniform illumination does not influence the recording parameters. The grating is not destroyed in darkness. After 72 hours the holographic grating was reconstructed with the same diffraction efficiency.

The observed behaviour of the characteristics of holographic recording in PST crystal give grounds to suggest that it is due to orientation and reorientation of polarized domain structures in the bulk of the crystal.

5. Conclusions

The refractive index of PST crystalline material was measured at 10 wavelenght points. A transmission window in the visible and infrared range was taken and the bandgap width was determined to be 3.2 eV.

PST crystal represents a new holographic recording material with diffuse phase transition at room temperature. η_{max} around 20% was achived, as the η and the time for reaching and erasure depend strongly on the preliminary treatments of the crystal. We suggest that the forming of the holographic grating is provoked by amplitude and phase gratings, such the last is combination from phase gratings due to photorefractive effect and changes typical of this diffuse transition. At present, the nature of holographic recording in this material is not yet clear and further investigations will be carried out to understand the physical mechanism.

Acknowledgements: We acknowledge the finansial support of the Bulgarian National Science Foundation under contract MU- F-12/96.

REFERENCES

1. Kang, Z., Caranoni. C., Siny. I., Nihoul. G., and Boulesteix C. (1990) Study of the ordering of Sc and Ta atoms in Pb_2ScTaO_6 by X-ray diffraction and High Resolution Electron Microscopy, *J. of Solid State Chem.* **87**, 308-320

2. Baba-Kishi, B. and Barber, D. (1990) Transmission Electron Microscope Studies of Phase Transitions in Single Crystals and Ceramics of Ferroelectric $Pb(Sc_{1/2}Ta_{1/2})O_3$, *J. Appl.Cryst.* **23**, 43-54

3. Rajibenbach, H., Bann., S. and Huignard, J-P. (1992) Long-term readout of photorefractive memories by using a storage /amplification two-crystal configuration, *Opics Letters* **17**, 1712-1714

4. Tonchev, D., Zhivkova, S. and Miteva, M. (1990) Holographic interferomeric microscope on the basis of a $Bi_{12}TiO_{20}$ crystal, *Applied Optics* **29**, 4753-4756

5. Fenberg, J. and Hellwarth, R. (1980) Phase-conjugating mirror with continuous-wave gain , *Optics Letters* **5**, 519-521

6. Duelli, M., Pourzand, A., Collings, N. and Dandliker, R (1997) Pure phase correlator with photorefractive filter memory, *Optics Letters* **22**, 87-89

7. Yue, X., Xu, J., Mersch, F., Rupp, R. and Kratzig, E. (1997) Photorefractive properties of $Bi_4Ti_3O_{12}$, *Physical Review B* **55**, 9495-9501

8. Sugg, B., Gratchenko, S. an d Rupp, R. (1996) Holographic recording at 633 nm in the garnet $Ca_3Mn_2Ge_3O_{12}$, *J. Opt. Soc. Am B* **13**, 2662-2670

9. Sainov, S. (1992) Differential laser microrefractometer, *Applied Optics* **31**, 6589-6591

10. Sainov, S. (1991) Laser Microrefractometer , *Rev. Sci. Instrum.* **62**, 3106-3107

NONLINEAR EFFECTS IN BULK SEMICONDUCTOR WAVEGUIDE SWITCHES

F. BERTRAND, R. EL BERMIL, N. PARAIRE, N. MORESMAU, P. DANSAS

CNRS URA22, Université Paris -Sud

Institut d'Electronique Fondamentale, Bat 220, 91405- Orsay Cedex FRANCE

1. Framework

In the following, we consider structures made of a resonator filled with a nonlinear material. The resonator is made of a waveguide with a diffraction grating coupler which allows excitation of guided waves in the film. Then reflection (R), transmission (T) and absorption (A=1-R-T) coefficients of the resonator exhibit resonances near guided mode characteristics (λ_m, θ_m). If such a structure is filled with a nonlinear material, it can be used as a switch or a bistable device when it is excited close to resonance[1].

Two such structures have been studied (Fig.1), made of III-V materials :

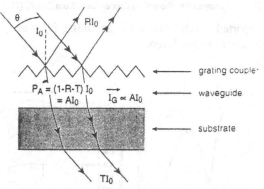

Figure 1 : Diagram of the guiding structures under study

- structure A, due to operate in transmission is made of a InGaAsP film on a transparent InP substrate. The alloy composition is chosen to have a bandgap edge around $0.98\mu m$

- structure B, due to operate in reflection is made of a GaAs film on a Bragg reflector (15.5 AlAs/GaAs pairs).

A. D. Boardman et al. (eds.),
Advanced Photonics with Second-order Optically Nonlinear Processes, 423–426.
© 1999 *Kluwer Academic Publishers. Printed in the Netherlands.*

424

2. Linear regime operation

In semiconductor materials, nonlinearities are associated to e-h photogeneration : absorption must then be present. In other respects, resonances are associated to guided wave propagation and the weaker the damping,the sharper the resonances.So, choosing operating conditions ($\lambda_o \approx \lambda_G$) and optimizing a device (via its grating coupler characteristics) necessitates a good knowledge of the guiding layer parameters, which are measured in the linear regime.

In Fig.2, transmissivity of structure A is reported versus λ, which allows to calculate the imaginary part of the film refractive index $n''(\lambda)$, and its real part (via Kramers Krönig relation) .

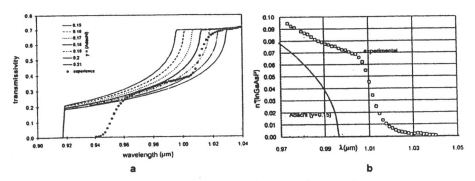

Figure 2 : a- Transmissivity of structure A : $T(\lambda)$,b- absorption of the guiding layer $n''(\lambda)$. Experimental results (dots)are compared to theoretical ones deduced from Adachi [2]

In Fig.3, $R(\lambda)$ is reported : this allows to determine $n(\lambda)$ and, in addition, the thicknesses of the structure various layers.

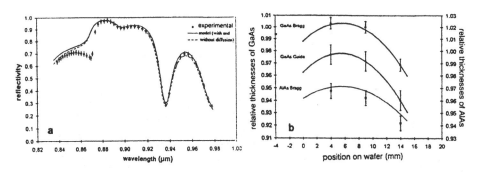

Figure 3 : a- Reflectivity of structure B : $R(\lambda)$, b- cartography of the thicknesses of the wafer various layers

These two figures show that experimental parameters are quite different from those expected. For instance, operating wavelengths will be respectively $\lambda_o \approx 1.04 \mu m$ and $\lambda_o \approx 0.88 \mu m$.

3. Nonlinear regime operation

In semiconductors, nonlinearities associated to absorption present two aspects [3,4]:

- electronic effects associated with e-h photogeneration (with a density δN) and various saturation phenomena. As a result, the material refractive index change is $\delta n_e = - K_e \delta N$, the dynamics of which obeys :

$$\partial(\delta N)/\partial t = \eta(N)I_G - K_1 \delta N - K_2 \delta N^2$$

the first term of the second member of the equation represents e-h photogeneration and is proportional to the guided light intensity I_G, the other terms are recombination ones.

- thermal effects associated to e-h recombination with phonon emission and temperature raise : $\delta n_T = K_T \delta T$ with

$$\partial(\delta T)/\partial t = \zeta \, \delta N - D\Delta\delta T$$

the last term of the expression second member represents thermal diffusion.It is clear that each contribution (δn_e and δn_T) has its own dynamics, so, nonlinearities cannot be described by a stationnary coefficient (as in Kerr effect...).Indeed, according to the relative importance of the 2 effects and the time domain under study, different phenomena can be observed.

Experimentally, we have used submicrosecond light pulses delivered by a Titanium-Sapphire CW laser together with an acousto-optic modulator. In Fig.4a, we see the incident pulse and that transmitted by structure A : bistability (2 switchings during excitation) is observed, due to predominant electronic effects (thermal effects only occur after a few milliseconds).On the contrary, Fig.4b shows light intensity reflected by structure B : only 1 switching can be seen.

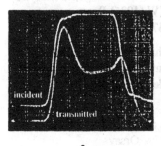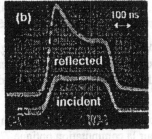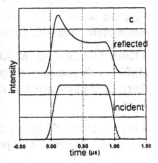

Figure 4 :Nonlinear responses of structures A(a): transmission and B(b): reflection, under square pulse excitation(a few 100ns) at λ_o. c- Modelling of the reflected signal.

426

This can be explained by modelling. Indeed, the global refractive index change can be described by :

$$n(t) = n_o + \delta n_e + \delta n_T = n_o - \gamma A(t)I_o(t) + \beta \int A(u)I_o(u)du$$

where $I_o(t)$ is the incident intensity, A the device absorptance, γ and β electronic and thermal nonlinear coefficients.

Using this relation, one can adjust the nonlinear coefficients to describe experimental results [5] : Fig.4c properly represents Fig.4b, with predominant thermal effects.

4. Conclusions

Nonlinear waveguide structures with grating couplers may operate as sensitive and contrasted alloptical switches (with threshold refractive index changes as low as $\approx 10^{-3}$). III-V materials exhibit high nonlinearities around their bandgap edge and are good candidates for optical switching. However, as these material absorption (n") and nonlinear coefficients vary very rapidly near their bandgap edge, it is important to measure them in-situ to define operating conditions and coupler characteristics. In these materials, various nonlinear phenomena can be observed, which depend on the material and device properties and on the time scale under study.

5. References

1.Paraire,N.,Dansas,P.,Koster,A.,Rousseau,M. and Laval,S.(1991) Sensitivity and switching contrast optimization in an optical signal processing waveguide structure, *Workshop on Optical Information Technology*, Springer-Verlag,Berlin,350-356
2.Adachi,S.,(1989)Optical properties of $In_{1-x}Ga_xAs_yP_{1-y}$ alloys, *Phys.Rev.*B39,12612-12620
3.Lee,Y.H.,Chavez-Pirson,A.,Koch,S.W.,Gibbs,H.M.,Park,S.H.,Morhange,J.,Jeffery,A. Peyghambarian,N.,Banyai,L.,Gossard,A.C.and Wiegmann,W.(1986) Room temperature optical nonlinearities in GaAs,*Phys.Rev.Lett.*57,2446-2449

4.Kisting,S.R.,Bohn,P.W.,Andideh,E.,Adesida,I.,Cunningham,B.T.and Stillman,G.E. (1990) High precision temperature- and energy-dependent refractive index of GaAs determined from excitation of optical waveguide eigenmodes, *Appl.Phys.Lett.*57, 1328-1330

5.Bertrand,F.(1997) Etude expérimentale et modélisation de structures guide d'ondes en semi-conducteurs III-V pour la commutation optique,*Thèse* Orsay

NONLINEAR TRANSMISSION OF ULTRASHORT LIGHT PULSES BY A THIN SEMICONDUCTOR FILM FOR THE CASE OF TWO-PHOTON BIEXCITON EXCITATION

P.KHADZHI*, S.GAIVAN*, O.TATARINSKAYA

State University of Moldova, Kishinev
* Institute of Applied Physics, Academy of Sciences of Moldova, Kishinev

We shall report a theoretical study of transient nonlinear transmission and reflection of ultrashort laser pulses by a thin semiconductor film (TSF) under the conditions of resonant two-photon excitation of biexcitons from the ground state. Since typical times of interaction between optical pulses and media are determined by oscillator strength of specific optical transitions, it is feasible that two-photon excitation of biexcitons from the ground state may facilitate our understanding of transient nonlinear transmission of ultrashort pulses through the TSF.

Suppose that two monochromatic laser pulses with electric field envelopes $E_{01}(t)$ and $E_{02}(t)$ and frequencies ω_1 and ω_2, respectively, are incident on the TSF in vacuum. Under these conditions the interaction of light with semiconductor corresponds to a process described by $\chi^{(2)}$. We assume that the thickness L of TSF is much less than the wavelength λ of the incident pulse and the pulse duration T is much less than the relaxation time of the medium. The durations of the pulses are much smaller than the biexciton relaxation time in the film, and the envelopes are slow functions of time. Suppose that the frequency of each pulse is resonant with neither the excitonic nor M-band exciton–biexciton transitions [1], but that the sum of the photon energies equals to that of the ground state to biexciton transition.

Using the relevant Hamiltonian [2] and the conditions of continuity of the tangential field components at the vacuum–film interface we have derived the following set of equations for determination of amplitudes of transmitted (E_1, E_2) pulses, when the amplitudes of incident pulses, (E_{01}, E_{02}) are given:

$$E_1 = E_{01} - \alpha_1 v E_2, \qquad E_2 = E_{02} - \alpha_2 v E_1, \qquad (1)$$

427

A. D. Boardman et al. (eds.),
Advanced Photonics with Second-order Optically Nonlinear Processes, 427–428.
© 1999 *Kluwer Academic Publishers. Printed in the Netherlands.*

$$\frac{dv}{dt} = \mu \frac{(E_{01} - \alpha_1 v E_{02})(E_{02} - \alpha_2 v E_{01})}{(1 - \alpha_1 \alpha_2 v^2)^2}. \tag{2}$$

where v is a imaginary component of the biexciton wave; $\alpha_1 = 2\pi\hbar\omega_1\mu L/c$, $\alpha_2 = 2\pi\hbar\omega_2\mu L/c$, μ is the constant of the two-photon generation of biexcitons from the ground state. The real component of the medium polarization is equal to zero under the condition of vanishing detuning from the resonance. The typical time τ_0 over which the amplitude of the transmitted pulse changes is determined by the incident radiation intensity:

$$\tau_0^{-1} = \sqrt{\alpha_1\alpha_2}\mu E_{01} E_{02}. \tag{3}$$

Therefore, the thin film easily transmits weak pulses and shortens those of higher intensity. It follows from Eqn (2) that the steady-state amplitudes of the transmitted pulses depend on those of the incident pulses, so that the film transmits a pulse with a high amplitude, whereas a less intense pulse is totally reflected.

The peculiarities of the transmission are studied in detail for the incident pulses with various envelopes, in particular with soliton-like profile. Several nonlinear effects are predicted for the transmission of ultrashort pulses by the film: intensity discrimination, compression and transformation of the profile of pulses incident on the film, and the possibility of "superradiative" amplification of a short pulses traversing a previously excited crystal. Specific examples are given to demonstrate various possibilities of controlling the transmission of both incident pulses by changing the amplitude of one of them. We predict that this technique can be widely used to control the transmission of both pulses by varying the amplitude of one pulse, and therefore can be employed in fast optical integrated circuits.

References

1. Khadzhi, P.I. (1988) *Nonlinear Optical Processes in a System of Excitons and Biexcitons in Semiconductors*, Ştiinţa, Kishinev.
2. Khadzhi, P.I. (1977) *Kinetics of Recombination Radiation from Excitons and Biexcitons in Semiconductors*, Ştiinţa, Kishinev.

THE OPTICAL AND STRUCTURAL PROPERTIES OF $H_xLi_{1-x}NbO_3$ PHASES, GENERATED IN PROTON EXCHANGED $LiNbO_3$ OPTICAL WAVEGUIDES

YU.N.KORKISHKO and V.A.FEDOROV
Moscow Institute of Electronic Technology (Technical University)
103498 Moscow, Zelenograd, Russia

S.M.KOSTRITSKII
Kemerovo State University, Department of Physics, 650043,
Kemerovo, Russia

Seven crystallographic phases $H_xLi_{1-x}NbO_3$ generated in proton exchanged $LiNbO_3$ optical waveguides have been identified and characterized. Correlation between the crystal structure, refractive indices, hydrogen concentration and ferroelectric properties has been experimentally determined.

This work allowed identifying and characterization different crystallographic phases of the proton exchanged (PE) $LiNbO_3$ waveguides, considering that there is a phase jump when gradually varying the hydrogen concentration in the exchanged layer, causes a sudden variation of the cell parameters and chemical bonds, even if the crystallographic system is conserved.

A seven distinct phases (α, κ_1, κ_2 and β_1 to β_4) have been observed in PE $LiNbO_3$ waveguides [1,2]. The κ_1 and κ_2- $H_xLi_{1-x}NbO_3$ phases can exist in high temperature (above 380°C) and low temperature (below 330°C) modifications. The α and β_i phases can be obtained by direct exchange, while the κ_i–phases can be formed by postexchange annealing only.

To realise the PE $LiNbO_3$ waveguides, one can use either a simple exchange, varying the temperature and the acidity of the bath and the duration of the exchange to modify the parameters of the waveguide, or a two or three step process, where the exchange is followed by annealing that duration and temperature further modify the waveguide parameters (Table 1).

The crystallographic $H_xLi_{1-x}NbO_3$ phases generated in proton exchanged $LiNbO_3$ waveguides have been studied by using Double Crystal X-ray Diffraction, IR reflection, Attenuated Total Reflection (ATR), and prism coupling methods. Essential changes of the crystal structure and the lattice vibrational spectrum at the phase boundaries were observed. Strong decrease of the LO-TO splitting for polar mode assignated to NbO_6 octahedra vibrations (800-900 cm^{-1}) was detected. The LO - TO splitting, which is proportional to width of main band ranged from 580 to 900 cm^{-1} characterised by a very different specific values for different phases [3].

A. D. Boardman et al. (eds.),
Advanced Photonics with Second-order Optically Nonlinear Processes, 429–432.
© *1999 Kluwer Academic Publishers. Printed in the Netherlands.*

Both, Double Crystal X-ray diffraction method and ATR methods indicated that when different $H_xLi_{1-x}NbO_3$ phases are generated in proton exchanged $LiNbO_3$ waveguides, these phases are organised as an individual layers, but not as a mixture of different phases.

TABLE 1. Fabrication conditions of studied samples.

Sample/ phase on the surface	Proton exchange Proton-exchange source	Tempera-ture (°C)	Annealing Time (h)	Tempe-rature (°C)	Time (h)
Z-1/β_4	$NH_4H_2PO_4$	235	1.6	-	-
Z-3/β_1	glycerine + $KHSO_4$ (C=2g/l)	230	23	-	-
Z-5/β_2	-"- (C=8g/l)	230	23	-	-
Z-6/β_1	pure glycerine	230	62	-	-
Z-7/β_3	$NH_4H_2PO_4$	230	4	330	0.3.
Z-8/κ_1^{HT}	glycerine + $KHSO_4$ (C=1g/l)	230	18	400	1.5
Z-9/κ_2^{LT}	-"- (C=1g/l)	230	18	325	13
Z-10/κ_1^{LT}	-"- (C=1g/l)	230	18	325	30
Z-11/κ_2^{HT}	-"- (C=1g/l)	230	18	400	0.3.
Z-12/α	-"- (C=1g/l)	230	6	325	60

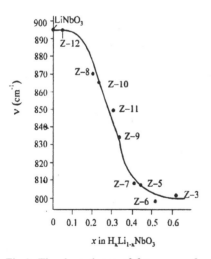

Fig.1. The dependence of frequency of LO phonon for NbO_6 vibrational mode polarised perpendicular to optical axis c versus calculated proton concentration.

Fig.1 shows the results of Kramers-Kroning analysis of IR reflection spectra, i.e. dependence of fundamental value of ν_{LO} versus hydrogen concentration x. The strong decrease of the LO-TO splitting for polar mode assigned to NbO_6 octahedra vibrations was observed for $H_xLi_{1-x}NbO_3$ phases with high hydrogen concentration.

The spontaneous polarization, electrooptic and nonlinear coefficients have been estimated for different $H_xLi_{1-x}NbO_3$ phases (Table 2). The α-phase PE $LiNbO_3$ waveguides characterised by very small degradation (less then 5%) of ferroelectric properties. It was found that the spontaneous polarization, electrooptic and nonlinear coefficients decrease with increasing hydrogen concentration and these values are less then 20% for the β_4 phase.

Only the β_1 -phase can be presented alone on top of the structure, the exchanged layer is then homogeneous, the rocking curves present a clear thin peak for the exchanged layer and the index profile has a step like form. The β_2 and β_3 phases can be found only in combination with the β_1 phase, organized as a sublayer. The β_4 phase can be found only in combination with two lower proton concentrated phases β_3 and β_1. The exchanged layer is then stratified in two or three single phase layers presenting two or three different indices.

TABLE 2. Calculated spontaneous polarization and electrooptical coefficients for some $H_xLi_{1-x}NbO_3$ phases.

Phase at the surface	r_{33}/r^0_{33} or r_{13}/r^0_{13}
κ_1^{HT}	0.925
κ_1^{LT}	0.91
κ_2^{HT}	0.88
κ_2^{LT}	0.83
α	0.99
β_1	0.61*
β_1	0.59*
β_3	0.61*
β_2	0.58*
β_1	0.19-0.28*

* - estimated values.

By annealing it is possible either to involve protons in a given phase or to go from a high concentration phase to a low concentration phase. For example, starting with the β_2-phase one can reach the β_1-phase. By further annealing at temperatures up to 330°C we were able to observe two layers in the crystallographic phases we call here κ_1^{LT} and κ_2^{LT}- phases (low temperature modifications) which cannot be observed after low temperature ($T_{PE}<350°C$) direct proton exchange. By annealing at temperatures above 380°C, a high temperature modification of these phases, called κ_1^{HT} and κ_2^{HT}, were observed. The κ_2 phases can be found only in combination with κ_1 phase, organized as a sublayer:

Note, that exchanged layers of different crystalline phases can generate similar index profiles but differ considerably in structural and other properties such as electrooptical and non-linear coefficients. The propagation losses, which are higher than 1 dB/cm when the β_2-phase is present, are of the order of 0.5 dB/cm only in the β_1-phase. Waveguides, containing the κ_1 and κ_2 phases are characterized by high losses (more than 10 dB/cm). Actually, most of the published results regarding high quality optical waveguides with optical losses less then 0.3 dB/cm were obtained using heavily annealed α-phase PE LiNbO$_3$ waveguides.

Using the early proposed method [4] we have reconstructed the unit cell (lattice parameters a and c) of the all unstrained $H_xLi_{1-x}NbO_3$ phases by examining the corresponding strained waveguiding layers. Ordinary and extraordinary refractive indices were obtained as a function of lattice parameters a and c .

We have proposed a simple and accurate technique to determine the hydrogen concentration in the PE LiNbO$_3$ waveguides for a wide range of fabrication conditions under which a different $H_xLi_{1-x}NbO_3$ phases can be generated [5]. By analyzing the areas under the evolving refractive-index curves during annealing of proton exchanged LiNbO$_3$ waveguides the relationship between extraordinary and ordinary index change and proton concentration have been obtained for all phases (Fig.2).

The plotted $H_xLi_{1-x}NbO_3$ phase diagram allows one to explain some optical phenomena, such as the refractive index increasing after annealing of PE waveguides in LiNbO$_3$, the step-like index profiles, specific features of proton exchange kinetics and some others. Knowledge of $H_xLi_{1-x}NbO_3$ single crystal phase diagram, which are different from known for powders (unstressed state) enables one to predict a properties of the waveguides and to optimize the fabrication process.

432

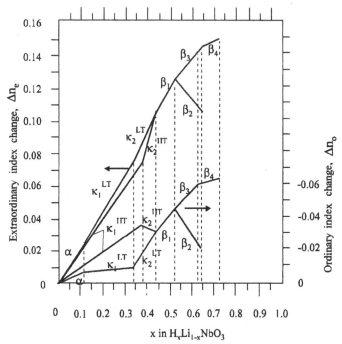

FIG.2. Extraordinary and ordinary refractive indices vs. hydrogen concentration for the different $H_xLi_{1-x}NbO_3$ phases.

References

1. Korkishko Yu. N., Fedorov V.A., De Micheli M.P., Baldi P., El Hadi K., and Leycuras A., (1996) Relationships between structural and optical properties of proton-exchanged waveguides on Z-cut lithium niobate, *Appl. Optics*, **35**, 7056-7080.

2. Korkishko Yu. N., Fedorov V.A., (1996) Structural Phase Diagramm of $H_xLi_{1-x}NbO_3$ waveguides: The Correlation Between Structural and Optical Properties, *IEEE Journal of Selected Topics on Quantum Electronics*, **2**, 187-196 .

3. Korkishko Yu. N., Fedorov V.A., and Kostritskii S.M., (1997) Optical and X-ray Characterization of $H_xLi_{1-x}NbO_3$ Phases Generated in Proton Exchanged LiNbO_3 Optical Waveguides, *submitted to JOSA B*.

4. Korkishko Yu. N.,. Fedorov V.A., (1994) Crystal structure and optical properties of proton-exchanged LiTaO_3 waveguides *Ferroelectrics*, **160**, 185-208.

5. Korkishko Yu. N., Fedorov V.A., (1997) Relationship Between Refractive Indices and Hydrogen Concentration in Proton-Exchanged LiNbO_3 Waveguides, *J. Applied Physics*, **82**, 171-183.

OPTICAL PROPERTIES OF $Bi_{12}TiO_{20}$ PHOTOREFRACTIVE CRYSTALS DOPED WITH Cu AND Ag

V. MARINOVA, V. TASSEV*, V. SAINOV

Central Laboratory of Optical Storage and Processing of Information,
Bulgarian Academy of Sciences, Sofia, Bulgaria
*Institute of Solid State Physics, Bulgarian Academy of Sciences,
Sofia, Bulgaria

1. Abstract

Photorefractive $Bi_{12}TiO_{20}$ crystals doped with Cu and Ag were grown by Czochralski method. The dopands (Cu, Ag) were introduced into the melt in the form of oxides. Transmission and reflection spectra were made in the visible range and absorption coefficients were determined. The optical rotatory power was measured in the spectral range (480-1180 nm) and was established , that doping with Ag and Cu decrease optical activity in comparison with undoped $Bi_{12}TiO_{20}$.

Keywords: photorefractive crystals, absorption coefficient, optical rotatory

2. Introduction

The sillenites with structure Bi_{12} M(=Si,Ge,Ti) O_{20} are well-known photorefractive materials. In the resent years they have attracted increasing attention because of the possibility to be used in reversible holographic recording elements, holographic interferometry, amplification of weak light signals, phase-conjugating mirror , etc [1,2,3].
The spreading of solitons in the sillenites is an interesting phenomenon for the scientists [4].
There is a well-developed technology for growth of large crystals with the high optical quality and reproductible physical parameters. By doping with suitable dopants the concentration of the photosensitive centres increases and the spectral sensitivity of the optical elements extends.
$Bi_{12}TiO_{20}$ (BTO) have some advantages compared to $Bi_{12}SiO_{20}$ and $Bi_{12}GeO_{20}$ crystals:
- higher linear electrooptical coefficients
- higher photosensitivity;
- higher sensitivity of the hologrphic recording in the red spectral region;
- lower optical rotatory power.

A. D. Boardman et al. (eds.),
Advanced Photonics with Second-order Optically Nonlinear Processes, 433–436.
© 1999 *Kluwer Academic Publishers. Printed in the Netherlands.*

The possibility by doping with different dopants the main optical parameters (as the spectral sensitivity and the absorption coefficients) to be modified is a factor with special importance at the produce of optical elements. Using different dopants the properties of BTO can be modified in order to be found the most suitable media for photorefractive applications.

3. Samples and methods of investigations

$Bi_{12}TiO_{20}$ crystals are grown by the high temperature solution method from Bi_2O_3. The doping elements are added in the melt as oxides with the following concentrations:

Dopants	moll %
Ag	0.1
Cu	0.2

The optical transmission and reflection spectra are measured at room temperature in the spectral range 0.40-0.85 μm by spectrophotometer "Perkin Elmer 330".
The optical rotatory power of the samples is measured by a 90°- optical scheme.

4. Results and discussion

The absorption coefficient α was calculated according to the formula [5]

$$\alpha = -\frac{\ln T}{d} - \frac{2R}{d}\left[1 + \frac{R}{2}\left(1 - \frac{T^2}{(1-R)^4}\right)\right] \tag{1}$$

where T is the transmission , R is the reflection and d is the specimen thickness.

Doping of $Bi_{12}TiO_{20}$ crystals with Ag and Cu results in strong changes in the visible optical spectrum. For instance, Ag causes slight bleaching of the crystals, whereas Cu shifts the transmission spectrum to the red spectral region. Cu-doped crystals exhibit a strong photochromic effect as the white light changes their colour from yellow-brown to dark-red. Fig. 1 shows the dependence between the absorption coefficient α (cm^{-1}) and the wavelength for doped and undoped BTO crystals.

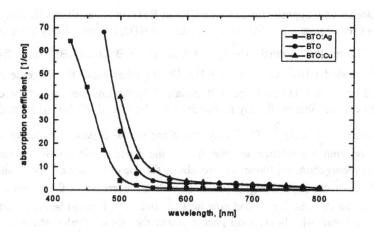

Figure.1. Absorption coefficient α (cm⁻¹) as function of wavelength
for undoped and doped BTO crystals with Ag and Cu.

Because of the fact that the high optical activity (OA) is an undesirable factor at the photorefractive applications [6] of the sillenites, our efforts were directed to be found dopands, that decrease the optical rotatory power (ORP-ρ).

The measurements in the range 0.48 - 1.18 μm show that OA of BTO is much lower than that one of BSO [7]. Cu and, especially, Ag are too suitable dopands because of they strongly decrease ORP of BTO: Cu - with about 31, 29, 21 and 8 % at 480, 522, 756 and 1018 nm, respectively, and Ag up to 70 % in the spectral range 0.48-0.75 μm (Fig.2).

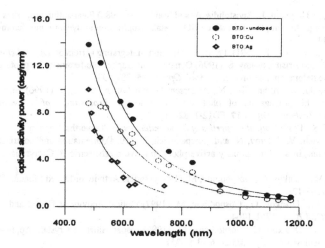

Figure 2. Optical rotatory power of BTO, BTO:Ag and BTO:Cu.

Cu is a dopant with opposite influence on ORP of BTO (in comparison with BSO: Cu). As far as Cu increases ρ of BSO and decreases ρ of BTO, it is most likely Cu to displace mainly Si from the "normal" $[SiO_4]^{4-}$ tetrahedra in BSO and Bi^{3+}-ions from the "defect" tetrahedra $[Bi^{3+}O_3h_0]^{3-}$ in BTO. On the other hand, the decrease of ORP provoked by Cu in BTO could be well explained by the lower energy of the electron transition t_1 - $2e$ (that is directly responsible for the OA of $Bi_{12}MO_{20}$) located at the defect tetraheron $[CuO_4]^{2-}$. Obviously, the place of the dopand occupation strongly depends on both the valence and the size of the dopant ionic radius. As some our previous investigations [8] shown the probability the dopants to occupy the M-site in the "defect" tetrahedra $[Bi^{3+}O_3h_0]^{3-}$ increases with the decrease of the valence. It was established for Al^{3+}, Sc^{3+}, Fe^{3+} and now for Cu^{2+} and Ag^{1+}. It could be explained by the fact, that the ions with larger radii occupy easier the Bi-thetrahedra where an O(3) is already absent. That decreases the number of the Bi-tetrahedra whose optical rotation is the same as that one of the Bi-polyhedra [9] that are the main reason for the BTO optical activity.

Acknowledgement:

We acknowledge the financial support of the Bulgarian National Science Foundation under contracts MU-F-12/96 and F-441.

References:

1. Herriau, J., Huignard, J., Apostolidis, A. and Mallick, S. (1985) Polarization properties in two wave mixing with moving grating in photorefractive BSO crystals. Application to dynamic interferometry, *Opt. Commun.* **56**, 141-144

2. Hong, J. and Chang, T. (1991), Photorefractive time-integrating correlator, *Optics Letters* **16**, 333-335

3. Pouet, B. and Krishnaswamy, S. (1996) Dynamic holographic interferometry by photorefractive crystals for quantitative deforation measurements, *Appl. Optics.* **35**, 787-793

4. Krolikowski, W., Akhmediev, N., Andersen, D. and Luther-Davies, B. (1996) Effect of natural optical activity on the propagation of photorefractive solotons, *Proceedings of Non-linear Cuided-Wave Phenomena, Technical Digest* **17**, FD12-1, 82-84

5. Moss, T. S. (1959) *Optical properties of Semiconductors*, Butterworths, London

6. Shepelievich. V., Egorov, N. and Shepelievich, V. (1994),Orientation and polarization effects of two-beam coupling in a cubic optically active photorefractive piezoelectric BSO crystals, *JOSA (B)* **11**, 1394-1402

7. Tassev, V., Diankov, G., Gospodinov, M. (1995) Optical Activity of Doped Sillenite Crystals, *Mater. Res. Bull.* **30**, 1263-1267

8. Tassev, V., Diankov, G., Gospodinov, M. (1997) Optical Activity of Doped and Co-Doped $Bi_{12}SiO_{20}$ Crystals, *JOSA B* **14**, 1761-1764

9. Tassev, V., Diankov, G., Gospodinov, M. (1996) Doped Sillenite Crystals Applicable for Fiber-Optic Magnetic Field Sensors, *Opt. Mater.* **6**, 3 47-351

COHERENT AND INCOHERENT OPTICAL PROCESSES AND PHASE SENSITIVE ADIABATIC STATES.

I. G. KOPRINKOV,

*Technical University of Sofia, Institute of Applied Physics,
8, Kliment Ochridski blvd., 1156 Sofia, Bulgaria.*

1. Introduction

The interaction of a quantum system (QS) with an electromagnetic field (EMF) represents a basical phenomenon for wide range of physical processes. In the optical phenomena, the main attention is normally paid to the evolution of the parameters of the optical fields. Among these, the optical phase plays a basical role when the optical processes are coherent. Since the field and matter (that mediates the interaction between the optical fields) participate in an apparently equivalent way in the interaction, it is naturally to rise up the question "What happens with the material phase?". It seems that the answer of this question is predetermined by the orthodoxal quantum mechanics, which gives physical meaning only to the modulus of the state vector. Thus, despite that the Schrödinger's equation actually determines the evolution of the entire state vector, the physical state of the QS remains determinate up to an arbitrary, constant (in configurational space) phase factor with unit modulus. This phase factor is irrelevant to the physical processes, that is - unobservable [1, 2].

The material phase has two contributions, one from the dynamical phase and other - from the geometric (Berry's) phase [3]. As summarized in [4], the Berry's phase has observable consequences in wide range of physical processes. A phenomenological appearance of the material (dynamical) phase may be distinguished in recent experiments with material wave-packets within atoms and molecules [5, 6], as well as with atomic de Broglie waves in the realization of the Feynman's *gedanken* experiment [7]. A dramatic change in the observed results has been obtained by shifting the phase with a constant value [5]. These facts, despite not considered in the sense discussed bellow, stimulated the present work.

The main subject of the present work is to consider how the material phase behaves during the formation of the perturbed states of the QS, the so called phase sensitive adiabatic states (ASs) [8], and how different optical processes (coherent and incoherent) can be related with the phase sensitive ASs. From

A. D. Boardman et al. (eds.),
Advanced Photonics with Second-order Optically Nonlinear Processes, 437–440.
© *1999 Kluwer Academic Publishers. Printed in the Netherlands.*

purely qualitative point of view, the ASs (semiclassical dressed states), being eigenstates of the perturbed QS's Hamiltonian, $\hat{H} = \hat{H}_o + \hat{H}'$, represent more adequate basis for consideration of the field-matter interaction problem than the states of the bare QS.

2. Phase Sensitive Adiabatic States

To derive the phase sensitive ASs, the quantum equations of motion for a two-level QS driven adiabatically by a quasimonochromatic EMF have been solved, using adiabatic and rotating-wave approximations. All phase factors (even the constant one) in the initial ground $|g\rangle$ and excited $|e\rangle$ eigenstates of the bare QS (switched-off EMF, E=0), shown in Fig.1, have been considered.

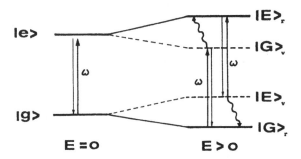

Figure 1. The states of the bare QS (E = 0) and the respective adiabatic states (E > 0).

The ground $|G\rangle$ and excited $|E\rangle$ orthonormal ASs can be represent as

$$|E\rangle = \cos(\vartheta/2)|E\rangle_r - \sin(\vartheta/2)|E\rangle_v$$
$$|G\rangle = \sin(\vartheta/2)|G\rangle_v + \cos(\vartheta/2)|G\rangle_r \qquad (1)$$

where indexices "r" and "v" stand for their real and virtual components.

At ground state initial conditions, (i.e., with certainty, the QS is initially in the ground state $|g\rangle$) the real and virtual components of the ASs (Fig.1) are [8] :

$$|G\rangle_r = |g\rangle \exp(-i\varphi_g)\exp(-i\omega_G t)$$
$$|G\rangle_v = |e\rangle \exp(-i\varphi_g)\exp[-i(\omega_G + \omega)t]$$
$$|E\rangle_r = |e\rangle \exp(-i\varphi_g)\exp(-i\omega_E t) \qquad (2)$$
$$|E\rangle_v = |g\rangle \exp(-i\varphi_g)\exp[-i(\omega_E - \omega)t]$$

where $\omega_G = \omega_g + \delta_-$ and $\omega_E = \omega_e - \delta_-$ are the Stark-shifted frequencies of the real ground and excited state, respectively, $\delta_\pm = \frac{1}{2}[\Delta\omega \pm (\Delta\omega^2 + \Omega^2)^{1/2}]$, $\Delta\omega = \omega_e - \omega_g - \omega$ is zero-order frequency detuning, $\cos(\vartheta/2) = (\delta_+/\Omega')^{1/2}$ and $\sin(\vartheta/2) = sgn(\Delta\omega)(-\delta_-/\Omega')^{1/2}$ are intensity dependent "weight" factors of the ASs components, $\Omega = \mu E/\hbar$ is the resonance Rabi frequency, and $\Omega' = (\Delta\omega^2 + \Omega^2)^{1/2}$ is the off-resonance Rabi frequency.

The real components of the ASs derive from the quasistationary eigenstates of the QS by their continuous evolution under the influence of the EMF (dynamic Stark-shift). The virtual components of the ASs have not analog among the initial quasistationary states. However, they are considered as physically real state to the very same extent, to what extent the entire adiabatic states are physically real.

The peculiarity of the solutions (2) is that the ground state constant material phase φ_g appears in all ASs components, whereas the excited state material phase φ_e totally disappears. At excited state initial conditions similar solutions take place, but now only the excited state material phase φ_e appears in all ASs components, whereas φ_g totally disappears. Thus, the material phase behaves as a traceable quantity in the formation of the ASs, which derive by a continuous and causal evolution from the initial quasistationary states. To the best of our knowledge, such a feature is formulated for the first time and we call it material phase tracking (MPT). Such a physically sensible behavior of the material phase, namely the MPT, and, especially, the experimentally observed manifestation of the material phase [5, 6, 7] means, from our point of view, that the material phase is causally involved in the dynamics of the QS.

3. Coherent and Incoherent Optical Processes in Terms of the Phase Sensitive Adiabatic States.

The huge variety of processes due to the interaction of the QSs with the EMF may be classified either as coherent processes (CPs) or as incoherent processes (ICPs). According to [9], this depends on the coherence or incoherence of the excitation (i.e., whether exist or does not exist a definite phase relationship between the induced polarization and the driving EMF), whereas, according to [10] - on adiabaticity versus nonadiabaticity of the excitation.

On the basis of the announced dynamical implication of the material phase, we relate the CPs and ICPs to the intimate phase behavior of the ASs. Due to the MPT, the real and virtual components of given AS become phase-coupled components (states). The dipole moments of the individual QSs of given ensemble, calculated by means of ASs (1) and (2), become phased to the local adiabatic EMF. According to the adiabatic theorem of quantum mechanics, the QS will remain in a given AS if the excitation is totally adiabatic. Transitions between different ASs may occur under the influence of nonadiabatic factors (zero-point vacuum fluctuations, nonadiabaticity of the forcing EMF, or collisions). This leads to the ASs and dipole moments' phase randomization. Consequently, if the process is adiabatic, the individual QSs remain within given AS, in which a definite phase relationship between their dipole moments and the driving local EMF exists. From the other side, if the process is nonadiabatic, this causes transition between different adiabatic states, which destroys the phase coherence between the dipole moments and the local EMF. Thus, a formal equivalence between both approaches [9] and [10] to the CPs and ICPs can be revealed, considering phase sensitive ASs. This allows to summarize that the CPs are those which develop within given single photon or multiphoton AS, whereas the ICPs are those which involve transitions between different ASs.

4. References

1. von Neumann, J. (1932) *Mathematische Grundlagen der Quantenmechanik*, Verlag von Julious Springer, Berlin, p.19, (in Russian).
2. Landau, L. D., Lifshitz, E. M. (1989) *Quantum Mechanics, nonrelativistic theory*, Nauka, Moscow, p.20 (in Russian).
3. Berry, M. V., (1984) Quantal phase factors accompanying adiabatic changes, Proc. R. Soc. Lond., **A392**, 45 - 57.
4. Anandan, J. (1992) The geometric phase, *Nature*, **360**, 307 - 313, and the references cited therein.
5. Noel, M. W. and Stround, Jr., C. R (1995). Young's double-slit interferometer within an atom, Phys. Rev. Lett., **75**, 1252 -1255.
6. Scherer, N. F. et al. (1991) Fluorescence-detected wave packet interferometry: Time resolved molecular spectroscopy with sequences of femtosecond phase-locked pulses, J. Chem. Phys., **95**, 1487 - 1511.
7. Chapman, M. S. et al. (1995) Photon scattering from atoms in an atom interferometer: coherence lost and regained, Phys. Rev. Lett., **75**, 3783 - 3787.
8. Koprinkov, I. G. (1997) Phase sensitive adiabatic states, Lasers'97 Conference, New Orleans, 1997, (accepted).
9. Grischkowsky, D. (1976) Coherent excitation, incoherent excitation and adiabatic states, Phys. Rev., **14**, 802 - 812.
10. Courtens, E. and Szöke, A. (1977) Time and spectral resolution in resonance scattering and resonance fluorescence, Phys. Rev., **A15**, 1588 - 1603.

HIGH AVERAGE POWER TUNABLE DEEP UV GENERATION USING CASCADING SECOND - ORDER NONLINEAR OPTICAL CONVERSIONS

I. G. KOPRINKOV*, G. A NAYLOR, J.-P. PIQUE
*Laboratoire de Spectrometrie Physique, Universite Joseph -
Fourier Grenoble-1. B.P. 87. 38402 Saint-Martin-d'Heres,
Cedex, France*
*)*Technical University of Sofia, Institute of Applied Physics,
8. Kliment Ochridski blvd.. 1156 Sofia. Bulgaria.*

1. Introduction

The nonlinear optical interactions represent a rich source of phenomena, that has scientific and practical importance. With increasing of the order of nonlinearity, the conversion efficiency strongly falls down. That is why, only the lowest order $\chi^{(2)}$ or $\chi^{(3)}$ processes have a satisfactory efficiency for real practical application. The second and third order nonlinear conversions in nonlinear crystals are now a mature technology for short wavelength generation in the range down to the vacuum UV, which, in particular, is very interesting for variety spectroscopy studies.

The high resolution linear spectroscopy imposes several requirements on the optical sources, i.e., to be narrow band, widely tunable, with good amplitude stability, good average power (so as to facilitate the acquisition and processing of the optical spectra) and enough peak power (for subsequent nonlinear conversions). To find a primary laser pump source that satisfies all of the above requirements is not an easy problem.

In this work we demonstrate a convenient and reliable way of generation of deep UV radiation, by tripling the emission of a 6.5 kHz repetition rate copper vapor laser (CVL) pumped dye laser (DL). This source is specially designed for spectroscopy of high-lying states. The nonlinear tripling can be done either by direct third harmonic generation (THG) in a single crystal or by two cascading second order processes - second harmonic generation (SHG) and sum frequency mixing (SFM). In principle, the cascading way takes place in single crystal, but using separate crystals allows independent optimization of both processes. A proper optimization of the cascading tripling scheme is necessary, because the peak power of the CVL pumped DL is relatively low. The cascading SHG and SFM in two separate crystals, chosen here, allow independent phase matching of both processes, which is impossible in the same crystal [1, 2].

A. D. Boardman et al. (eds.),
Advanced Photonics with Second-order Optically Nonlinear Processes, 441–444.
© *1999 Kluwer Academic Publishers. Printed in the Netherlands.*

2. Experimental arrangement

The CVL pumped DL and the nonlinear conversion schemes are shown in Fig.1. The CVL is an Oxford Laser Cu60. The DL consists of an oscillator and two amplifiers A1 and A2. The DL oscillator has an Littman type cavity, which consists of a grazing incidence grating (G) (11cm long, 2400 l/mm), prism beam expander (PBE), an output coupling mirror (OM) and a tuning mirror (TM). The oscillator dye was DCM (10^{-3} M/l in methanol).

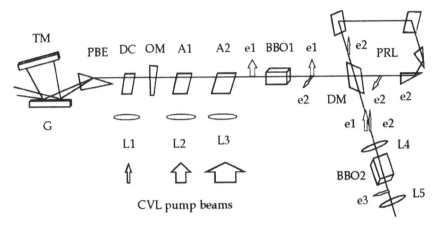

Fig. 1. Copper vapor laser pumped dye laser oscillator-amplifiers system and the nonlinear tripling scheme.

The nonlinear conversion scheme is also shown in Fig.1. The SHG is generated in a 9 mm long BBO crystal using type I phase-matching. The SFM of the second harmonic and the non-converted part of the fundamental is performed in a second 8 mm long BBO crystal cut for type I phase-matching, too. To ensure right polarization of the fundamental and the second harmonic, we separate both beams by a dichroic mirror DM. The polarization of the fundamental was rotated 90^0 by a prism type polarization rotation loop (PRL), which is wavelength independent. The fundamental and the second harmonic were focused into the second BBO crystal and subsequently recolimated using lenses L4 and L5.

3 Results and discussions

The typical parameters of the DL oscillator-amplifier system in this experiment were: 2.3 W average power, 20 kW peak power and average bandwidth 3GHz, or $0.1 cm^{-1}$, with long term fluctuations of only of few percent. The average power of the generated second harmonic was about 250mW. At these conditions, the present concept, based on a combination of cascading scheme for the nonlinear tripling and a copper vapor laser as a pump source, allows to reach, at the same time, several important spectroscopic features for the generated deep uv radiation: wide tunability, 207-220 nm; narrow bandwidth, <3 GHz, (or bellow 0.1 cm^{-1}); deep uv wavelength, around 213 nm at the maximum output power; high repetition rate, 6.5 kHz; and high average power, 10 mW. With some expense of the average power, the bandwidth of the emission can be reduced bellow 1GHz. For comparison, some representative results using different concepts of tunable deep uv generation in nonlinear crystals are shown in Table 1

TABLE 1. Comparison between different nonlinear conversion schemes.

Ref. No:	pump laser operation mode & parameters	nonlinear conversion scheme	parameters of the deep uv emission
1	mode-locked, single ps-pulses I_p*=50GW/cm^2	direct THG	wideband, nontunable $\eta = 0.8\%$
2	mode-locked, high RR ps-pulses I_p=1.5GW/cm^2	direct THG	wideband, $\eta = 7 \times 10^{-5}$
3	CW, narrowband, tunable	cascading SHG & SFM, DIWs, NLCEC	narrowband, tunable, 194nm, 16µW
4	CW, narrowband	cascading SHG & SFM, DIWs, NLCEC	narrowband, tunable 194nm, 2mW
5	CW, mode-locked wideband, tunable	cascading SHG & SFM, SIW, no NLCEC	wideband, tunable, 210nm, 10mW
this work	pulsed, free running narrowband, tunable	cascading SHG & SFM, SIW, no NLCEC	narrowband (3GHz), tunable,213nm,10mW

*) Pump intensity (Ip), repetition rate (RR), single initial wavelength (SIW), double initial wavelengths (DIWs), nonlinear conversion enhancement cavity (NLCEC)

The total conversion efficiency of the whole tripling process, $\eta = P_{3\omega}/(P_{2\omega}P_{\omega})^{1/2}$, is 1.3 %, based on the average power. As must be expected, this is lower than the results obtained using high peak power lasers. The average power, however, is among the highest reported for tunable narrow-band radiation in this spectral range [6], and exceeds by several orders of magnitude that, obtained by direct phase-matched THG [1, 2]. It is important to note that above results have been achieved using only convenient and reliable experimental solutions: i) single initial wavelength scheme, thus avoiding complications with the time synchronization and space overlapping of two independent laser beams, ii) no external enhancement cavities for the nonlinear crystals, that compensate insufficient peak pump power, but complicate the system as a whole.

An additional advantage of this system is that the average power is spread over time at 6.5 kHz repetition rate, reducing the peak power. High peak power is not suitable for high resolution linear spectroscopy, because this can lead to a number of undesired concomitant effects, i.g., Stark-shift of the respective levels, multiphoton ionization, and variety of laser type or multiphoton stimulated processes. All these may be a source of errors and confusion.

The shortest wavelength generated by this DCM based dye laser system (chosen for specific spectroscopic applications) was around 207nm. Wavelengths around or even slightly bellow 200nm still can be generated using this single initial wavelength scheme, simply changing the DCM dye by a shorter wavelength one, for example, Rhodamin 6G whose efficiency is even higher. To reach the transparency limit of BBO (~190nm), two widely different initial wavelengths are necessary for phase-matched SFM [3, 4].

In conclusion, we demonstrate a convenient and reliable way of generation of deep uv emission, using cascading second order nonlinear conversion, and a copper vapor laser based system. The generated emission satisfies in various ways the main requirements of the high resolution linear spectroscopy of high-lying atomic and molecular states.

4. References

1. Qiu, P.and Penzkofer, A. (1988) Picosecond third-harmonic light generation in $\beta - BaB_2O_4$, *Appl. Phys.* B45, 225 - 236.
2. Tomov, I. V, Wanterghem, B. V.. and Rentzepis. P. M. (1992), *Appl. Opt.* **31**, 4172.
3. Watanabe, M., Hayasaka, K., Imajo, H., and Urabe, S. (1992) Continuous-wave sum-frequency generation near 194 nm in $\beta - BaB_2O_4$ crystals with an enhancement cavity, *Opt. Lett.*. **17**, 46 - 48.
4. Berkeland, D. J.. Cruz. F. C., and Bergquist, J. C. (1997) Sum-frequency generation with beta-barium borate pushes ultraviolet output to 194 nm, *Laser Focus World*, **33**, No 8, p.9.
5. Nebel, A., and Beigang, R. (1991) External frequency conversion of cw mode-locked $Ti:Al_2O_3$ laser radiation , *Opt.. Lett.,* **16**. 1729 - 1731.
6. Koprinkov, I. G., Naylor, G. A., and Pique, J.-P. (1994) Generation of 10 mW tunable narrowband radiation around 210 nm using 6.5 kHz repetition rate copper vapour laser pumped dye laser, *Opt. Commun.,* **104**, 363 - 368

OPTICAL FILTERS AND SWITCHES USING PHOTONIC BANDGAP STRUCTURES

P. DANSAS, N.A. PARAIRE

CNRS URA22, Université Paris Sud

Institut d'Electronique Fondamentale, Bat 220, 91405- Orsay Cedex FRANCE

1. Introduction

Photonic crystals are artificial media which exhibit a 3D periodicity, with periods of the order of the electromagnetic (EM) wavelength under study. They behave, for EM waves as semiconductors for electrons : they present frequency ranges where EM waves can propagate with a dispersion relation $\omega(k)$ and frequency bandgaps (BG). An EM wave incident on a photonic crystal with a frequency included in a BG is totally reflected (T=0).If a defect is inserted in such a structure, propagation modes can appear in BG and the material behaves as a filter.

In the following, we present theoretical studies concerning 2D photonic structures made of dielectric rods in air.Two types of structures are studied : regular stacks of periodic grids of rods, which behave as perfect reflectors for specific spectral bands,and stacks exhibiting a central defect plane which can operate as filters or switches.

The structure physical parameters are chosen to obtain devices operating around $\lambda=1.55\mu m$.

2. Description of the photonic structures under study

Our approach is the following : we consider crystals made of N grids of rods (each grid is a 1D diffraction grating) separated by homogeneous layers. We calculate the various diffraction coefficients (R, T, and A = 1-R-T) of such structures, ie quantities which can be reached experimentally. These quantities are modeled using a method we have developed based on a rigorous differential theory of diffraction gratings [1], the stack optical properties being calculated by means of a numerically stable S-matrix algorithm [2].

445

A. D. Boardman et al. (eds.),

Advanced Photonics with Second-order Optically Nonlinear Processes, 445–448.

Regular structures are made of N grids (N odd), and when a structure presents a defect, it is a central one so that, if p is the number of grids before the defect plane, N = 2p+1.

With this approach, we define « effective BG » (ie λ ranges where the transmission coefficient T is smaller than a pre-requisite value T_L). A BG is characterized by a central wavelength λ_c and bandwidth $\Delta\lambda$, quantities which depend on N and on experimental conditions [3]. We also determine the characteristics of the implemented filters (central wavelength λ_D, passband $\delta\lambda$.and change in R and T coefficients as λ passes through λ_D) versus various parameters (defect size c, refractive index n and absorption n", p, BG number, ligth polarization...). We can then determine switching conditions : the index change necessary to induce switching and the EM field enhancement location.

3. Regular structures : effective BG determination

We concentrate on square lattices made of stacks of square dielectric rods. These assumptions are made to minimize computation time : indeed, a grid of rods of any shape can be considered as a succession of slabs of rectangular gratings. a is the rod size, c the defect rod size, Λ the grating period and λ the incident wavelength. The filling factor is defined as $f = (a/\Lambda)^2$. In the following, we have fixed : $\Lambda = 0.6\mu m$, f = 0.125 and rod refractive index $n_r = (2.85,0.)$.Structures are defined within a scaling factor : so, if f is unchanged, λ_c/Λ and $\Delta\lambda/\lambda_c$ are unchanged too.

λ is varied from Λ to 5μm. In this range, we have reported in fig 1 T(λ) for various values of N . For TE polarization, one can see 2 BG, the larger one including λ=1.55mm.. In fig.1b, transmission curves are reported for TM polarization : 3BG can be seen.The minimum value T_m of T and $\Delta\lambda$ strongly vary with N and the BG number : indeed, for TM BG2, $T_m<T_L = 10^{-5}$ is reached with only 5 rod grids, which corresponds to a structure which can be carried out relatively easily.

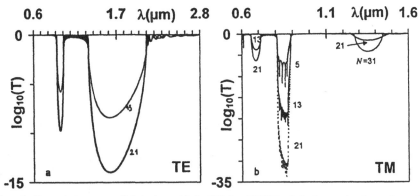

Figure 1 : Transmission curves of photonic crystals for TE(a) and TM(b) polarizations and various values of N

4. Structure with a defect plane

4.1 INFLUENCE OF THE DEFECT REFRACTIVE INDEX

The defect plane rods may not have the same refractive index ($n \neq n_r$)or size ($c \neq a$) as the other rods. If the defect has appropriate characteristics, it is associated to a defect mode yielding a transmission/reflection peak in every BG. For instance, for $c = a$, as n varies from 1 to 4.5 (in order to model feasible devices), λ_D varies and passes through the various BG : the structure acts as a filter with adjustable narrow band ($\lambda_D, \delta\lambda$). As TE band 2 is very large, we have assumed $c = 2a$ (see fig.2). Then, 2 values of λ_D are simultaneously present in the gap. In fig 2b, the dispersion curve $\lambda_D(n)$ of one defect mode is reported, and so is its linewidth. This quantity is minimum ($\delta\lambda \approx .39$Å)for n=2.7and λ_D =1.6444µm. Around these values, defect mode dispersion is maximum ($d\lambda_D/dn=3.10^3$Å), and a change in n of the order of 10^{-4} allows a switching from a low to a high transmission state.Then, if the defect is made with a nonlinear material, and if the EM field is enhanced in the defect plane, a change in the incident beam intensity will drive a switching of the device.

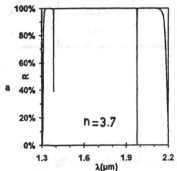
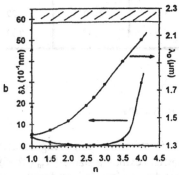

Figure2 Defect structure : a-reflection coefficient for TE BG2-b- dispersion curve $\lambda_D(n)$ of the filter

4.2 INFLUENCE OF THE DEFECT ABSORPTION (N")

Calculations reported here are performed for TE band 2, c=2a and N=6+1+6.

Experimentally, damping cannot be excluded : we have then introduced n" in the -defect plane.Absorption occurs in this plane, characterized by an absorption coefficient A=1-R-T, and A/$\delta\lambda$ is a measure of the EM field enhancement in the plane [3]. Moreover, as soon as absorption is present, the filter performances ($\Delta T = T(\lambda_D)$-Tmin and ΔR = Rmax-R(λ_D))are modified.We report in fig.5 the changes in the filter efficiency (characterized by ΔT or ΔR) and the switch sensitivity (characterized by A/$\delta\lambda$) versus n": they don't exhibit the same variations, and so a trade-off must be found between these quantities.

448

Indeed, efficiency and sensitivity depend simultaneously on p and n" (see Table 1) which must be chosen accordingly.

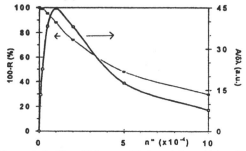

Figure3 : Changes in the filter efficiency and in the switch sensitivity with damping n"

Table 1

p	n"	A/δλ	R_{min}%	ΔT%	ΔR%
6	5×10^{-5}	38.37	4.73	59.11	95.27
6	2×10^{-4}	38.3	25.94	24.18	74.06
5	2×10^{-4}	9.29	3.92	64.48	96.08
6	10^{-3}	7.77	70.16	2.64	29.84

5. Conclusions

Photonic structures consisting of dielectric columns in air, with suitable physical parameters exhibit a stable ($T_m << 10^{-5}$) bandgap around $\lambda = 1.55 \mu m$, using N = 13. If a central defect plane is introduced in such a structure, it can operate as a filter at any chosen wavelength for small grid numbers N. For specific BG and wavelengths, the EM field is enhanced in the defect plane and allows switching operation if the defect is made of a nonlinear material. Using Kerr nonlinearities, which don't induce absorption seems appropiate to obtain efficient devices.

6. References

1. Chateau,P.and Hugonin, JP (1994)Algorithm for the rigorous coupled wave analysis of grating diffraction,*J.Opt.Soc.Am.***A11**,1321-1331
2. Li,L(1996) Formulation and comparison of two recursive matrix algorithms for modelling layered diffraction gratings,*J.Opt.Soc.Am.***A13**,1024-1035
3. Dansas,P. and Paraire, N.A. (1997) Feasibility of optical filters and switches using plastic photonic bandgap structures,*SPIE Proc.N°313527*

THIN LAYER MODIFICATION OF P.V.D.F. WITH COPPER SULFIDE

T. ANGELOV, A. KALT
Laboratoire de Chimie Textile, E.N.S.C. de Mulhause,
3, rue Alfred Werner - F-68093 Mulhause Cedex, France

D. LOUGNOT
Laboratoire de Photochimie Generale, E.N.S.C. de Mulhause, France
3, rue Alfred Werner - F-68093 Mulhause Cedex, France

The polyvinylidene flourede is known for its piezo- and pyroelectric properties. At the other hand the non-linear optical properties of nanocrystals of semi-conductors inbeded in polymer matrixes are of great interest. We studied the PVDF-Cu_xS couple. Our aim was to obtain good adhesion of a PVDF film with thickness of 25 mm, by a surface modification of consecutive deflouridation and one aclrylat monomer annexation. Cu_xS was adjourned from a solution with the help of a oxidation-reduction process, 1 μm thick and a resistivity of 1.3 kΩ.m measured with two point method.

The acquired films were analysed with XPS and in UV and IR regions of electromagnetic spectrum under different angles, where full internal reflection was observed.

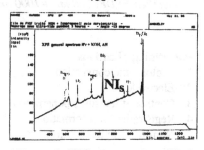

Figure 1. FTIR spectra of "Kynar 401".

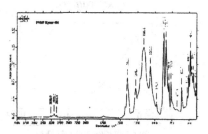

Figure 2. XPS spectra of modified PVDF with AN.

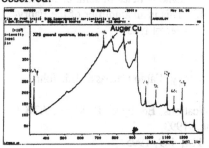

Figure 3. XPS spectra of modified PVDF with AN and Cu_xS

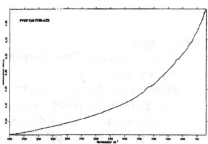

Figure 4. FTIR - ATR spectra.

A. D. Boardman et al. (eds.),
Advanced Photonics with Second-order Optically Nonlinear Processes, 449–450.
© 1999 *Kluwer Academic Publishers. Printed in the Netherlands.*

450

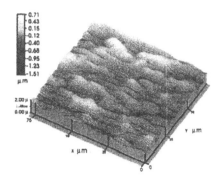

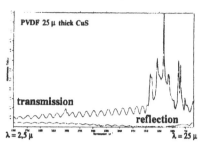

Figure 6. FTIR spectra with microscope "Bruker" - reflection, transmition.

Figure 5. Image of PVDF surface after modifications.

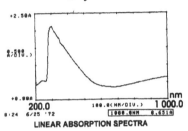

LINEAR ABSORPTION SPECTRA

Figure 7. UV and visible spectras of modified PVDF: 200 nm ÷ 1000 nm

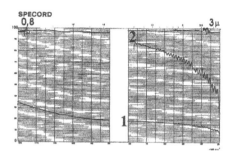

Figure 8. Near IR transmition spectra of modified PVDF: 0,8 μ ÷ 3 μ.
1 - normal spectra. 2 - 20% increase 1.

Potential applications
1. Phototermic conversion
2. Electrodes
3. Control of the sun's penetration
4. Registration of information
5. Non-linear optics (optional)

Conclusions
1. We modified the surface of PVDF with a "p" semi-conductor.
2. We observed total internal reflection.
3. We combined piezo- and pyroelectric properties with nanoparticles of semi-conductors.
4. This is an amplifier of the Fabry-Perrot type

Bibliography
1. Polimere surfaces. From Physics to Technology, Fabio Carbassi, Ed. John Wiley & Sons, 1994.
2. Polymers for electronic and photonic applications, Ed. by C. P. Wang, AT&T, Lab. Prinston, N. Jersey, Acad. Press Inc., 1993.
3. Ivan Grozdanov J. of Solid State Chem., 114, 469, 1995.
4. Artemyev M. V., J. Appl. Phys., 80, 12, 1996.

QUADRATIC SOLITONS: PAST, PRESENT, AND FUTURE

YURI S. KIVSHAR

Australian Photonics Cooperative Research Centre
Research School for Physical Sciences and Engineering
Optical Sciences Centre, The Australian National University
ACT 0200 Canberra, Australia

Abstract. The theory of two-wave solitary waves supported by parametric wave mixing in an optical medium with second-order nonlinear responce is presented. This includes a brief discussion of the existence and stability of bright solitons due to second-harmonic generation, dark solitons, the effect of competing nonlinearities, self-focusing in quasi-phase-matched materials, etc. One of the reasons for the current high level of interest is the web of connections with other types of non-Kerr solitary waves and relatively low switching power expected from the cascaded nonlinearities. A summary of important unsolved problems and hot topics in this field is also included.

1. Introduction

Recent years have shown increased interest from different experimental and theoretical groups in the study of self-guided optical beams that propagate in slab waveguides or bulk nonlinear media without supporting waveguide structures. Such beams are commonly referred to by physicists working in nonlinear optics as *spatial optical solitons*.

Simple physics explains the existence of spatial solitons in a self-focusing nonlinear medium. First, we recall the physics of optical waveguides. Optical beams have an innate tendency to spread (diffract) as they propagate in a homogeneous medium. However, the beam's *diffraction* can be compensated for by beam *refraction* if the refractive index is increased in the region of the beam. An optical waveguide is an important mean to provide a balance between diffraction and refraction if the medium is uniform in the direction of propagation. The corresponding low-intensity propagation of light is confined in the transverse direction of the waveguide, and it is de-

A. D. Boardman et al. (eds.),
Advanced Photonics with Second-order Optically Nonlinear Processes, 451–475.
© 1999 *Kluwer Academic Publishers. Printed in the Netherlands.*

scribed by the so-called linear *guided modes*, spatially localized eigenmodes of the electric field in the waveguide.

As was discovered long time ago [1], the similar effect, i.e. suppression of diffraction through a local change of the refractive index, can be produced solely by nonlinearity. As has been established in a number of experimental research [2], some materials can display considerable optical nonlinearities when their properties are modified by the light propagation. In particular, if the nonlinearity leads to a change of the refractive index of the medium in such a way that it generates *an effective positive lens* to the beam, the beam can become *self-trapped* and it propagates unchanged without any external waveguiding structure. Such stationary self-guided beams are called these days *spatial optical solitons*, they exist with profiles of certain form allowing a local compensation of the beam diffraction by the nonlinearity-induced change in the material refractive index.

Until recently, the theory of spatial optical solitons has been based on the nonlinear Schrödinger (NLS) equation with a cubic nonlinearity, or its generalizations describing the optical media with a non-Kerr nonlinear response (see, e.g., an extended list of references in the review papers [3, 4]). However, some other models supporting solitary waves have been known for a long time. In particular, one of the important mechanisms to enhance nonlinear properties of optical materials is a resonant, phase-matched interaction between the modes of different frequencies. In this latter case, the mutual beam coupling between different frequencies can modify drastically the properties of a single beam leading to the parametric beam self-trapping and multi-component solitary waves, as it occurs in the case of *quadratic solitons* [5, 6] supported by the so-called *cascaded nonlinearities* [7].

The main purpose of this chapter is to discuss, starting from the basic physical models, the most recent advances in the theory of self-guided beams supported by quadratic nonlinearities, and to provide a comprehensive list of open problems and future directions in this field.

2. Solitary Waves in $\chi^{(2)}$ Media

2.1. MODELS FOR QUADRATIC SOLITONS

When the fundamental frequency becomes phase matched with one of its harmonics, the model described by the NLS equation fails and the wave propagation should be described by some other models. In the case of quadratic (or the so-called $\chi^{(2)}$) nonlinearity, the simplest effect of resonant wave interaction is the generation of the second harmonics 2ω by the fundamental w with the frequency ω, the process being a particular case of a more general process of *three-wave mixing*. Such an interaction is efficient provided the matching conditions between the wave propagation constants are satisfied. This phenomenon is well known in the theory of second-harmonic generation (SHG) (see, e.g., Refs. [8, 9]). For example, in an anisotropic medium, for any wave vector direction $\mathbf{k}/k$, two *different* corresponding values of $k(\omega)$ can be found. In other words, for any direction of propagation there are two normal waves (which are called *ordinary* and *extraordinary* waves) which have different polarizations and travel with *different* phase velocities. For the ordinary wave the direction of wave vector $\mathbf{k}$ coincides with the direction of the Poynting vector $\mathbf{s}$ (i.e. with the direction of energy flow), whereas for the extraordinary wave the directions of $\mathbf{k}$ and $\mathbf{s}$ do not coincide.

To derive the model of three-wave mixing in a diffractive medium, one should consider parametric interaction between three stationary quasi-plane monochromatic waves with the envelopes E_j (where $j = 1, 2, 3$) and assume $\omega_1 + \omega_2 = \omega_3$ with the corresponding wave vectors to be nearly matched, i.e. $k_1(\omega_1) + k_2(\omega_2) - k_3(\omega_3) = \Delta k$, where $\Delta k \ll k_3$. If all three vectors $\mathbf{k}_j$ have the same direction, there is no phase velocity walk-off. However, if some of three waves are extraordinary then their energy flows diverge and this should be taken into account in the structure of the slowly varying envelopes E_j. Choosing the z-axis as the direction of $\mathbf{k}_j$, and the x-axis being in the plane defined by $\mathbf{k}_j$ and the direction of the energy walk-off, we consider the electric field as a sum of three fields of the resonantly interacting frequencies. As a result, in the approximation of slowly varying envelopes, we can derive a system of equations that describes the type II SHG,

$$2ik_1 \frac{\partial E_1}{\partial z} - 2ik_1\rho_1 \frac{\partial E_1}{\partial x} + \frac{\partial^2 E_1}{\partial x^2} + \frac{\partial^2 E_1}{\partial y^2} + \frac{8\pi\omega_1^2}{c^2} \tilde{\chi}_1^{(2)} E_3 E_2^* e^{-i\Delta kz} = 0,$$

$$2ik_2 \frac{\partial E_2}{\partial z} - 2ik_2\rho_2 \frac{\partial E_2}{\partial x} + \frac{\partial^2 E_2}{\partial x^2} + \frac{\partial^2 E_2}{\partial y^2} + \frac{8\pi\omega_2^2}{c^2} \tilde{\chi}_2^{(2)} E_3 E_1^* e^{-i\Delta kz} = 0,$$

$$2ik_3 \frac{\partial E_3}{\partial z} - 2ik_3\rho_3 \frac{\partial E_3}{\partial x} + \frac{\partial^2 E_3}{\partial x^2} + \frac{\partial^2 E_3}{\partial y^2} + \frac{8\pi\omega_3^2}{c^2} \tilde{\chi}_3^{(2)} E_1 E_2 e^{i\Delta kz} = 0.$$

$$(1)$$

These equations describe the case when the spatial walk-off of all waves occurs in the same plain. Formally, this is true only for single-axis crystals, but the corresponding generalization is trivial. For a slab waveguide, the structure of the linear guided modes in the direction of the trapping provided by the waveguide is known. Then, using an approximate separation of variables, E_j, $E_j(x, y, z) = F_j(y)\mathcal{E}_j(x, z)$, and integrating out the dependencies in y, we can obtain the similar system of nonlinear coupled equations but with the normalized (scaled) coefficients.

In the limiting case, when $\omega_1 = \omega_2 = \omega_3/2$, only one characteristic frequency $\omega_0 \equiv \omega_1$ is involved. This requires only one source of coherent radiation at the fundamental frequency ω_0 and a wave of the double frequency $2\omega_0$ is generated due to SHG phenomenon (type I SHG). In this case, we put $E_1 = E_2$ and, therefore, the spatial solitons due to type I SHG in a $\chi^{(2)}$ slab waveguide are described by the system of two coupled equations:

$$2ik_0 \frac{\partial E_1}{\partial z} + \frac{\partial^2 E_1}{\partial x^2} + \frac{8\pi\omega_0^2}{c^2} \tilde{\chi}_1^{(2)} E_3 E_1^* e^{-i\Delta k z} = 0,$$

$$4ik_0 \frac{\partial E_3}{\partial z} - 4ik_0\rho_3 \frac{\partial E_3}{\partial x} + \frac{\partial^2 E_3}{\partial x^2} + \frac{32\pi\omega_0^2}{c^2} \tilde{\chi}_3^{(2)} E_1^2 e^{i\Delta k z} = 0,$$

$$(2)$$

where ω_0, k_0 and E_1 are the frequency, wave number and the electric field intensity of the first harmonic wave, respectively; E_3 and ρ_3 are the electric field intensity and the walk-off angle for the second harmonic wave, and $\Delta k = 2k_0 - k_3$ is the wave vector mismatch.

In the case of spatial solitons, we normalize Eqs. (2) measuring the transverse coordinate x in units of the input beam size r_0, and the propagation coordinate z, in units of the diffraction length $z_d \equiv r_0^2 k_2 = r_0^2/\gamma_1$. Introducing the dimensionless fields, $E_1 = (v\sqrt{|\gamma_1\gamma_2|}/\sqrt{2\chi_1\chi_2 r_0^4})\exp(i\beta z)$ and $E_2 = (w|\gamma_1|/\chi_1 r_0^2)\exp[i(2\beta + \Delta)z]$, we finally obtain

$$i\frac{\partial v}{\partial z} + r\frac{\partial^2 v}{\partial x^2} - \beta v + wv^* = 0,$$

$$i\sigma\frac{\partial w}{\partial z} - i\delta\frac{\partial w}{\partial x} + s\frac{\partial^2 w}{\partial x^2} - \sigma(2\beta + \Delta)w + \frac{v^2}{2} = 0,$$

$$(3)$$

where $\Delta \equiv z_d\Delta k$; $\delta \equiv \delta_2 r_0/|\gamma_2|$; $r \equiv \text{sign}(\gamma_1)$; $s \equiv \text{sign}(\gamma_2)$ and $\sigma \equiv |\gamma_1/\gamma_2|$. The dimensionless parameter β is proportional to the nonlinearity-induced phase velocity shift.

2.2. SOLITONS IN THE CASCADING LIMIT

In the simplest case of no walk-off, the system (3) can be further normalized to scale out the propagation constant β,

$$i\frac{\partial v}{\partial z} + r\frac{\partial^2 v}{\partial x^2} - v + wv^* = 0,$$

$$i\sigma\frac{\partial w}{\partial z} + s\frac{\partial^2 w}{\partial x^2} - \alpha w + \frac{v^2}{2} = 0,$$

$$(4)$$

where the dimensionless parameter $\alpha \equiv 2\sigma + \sigma\Delta/\beta$ includes the mismatch parameter. Equations (4) is the generic model of two-wave $\chi^{(2)}$ solitons in the absence of walk-off. Its solutions have been recently analyzed by many authors, in the (1+1) dimensional case (see, e.g., Refs. [10, 11, 12, 13] to cite a few), and in a more general (2+1) dimensional case [14, 15]. Solitary waves in the case of more general, nondegenerated three-wave mixing have been also investigated (see, e.g., Refs. [16, 17]). Experimental observation of quadratic solitons has been reported for both waveguide geometry [18] and for beam propagation in a bulk [19, 20] (see also a review paper [7]).

It is straightforward to see why (and when) we expect to find spatially localized solutions of Eqs. (4). Indeed, let us consider the limit of large α, which corresponds to large positive values of the mismatch Δk. In this case, the second equation of the system (4) can be approximately reduced to the form $w \approx v^2/(2\alpha)$. The substitution of this expression into the first equation of the system (4) results in the standard NLS equation for the first harmonic,

$$i\frac{\partial v}{\partial z} + r\frac{\partial^2 v}{\partial x^2} - v + \frac{1}{2\alpha}|v|^2 v = 0. \tag{5}$$

The NLS equation possesses stable bright (at $r = +1$) or dark (at $r = -1$) soliton solutions. We will call the limit of large α *the cascading limit*. In this limit the effective Kerr-like behaviour due to cascaded $\chi^{(2)}$ effects is clearly seen and the second harmonic component w is much weaker than the first harmonic v.

Using such a simple reduction to the NLS equation, we may look for a stationary (i.e. z-independent) localized solutions of Eqs. (4) in the form of an asymptotic series in the parameter α^{-1} and find the real functions $v(x)$ and $w(x)$ in the form of asymptotic series,

$$v(x) = 2\alpha^{1/2}\operatorname{sech} x + 4s\alpha^{-1/2}\tanh^2 x \operatorname{sech} x + \ldots,$$
$$w(x) = 2\operatorname{sech}^2 x + s\alpha^{-1}(16\operatorname{sech}^2 x - 20\operatorname{sech}^4 x) + \ldots, \tag{6}$$

for bright solitons at $r = +1$, and

$$v(\tau) = \sqrt{2}\alpha^{1/2}\tanh\tau + \sqrt{2}s\alpha^{-1/2}(\tau\operatorname{sech}^2\tau - \tanh\tau\operatorname{sech}^2\tau) + \ldots$$
$$w(\tau) = \tanh^2\tau + s\alpha^{-1}(2\tau\tanh\tau\operatorname{sech}^2\tau - 4\operatorname{sech}^2\tau + 5\operatorname{sech}^4\tau) + \ldots, \tag{7}$$

where $\tau \equiv x/\sqrt{2}$, for dark solitons at $r = -1$.

Properties of Kerr solitons of Eq. (5) are well known. Existence of the asymptotic solutions (6) and (7) obtained in the cascading limit suggests that for $\alpha \gg 1$ the system (4) should have stable bright solitons for $r = +1$, $s = \pm 1$, and stable dark solitons for $r = -1$, $s = \pm 1$, similar to Eq. (5). However, this conclusion does not provide a complete answer, because: (i) formal asymptotic solutions (6) and (7) can be nonstationary for the system (4) due to their resonance with linear waves (due to asymptotic terms 'beyond all orders'); (ii) in the case of dark solitons, the solutions (7) can also be unstable due to *parametric modulational instability*, as has been first demonstrated in Refs. [21, 22].

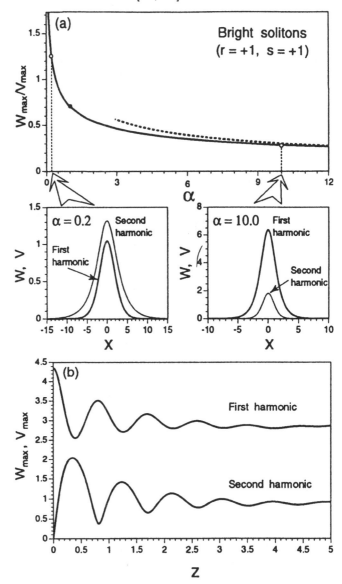

Fig. 1. (a) Two-wave bright solitons at $r = +1$ and $s = +1$ for different values of the dimenionless parameter α. The filled circle at $\alpha = 1$ corresponds to the exact solution (8). (b) Self-trapping of an initial sech-profile input into a slightly oscillating two-wave bright soliton. Shown are the peak intensities of the fundamental and second harmonics.

2.3. FAMILIES OF BRIGHT SOLITONS

As follows from the NLS limit valid for large α, bright solitons of two coupled equations (4) should exist for $r = +1$ in the form of one-hump localized profiles for the real functions $v(x)$ and $w(x)$. Such solutions have been first found by Buryak and Kivshar [11, 13] using the numerical shooting technique, for any positive value of α.

Examples of two-wave localized solutions of Eqs. (4), found numerically, are presented in Fig. 1(a) for $\alpha = 0.2$ and $\alpha = 10$. For $\alpha \gg 1$ the maximum amplitude of the fundamental component v_{max} is much larger than the similar value w_{max} for the second harmonic component. This case corresponds to the asymptotic solution (6), $v \approx \pm 2\sqrt{\alpha}\,\text{sech}\,x$, $w \approx 2\,\text{sech}^2 x$. The ratio w_{max}/v_{max} characterizing the whole family is plotted in Fig. 1(a) where the filled circle corresponds to the exact solution [5]

$$v(x) = \sqrt{2}\,w(x) = \frac{3\sqrt{2}}{2}\text{sech}^2(x/2), \tag{8}$$

that exists at $\alpha = 1$, and the asymptotic (dashed) curve corresponds to the NLS limit, for which $w_{max}/v_{max} \approx 1/\sqrt{\alpha}$.

The numerical analysis of the stability of this soliton family based on the direct integration of Eqs. (4) and the analysis of the corresponding linearized eigenvalue problem indicated that both stable and unstable two-wave solitons exist, depending on the values of system parameters α and σ [23]. In the cascading limit $\alpha \gg 1$, the solitons shown in Fig. 1(a) are stable, whereas in the other limit ($\alpha \to 0$) solitons become unstable [e.g. left-bottom soliton of Fig. 1(a) is unstable for $\sigma = 2$]. In spite of the instability in the limit of small α, the parametric solitons are stable for $\alpha \sim 1$ even under the action of very strong perturbations. These two-wave solitons can be generated from a rather broad class of initial conditions. Figure 1(b) shows an example of the soliton generation from a first harmonic sech-form input pulse. Due to diffraction, the input pulse becomes broader, but then it generates the second harmonics and, after a rather short transition period, a two-component bright soliton emerges, with a profile oscillating due to excitation of a soliton internal mode. This kind of behaviour is possible only due to the existence of the continuous family of stable bright solitons.

Additionally to one-hump localized solutions, the numerical analysis indicates the existence of continuous (in α) families of two-hump (and even multi-hump) bright solitons, which can be treated as bound states of one-hump solitons [24, 25]. These solitons exist only for $0 < \alpha < 1$. At $\alpha \to 1$ the distance between the neighboring solitons increases to infinity. Numerical stability analysis indicates that all these multi-hump bright solitons are unstable and either split into partial stable solitons or disintegrate completely, for sufficiently small values of α where stable single solitons do not exist.

In spite of the fact that in the cascading limit the effective NLS equation (5) does not depend on the sign s, localized solutions of Eqs. (4) are very different for $s = +1$ and $s = -1$. A simple analysis of the soliton tails indicates that for $s = -1$ one-hump solitons do not exist due to the resonance with linear waves. However, the numerical analysis still allows to find bound states of such solitons existing as discrete sets of two- (and multi-) soliton radiationless states where radiation is suppressed outside, but exists between the solitons in the form of trapped oscillations. Such bound states of solitons in the presence of radiation may occur in other nonlinear systems where single solitons do not exist . It is interesting, that in this case one solution is also known in an explicit analytical form [26]. It exists at $\alpha = 2$ and has the form,

$$v(x) = 6\sqrt{2}\,\tanh x\,\mathrm{sech} x, \quad w(x) = 6\,\mathrm{sech}^2 x. \tag{9}$$

Solution (9) represents one member of the family of two-soliton bound states of an integer order. Because of a delicate balance between the solitons and radiation for such stationary solutions to exist, all these bound states are unstable, they either split into single radiating solitons, or disintegrate in a more complicated fashion.

2.4. FAMILIES OF DARK SOLITONS

Following the preliminary results of the cascading limit when the effective NLS equation (5) is valid, we expect to find dark solitons in the case $r = -1$, that corresponds to a defocusing effective cubic nonlinearity. Indeed, the numerical results obtained by Buryak and Kivshar [11, 13] indicate that single dark radiationless solitons exist for $r = -s = \pm 1$ as localized solutions of Eqs. (4). In the case $r = -s = -1$, a continuous family of parametric dark solitons exists for $0 < \alpha < \infty$ and in the interval $0 < \alpha < 8$ these solitons have nonmonotonic radiationless oscillatory tails. Examples of these two-wave dark solitons are presented in Fig. 2 for $\alpha = 1.0$ (nonmonotonic tails) and $\alpha = 10.0$ (monotonic tails).

For the cascading limit ($\alpha \gg 1$) the solution can be presented in the asymptotic form (7). When α is not large, the asymptotic solution (7) fails (e.g. it does not describe oscillatory tails for $\alpha < 8$), but the families of dark solitons still exists for $\alpha > 0$, and it can be characterized, e.g., by the minimum amplitude of the second harmonic w_{min}. The dependence of w_{min} versus α is shown in Fig. 2. For large α it approaches the asymptotic dashed curve $w_{min} \approx 1/\alpha$ which corresponds to the NLS solitons obtained in the cascading limit. Dark solitons of Fig. 2 are stable if their backgrounds are modulationally stable (i.e. the stability domain is $2 < \alpha < \infty$, and it does not depend on σ).

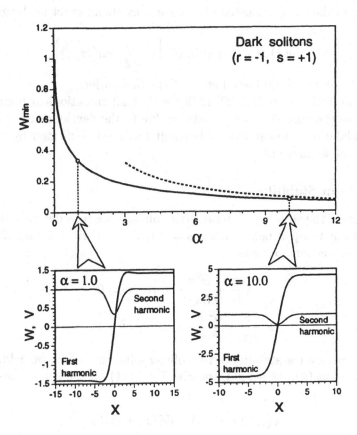

Fig. 2. Two-wave dark solitons for $r = -1$ and $s = +1$ for different values of α. Oscillating tails appear for $\alpha < 8$.

Due to existence of decaying oscillating tails, a dark soliton can trap another dark soliton to form a bound state, *a twin-hole dark soliton* [22]. For $\alpha \to 8$ the distance between the neighboring dark solitons in a bound state increases to infinity. To the best of our knowledge, this is the first example when stable twin-hole dark solitons have been identified.

The dark solitons presented above exist for the case $r = -s = -1$ in Eqs. (4). Recently other continuous families of dark solitons have been found for $r = -s = +1$ in the interval $0 < \alpha < 2$ [25]. These dark solitons also possess oscillating tails, and thus they can form bound states. Stability of these solutions is still an open problem.

Similar to the case of bright solitons, for $r = s = -1$ in Eqs. (4) single dark radiationless solitons do not exist due to a resonance with linear waves. However, a discrete sets of two- (and multi-) soliton radiationless bound states can be found, these solutions appear due to trapping of radiation. Similar to the cases discussed above, there exists an exact analytic solution [27]

$$v(x) = \sqrt{2}\,w(x) = \sqrt{2}\left[1 - \frac{3}{2}\mathrm{sech}^2(x/2)\right], \tag{10}$$

which is a two-soliton bound state of the first order.

Analytical results [21, 22] indicate that all radiationless bound states of radiative dark solitons are unstable due to the development of parametric modulational instability. This result has been confirmed by the direct numerical simulations.

3. Soliton Stability

To analyze the stability of parametric solitons, we consider only the case of spatial self-trapped beams when $s = +1$, $r = +1$. In this case the dimensionless equations become

$$i\frac{\partial v}{\partial z} + \frac{\partial^2 v}{\partial x^2} - v + v^*w = 0,$$
$$i\sigma\frac{\partial w}{\partial z} + \frac{\partial^2 w}{\partial x^2} - \alpha w + \frac{1}{2}v^2 = 0. \tag{11}$$

To analyze the stability of the soliton solutions, which we define here as $v_0(x)$ and $w_0(x)$, we should linearize Eqs. (11) on the soliton background as the following,

$$v(x, z) = v_0(x) + [V_r(x) + iV_i(x)]e^{\lambda z},$$
$$w(x, z) = w_0(x) + [W_r(x) + iW_i(x)]e^{\lambda z},$$

and investigate the corresponding linear eigenvalue problem for the corrections (V_r, V_i) and (W_r, W_i). This problem can be solved by the asymptotic method for *nonzero but small* λ. Near the instability threshold, such solutions are expected to exist only for special values of the parameters α near the critical curve $\alpha = \alpha_c(\sigma)$. The instability threshold, as well as the general dependence $\lambda(\alpha; \sigma)$, can be found from the corresponding solvability conditions to the linear problem. This analysis was recently presented in Ref. [23], and it was shown that the instability threshold is given by the generalized Vakhitov–Kolokolov criterion, $\partial\tilde{P}/\partial\beta = 0$, where $\beta(\alpha; \sigma) = (2\sigma - \alpha)^{-1}$ is the renormalized soliton propagation constant and $\tilde{P} = (2\sigma - \alpha)^{-3/2}P$,

$$P(\alpha; \sigma) = \frac{1}{2}\int_{-\infty}^{+\infty}\left[v_0^2(x; \alpha) + 2\sigma w_0^2(x; \alpha)\right]dx$$

is the renormalized energy (Menley-Rowe) invariant of Eqs. (11). In all these formulas the parameter σ is considered as an arbitrary parameter. Using this criterion and the numerical results on the stationary soliton solutions, the instability threshold curve $\alpha = \alpha_c(\sigma)$ was calculated in Ref. [23]. Moreover, it was shown that the two-wave parametric solitons are unstable for $\alpha < \alpha_c$.

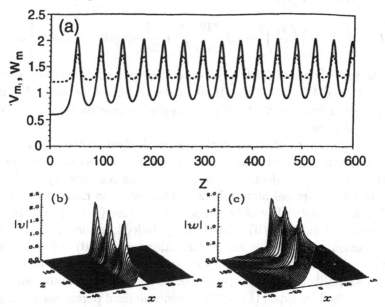

Fig. 3. Periodic oscillations of the two-wave soliton of the model (11) after the development of instability ($\sigma = 2$, $\alpha = 0.05$). (a) Evolution of the harmonics amplitudes $v_m \equiv |v(0,z)|$ (solid) and $w_m \equiv |w(0,z)|$ (dashed). (b),(c) Propagation of the soliton components v and w, respectively.

The important physical question is the development of linear instability in the subsequent dynamics of the two-wave solitons. In order to describe this analytically, we should take into account *nonlinear effects*. Following the original paper [23], we select $\alpha \equiv \alpha_0$ close to α_c, so that the small parameter ϵ characterizes the deviation $(\alpha_0 - \alpha_c) \sim O(\epsilon^2)$. Then, it follows from the linear theory that the growth rate has the order $O(\epsilon)$ and, therefore, the unstable linear perturbations grow on the 'slow' scale $Z = \epsilon z$. This allows us to introduce the slowly varying complex phase $S = S(Z)$ and look for the perturbed solutions of Eqs. (11) in the form of asymptotic series $v = [v_0(x;\alpha) + \epsilon^2 \Omega V_r(x;\alpha) + O(\epsilon^3)]e^{2i\epsilon S}$, $w = [w_0(x;\alpha) + \epsilon^2 \Omega W_r(x;\alpha) + O(\epsilon^3)]e^{i\epsilon S}$, where the functions V_r and W_r are the solutions of the linearized problem, and $\Omega = (2\sigma - \alpha)dS/dZ$ describes a correction to the soliton propagation constant. Using the asymptotic multi-scale technique we find the nonlinear equation for the function Ω,

$$\tilde{M}_s(\alpha_c)\frac{d^2\Omega}{dZ^2} + \frac{1}{\epsilon^2}\left(\frac{\partial\tilde{P}}{\partial\beta}\right)_{\alpha=\alpha_0}\Omega + \frac{1}{2}\left(\frac{\partial^2\tilde{P}}{\partial\beta^2}\right)_{\alpha=\alpha_c}\Omega^2 = 0, \qquad (12)$$

where the renormalized soliton mass is always positive, and it is defined as $\tilde{M}_s = (2\sigma - \alpha)^{-1/2}M_s(\alpha;\sigma)$,

$$M_s = \int_{-\infty}^{+\infty}\left[v_0^2\left(\frac{V_i}{v_0}\right)_x^2 + w_0^2\left(\frac{W_i}{w_0}\right)_x^2 + \frac{1}{2w_0}(v_0W_i - 2w_0V_i)^2\right]dx,$$

where the index x stand for the corresponding partial derivatives, and the functions V_i and W_i are the solutions of the linearized problem. We mention that the derivative $(\partial^2\tilde{P}/\partial\beta^2)_{\alpha=\alpha_c}$ is always positive for the two-wave parametric solitons.

It turns out that the asymptotic theory applied to the two-wave $\chi^{(2)}$ solitons gives, in a renormalized form, essentially the same equation (12) of motion of the equivalent particle for the (one-component) bright solitons in non-Kerr optical materials [28]. This equation describes two main scenarios of the instability of bright solitons, either (i) long-lived periodic amplitude oscillations or (ii) soliton decay. Indeed, according to Eq. (12), the exponential growth of linear perturbations with $\Omega(0) > 0$ is stabilized by nonlinearity leading to oscillations around a novel stable equilibrium state. This equilibrium state corresponds to a stationary soliton which can be also described by Eqs. (11) but for a renormalized parameter α lying *inside* the stability region. Therefore, for a slightly increased amplitude of an unstable soliton our analytical model (12) predicts *in-phase* pulsations of the fundamental and second harmonics around a novel stable soliton, and this exactly corresponds to the evolution observed numerically and presented in Fig. 3. For $\Omega(0) < 0$, according to Eq. (12), such a stabilization is not possible and, as a result, a slightly decreased amplitude of an unstable soliton gradually decreases further (see some other examples in Refs. [28]).

4. Quadratic Solitons in QPM Systems

In the context of second-harmonic generation the quasi-phase-matching (QPM) technique is known as an attractive way to achieve good phase-matching, and it has been studied intensively (see Ref. [29] for a review). The QPM technique relies on the periodic modulation of the nonlinear susceptibility and/or refractive index, by which an additional (grating) wavevector is introduced, that can compensate for the mismatch between the wavevectors of the fundamental and second-harmonic waves. With the QPM technique, phase-matching becomes possible at ambient temperatures, and does not introduce spatial walk-off; the polarization with the largest nonlinearity can be used, and materials with strong nonlinearities can be explored, that are not phase matchable by angle or temperature tuning. The physics of QPM has been known since 1962 [30], but only recently have the experimental difficulties been overcome and stable techniques been developed, such as domain inversion in ferroelectric materials [31], proton exchange [32], and etching and cladding [33], to mention a few.

Recent studies [34, 35] have addressed the question if QPM can be employed to achieve self-trapping of light and to support spatial solitary waves in quadratic materials. The answer is not obvious, because resonances between the domain length of the periodic structure and the beam characteristic length might induce instability. In particular, Clausen *et al.* [34] derived effective average equations that include both quadratic and periodicity-induced *cubic* nonlinearities which allowed to predict the existence of a *novel class of solitary waves, QPM solitons.*

To demonstrate the origin of the induced cubic nonlinearities, we follow the original paper by Clausen *et al.* and consider the interaction of a cw beam with the fundamental frequency ω and its second harmonic (2ω), propagating in a QPM $\chi^{(2)}$ slab waveguide, where only the nonlinear susceptibility is modulated. Assuming nonlinearity to be of the same order as diffraction, the evolution of the slowly varying beam envelopes is governed by the normalized equations

$$i\frac{\partial W}{\partial z} + \frac{1}{2}\frac{\partial^2 W}{\partial x^2} + d(z)W^*V e^{-i\beta z} = 0,$$
$$i\frac{\partial V}{\partial z} + \frac{1}{4}\frac{\partial^2 V}{\partial x^2} + d(z)W^2 e^{i\beta z} = 0,$$

(13)

where $W(x,z)$ and $V(x,z)$ are the envelopes of the fundamental and the second harmonic, respectively. The parameter $\beta = \Delta k |k_\omega| x_0^2$ is proportional to the phase mismatch $\Delta k = 2k_\omega - k_{2\omega}$, k_ω and $k_{2\omega}$ being the wave numbers at the two frequencies. The normalization parameter x_0 is equal to the input beamwidth. Spatial walk-off is neglected; it will usually not be present in QPM materials, since perpendicular or parallel polarization states can be employed. The transverse coordinate x is measured in units of x_0, and the propagation coordinate z is measured in units of the diffraction length $l_d = x_0^2 |k_\omega|$. The spatial, periodic modulation of the nonlinear susceptibility $\chi^{(2)}$ is described by the QPM-grating function $d(z)$, whose amplitude is normalized to 1, and whose domain length we define as π/κ. In general, the periodic function $d(z)$ can be expanded in a Fourier series $d(z) = \sum_n d_n e^{in\kappa z}$, where the summation is over all n from $-\infty$ to ∞. In many physical applications the QPM grating can be well approximated by the square function for which the Fourier series contain only odd harmonics, $d_{2n} = 0$ and $d_{2n+1} = 2/i\pi(2n+1)$.

Inserting $d(z)$ into Eqs. (13) and making the transformation $W(x, z) = w(x, z)$ and $V(x, z) = v(x, z) \exp(i\tilde{\beta}z)$, where $\tilde{\beta} = \beta - m\kappa$ is the effective phase-mismatch parameter for QPM of the mth order, we assume that the QPM period is well controlled, so that $\beta \approx m\kappa$. This means that $\tilde{\beta}$ is of the order of one or less (ideally 0), even though β might be large itself.

The equation that follow include coefficients that are periodically varying with the period $2\pi/\kappa$. If κ is sufficiently large, the dynamics could therefore be adequately described by averaged equations. Physically, $m\kappa \approx \beta \gg 1$ means that the coherence length $l_c = 2\pi/\Delta k$ is much smaller than the diffraction length l_d, since $\beta = 2\pi l_d / l_c$. To derive these equations, Clausen et al. [34] used an approach based on the asymptotic expansion technique, which was successfully applied in many types of soliton problems [36].

Let us consider the case where $\kappa \gg 1$ and expand the functions $w(x, z)$ and $v(x, z)$ in a Fourier series $w = \sum_n w_n e^{inm\kappa z}$ and $v = \sum_n v_n e^{inm\kappa z}$, where $w_n(x, z)$ and $v_n(x, z)$ are assumed to vary slowly compared with $\exp(i\kappa z)$. This gives the equations for the coefficients w_n and v_n. Now, we assume that the higher harmonics are of order of $1/\kappa \ll 1$ or smaller, compared to the averages w_0 and v_0. Taking into account only the lowest order terms in the equations for the harmonics, we arrive at the average equations [34]

$$
\begin{aligned}
&i\frac{\partial w_0}{\partial z} + \frac{1}{2}\frac{\partial^2 w_0}{\partial x^2} + d_m w_0^* v_0 + (\gamma|w_0|^2 + \rho|v_0|^2)w_0 = 0, \\
&i\frac{\partial v_0}{\partial z} + \frac{1}{4}\frac{\partial^2 v_0}{\partial x^2} - \tilde{\beta}v_0 + d_{-m}w_0^2 + 2\eta|w_0|^2 v_0 = 0,
\end{aligned}
\tag{14}
$$

where γ, ρ, and η are all of the order of $1/\kappa$ and given by

$$
\gamma = \frac{1}{m\kappa}\sum_{n\neq 0}\frac{\gamma_n}{n}, \quad \rho = \frac{1}{m\kappa}\sum_{n\neq 0}\frac{\rho_n}{n}, \quad \eta = \frac{1}{m\kappa}\sum_{n\neq 0}\frac{\eta_n}{n},
$$

with $\gamma_n = d_{m(n-1)}d_{m(1-n)}$, $\rho_n = d_{m(n+1)}^* d_{m(n+1)}$, and $\eta_n = d_{m(n+1)}d_{m(-n-1)}$. From Eqs. (14) follows the important result that the QPM grating introduces an effective cubic nonlinearity in the form of self- and cross-phase modulation terms. However, the self-phase modulation does not appear for the second harmonic, making the localized solutions and the system dynamics be different from the earlier analyzed case of competing nonlinearities [37]. Thus, the so-called V-soliton [37], where $w_0 = 0$ and v_0 is a nonlinear Schrödinger soliton, does not exist in Eqs. (14).

Let us consider the most efficient QPM of first order, $m=1$, and the square grating. Then, the expansions for w and v involve only even components, and the coefficients γ, ρ, and η become real, Eqs. (14) reduce to the following

$$i\frac{\partial w_0}{\partial z} + \frac{1}{2}\frac{\partial^2 w_0}{\partial x^2} - i\chi w_0^* v_0 + \gamma(|w_0|^2 - |v_0|^2)w_0 = 0,$$

$$i\frac{\partial v_0}{\partial z} + \frac{1}{4}\frac{\partial^2 v_0}{\partial x^2} - \tilde{\beta}v_0 + i\chi w_0^2 - 2\gamma|w_0|^2 v_0 = 0,$$

(15)

where both the quadratic and cubic nonlinearity coefficients are calculated in an explicit form, $\chi=2/\pi$, $\gamma=\kappa^{-1}(1\text{-}8/\pi^2)$. Note the $\pi/2$ phase-shift in front of the quadratic terms and the opposite signs of the cubic self- and cross-phase nonlinear terms.

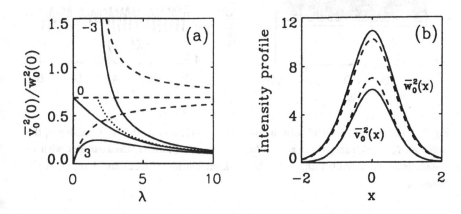

Fig. 4. Soliton families of the QPM-system (15) for $\kappa=10$ (solid curves) and the zero-order approximation ($\gamma=0$, dashed curves). (a) Ratio between the peak intensity of the second harmonic, $\bar{v}_0^2(0)$, and the fundamental, $\bar{w}_0^2(0)$, vs. λ. The value of $\tilde{\beta}$ is indicated at each pair of curves. The dotted curve shows the asymptotic result $\chi^2/18\gamma\lambda$. (b) Profiles $\bar{w}_0^2(x)$ and $\bar{v}_0^2(x)$ for $\tilde{\beta}=0$ and $\lambda=1$.

Stationary solutions of Eqs. (15) have the form

$$w_0(x, z) = \bar{w}_0(x)e^{i\lambda z}, \quad v_0(x, z) = i\bar{v}_0(x)e^{2i\lambda z},$$

(16)

where the real and localized profiles $\bar{w}_0(x)$ and $\bar{v}_0(x)$ are determined by the set of ordinary differential equations.

Analysis shows that localized solutions (16) exist only for positive values of the wave number λ, satisfying $\lambda > \max\{0, -\tilde{\beta}/2\}$. Note again the $\pi/2$ phase-shift appearing in the definition of the stationary solutions (16), in order for $\bar{w}_0(x)$ and $\bar{v}_0(x)$ to be real. These localized solutions were found numerically for any allowed value of λ and for the coefficients that correspond to the square grating [34]. Figure 4 shows some of the properties of these numerical solutions for $\kappa=10$, e.g., the ratio of peak intensities vs. λ and the characteristic profiles of the solutions for $\lambda=1$, compared with the corresponding results for the zero-order approximation ($\gamma=0$, dashed curves).

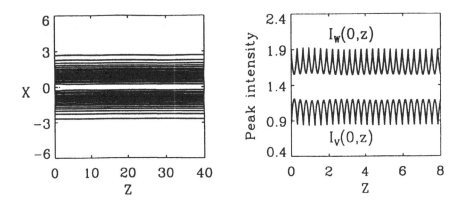

Fig. 5. Excitation of a QPM soliton with $\tilde{\beta}=0$, $\kappa=10$, and $\lambda=0.4$ (a) Intensity of the fundamental, $I_w(x, z)$, sampled at $z=n4\pi/\kappa$, where n is an integer. Contour plot with 10 equidistant levels between 0.1 and 1.92. (b) Peak intensities, $I_w(0, z)$ and $I_v(0, z)$, of the two harmonics.

Looking at the ratio of the peak intensities shown in Fig. 4(a), the cubic correction terms are seen to have a significant effect. In the zero-order approximation ($\gamma=0$), this ratio is a constant for $\tilde{\beta}=0$, which we find numerically to be 0.6865. However, in the QPM system with the induced Kerr effects ($\gamma\neq0$), this ratio tends to $\chi^2/18\gamma\lambda$ for $\lambda\gg1$, shown as a dotted curve.

For negative $\tilde{\beta}$ there is a power threshold for the existence of solitons, which occurs close to the cut-off at $\lambda = \max\{0, -\tilde{\beta}/2\}$. The induced Kerr effects are seen to *increase* this threshold power. However, for positive $\tilde{\beta}$ the Kerr effects *decrease* the power required for generating a solitary wave with a certain propagation constant λ.

In order to test these asymptotic results, Clausen *et al.* [34] used a QPM soliton as initial condition in Eqs. (13), which were solved numerically for the square grating. The results for $\tilde{\beta}=0$ and $\kappa=10$ are plotted in Fig. 5 and show clearly that the soliton propagates undistorted along z, oscillating with the period π/κ.

It is important to note that the averaged equations seem to be applicable even when the diffraction length and the coherence length are of the same order, $l_d \sim l_c$.

The average QPM soliton in quadratic media can be regarded as *a spatial analogue* of the *guided-center soliton* [38], known from the theory of pulse propagation in nonlinear optical fibers with periodic amplification and dispersion management. However, unlike the guided-center soliton, the periodic modulation of the quadratic nonlinearity does not alter the existing nonlinearity, it induces effective *higher order* nonlinearities, which can significantly modify the solitary waves.

5. Quadratic Solitons in Higher Dimensions

In a medium with quadratic nonlinear response, parametric interaction between harmonics has been shown to support stable solitary waves even in higher dimensions and the instability, if it occurs, does not display the wave collapse dynamics [6, 39].

Mutual focusing of the fundamental and second-harmonic beams in a diffractive quadratic medium and the physical mechanism of the self-guided beam generation was first discussed by Karamzin and Sukhorukov [5]. Nonlinear refraction of intense light beams in a diffractive medium has been a subject of many papers discussing the efficiency of the SHG in crystals (see, e.g. Refs. [40] and references therein), however not specifically dealing with the soliton generation. In particular, the self-action of light during the SHG generation and nonlinear-induced phase shift have been observed experimentally in a crystal of barium-sodium niobate by Belashenkov *et al.*[41] for a power density of the order of 1 MW/cm^2.

Multidimensional spatial solitons have been discussed first for parametric two-wave or type I SHG process [5] and then for three-wave parametric interaction [6] for the special case of the zero phase matching. Kanashov and Rubenchik [6] were first who demonstrated the possibility of noncollapsing stable solitary waves of higher dimensions in the case of parametric wave mixing in quadratic media.

The next effort to analyze multidimensional solitary waves was attempted by Hayata and Koshiba [42] who not only rediscovered once more again the exact analytical solution of Karamzin and Sukhorukov [5] but also applied a kind of Hartree-like approach to construct an approximate multidimensional solitary wave of the radial symmetry for a special value of the phase matching parameter. Numerical simulations demonstrating the generation of two-dimensional self-trapped beams due to parametric interaction at a nonzero phase mismatch and in the presence of walk-off have been reported by Torner *et al.*[14] whereas the beam collisions and steering have been analyzed much later [43, 44].

Buryak *et al.*[15] demonstrated the existence of a two-parameter class of (2+1)-dimensional stable solitary waves in a bulk quadratic medium, the parameters are the dimensionless phase mismatch and the amplitude of one of the fields. Importantly, the approximate solution found by Hayata and Koshiba [42] was shown to be just a single point of this numerically found family of localized waves. Two-dimensional solitary waves has been also described by means of the variational approach [45], and their existence and properties were independently confirmed by other researchers [46].

Recently the theory of multidimensional solitary waves has been advanced by the study of the beam self-focusing in the presence of both cubic and quadratic nonlinearities [47], parametric solitons in a defocusing Kerr medium [48], and also walking solitons in a bulk medium [49].

Higher-dimensional generalizations may include not only one more transverse dimension [for the (2+1)-dimensional solitary waves in a bulk medium] but also a temporal variable (spatio-temporal beam self-trapping, or *light bullets*). The general analysis of spatio-temporal localized structures is difficult because the equations for the fundamental and second harmonics are nonsymmetrical, not allowing to introduce an effective radial coordinate. More recently, however, a rather general analysis of the symmetry-breaking instabilities of quadratic solitons has been initiated (see [50, 51]) and even the first stone into the theory of $\chi^{(2)}$ *light bullets* was put down [52, 53]. Nevertheless, this important topic is still far from completeness and many other efforts would be required.

6. Future Directions and Open Problems

As has been mentioned above, the important achievement of the theory during last years was to recognize that parametric wave mixing can support solitary waves in diffractive (and dispersive) optical media with the properties similar to those of the solitary waves in non-Kerr optical media existing due to a nonlinearity-induced change of the refractive index. This, in turn, allows to have a new, "fresh" look at the properties of non-Kerr solitary waves, not described by exactly integrable nonlinear models. Moreover, the theoretical activity in this field has been enhanced drastically by the experimental discovery of the parametric $\chi^{(2)}$ solitons in waveguides and a bulk. This led to much deeper understanding of many properties of solitary waves in nonintegrable models not previously addressed in nonlinear physics and, in particular, nonlinear optics. As a consequence, a series of novel fundamental results have been obtained for solitary waves in nonintegrable models, such as the generalized stability criterion for multiparameter solitary waves, the existence criterion for the soliton internal modes, analytical models for the instability-induced dynamics of solitary waves and their interactions, etc.

Different aspects of the physics of nonlinear effects, instabilities, and spatial solitons in a quadratic medium mentioned above demonstrate a number of very interesting features of this type of nonlinear waves which, as we believe, still require some additional efforts to be completely understood. This is an exciting forefront area of modern nonlinear optics that is closely related to technological applications. Therefore, it is hard to predict how this topic will develop in the near future. If the further experimental demonstrations will be successful, more applied and technological aspects of this field, especially those closely related to the design and principles of nonlinear all-optical switching devices, will rapidly develop. However, we would like to mention some important problems, especially in the theory of two- and three-wave solitary waves in $\chi^{(2)}$ media, which still remain unsolved or less developed. We believe, active researchers will find this list useful to work on completing the picture of parametric solitary waves.

Spatial walk-off in diffractive three-wave mixing. Effect of the spatial walk-off (or group-velocity dispersion) on three-wave parametric mixing in diffractive (dispersive) media has been not completely understood. Indeed, it is well known that wave mixing in a quadratic medium can support three-wave bright solitary waves even without diffraction (or dispersion) taken into account. This case is described by the model known to be exactly integrable by the inverse scattering transform (see, e.g., [54]). This may be a good case of temporal solitons when there always exists group-velocity mismatch between the parametrically interacting waves (see also [55, 56]; and references therein). Clearly, when pulses become shorter, the effect of the group-velocity dispersion should be taken into account leading to the model of three wave mixing in the presence of both walk-off and dispersion. The first effort in this direction has been made by Mihalache et al.[57], however a detailed consideration, including the proof of the stability criterion is still an open problem. A simpler problem is to understand the influence of the group-velocity mismatch on modulational instability. Because the phenomenon of modulational instability is closely related to the existence of solitary waves, this will be the first step towards understanding a more general class of three-wave solitary waves.

Classification and stability of grey solitons. All the solutions describing dark solitary for parametric wave mixing waves discussed up to now belong to a very narrow class of the so-called 'black' solitons, wich are characterized by the zero steering velocity relative to the background. Extending this class of localized waves is an actual problem for the future experimental verification of dark solitons. This, from the other hand, will allow to answer an important question if the remormalized momentum (see, e.g., [4] for a recent review of dark solitons) can serve for the stability criterion. This problem will require numerical solutions of three coupled equations because grey solitons are described by a complex envelope with nonzero boundary conditions.

Dark solitons in $\chi^{(2)}$ media have been shown to be *unstable* for the case when the mode dispersions have the same sign. This is also the important case of spatial dark solitons. *Can parametric interactions in a $\chi^{(2)}$ medium support stable spatial dark solitons ?* A very recent results by Alexander et al.[58] indicate that the effect of the next order (i.e. defocusing cubic) nonlinearity on dark solitons may suppress this instability and finally lead to suggestions for experimental verifications. Other possibilities are to consider nondegenerated wave mixing or the effect of the induced cubic nonlinearity in QPM geometry.

Generalized cascading effects. Recent results demonstrating cascading effects in materials with purely cubic nonlinearity for all-optical switching [59] and non-Kerr solitary waves and soliton multistability due to third harmonic generation [60] suggest another wide class of problems not analyzed in detail yet, the cascaded effects due to four-wave mixing in materials with conventional (cubic or Kerr) nonlinearities. One of the important physical idea not emphasized up to now is the enhancement of nonlinear properties of Kerr materials near the phase matching. This effect is expected to modify the properties of solitary waves near resonances, inducing *effective non-Kerr nonlinearities*. However, many properties of nonlinear four-wave mixing in diffractive and dispersive media have been not even properly discussed, and many problems are still open. It is worth to mention recent papers by Lundquist et al.[61, 62] which indicate the existence of a complicated structure of solitary waves and the specific properties of their stability. Nondegenerate four-wave mixing as well as the effect of walk-off are of a great importance to be addressed for the success of the future experimental verifications.

On the other hand, also cascaded quadratic interactions might result into mimicked overall four-wave mixing, which has been observed either in the frequency-degenerate noncollinear case [63, 64] or in the frequency nondegenerate case. The possibility to sustain solitons in these interactions has not been considered yet.

Higher-order dispersion. In the main approximation, the equations for *temporal* and *spatial* quadratic solitons are similar. However, taking into account higher-order effects makes them very different (see, e.g., Ref. [65]). Indeed, for spatial solitons, the most important higher-order effect is the cubic nonlinearity, the resulting equations describe a competition between two types of self-focusing, already well understood (see, e.g., Ref. [66] and references therein). In contrast, for the ultrashort pulses (temporal solitons) the most important effect of the next order is the third-order and nonlinear dispersion [65]. The simplest analyzed solution of the corresponding generalized equations have been recently found by Mihalache *et al.*[67]. Nothing else is known so far.

Vectorial wave-mixing interactions. Besides usual type I and type II SHG, more general polarization configurations involve multicomponent mixing (see, e.g., [68, 69]. For SHG, the analysis of a full vectorial interaction between fundamental and second-harmonic waves (see, e.g., an example of magneto-oticals interaction [70]) leads to four coupled equations for two polarizations of each of the harmonics. This class of equations has been not investigated in detail yet, and only the analysis of modulational instability [71] has been completed, showing many novel features introduced by this kind of a combined type I and type II parameteric SHG wave mixing.

Light guiding light concept. The use of spatial solitons for *light guiding light* is an important theoretical concept that is also attractive for physical applications. However, the analysis of the soliton-based optical devices for materials with quadratic nonlinearities is still an open area awaiting more active researchers. This topic is closely related to the use of quadratic solitons for all-optical switching devices (see, e.g. some preliminary studies on directional couplers by Mak *et al.*[72]).

Coherent interaction with a resonant medium. Important concepts closely related to the theory of quadratic solitons appear in the problem of degenerate and nondegenerate coherent interaction of light with a resonant medium (see, e.g., a review paper [73]). Under the condition of two-photon resonance, the Maxwell-Bloch equations for the field interacting with a two-level medium can be reduced to a special case of the model for quadratic solitons, for a component of the Bloch vector and the electric field envelope [74]. As has been recently shown by Zabolotskii [75], the inclusion of diffraction into the equation for the electric field leads to the formation of stable localized structures, or quadratic solitons. However, a more detailed analysis of the properties of this model is still missing, as well as the study of nondegenerate interactions in the presence of diffraction. .

Surface modes and interactions with interfaces. Nonlinear guided modes and their stability in cubic nonlinear media have been an important topic of extensive theoretical and experimental research in the past (e.g., a review paper [76]). It is natural to extend this concept to the case of quadratic nonlinearities, when the beam interaction with interfaces is driven by resonant quadratic nonlinearities. Some papers recently published on this subject indicate the possibility of novel phenomena including the pseudo-metal reflection at the interface between linear and quadratic nonlinear materials [77] and the power-dependent reflection and refraction of the solitary wave at the interface between two quadratic nonlinear media [78]. This seems to be an important direction to explore some future applications of quadratic materials for all-optical switching devices.

Parametric vortex solitons. Recently reported second-harmonic generation by the use of Laguerre-Gaussian laser modes [79] and the sum-frequency
generation of light beams with vortices [80] open a simple way to construct
vortices of different topological charges and different frequencies. However,
most of those experiments have been performed *at low input intesities.*
We believe that similar techniques will be useful to generate doughnut
and vortex-like modes and observe their instability due to parametric self-
focusing as predicted theoretically [81, 82], and also to generate stable two-
wave vortex solitons recently predicted by Alexander *et al.* [58].

Solitons in $\chi^{(2)}$ gratings. In spite of the theoretical results concerning
parametric gap solitons, the physics of the propagation in $\chi^{(2)}$ gratings remains to be fully understood. Even in the cw limit the integrability and
stability of the coupled-mode ordinary differential equations is not known.
In long-wavelength gratings solitons of genuine parametric nature (i.e., non
Kerr equivalent) remain to be found, whereas in Bragg gratings the rich-
ness of localized solutions still demands for a complete classification, while
properties such as soliton stability, multidimensional localization, and self-
pulsing or modulational instabilities are likely to be investigated in the near
future. More general and complex interactions involving three waves might
also have interest in view of possible experiments.

Trapping in the presence of dc fields. The parametric solitons considered so far are sustained by degenerate or nondegenerate mixing of two
or three waves at optical frequencies. It is well known, however, that in
quadratic media also dc fields arise due to nonlinear optical rectification
of the involved optical frequencies. The dc fields can in turn contribute to
the nonlinear index through the linear electro-optic tensor, resulting in a
nonnegligible effect as pointed out by Bosshard *et al.*[83] (see also [84]).
The effect of the dc fields on the propagation and trapping of envelopes is
not known, and only recent attempts to obtain the governing equations for
propagating envelopes have been reported to date [85].

Quantum effects in dispersive $\chi^{(2)}$ media. It is well known that parametric mixing exhibits nonclassical features related to the quantum nature
of light. The research on classical parametric solitons will definitely have
impact over the physical understanding of quantum propagation in dispersive $\chi^{(2)}$ media. Preliminary results concern the calculations of the SHG
quadrature squeezing spectrum due to modulational instabilities [86], and
the existence of two-photon quantum bound states with large binding energy that in analogy with a quark model of the meson have been termed
optical mesons [87].

7. Conclusion

I have presented a brief overview of the theory of two-wave solitary waves in diffractive nonlinear optical media with a quadratic responce, including the discussion of the classes of bright and dark solitary waves and stability of the fundamental (no nodes) solutions. Several related issues, such as solitary waves in quasi-phase-matched systems and solitons supported by competing nonlinearities, have been mentioned as well. I have provided an extended list of the future directions and unsolved problems in this field which, as I believe, can be considered as a guide for those researchers who are willing to contribute into this field and to develop applications of cascaded nonlinearities for all-optical switching devices. Moreover, I expect that the analytical methods recently developed for the quadratic solitons will be also useful for other nonintegrable nonlinear models of various physical context.

Acknowledgements

I would like to thank my colleagues and students at the Optical Sciences Centre, Australian National University, who contributed enormously into my understanding of the field of cascaded nonlinearities and parametric solitons and the results presented here. An extended version of this paper (with many other results) will appear as a review paper in Reviews of Modern Physics, co-authored with A.V. Buryak, S. Trillo, and W. Torruellas.

References

1. R.Y. Chiao, E. Garmire, and C.H. Townes, Phys. Rev. Lett. **13**, 479 (1964).
2. For the most recent overview of experimental observations of different types of spatial optical solitons see G.I. Stegemen, Optica Aplicata **26**, 240 (1996), and M. Segev, J. Opt. Quantum Electron. (1998) in press.
3. A.D. Boardman and K. Xie, Radio Sci. **28**, 891 (1993).
4. Yu.S. Kivshar and B. Luther-Davies, Phys. Rep. **298**, 81 (1998).
5. Yu.N. Karamzin and A.P. Sukhorukov, Zh. Eksp. Teor. Fiz. **68**, 834 (1975) [Sov. Phys. JETP **41**, 414 (1976)].
6. A.A. Kanashov and A.M. Rubenchik, Physica D 4, 122 (1981).
7. See, for example, the recent review paper on cascaded effects in $\chi^{(2)}$ materials, G.I. Stegeman, D.J. Hagan, and L. Torner, J. Opt. Quantum Electron. **28**, 1691 (1996).
8. D.L. Mills, *Nonlinear Optics: Basic Concepts* (Springer-Verlag, Berlin, 1991).
9. R.W. Boyd, *Nonlinear Optics* (Academic Press, San Diego, 1992).
10. R. Schiek, J. Opt. Soc. Am. B **10**, 1848 (1993).
11. A.V. Buryak and Yu.S. Kivshar, Opt. Lett. **19**, 1612 (1994); **20**, 1080 (1995).
12. L. Torner, C.R. Menyuk, and G.I. Stegeman, Opt. Lett. **19**, 1615 (1994).
13. A.V. Buryak and Yu.S. Kivshar, Phys. Lett. A **197**, 407 (1995).
14. L. Torner, C.R. Menyuk, W.E. Torruellas, and G.I. Stegeman, Opt. Lett. **20**, 13 (1995).
15. A.V. Buryak, Yu.S. Kivshar, and V.V. Steblina, Phys. Rev. A **52**, 1670 (1995).
16. A.V. Buryak, Yu.S. Kivshar, and S. Trillo, Phys Rev. Lett. **77**, 5210 (1996).

474

17. A.V. Buryak and Yu.S. Kivshar, Phys. Rev. Lett. **78**, 3286 (1997).
18. R. Schiek, Y. Baek, G.I. Stegeman, Phys. Rev. E **53**, 1138 (1996).
19. W.E. Torruellas, Z. Wang, D.J. Hagan, E.W. VanStryland, G.I. Stegeman, L. Torner, and C.R. Menyuk, Phys. Rev. Lett. **74**, 5036 (1995).
20. R.A. Furst, D.M. Baboiu, B.L. Lawrence, W.E. Torruellas, G.I. Stegeman, S. Trillo, and S. Wabnitz, Phys. Rev. Lett. **78**, 2756 (1997).
21. P. Ferro and S. Trillo, Phys. Rev. E **51**, 4994 (1995).
22. A.V. Buryak and Yu.S. Kivshar, Phys. Rev. A **51**, R41 (1995).
23. D.E. Pelinovsky, A.V. Buryak, and Yu.S. Kivshar, Phys. Rev. Lett. **75**, 591 (1995).
24. A.D. Boardman, K. Xie, and A. Sangarpaul, Phys. Rev. A **52**, 4099 (1995).
25. H. He, M.J. Werner, and P.D. Drummond, Phys. Rev. E **54**, 896 (1996).
26. M.J. Werner and P.D. Drummond, J. Opt. Soc. Am. B **10**, 2390 (1993).
27. K. Hayata and M. Koshiba, Phys. Rev. A **50**, 675 (1994).
28. D.E. Pelinovsky, Yu.S. Kivshar, and V.V. Afanasjev, Phys. Rev. E **53**, 1940 (1996); Phys. Rev. E **54**, 2015 (1996).
29. M.M. Fejer, G.A. Magel, D.H. Jundt, and R.L. Byer, IEEE J. Quantum Electron. **28**, 2631 (1992).
30. J.A. Armstrong, N. Bloembergen, J. Ducuing, and P.S. Pershan, Phys. Rev. **127**, 1918 (1962); P.A. Franken and J.F. Ward, Rev. Mod. Phys. **35**, 23 (1963).
31. E.J. Lim, M.M. Fejer, and R.L. Byer, Electron. Lett. **25**, 174 (1989).
32. K. Mizuuchi, K. Yamamoto, and T. Taniuchi, Appl. Phys. Lett. **58**, 2732 (1991).
33. T. Fujimura, T. Suhara, and H. Nishihara, Electron. Lett. **27**, 1207 (1991).
34. C.B. Clausen, O. Bang, and Yu.S. Kivshar, Phys. Rev. Lett. **78**, 4749 (1997).
35. L. Torner and G.I. Stegeman, J. Opt. Soc. Am. B **14**, 3127 (1997).
36. Yu.S. Kivshar, N. Grønbech-Jensen, and R. Parmentier, Phys. Rev. E **49**, 4542 (1994); Yu.S. Kivshar and K.H. Spatschek, Chaos, Solitons, and Fractals **5**, 2551 (1995).
37. A.V. Buryak, Yu.S. Kivshar, and S. Trillo, Opt. Lett. **20**, 1961 (1995); Opt. Commun. **122**, 200 (1996).
38. A. Hasegawa and Y. Kodama, Phys. Rev. Lett. **66**, 161 (1991); Opt. Lett. **16**, 1385 (1991).
39. L. Bergé, V.K. Mezentsev, J.J. Rasmussen, and J. Wyller, Phys. Rev. A. **52**, R28 (1995).
40. P. Pliszka, and P.P. Banerjee, J. Opt. Soc. Am. B. **10**, 1810 (1993); J. Mod. Opt. **40**, 1909 (1993).
41. N.R. Belashenkov, S.V. Gagarskii, and M.V. Inochkin, Opt. Spect. (USSR), **66**, 806 (1989).
42. Hayata, K., and M. Koshiba, Phys. Rev. Lett. **71**, 3275 (1993).
43. G. Leo, G. Assanto, and W.E. Torruellas, Opt. Lett. **22**, 7 (1997); Opt. Comm. **134**, 223 (1997).
44. G. Leo, and G. Assanto, Opt. Lett. **22**, 1391 (1997).
45. V.V. Steblina, Yu.S. Kivshar, M. Lisak, and B.A. Malomed, Opt. Comm. **118**, 345 (1995).
46. L. Torner, D. Mihalache, D. Mazilu, E.M. Wright, W.E. Torruellas, G.I. Stegeman, Opt. Comm. **121**, 149 (1995).
47. L. Bergé, O. Bang, J.J. Rasmussen, and V.K. Mezentsev, Phys. Rev. E. **55**, 3555 (1997).
48. O. Bang, Yu.S. Kivshar, and A.V. Buryak, Opt. Lett. **22**, 1680 (1997).
49. Mihalache, D., D. Mazilu, L.-C. Crasovan, and L. Torner, Opt. Commun. **137**, 113 (1997).
50. A. De Rossi, S. Trillo, A.V. Buryak, and Yu.S. Kivshar, Opt. Lett. **22**, 868 (1997).
51. A. De Rossi, S. Trillo, A.V. Buryak, and Yu.S. Kivshar, Phys. Rev. E **56**, R4959 (1997).
52. B.A. Malomed, P. Drummond, H. He, A. Berntson, D. Anderson, and M. Lisak, Phys. Rev. E **56**, 4725 (1997).

53. D.V. Skryabin, and W.J. Firth, Opt. Commun. **148**, 79 (1998).
54. D.J. Kaup, A. Bers, and A. Rieman, Rev. Mod. Phys. **51**, 275 (1979)
55. A. Stabinis, G. Valiulis, and E.A. Ibragimov, Opt. Comm. **86**, 301 (1991).
56. E. Ibragimov, and A. Struthers, Opt. Lett. **21**, 1582 (1996); J. Opt. Soc. Am. B **14**, 1472 (1997).
57. D. Mihalache, D. Mazilu, L.-C. Crasovan, and L. Torner, Phys. Rev. E, **56**, R6294 (1997).
58. T.J. Alexander, A. V. Buryak, and Yu. S. Kivshar, Opt. Lett. **23**, 670 (1998).
59. S. Saltiel, S. Tanev, and A.D. Boardman, Opt. Lett. **22**, 148 (1997).
60. R.A. Sammut, A.V. Buryak, and Yu.S. Kivshar, Opt. Lett. **22**, 1385 (1997).
61. P.B. Lundquist, and D.R. Andersen, J. Opt. Soc. Am. B. **14**, 87 (1997).
62. P.B. Lundquist, D.R. Andersen, and Yu.S. Kivshar, Phys. Rev. E **57**, 3551 (1998).
63. R. Danielius, P. Di Trapani, A. Dubietis, A. Piskarskas, and D. Podenas, Opt. Lett. **18**, 574 (1993).
64. R. Danielius, A. Dubietis, and A. Piskarskas, Opt. Lett. **20**, 1521 (1995).
65. O. Bang, J. Opt. Soc. Am. B. **14**, 51 (1997).
66. O. Bang, Yu. S. Kivshar, A.V. Buryak, A. De Rossi, and S. Trillo, Phys. Rev. E **58** September (1998).
67. D. Mihalache, L.-C. Crasovan, and M.C. Panoiu, J. Phys. A **30**, 5855 (1997).
68. G. Assanto, I. Torelli, and S. Trillo, Electr. Lett. **30**, 773 (1994); Opt. Lett. **19**, 1720 (1994).
69. G. Assanto, G.I. Stegeman, M. Sheik-Bahae, and E. Van Stryland, J. Quantum Electron. **31**, 673 (1995).
70. A.D. Boardman and K. Xie, Phys. Rev. E **55**, 1899 (1997)
71. A.D. Boardman, P. Bontemps, and K. Xie, J. Opt. Soc. Am. B **14**, 3119 (1997).
72. W.C.K. Mak, B.A. Malomed, and P. Chu, Phys. Rev. E **57**, 1092 (1998).
73. A.I. Maimistov, A.M. Basharov, S.O. Elyutin, and Yu.M. Sklayrov, Phys. Rep. **191**, 1 (1991).
74. H. Steudel, Physica D **6**, 155 (1983).
75. A.A. Zabolotskii, Optika Spektr. **83**, 782 (1997) [Optics and Spectroscopy **83**, 721 (1997)].
76. G.I. Stegeman, In: *"Nonlinear Optics and Optical Physics"*, Eds. I.C. Khoo *et al.*(World Scientific, Singapore), Chap. 10 (1994).
77. J. Martorell, R. Vilaseca, and R. Corbalán, Opt. Comm. **144**, 65 (1997).
78. A.D. Capobianco, C. De Angelis, A. Laureti-Palma, G.F. Nalesso, J. Opt. Soc. Am. B **14**, 1956 (1997).
79. A., Beržanskis, A. Matijošius, A. Piskarskas, V. Smilgevičius, and A. Stabinis, Opt. Commun. **140**, 273 (1997).
80. K. Dholakia, N.B. Simpson, M.J. Padgett, and L. Allen, Phys. Rev. A **54**, R3742 (1996).
81. W.J. Firth, and D. V. Skryabin, Phys. Rev. Lett. **79**, 2450 (1997).
82. L. Torner, and D.V. Petrov, Electron. Lett. **33**, 608 (1997); J. Opt. Soc. Am. B **14**, 2017 (1997).
83. Ch. Bosshard, R. Spreiter, M. Zgonik, and P. Günter, Phys. Rev. Lett. **74**, 2816 (1995).
84. M. Zgonik, and P. Günter, J. Opt. Soc. Am. B **13**, 570 (1996).
85. M.J. Ablowitz, G. Biondini, and S. Blair, Phys. Lett. A **236**, 520 (1997).
86. T.A.B. Kennedy, T. A. B. and S. Trillo, Phys. Rev. A. **54**, 4396 (1996).
87. P.D. Drummond and H. He, Phys. Rev. A **56**, R110 (1997).

Index